Computational Physics with R

Online at: https://doi.org/10.1088/978-0-7503-2632-2

Computational Physics with R

James Foadi
Department of Mathematical Sciences, University of Bath, Bath, UK

IOP Publishing, Bristol, UK

ISBN 978-0-7503-2632-2 (ebook)
ISBN 978-0-7503-2630-8 (print)
ISBN 978-0-7503-2633-9 (myPrint)
ISBN 978-0-7503-2631-5 (mobi)

DOI 10.1088/978-0-7503-2632-2

Version: 20260101

IOP ebooks

British Library Cataloguing-in-Publication Data: A catalogue record for this book is available from the British Library.

Published by IOP Publishing, wholly owned by The Institute of Physics, London

IOP Publishing, No.2 The Distillery, Glassfields, Avon Street, Bristol, BS2 0GR, UK

US Office: IOP Publishing, Inc., 190 North Independence Mall West, Suite 601, Philadelphia, PA 19106, USA

To my father, Hushyar Foady, and to the memory of my mother, Amalia Carraro.

Contents

Part IV Appendices

Preface

While there exist several excellent books on computational physics, none of them, to the knowledge of the author, uses R as its associated programming language. I have written this book primarily to fill this gap in the market and also because I am very passionate about R and its potential as a tool for physicists. R is widely recognised in statistics and data science, but less so in physics; my hope is that this book will encourage students and colleagues to discover its power and versatility.

The book is organised in three parts. Part I introduces the subject of computational physics itself: what it is, why it matters, and how it fits into the broader scientific landscape. It also provides some background on how numbers and functions are represented in a computer, together with a gentle introduction to R for those new to the language. Part II forms the core of the book. Here the main numerical methods used in physics are developed: differentiation, integration, and the solution of ordinary differential equations. The emphasis is on understanding the underlying mathematics, implementing the algorithms in R, and reinforcing learning through examples and exercises. Part III shifts perspective and shows how R, with its wide range of packages, can be used as a modern computational toolbox. Instead of building algorithms from scratch, the focus here is on solving physics problems directly with existing code, illustrating the workflows that a researcher might use in practice.

This book is written with several audiences in mind. Undergraduates with some exposure to programming will find it accessible and structured for learning. Graduate students will find it a flexible and comprehensive resource. Colleagues who are curious about the use of R in physics may also find it a useful reference and a source of ideas.

The pedagogical approach is guided by two principles. First, R provides a natural bridge to statistics, a field in which physics students often receive little training beyond the basics. Working with R offers an opportunity to refresh and expand that knowledge while learning computational physics. Second, concepts are always illustrated, either with formulas or with coding examples. The mathematics is not always rigorous, in order to keep the exposition accessible, but the quantitative foundation is solid. Proofs and more advanced material are collected in appendices, allowing interested readers to pursue them as needed.

There are different ways of using this book. If you want a quick introduction to R, chapter 2 on its own may be enough. If you are after a classic introduction to computational physics, then combining chapters 1 and 2 with Part II provides a solid one- or two-semester course, particularly suitable for undergraduates. If you already have experience in computational physics but are new to R, you may prefer to start with chapter 2 and then explore Part III.

Some areas of computational physics are not covered. Partial differential equations, spectral methods (notably Fourier transforms), molecular dynamics, computational fluid dynamics, and parallel or high-performance computing lie beyond the scope of this book. The addition of these topics would have turned an already sizeable book into a reference manual, which was never the author's intention. The goal here is a focused, pedagogical text, not an encyclopaedia.

The section on exercises deserves a special mention. Exercises are central to the learning process, and in this book solutions to *all* exercises are provided. This is extraordinary, since very few textbooks offer such complete support. Having full solutions allows readers to check their progress, confirm their understanding, and compare their own approaches with those suggested here. Furthermore, some extra, advanced insights not present in the text, are presented in the guided solutions to the exercises. Instructors may also find them especially valuable when adopting the book for teaching. A link to all solutions and computer code[1] available to download from my website, is provided in appendix K.

Bath, September 2025
James Foadi

[1] Most computer code is related to the demonstrations printed in the book, but other code is also included like, for example, all the code related to the nuclear reactor Monte Carlo simulation.

Acknowledgements

Although this book was written largely on my own, I wish to express my gratitude to those who have shaped my journey in science and inspired my work. I am indebted to the late Professor Michael Mark Woolfson of the University of York, whose supervision during my PhD sparked my enduring passion for algorithms and computing, and to Professor Garib Murshudov of the MRC Laboratory of Molecular Biology, who first introduced me to the R language.

I am also deeply thankful to two of my students, Sofia Moliner Bobo and Federico Colio, who kindly assisted me in reviewing the proofs. Seeing their professionalism, dedication, and growth has been one of the most rewarding aspects of my academic life. Their help not only improved the final manuscript, but also reminded me of the joy of teaching and learning through others.

I would also like to thank the editors who supported this project. John Navas, who first approached me with the idea of writing a book for the Institute of Physics, and Phoebe Hooper, who patiently oversaw my progress and offered invaluable guidance in finalising and polishing the manuscript. I am deeply grateful to them both.

I owe a deep debt of gratitude to my uncle, Ing. Fahim Avaregan, who with great patience and methodical teaching instilled in me a love of mathematics from early childhood. Without his guidance, I might never have found my way into science at all. Thank you from the bottom of my heart, Fahim!

Finally, I thank my wife Sonia and my daughters Sophie and Nadine for their endless patience and understanding during the many hours I spent secluded in my study. Their support and forbearance made this book possible.

Author biography

James Foadi

Dr James Foadi is a mathematical and computational physicist with over 20 years of experience in the development of mathematical and statistical methods for structural biology. He has researched and taught in various universities across the UK and held a research position at the Diamond Light Source synchrotron for 10 years. Dr Foadi is currently a lecturer in the Department of Mathematical Sciences at the University of Bath.

Part I

Introduction to computational physics and the R platform

This first part lays the groundwork for the rest of the book by introducing both the motivations behind computational physics and the computational environment in which the methods will be developed: the R platform.

The chapters in this part guide the reader through the nature of computational physics as a discipline and explore the structure and limitations of numerical computation, with particular attention to floating point arithmetics and sources of error. The basic use of R is introduced gradually, alongside tools for plotting, programming structures, and handling vectors and matrices, skills that are fundamental for all later implementations.

This part also prepares the reader to write clear, functional, and reproducible code in R, with a focus on best practices and simplicity. It is intended for readers with a background in mathematics or physics but with little or no prior experience in numerical computing or R.

IOP Publishing

Computational Physics with R

James Foadi

Chapter 1

Introduction to computational physics

It is pedantic to explain why computational physics is part of a physicist's training, given how evident is digital technology in modern times. Even experimental physicists in a laboratory need to operate computers attached to the equipment and write numerous lines of code to activate and complete various essential operations. And it has become nowadays common to implement some basic computer simulations of certain phenomena, before embarking on any time-consuming physical study of the same. At a pedagogical level one could say that computational physics is very formative for undergraduates as it forces them to think in terms of constituent and elementary processes, represented by algorithms. From a practical point of view, to take part in an introductory computational physics course will also be beneficial because it enables one to learn one or more programming languages.

In this chapter the main parts of modern computational physics and its specific fields of application will be sketched. The picture that emerges is not complete and certainly biased by the author's professional formation and preferences, but it contains a sizable subset of topics commonly taught in modern computational physics courses. The chapter also provides details on the binary floating point system used in all computer calculations and explains the main sources of error when numerical algorithms are developed, truncation and round off errors. It is important to learn this topic at the outset as one should consider the risks of neglecting these errors before 'transforming any algorithm into code'.

1.1 The landscape of computational physics

Computational physics has grown into a broad and diverse discipline. Its scope varies depending on the author, the institution, and the intended audience, but a few common threads emerge. What unites them is the attempt to capture physical processes through numerical algorithms, when exact analysis is impossible and experiment alone cannot provide sufficient insight. In this sense, computational

doi:10.1088/978-0-7503-2632-2ch1

physics sits between theory and experiment, both borrowing from and contributing to each. Across the modern landscape one can identify several recurring areas:

- **Approximation methods:** numerical differentiation, quadrature, interpolation, and curve fitting. These topics are essential because they represent the simplest way of turning continuous mathematics into computable steps.
- **Differential equations:** methods for solving ordinary differential equations, both initial and boundary value problems, and algorithms for partial differential equations such as diffusion, wave, or Poisson-type equations. This area is central because so much of physics rests on differential formulations.
- **Stochastic and statistical methods:** Monte Carlo simulation, random number generation, random walks, and modelling of many-body systems. These tools are indispensable when dealing with randomness or high-dimensional problems.
- **Numerical linear algebra:** solvers for systems of linear equations and eigenvalue problems, which underpin discretised models of continuous physics and large-scale computations.
- **Specialised applications:** molecular dynamics, percolation and lattice models, computational fluid dynamics, electronic structure calculations, and high-performance or parallel computing. These topics demonstrate the reach of computational physics into advanced and interdisciplinary areas.

Not every course or textbook covers all these parts. Some choices emphasise numerical methods for quantum mechanics, others focus on heat conduction or fluid dynamics, while others build extensive simulations of many-body systems. The present book belongs firmly to the pedagogical tradition. Its purpose is not to give a comprehensive survey of all possible topics, but to provide students with the essential building blocks of computational thinking in physics. The central aim is that students should learn to instruct the computer, by writing their own programs in R, rather than relying on pre-packaged routines. In this way the computer becomes a laboratory where one experiments with algorithms, observes their behaviour, and develops intuition about both mathematics and physics.

In concrete terms, the book emphasises: approximation of functions and integrals; the solution of ordinary differential equations; Monte Carlo methods for integration and simple simulations; and a careful discussion of numerical error, floating-point arithmetic, and computational efficiency. A short excursion into data-driven models and machine learning is also included, to connect with current practice. Deliberately excluded are topics such as Fourier and spectral methods, large-scale molecular dynamics, or high-performance computing, which lie beyond the intended scope. This selection is not a limitation but a choice: by focusing on a core set of methods and by insisting on direct programming experience, the book aims to make computational physics accessible, formative, and useful to undergraduates who are beginning their journey into the subject.

1.2 Binary representation of numbers

While humans have evolved a numeral system based on the powers of 10, for computers' memory it is better to store and process numbers using a base 2 system. Consider, for instance, 182. In the base 10 system this is represented as '182' because

$$182 = \mathbf{1} \times 10^2 + \mathbf{8} \times 10^1 + \mathbf{2} \times 10^0.$$

In base 2 the representation is different and 182 becomes '10110110' because

$$\begin{aligned}10110110 = &\mathbf{1} \times 2^7 + \mathbf{0} \times 2^6 + \mathbf{1} \times 2^5 + \mathbf{1} \times 2^4 \\ &+ \mathbf{0} \times 2^3 \mathbf{1} \times 2^2 + \mathbf{1} \times 2^1 + \mathbf{0} \times 2^0.\end{aligned}$$

In the base 10 representation, numbers are composed of the 10 symbols 0, 1,…,9. In base 2, numbers are composed of the 2 symbols 0, 1.

To transform a base 10 into a base 2 number, we keep dividing by 2 and store the remainder; the binary representation is given by the succession of remainders, taken in reverse order. An example, still using 182, can clarify the procedure. In the following table, divisions by 2 are kept on the left and remainders on the right.

182	÷	2	0
91	÷	2	1
45	÷	2	1
22	÷	2	0
11	÷	2	1
5	÷	2	1
2	÷	2	0
1	÷	2	1
0			

By reading the remainders bottom to top, we obtain the binary representation of the given number, which is '10110110'.

Example 1.1 *Let us transform into binary form the numbers 71 and 380. Using the same tabular arrangement used earlier we get*:

71	÷	2	1
35	÷	2	1
17	÷	2	1
8	÷	2	0
4	÷	2	0
2	÷	2	0
1	÷	2	1
0			

Therefore, 71='1000111'. For the other number we have:

380	÷	2	0
190	÷	2	0
95	÷	2	1
47	÷	2	1
23	÷	2	1
11	÷	2	1
5	÷	2	1
4	÷	2	0
2	÷	2	0
1	÷	2	1
0			

Thus, 380='10011111100'.

Fractional numbers, too, can be represented in binary form. This time the conversion is obtained via a multiplication, rather than division, by 2. The result is a succession of zeros and ones read from top to bottom. As an example, the transformation of 0.875 from base 10 to base 2 is here illustrated via a table.

0.875	×	2	=	1.750	1
0.750	×	2	=	1.500	1
0.500	×	2	=	1.000	1
1.000					

The decimal representation of 0.75 is therefore '0.111'. Indeed we have

$$0.875 = \mathbf{0} \times 2^{0} + \mathbf{1} \times 2^{-1} + \mathbf{1} \times 2^{-2} + \mathbf{1} \times 2^{-3}.$$

The procedure for decimals, as seen above, stops when the product gives 1.

Example 1.2 *Let us find the binary representation of 0.2. The procedure is interesting because here we realise that it does not end. After the first four digits, the values repeat themselves.*

0.2	×	2	=	0.4	0
0.4	×	2	=	0.8	0
0.8	×	2	=	1.6	1
0.6	×	2	=	1.2	1
0.2	×	2	=	0.4	0
			...		

We are sure that the first four digits, '0011', repeat themselves because the next number to multiply by 2 is 0.2, which is the number from which the procedure started. Therefore we conclude that 0.2 is represented by a periodical, '$0.\overline{0011}$'.

There are numbers, in the base 10 system, which cannot be represented accurately like, for instance, $\sqrt{2}$ or $1.\bar{3}$ (periodical 3). The reason is that one would need an infinite stream of decimals after the point, to give the correct representation. In cases like these, the number is approximated by 'rounding' the last decimal to the closest digit. So, when using an approximation to two decimals, $\sqrt{2}$ becomes 1.41, while $1.\bar{3}$ becomes 1.33. In a similar fashion, there are numbers that cannot be accurately represented in binary form, as shown in the last example. The unfortunate (perhaps mostly pedagogical!) drawback of this is that they often correspond to numbers which have an exact representation in base 10. A well-known example is 0.1, which in binary format is $0.0\overline{0011}$. Numbers like this will be stored with a finite amount of digits in the computer's memory, providing an approximation, commonly known as *round off error* (or *rounding error*).

1.3 The IEEE standard

Most computers store and process numbers in a binary format known as the *IEEE 754 standard* [1]. The actual standard, which handles 32-bits (*single precision*, with 7 significant digits) and 64-bits (*double precision*, with 16 significant digits), is illustrated in appendix D. Here we would like to explain in some detail how numbers are represented, but using a *toy IEEE standard*, which consists only of 8 bits. All numbers represented by the 8-bits system will be smaller, but enable demonstrations and explanations without sacrificing clarity. This system will be used throughout for various explanations, to the end of the chapter

A binary floating point number is represented by a string of bits, subdivided into three segments called *sign*, *exponent* and *significand* (or *mantissa*). More specifically, in the 8-bits standard adopted, any number appears as follows:

$$1.\, d_1 d_2 d_3 d_4 \times 2^p,$$

where d_1, d_2, d_3, d_4 can assume only values 0, 1, and where the exponent p is an integer such that $-2 \leqslant p \leqslant 3$. The only number to the left of the floating point is 1. A number in this form is known as *normalised*. In general, there are m digits d_i in standardised systems, like the 32-bits system, where $m = 23$ (and $-126 \leqslant p \leqslant 127$), or the 64-bits system, where $m = 52$ (and $-1022 \leqslant p \leqslant 1023$). In the 8-bits system, the string is structured as follows:

1. Bit 1. It stores the sign. A 0 means '+' and a 1 means '−'.
2. Bits 2–4. These bits are reserved for the exponent. They store an integer between 0 and 7. Using an *offset* equal to 3 (the value 3 comes from $2^{3-1} - 1 = 3$, where the 3 of 2^{3-1} counts how many bits are in the exponent), these 8 integer values correspond to powers of 2 in the range between $0 - 3 = -3$ and $7 - 3 = 4$. For practical reasons, the powers -3 and 4 (the outmost values in the range) are reserved for special cases. So the used range is between -2 and 3, corresponding to $2^{-2} = 0.25$ and $2^3 = 8$.

3. Bits 5–8. These are reserved for the significand and are responsible for the number's precision. As shown in the following examples, the mantissa is normalised to have always 1 before the floating point. As this never changes, the information can be omitted and the significand includes only the bits after the floating point.

The special cases (exponents 000 or 111) can be thus summarised:

Exponent value	Significand	Represents
111	All zeros	Infinity ($\pm\infty$)
111	Not all zeros	Not a number (NaN)
000	All zeros	Zero (0)
000	Not all zeros	Subnormal numbers (very small)

Given the different number of bits in the exponent, the special cases for the 32-bits system will have exponent equal to 00000000 or 11111111, while those for the 64-bits system will have exponent equal to 00000000000 or 11111111111.

In relation to the last row of the table, even though it is not possible to represent numbers smaller than a given value, the IEEE standard allows a gradual shift towards zero by enabling the representation of so-called *subnormal numbers* (see example 1.7). In general, using a base β numeral system with m digits for the significand and a $[e_{\min}, e_{\max}]$ range for the exponent, the largest and smallest normal numbers representable are:

$$\text{Largest}: \quad (\beta - \beta^{-m})\beta^{e_{\max} - 1}, \qquad \text{Smallest}: \quad \beta^{e_{\min} + 1} \tag{1.1}$$

Example 1.3 *Let us transform 00110011, written in the IEEE 8-bit toy system, into a decimal number. The first bit, 0, means that the number is positive. The next three bits, 011, correspond to the binary form of the integer 3. Using the offset* -3*, this means that the exponent is* $p = 3 - 3 = 0$*. Therefore, the next four bits, 0011, will be multiplied by* $2^0 = 1$*. As 1 is always supposed to be the number before the fixed point, we have*

$$00110011 \rightarrow 1.0011 \times 2^0 = 1.0011.$$

The transformation from binary to decimals, yields:

$$1.0011 = \mathbf{1} \times 2^0 + \mathbf{0} \times 2^{-1} + \mathbf{0} \times 2^{-2} + \mathbf{1} \times 2^{-3} + \mathbf{1} \times 2^{-4} = 1.1875.$$

This means that 00110011 represents exactly the decimal 1.1875.

Example 1.4 *To write* 5.25 *in the 8-bits system, this is first transformed into a base 2 number, thus becoming* 101.01*. Next comes normalisation, in which the floating point is shifted two positions to the left, leaving only a 1 before the point itself. To make up for the decrease in value, a* 2^2 *must be factorised. Thus:*

$$5.25 \rightarrow 101.01 \rightarrow 1.0101 \times 2^2$$

The first bit of the 8-bits string must be 0 because the number is positive, while the bits from 5 to 8 are 0101. *For the exponent, an offset 3 must be added to the exponent 2, so that the bits 2 to 4 in the string accommodate the integer 5, i.e.* 101. *The 8-bits string is, thus:*

$$5.25 \rightarrow 0 \quad 101 \quad 0101 \rightarrow 01010101$$

Example 1.5 *Let us write the largest positive number in the 8-bits system. As explained earlier, the largest usable power of 2 is 3 which, with the offset, becomes 6, i.e. the binary* 110. *The exponent part of the string is therefore* 110. *The largest significand has all 1's, i.e. it is* 1111. *This corresponds to floating point binary* 1.1111. *The number is, therefore*

$$01101111 \rightarrow 0 \quad 110 \quad 1111 \rightarrow 1.1111 \times 2^3 \rightarrow 1111.1$$

The binary 1111.1 *is* 15.5. *The largest positive number is thus* 15.5. *This is also in agreement with formula (1.1).*

Example 1.6 *Let us write the smallest negative number in the IEEE 8-bits standard. Negative means simply that the string starts with 1 instead of 0. For the rest, we can proceed as if the number were positive. The smallest usable power of 2 is* -2. *With offset 3, this becomes 1, and corresponds to string* 001. *The smallest significand is* 0000, *corresponding to binary* 1.0000, *which is 1, in base 10. Thus, the smallest negative floating number is*

$$10010000 \rightarrow 1 \quad 001 \quad 0000 \rightarrow -1.0000 \times 2^{-2} \rightarrow -0.01$$

which corresponds to -0.25, *in base 10. This is also in agreement with formula (1.1).*

Example 1.7 *We have mentioned that strings with exponent* 000 *and a significand different from* 0000, *correspond to subnormal numbers, which are smaller than the smallest regular positive,* 0.25, *but greater than 0. Here the convention is that a 0, rather than a 1, appears before the floating point. Also, even though the exponent part indicates* 000 *(an indication, indeed, of the number being subnormal), the significand is meant to be multiplied by the smallest power of 2 usable, which is* 2^{-2}. *There are therefore* 15 *positive subnormal numbers. The smallest is 00000001, corresponding to the binary* 0.0001×2^{-2} *(*$2^{-4} \times 2^{-2} = 2^{-6} = 0.015\,625$ *in base 10). The next smallest is 00000010, corresponding to* 0.0010×2^{-2}, *which is* 0.031 25 *in base 10. The largest subnormal number is 00001111, i.e.* 0.1111×2^{-2}, *or 0.001 111 (0.234 375, in base 10).*

Example 1.8 *Irrational or periodic numbers, and even some rational numbers like* 0.1, *cannot exactly be represented in binary notation. Consider* 0.3, *which is the periodic* $0.01\bar{0}0$. *To obtain the normalised expression we need to move the floating point of two* *bits to the right. Therefore we need to round the number to six bits, prior to normalising it. Using the IEEE rounding rules (see appendix E), we round down to* 0.010 011. *The normalised form yields* 1.0011×2^{-2}. *The 8-bits string is therefore*

$$0.3 \to 0.01\bar{0}\bar{0} \to 1.0011 \times 2^{-2} \to 0 \quad 001 \quad 0011 \to 00010011.$$

This number, as expected, is not really 0.3. *It is, in fact,* 0.010 011, *which corresponds to* 0.296 875. *This is a typical case of round off error, where* 0.3 *is approximated by* 0.296 875, *in the 8-bits binary system.*

1.4 Floating point arithmetics

In this section, addition and multiplication of binary, floating point numbers will be explored. We need, first, to understand how two floating point binary numbers are added together. The recipe is relatively simple.

Addition of two binary floating point numbers

If the two numbers are $x = \alpha_x \times 2^{n_x}$ and $y = \alpha_y \times 2^{n_y}$, and n_x and n_y are the two exponents, then:

1. If n_x is the largest exponent, transform y so to have exponent n_x and appropriate significand, without altering the number of digits, m. The transformed expression for y will be, say, $y = \alpha_y' \times 2^{n_x}$.
2. Add the two significands:

$$z = x + y = (\alpha_x + \alpha_y') \times 2^{n_x}.$$

3. Normalise the result obtained, not forgetting that m must be kept fixed.

Example 1.9 *Let us add up* $x = 0.5$ *and* $y = 0.3$. *The two numbers, in 8-bits format, become (see example 1.8)*

$$x = 1.0000 \times 2^{-1}, \qquad y = 1.0011 \times 2^{-2}.$$

The second number has smaller exponent and must be transformed into a number with exponent -1:

$$y = 1.0011 \times 2^{-2} \to 0.1010 \times 2^{-1},$$

where we have rounded up 1.0011 *to* 1.010, *following the IEEE rules for rounding (see appendix E). Now the two numbers can be added up*:

$$z = x + y = (1.0000 + 0.1010) \times 2^{-1} = 1.1010 \times 2^{-1}.$$

The number is already in normalised form. Therefore, the result obtained is the final sum. This corresponds to 0.8125, *which is not the correct* 0.8, *because of the round off in representing* 0.3.

Example 1.10 *Let us add up* $x = 4$ *and* $y = 2.25$. *In binary we have*

$$x = 1.0000 \times 2^2, \qquad y = 1.0010 \times 2^1.$$

The second exponent is smaller than the first, thus y *needs to be converted:*

$$y = 1.0010 \times 2^1 = 0.1001 \times 2^2.$$

It is now possible to add up the digits of the significand:

$$z = x + y = (1.0000 + 0.1001) \times 2^2 = 1.1001 \times 2^2.$$

The number is already in normalised form and it is equivalent to 6.25. *This is also exactly the value obtained by adding up* 4 *and* 2.25. *In this case there were no round off errors to start with and none turned up after addition.*

Example 1.11 *In this example, two numbers which are represented exactly as 8-bits binary numbers will give a sum which cannot be exactly represented because of an intermediate round off error. The two numbers are* $x = 2.5$, *exactly represented by* 1.1000×2^1, *and* $y = 0.4375$, *exactly represented as* 1.1100×2^{-2}. *As the first number has a higher exponent, the second one must be transformed:*

$$y = 1.1100 \times 2^{-2} \rightarrow 0.001\,110\,0 \times 2^1 \rightarrow 0.0100 \times 2^1,$$

where rounding up was necessary for the transformation as round-to-even (see appendix E). The addition can now be carried out:

$$z = x + y = (1.0100 + 0.0100) \times 2^1 = 1.1000 \times 2^1.$$

This value is equal to 3, *which is not the exact value,* 2.9375. *In this case, the transformation of the second number to match the higher exponent has introduced a round off error. This typically happens when one of the two numbers is noticeably smaller than the first.*

The subtraction of two numbers can be treated as addition, once the second number is transformed into its negative. The subtraction can then be carried out according to the following recipe:

Subtraction of two binary floating point numbers

The two numbers, x and y, are already normalised, and we will use here the same notation adopted for addition.

1. Transform the expression for y as $y = \alpha'_y \times 2^{n_x}$, to have the same exponent of x.
2. Now we need to *negate* y. All bits are flipped: the zeros become ones and the ones zeros; next, we add 1 to the last bit of the number so transformed, and proceed carrying through the sums, if needed, to the left of the significand. The result is the negative of the initial number, as prepared for the following addition.

3. Add the two significands, including the zeros and ones before the floating point.
4. If the resulting bit before the floating point is 0, then nothing else is needed. If it is 1, then proceed with the negation of the result obtained.

Example 1.12 *Demonstrate, using the 8-bits system, the subtraction* $10 - 7.5$. *The result is* 2.5, *which in normalised form becomes* 1.0100×2^1. *The starting normalised numbers are*

$$x = 10 = 1.0100 \times 2^3, \qquad y = 7.5 = 1.1110 \times 2^2.$$

The second one has smaller exponent and must be transformed into an equivalent (non-normalised) number with exponent 3:

$$y \to y = 0.1111 \times 2^3.$$

Next, this number must be negated. All bits are flipped and then 1 *is added to the last bit (after the floating point)*:

$$0.1111 \times 2^3 \to (1.0000 + 0.0001) \times 2^3 = 1.0001 \times 2^3.$$

Now we can carry out the addition:

$$\begin{array}{l} 1.0100 \times 2^3 + \\ 1.0001 \times 2^3 = \\ \hline 0.0101 \times 2^3, \end{array}$$

which is equivalent to the normalised 1.0100×2^1, *as expected. Note how we have discarded the carried* 1 *in the leftmost digit because outside the normalised format, and because the number obtained would have still been within the range of possible values (not an overflow). In the case just treated, as the digit before the floating point was* 0, *nothing else was needed. Let us then swap* 10 *with* 7.5 *and observe how* -2.5 *is, instead, obtained. This time the term to be negated is* $x = 10 = 1.0100 \times 2^3$. *The negation yields*:

$$1.0100 \times 2^3 \to 0.1011 + 0.0001 = 0.1100 \times 2^3.$$

The addition is:

$$\begin{array}{l} 0.1111 \times 2^3 + \\ 0.1100 \times 2^3 = \\ \hline 1.1011 \times 2^3. \end{array}$$

Here the result starts by 1 *and this means that the number is negative. Its value can be found by flipping and adding* 1 *to the last bit (after the floating point)*:

$$1.1011 \times 2^3 \to (0.0100 + 0.0001) \times 2^3 \to 0.0101 \times 2^3 \quad \Leftrightarrow 2.5.$$

The final result, considering the negative sign (represented by the 1 *before flipping), is* -2.5, *as expected.*

Example 1.13 *The subtraction of a number from itself should yield* 0. *Let us choose any number, say* 12.25. *In binary this is* 1100.01, *but when transformed into the normalised 8-bits format, round off occurs and the number becomes* 1.1001×2^3, *which corresponds to* 12.5. *This is not a problem for the operation we are describing, as we will subtract the approximated numbers. The negative of* 1.1001×2^3 *is*

$$1.1001 \times 2^3 \rightarrow (0.0110 + 0.0001) \times 2^3 \rightarrow 0.0111.$$

The subtraction, therefore, yields

$$\begin{array}{r} 1.1001 \times 2^3 + \\ 0.0111 \times 2^3 = \\ \hline 0.0000 \times 2^3, \end{array}$$

which corresponds to 0, *as expected.*

Example 1.14 *When a positive number must be subtracted from a negative number, like for example in* $-2 - (5)$, *there is no need to apply the rules for subtraction. One should simply perform addition between two positive numbers, like in* $2 + 5$, *and change sign to the result.*

One can carry out subtraction without bits flipping, if preferred. It suffices to remember that 1 can be *borrowed* from the adjacent bit, if needed. The reason being that while $1 - 0 = 0$, $0 - 1$ is not possible. For example, to calculate

$$1.0100 \times 2^3 - 0.1111 \times 2^3,$$

corresponding to $10 - 7.5$, we proceed as follows. First we align the two numbers:

$$\begin{array}{r} 1.0100 \times 2^3 - \\ 0.1111 \times 2^3 = \\ \hline \end{array}$$

The first subtraction (rightmost bits) is $0 - 1$, which cannot happen. Therefore, a 1 is borrowed from the adjacent bit on the left. But this is also a 0 and will, in turn, borrow from the adjacent 1 on the left. The situation is depicted here:

$$\begin{array}{rrrr} 1.00 & 1 & 10 & - \\ 0.11 & 1 & 1 & = \\ \hline & 0 & 1 & \end{array}$$

Borrowing happens, again for the next bit on the left (that meanwhile has become zero). This borrows, in turn, from the adjacent bit on the left which, again, borrows from the first bit, which is a 1.

$$
\begin{array}{l}
0.1\ 10\ 1\ 10\ - \\
0.1\ \ \ 1\ 1\ \ \ 1\ = \\
\hline
0.0\ \ \ 1\ 0\ \ \ 1
\end{array}
$$

To summarise:

$$
\begin{array}{l}
1.0100 \times 2^3\ - \\
0.1111 \times 2^3\ = \\
\hline
0.0101 \times 2^3
\end{array}
$$

The result is $0.0101 \times 2^3 = 10.1 = 2.5$. The complication brought about by 'borrowing' is the equivalent of the complication brought about by bits flipping.

Product of two binary floating point numbers

For the product, there is no need to transform one of the two numbers, in order to have a common power of 2. First, a normal product of the significand of the two numbers is carried out, remembering that

$$0 \times 0 = 0, \quad 0 \times 1 = 0, \quad 1 \times 0 = 0, \quad 1 \times 1 = 1.$$

For the exponent we simply add the powers of 2. The following two examples are enough to demonstrate the procedure.

Example 1.15 *Let us multiply* 2 *and* 3 *in base* 2. *The two numbers are, respectively*

$$2 \to 10 \to 1.0000 \times 2^1, \quad 3 \to 11 \to 1.1000 \times 2^1.$$

For the product of the significand we have

$$1.0000 \times 1.1000 = 1.1000.$$

The product of the exponents yields

$$2^1 \times 2^1 = 2^2.$$

Therefore

$$2 \times 3 \to 1.1000 \times 2^2 \to 110.0000 = 6,$$

as expected.

Example 1.16 *Let us multiply* 1.5 *and* 7.125 *in base* 2. *The two numbers are, respectively*

$$1.5 \to 1.1000 \times 2^0, 7.125 \to 1.1101 \times 2^2.$$

The second number does not correspond to an exactly representable number because of round off error, but to 7.25. *As a consequence, the product will not yield the correct* 10.6875, *but a value affected by that error,* 10.875. *Indeed, the product of the significand yields*

$$1.1000 \times 1.1101 = 10.101\ 110\ 00,$$

while the product of the exponents yields 2^2. *Therefore*

$$1.5 \times 7.125 \rightarrow 1010.1110 = 10.875,$$

as expected.

Division of two binary floating point numbers

The rules of the division are similar to those for the multiplication, with the difference that the powers of two have to be subtracted, while the actual division of significand (for example using the long-division technique) can be tedious. We will demonstrate this with just one simple example. To get a refresher of long divisions look at, for instance reference [2].

Example 1.17 *Let us divide* 3 *by* 2 *in base* 2. *The two numbers, in the 8-bit system, are* 1.1000×2^1 *and* 1.0000×2^1. *The exponent of the division will therefore be* $1 - 1 = 0$. *For the significand, we do not need to use the long division for this simple example because the divisor is just* 1:

$$1.1000 \div 1.0000 = 1.1000.$$

Thus

$$3 \div 2 \rightarrow 1.1000 \times 2^0 = 1.1000 = 1.5,$$

as expected.

1.5 Truncation and round off errors

The two most important errors affecting all calculations carried out with numerical methods with a computer are *truncation errors* and *round off errors.*

1.5.1 Truncation errors

These are essentially due to the nature of numerical methods, which very often try and calculate quantities using approximations. A well known approximation, for instance, is the one approximating the exponential e^x:

$$e^x = \sum_{k=0}^{\infty} \frac{x^k}{k!} = 1 + x + \frac{x^2}{2} + \cdots$$

The correct value of e^x, for any x, is reached only when an infinite number of terms is used, because the series is convergent. In practical terms one will have to limit the accuracy of e^x by considering a finite number of terms. For example, e can be approximated just by three terms, thus yielding

$$e = 1 + 1 + \frac{1}{2} = 2.5.$$

This is a very crude approximation of e. The error associated with this approximation is known as *truncation error*, because it is due to a truncation of the full series (which provides the correct value) to a few terms. Truncation errors constantly arise from the many existing numerical methods and can very rarely be eliminated. It is important to be aware of them quantitatively, so to estimate appropriate margins of error in all calculations.

1.5.2 Round off errors

Even when a truncation error is not present in a numerical calculation, errors can arise because of the way decimal numbers are represented in computers. As explained in section 1.2, these are known as *round off errors* and arise essentially due to the finite number of bits needed for computer storage. Most of the time, this fact has very little consequence because the 32- or 64-bits representation allows for many bits to be stored after the decimal point (see appendix D). But it can sometimes have unpleasant consequences. A piece of code could, for instance, include an `if` statement that is triggered by the equality of a variable with an expression containing round off errors. In such a circumstance, the equality might never be realised, even when it should, and this would potentially have catastrophic consequences for the code's execution. Another undesired effect of round off is the growth of the overall error due to the very many calculations (ten or even a hundred thousands of times) involved in routine algorithms. For these and related reasons, one should have clear in mind the risk associated with coded operations, and provide check and safety mechanisms to avoid such risks. An accessible, but thorough, description of round off errors is provided in the popular article by Goldberg [3].

1.6 Quantifying numerical errors

All errors need to be quantified in some way so that their magnitude can be evaluated, when considering different calculations. The two most commonly known types are *absolute errors* and *relative errors*. A numerical error arises when a number x is approximated by another number, $\tilde{x}$. The difference between these two numbers will be indicative of the error brought about by the approximation. More specifically:

- The *absolute error*, E_a, is defined as

$$E_a = |x - \tilde{x}| \tag{1.2}$$

- The *relative error*, E_r, is defined as

$$E_r = \frac{|x - \tilde{x}|}{|x|}. \tag{1.3}$$

While the absolute error gives a feeling of the actual numeric discrepancy, the relative error provides a rough estimate of the number of digits which x and its approximation have in common. There exists, in fact, a rule of thumb according to which

$$E_{\rm r} = \frac{|x - \tilde{x}|}{|x|} \approx 10^{-m},$$

and where m is roughly equal to the number of significant digits that x and $\tilde{x}$ have in common.

Example 1.18 *Consider a number,* $x = 1.234\ 56$, *and its approximation,* $\tilde{x} = 1.234\ 92$. *These two numbers have four significant digits in common,* 1234. *The relative error is*

$$E_{\rm r} = \frac{|1.234\ 56 - 1.234\ 92|}{|1.234\ 56|} \approx 0.0003 \approx 10^{-4}.$$

In this case $m = 4$ *which is exactly the number of common significant digits.*

What we just demonstrated is only a rule of thumb and in general the value of m is only approximately equal to the number of common significant digits.

An error often considered with the relative error is the *reciprocal relative error*, $E_{\rm rr}$, defined by

$$E_{\rm rr} = \frac{|\tilde{x} - x|}{|\tilde{x}|}. \tag{1.4}$$

In appendix A.8 it is proved that

$$E_{\rm rr} = \frac{|\tilde{x} - x|}{|\tilde{x}|} \leqslant \frac{E_{\rm r}}{1 - E_{\rm r}} \tag{1.5}$$

Example 1.19 *Using data from example 1.18, let us calculate* $E_{\rm rr}$ *and verify that it obeys inequality (1.5). The calculation of* $E_{\rm rr}$ *is straightforward*:

$$E_{rr} = \frac{|1.234\ 92 - 1.234\ 56|}{|1.234\ 92|} \approx 0.000\ 291\ 5169.$$

As we have used 10 *decimals for* $E_{\rm rr}$, *we should use also* 10 *decimals for* $E_{\rm r}$:

$$E_r = \frac{|1.234\ 56 - 1.234\ 92|}{1.234\ 56} \approx 0.000\ 291\ 6019.$$

We also have

$$\frac{E_{\rm r}}{1 - E_{\rm r}} \approx 0.000\ 291\ 6869$$

and it is clear that 0.000 291 6869 > 0.000 291 5169, *which indicates that* E_{rr} *and* E_r *satisfy relation (1.5).*

1.6.1 Relative errors and machine epsilon

A real number, x, is in general approximated by a binary floating point number, $\tilde{x}$. It is possible to fix an upper limit to the relative error due to this approximation. The following inequality holds,

$$E_r = \frac{|x - \tilde{x}|}{|x|} \leqslant \epsilon_{\mathrm{mach}}, \tag{1.6}$$

where ϵ_{mach} is known as *machine precision*, or *machine epsilon* (also shortened to *macheps*). This is defined as the distance between 1 (exactly representable in the IEEE system) and the next (larger) decimal exactly representable in the IEEE system. We can easily derive this number because we know how decimals are represented in the IEEE system. 1 is represented by $1.0 \ldots 0 \times 2^0$, where the significand has m zeros. The next decimal, exactly representable in the same system is

$$1.0 \ldots 0 \times 2^0 + 0.0 \ldots 01 \times 2^0 = 1.0 \ldots 0 \times 2^0 + 2^{0-m} = 1 + 2^{-m}.$$

The distance between these two exactly representable and adjacent numbers is therefore 2^{-m}:

$$\epsilon_{\mathrm{mach}} = 2^{-m} \tag{1.7}$$

Let us now try and prove inequality (1.6). The decimal x is placed between two adjacent numbers, $\tilde{x}_b$ and $\tilde{x}_u$, exactly representable in the IEEE system and for which we assume

$$\tilde{x}_b \leqslant x \leqslant \tilde{x}_u.$$

It is relatively straightforward to see that

$$|\tilde{x}_u - \tilde{x}_b| = \epsilon_{\mathrm{mach}} |\tilde{x}_b|.$$

Therefore, considering that x sits between $\tilde{x}_b$ and $\tilde{x}_u$, knowing that $x \geqslant \tilde{x}_b$, and taking into account that $\tilde{x}$ could be either $\tilde{x}_b$ or $\tilde{x}_u$, we have

$$E_r = \frac{|x - \tilde{x}|}{|x|} \leqslant \frac{|\tilde{x}_u - \tilde{x}_b|}{|x|} = \frac{\epsilon_{\mathrm{mach}} |\tilde{x}_b|}{|x|} \leqslant \frac{\epsilon_{\mathrm{mach}} |\tilde{x}_b|}{|\tilde{x}_b|} = \epsilon_{\mathrm{mach}},$$

as reported in inequality (1.6).

Example 1.20 *Let us check wether the round off representation of x in example 1.8 obeys equation (1.6). The calculation can be carried out using base* 10 *numbers, as the final result is equivalent. In the example,* $x = 0.3$, *while* $\tilde{x} = 0.296\ 875$. *Therefore,*

$$E_r = \frac{|0.3 - 0.296\ 875|}{|0.3|} = \frac{0.003\ 125}{0.3} \approx 0.010\ 4167.$$

In the example, $m = 4$, therefore the upper limit for the relative error is $2^{-4} = 0.0625$. *The value of E_r is, indeed, smaller than* 0.0625.

1.6.2 Taking errors into account

Two important reasons to consider errors in calculations are that an exact result is rarely obtained and that errors must be understood and taken care of. This is easier to do when errors arise from known mathematical formulas, like in the case of truncation errors. It is more complicated when they arise from the machine's precision or from random causes. There are many systematic and varied ways to go about calculating errors and containing them within given boundaries, but to master them takes a whole and lengthy different course of studies [4–6]. In the present textbook we suggest an ad hoc approach, where after a specific algorithm has been designed and implemented in code, tests are done to ascertain the presence of such errors and appropriate measures are taken to remove or contain them. More details on this are contained in section 2.25.

1.7 Exercises on the computer representation of numbers

Exercise 01

Transform the following numbers from base 2 to base 10:

1. 1000.101
2. 111.11

Exercise 02

Transform the following numbers from base 10 to base 2:

1. 32.3125
2. 5.375

Exercise 03

Find the base 10 expression of the three consecutive numbers, 01101001, 01101010, 01101011, written in the toy 8-bit IEEE system.

Exercise 04

The smallest normal positive number in the IEEE 8-bit toy system is 0.25. It is possible to represent exactly a handful of numbers smaller than 0.25, but greater than 0 within the IEEE system, the so-called subnormal numbers (see main text). List all 15 subnormal numbers representable with the IEEE 8-bit system.

Exercise 05

Calculate the smallest positive normal number and the largest positive subnormal number in the IEEE 32-bit system.

Exercise 06

Verify that there are $n = 2^m$ normal numbers between any two consecutive powers of 2, in the IEEE system, where the two consecutive numbers are represented as

$$2^p, 2^{p+1}, \qquad e_{\min+1} \leqslant p \leqslant e_{\max} - 1.$$

In particular, find out how many normal numbers exist between 2 and 4 in the IEEE 32-bit and 64-bit systems.

Exercise 07

Represent $\sqrt{2}$ in the IEEE 8-bit toy system. Calculate the absolute and relative errors, due to round off.

Exercise 08

Find the sum and difference of the two base 2 numbers, 110.011, 110.010, once they have been represented in the IEEE 8-bit toy system. Work out the absolute and relative errors of both the addition and subtraction.

Exercise 09

Calculate the absolute and relative error when $\sin(x)$ is replaced by the truncated expansion

$$\sin(x) \approx x - x^3/6,$$

for $x = 0.1$ and $x = \pi/6$ (all values in radians).

References

[1] Wikipedia 2022 IEEE 754 https://en.wikipedia.org/wiki/IEEE_754 (Accessed: 2022-09-13)

[2] WikiHow 2025 How to divide binary numbers https://www.wikihow.com/Divide-Binary-Numbers (Accessed: 2022-12-14)

[3] Goldberg D 1991 What every computer scientist should know about floating-point arithmetic *ACM Comput. Surv.* **23** 5

[4] Wilkinson J H 1994 *Rounding Errors in Algebraic Processes* (Dover)

[5] Muller J M, Brisebarre N, de Dinechin F, Jeannerod C P, Lefèvre V, Melquiond G, Revol N, Stehlé D and Torres S 2010 *Handbook of Floating-Point Arithmetic* (Birkhäuser)

[6] Higham N J 2002 *Accuracy and Stability of Numerical Algorithms* 2nd edn (SIAM)

IOP Publishing

Computational Physics with R

James Foadi

Chapter 2

Introduction to the R platform

R can be seen as a powerful platform for statistical computing, as a very refined tool for data analysis, as an extraordinarily flexible creator of sophisticated graphics, as a modern and fully fledged programming language and more. In fact, the R platform is one of the most used data analysis and computation tools in the academic and professional industry world, especially because users have found it easy to develop new code to perform many useful operations and tasks [1–5].

R has been around since the 90s and plenty of introductory and specialised material has been written by many authors in the most disparate areas of research. A small selection can be found in the bibliography [6–15].

This chapter is a self-contained, hands-on introduction to the R statistical platform and programming language. It is not meant to be comprehensive and the reader interested in deepening specific aspects of the language should refer to the bibliography suggested. The goal of this chapter is to enable learning and assimilation of basic and intermediate elements of the programming language for the acquisition of the know-how necessary to follow the chapters in parts II and III.

2.1 R and RStudio installation

Details related to installation specifics will be omitted in this book as they are abundantly provided in the official websites, both for R and RStudio. The main reference for R is the website maintained by the R Foundation [16]. The main reference for RStudio is the website maintained by the RStudio Team [17]. RStudio is an integrated development environment (IDE) for R. It has become so popular in recent years as to be considered the premier IDE for R. RStudio is the IDE chosen to manage the R code developed in this book. Other IDEs are widely used and present features that make them popular across different communities of code developers. Among them one can find Tinn-R [18], Emacs-ESS [19], RKWard [20], Eclipse StatET [21] and Rcommander [22].

doi:10.1088/978-0-7503-2632-2ch2

Needless to say, R installs smoothly on all major operating systems. The same goes for RStudio and for the other popular IDEs. Some of the IDEs will complain if R has not yet been installed in the system.

2.2 Running RStudio

When RStudio is started, a window like the one shown in figure 2.1 pops up. Initially, the window is divided into three visible sub-windows, one on the left and two on the right. We will describe the two windows on the right later. The window on the left is the *console*, a reactive text panel in which words or symbols, or more appropriately *commands*, are typed in and in which other words and symbols appear as a response. When RStudio is started, a greeting message appears automatically in the console, including among other things the running version of R and the message `Workspace loaded from /.RData`. This message reveals immediately an important aspect of the language. R is not a compiled language, but rather a scripting language; what is typed in the console is immediately *interpreted* by a program in the R system, called the *R interpreter*, which reads in the typed commands, translates them in code that the computer can act on and write to the console the reaction to the commands in human-readable form. An *R session* is whatever is typed in and answered out from the moment RStudio (or any other IDE) is started, to the moment RStudio is quitted. In any R session, numerical values and variables appear progressively in the console and are kept in the active computer memory. When RStudio is ended, these are all saved, by default, in the so-called *workspace*, which is a file (quite often hidden) in the specific directory RStudio pointed at when started. In the session just started one can see clearly that the workspace is the hidden file `.RData`. The option of having each session saved in a

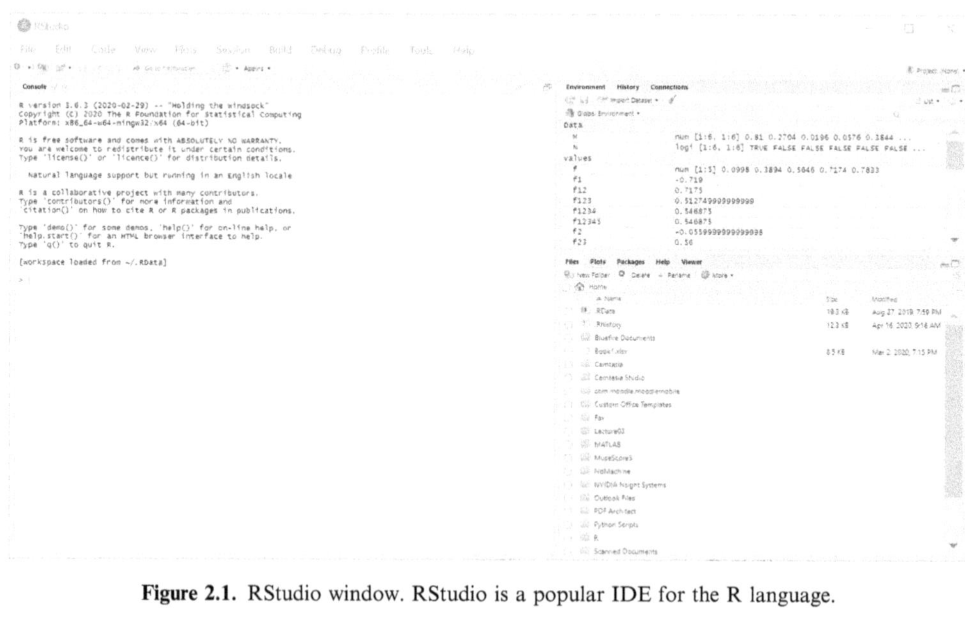

Figure 2.1. RStudio window. RStudio is a popular IDE for the R language.

separate file is a useful one because the user can suspend calculations at any time and reconsider them in a different R session, without losing all steps performed. A demonstration can illustrate the idea practically. The workspace is first filled with two objects containing the numbers 4 and π; then the session is quitted. A prompt is automatically raised by RStudio, asking to save the workspace to .RData. When 'yes' is answered, the two objects with the numbers stored will be automatically saved and are available for further work.

```
# When a line is preceded by the '#' symbol, it's considered
# a comment and it's not interpreted

# To check what's in the workspace, use the command ls()
ls()

## character(0)

# Create an object x and assign it the number 4
x <- 4

# Create an object y and assign it pi (equal 3.14...)
y <- pi

# Check the workspace again: now there are two objects
ls()

## [1] "x" "y"

# Quit with the R command q(), or ending RStudio.
# Save workspace image when prompted to do so.
```

The command `ls()` prompted the answer `character(0)` which means that no objects are stored in the workspace, initially. Then two new objects are created and assigned numeric values, using the *assignment operator* `<-` (a 'less-than' followed by a 'minus'). The = operator can also be used to assign numeric values; there is a slight difference between the two symbols that cannot be appreciated at this stage, but in what follows we will always use the `<-` operator as this characterises the R language in a distinctive way. The answer to the command `ls()` after the two objects `x,y` have been created clearly includes the objects' names.

After a new RStudio session is started, it is immediate to verify that the workspace content has not been lost, again using the command `ls()`.

```
ls()

## [1] "x" "y"
```

The workspace can be managed effectively according to the way users wish to organise their work. Multiple workspaces can be created within the same directory and loaded after RStudio is started. They do not need to be hidden and, in fact, it makes sense to give different specific names to different workspaces. A name different from the default `.RData` is given using the command `save.image (file="FileName.RData")`. The specific workspace can be loaded later using the command `load(file="FileName.RData")`. In the following code snippet the command `rm` to eliminate specific objects in the current workspace is also used.

```
# Save current workspace in a file named w01.RData
save.image(file="w01.RData")

# Eliminate x and y. Now the workspace is empty
rm(x,y)
ls()

## character(0)

# Create new objects
w <- "a"
z <- 2.1
s <- Inf # The symbol for infinity

# Check what's in the current workspace
ls()

## [1] "s" "w" "z"

# Save it in a new file, w02.RData
save.image(file="w02.RData")

# Eliminate the new objects
rm(w,z,s)

# Nothing in the current workspace
ls()
```

```
## character(0)
```

```
# Load objects from the first and second saved workspaces
load(file="w01.RData")
load(file="w02.RData")
ls()
```

```
## [1] "s" "w" "x" "y" "z"
```

2.3 R as a powerful calculator

There is no need to learn programming if one wants to use R only as a calculator. In fact, R is a very powerful calculator including hundreds of mathematical operations and functions. A sample is listed in tables 2.1 and 2.2. All the operators and functions provided can be used to obtain results of easy or complicated calculations straightaway, as shown in the following demonstration.

```
# Simple arithmetics on x1, x2, x3, x4
x1 <- 1
x2 <- 5
x3 <- 2
x4 <- 10

# The value of this operation is stored in x5
x5 <- x1/x2 + x3/x4

# Typing the object x5 displays its value
x5
```

Table 2.1. Some of the operators available in R.

Arithmetic operators		Relational operators		Logical operators	
+	Addition	<	Less than	!	Logical NOT
-	Subtraction	>	Greater than	&&	Logical AND
*	Multiplication	<=	Less than or equal to	&	Element-wise logical AND
/	Division	>=	Greater than or equal to	||	Logical OR
^	Exponent	==	Equal to	|	Element-wise logical OR
%%	Modulus	!=	Not equal to		
%/%	Integer Division				

Table 2.2. Some of the mathematical functions available in R.

Exponential and logarithm		Trigonometric and hyperbolic		Other	
`exp(x)`	e^x	`sin(x)`	$\sin(x) x$ in radians	`abs(x)`	$\lvert x \rvert$
`log(x)`	$\log(x)$ base e	`cos(x)`	$\cos(x) x$ in radians	`sqrt(x)`	$\sqrt{x}$
`log10(x)`	$\log_{10}(x)$ base 10	`tan(x)`	$\tan(x) x$ in radians	`floor(x)`	Maximum integer $<x$
`log(x,n)`	$\log_n(x)$ base n	`asin(x)`	$\sin^{-1}(x)$ value in radians	`ceiling(x)`	Minimum integer $>x$
		`acos(x)`	$\cos^{-1}(x)$ value in radians	`round(x)`	Round the value of x to an integer
		`sinh(x)`	$\sinh(x)$		
		`cosh(x)`	$\cosh(x)$		
		`tanh(x)`	$\tanh(x)$		
		`asinh(x)`	$\sinh^{-1}(x)$		
		`acosh(x)`	$\cosh^{-1}(x)$		
		`atanh(x)`	$\tanh^{-1}(x)$		

```
## [1] 0.4
```

```
# Results from simple relational operations
x5 > 0.4
```

```
## [1] FALSE
```

```
x5 >= 0.4
```

```
## [1] TRUE
```

```
x5 == 0.4
```

```
## [1] TRUE
```

```
# Results from relational operations can be
# stored and re-used with logical operations
x6 <- x5 > 0.4
x6
```

```
## [1] FALSE

x7 <- x5 == 0.4
x7

## [1] TRUE

!x6 # Returns TRUE

## [1] TRUE

x6 && x7 # Returns FALSE

## [1] FALSE

x6 || x7 # Returns TRUE

## [1] TRUE

# Use of mathematical functions

# sine and cosine of 60 degrees
a <- 60*pi/180 # from degrees to radians
sin(a)

## [1] 0.8660254

cos(a)

## [1] 0.5

# Well known trigonometric formula
(cos(a))^2+(sin(a))^2 == 1
```

```
## [1] TRUE
```

```
# Lorentz relativistic factor
c <- 299800000 #Speed of light (m/s)
v <- 1000 # First speed (small)
1/sqrt(1-(v/c)^2) # The factor is essentially 1
```

```
## [1] 1
```

```
v <- 1500000 # Second speed (large)
1/sqrt(1-(v/c)^2)
```

```
## [1] 1.000013
```

```
v <- c # Third speed (speed of light)
1/sqrt(1-(v/c)^2) # The factor becomes infinite
```

```
## [1] Inf
```

2.4 Saving code to repeat calculations. R scripts

Quite often, R is used to test ideas by inputting commands for a series of calculations. Once all steps are complete, for example because a numeric answer has been found, it is useful to store them away for a later re-enactment using different parameters, perhaps changing slightly some of the steps or simply for the purpose of illustrating one's ideas. In fact, all code typed in an RStudio session is stored in a hidden file called `.Rhistory`. When RStudio starts, the default mode is for all typed code included in `.Rhistory` to be available in the console. Each line of code can be re-created with the command `history()`. Using RStudio, the result of this command is displayed in the top part of the right hand side of the window, under the history tab. Specific parts of the code can then be re-generated in the console as shown in the following demonstration.

First, let us clear all previous history by pressing the *brush* symbol in the top, right hand part of the RStudio window; the symbol is shown in figure 2.2. Then let us create a few steps of a specific calculation.

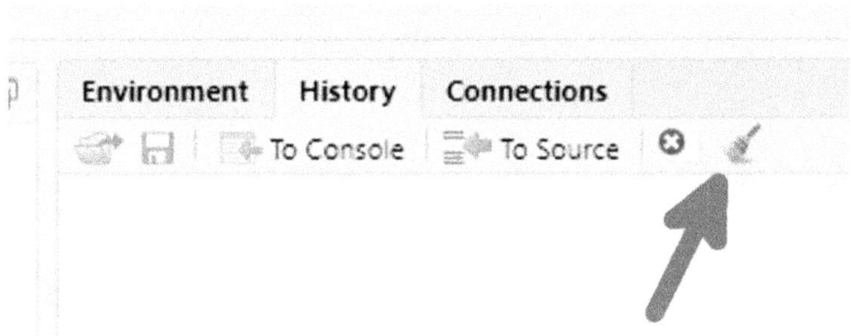

Figure 2.2. Clear-history button in the RStudio window.

```
# Trajectory path of a cannon ball without
# considering air resistance

# Gravity acceleration (metres/second^2)
g <- 9.8

# Initial speed (metres/second)
v <- 50

# Angle (degrees)
alpha <- 30

# Angle in radians
alpha <- alpha*pi/180

# Horizontal and vertical velocities
vx <- v*cos(alpha)
vy <- v*sin(alpha)

# Time of flight
tf <- 2*vy/g

# Range (metres)
D <- vx*tf
D

## [1] 220.9248
```

All instructions typed in are now clearly visible in the history tab (see figure 2.3).

Lines that need to be re-executed can be selected (they will be highlighted) and re-generated in the console using the *To Console* button, also available under the history tab (see figure 2.4). Other ways to re-generate typed commands from the .

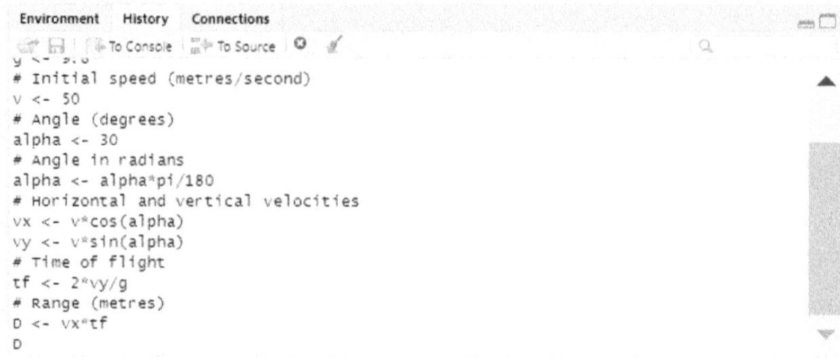

Figure 2.3. History content in the RStudio window.

Figure 2.4. To re-generate (and execute) commands available under the history tab, it suffices to highlight the command needed and, after the selection is complete, to push the *To Console* button.

Rhistory file are available when using different IDEs and they can be found out searching the specific IDE documentation.

Another, perhaps more popular and useful, way to re-generate commands previously typed is *via* the so-called *scripts*. Very simply, an *R script* is a collection of R commands in a single file. Traditionally, this file has extension `.R`. Commands stored in an R script are normally referred to as *R code*. An R script can be swiftly generated pushing the *Save* button (blue floppy-disk symbol) under the history tab. For example, once all lines of the history previously displayed are saved to a file named `cannon_ball.R`, its content will be:

```
# Trajectory path of a cannon ball without
# considering air resistance
# Gravity acceleration (metres/second^2)
g <- 9.8
# Initial speed (metres/second)
v <- 50
# Angle (degrees)
alpha <- 30
# Angle in radians
alpha <- alpha*pi/180
# Horizontal and vertical velocities
vx <- v*cos(alpha)
vy <- v*sin(alpha)
# Time of flight
tf <- 2*vy/g
# Range (metres)
D <- vx*tf
D
alpha <- 30
alpha <- alpha*pi/180
vx <- v*cos(alpha)
vy <- v*sin(alpha)
tf <- 2*vy/g
D <- vx*tf
D
```

To execute all code in an R script like `cannon_ball.R`, use the command `source` from within the console. Various options can be added to this command. For example, when `print.evaluated` is set to `TRUE`, the command `source` will also display the value of the given objects, when executed.

```
source(file="cannon_ball.R",print.eval=TRUE)
```

```
## [1] 220.9248
## [1] 220.9248
```

An R script can be edited before being executed in the console, with `source`. In RStudio this is done with the button indicated by the red arrow in figure 2.5. The file is then shown in a fourth panel on the left side of the RStudio window, ready to be edited and saved. We could, for example, change the second value of `alpha` to 45 (instead of 30). Once the script is saved and sourced in the console, this output is shown:

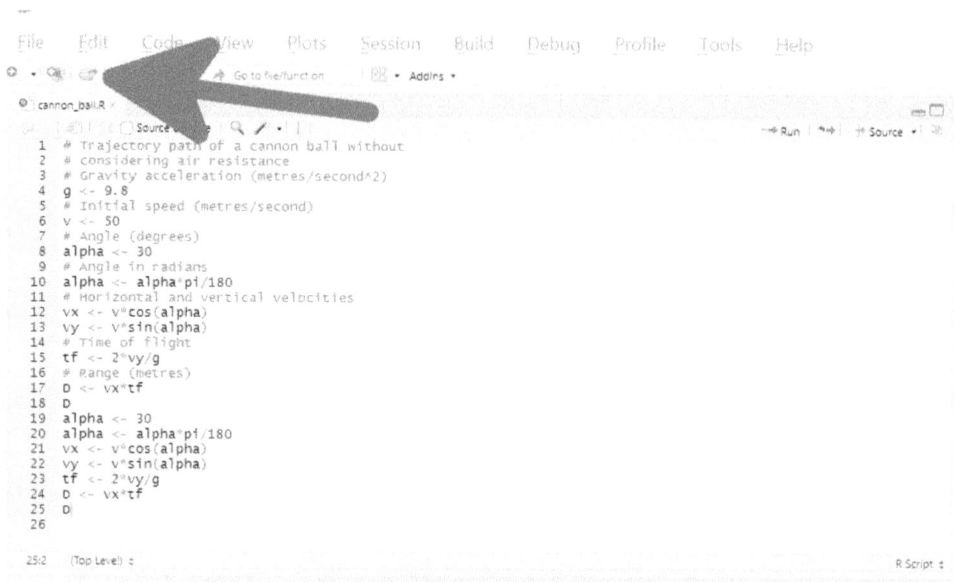

Figure 2.5. How to open and edit an R script in RStudio.

```
source(file="cannon_ball.R",print.eval=TRUE)
```

```
## [1] 220.9248
## [1] 255.102
```

The second printed value appears now different from the first. New R scripts to be edited *de novo* can be created with the third button on the left of the red arrow shown in figure 2.5.

2.5 Vectors

One of the most useful features of R is the possibility of carrying out the same set of operations on multiple objects at the same time with just one command. Such operations are supposed to be applied to so-called *vectors*, a special class of R objects containing the same data type multiple times. The easiest way to create a vector is via the `c()`, or *concatenation*, function. For instance, to create a vector containing five numbers, we type:

```
v <- c(1,-1,2.5,0,13)
v
## [1]  1.0 -1.0  2.5  0.0 13.0
```

A longer vector takes more time to be typed in and for this reason there exist several ways to generate long vectors with regular features. For instance, the command line `1:20` generates a vector of length 20 with components 1, 2, 3, …, 19, 20. This syntax works also backwards and with decimals. Vectors can also have characters (strings of letters) as components. The number of components of any vector (its length) can be found using the command `length`.

```
v <- 1:20

# The number at the start of each display line
# is the index of the first vector component
# being displayed on that line
v
```

```
##  [1]  1  2  3  4  5  6  7  8  9 10 11 12 13 14 15 16 17 18
## [19] 19 20
```

```
# Regular-patterned vectors can start from
# negative numbers and can be decimals
v <- -2:5
v
```

```
## [1] -2 -1  0  1  2  3  4  5
```

```
v <- -1.2:20
v
```

```
##  [1] -1.2 -0.2  0.8  1.8  2.8  3.8  4.8  5.8  6.8  7.8  8.8
## [12]  9.8 10.8 11.8 12.8 13.8 14.8 15.8 16.8 17.8 18.8 19.8
```

```
length(v)
```

```
## [1] 22
```

```
# Vectors can also filled in a backward fashion
v <- 20:-10
v
```

```
##  [1]  20  19  18  17  16  15  14  13  12  11  10   9   8   7
## [15]   6   5   4   3   2   1   0  -1  -2  -3  -4  -5  -6  -7
## [29]  -8  -9 -10
```

```
# Vectors can also be formed by characters (strings)
v <- c("my","name","is","James")
v
```

```
## [1] "my"    "name"  "is"    "James"
```

```
length(v)
```

```
## [1] 4
```

A flexible way to create vectors with regularly-spaced components (*grids*) is with the command `seq`. The advantage of using this command is that the selected start and end can be made to correspond to the first and last component of the vector.

```
# Regular grid between -1 and 1, including 11 points
x <- seq(-1,1,length=11)
x
```

```
##  [1] -1.0 -0.8 -0.6 -0.4 -0.2  0.0  0.2  0.4  0.6  0.8  1.0
```

```
# The same grid created using the step's width
x <- seq(-1,1,by=0.2)
x
```

```
##  [1] -1.0 -0.8 -0.6 -0.4 -0.2  0.0  0.2  0.4  0.6  0.8  1.0
```

Vectors with components having a same value (constant vectors) can be created with the function `rep`.

```
# Null 3D vector
x <- rep(0,times=3)
x
```

```
## [1] 0 0 0
```

```
# Vector with 24 "pi's"
x <- rep(pi,times=24)
x
```

```
##  [1] 3.141593 3.141593 3.141593 3.141593 3.141593 3.141593
##  [7] 3.141593 3.141593 3.141593 3.141593 3.141593 3.141593
## [13] 3.141593 3.141593 3.141593 3.141593 3.141593 3.141593
## [19] 3.141593 3.141593 3.141593 3.141593 3.141593 3.141593
```

In fact, a complicated numerical pattern can be created with an appropriate use of both `seq` and `rep` commands, as shown here.

```
# Alternate -1 and 1 five times
x <- rep(c(-1,1),times=5)
x
```

```
##  [1] -1  1 -1  1 -1  1 -1  1 -1  1
```

```
# Repeat 2 four times and 3 two times
x <- rep(c(2,3),times=c(4,2))
x
```

```
## [1] 2 2 2 2 3 3
```

```
# Two identical grids
x <- rep(seq(0,1,by=0.1),times=2)
x
```

```
##  [1] 0.0 0.1 0.2 0.3 0.4 0.5 0.6 0.7 0.8 0.9 1.0 0.0 0.1 0.2
## [15] 0.3 0.4 0.5 0.6 0.7 0.8 0.9 1.0
```

```
# Each element of the grid is repeated twice
# Notice, the second rep has times=11 because the
# grid has length 11
x <- rep(seq(0,1,by=0.1),times=rep(2,times=11))
x
```

```
##  [1] 0.0 0.0 0.1 0.1 0.2 0.2 0.3 0.3 0.4 0.4 0.5 0.5 0.6 0.6
## [15] 0.7 0.7 0.8 0.8 0.9 0.9 1.0 1.0
```

Vector components can be accessed using the vector name, followed by bracketed, `[]`, values. Several extraction patterns are possible. They all consider the components position as the component of an index vector. An interesting feature of the mechanism is that an index preceded by a 'minus' sign means exclusion of that specific index from the extraction.

```
# Grid of length 11
x <- seq(0,1,length=11)
x
```

```
##  [1] 0.0 0.1 0.2 0.3 0.4 0.5 0.6 0.7 0.8 0.9 1.0
```

```
# Access 5th component
x[5]
```

```
## [1] 0.4
```

```
# Access 5th and 8th component
x[c(5,8)]
```

```
## [1] 0.4 0.7
```

```
# Access all components from the second to the sixth
x[2:6]
```

```
## [1] 0.1 0.2 0.3 0.4 0.5
```

```
# Extract three central components and save them to a new vector
y <- x[5:7]
y

## [1] 0.4 0.5 0.6

length(y)

## [1] 3

# Omit second component
x[-2]

##  [1] 0.0 0.2 0.3 0.4 0.5 0.6 0.7 0.8 0.9 1.0

# Omit first and last component
x[-c(1,length(x))]

## [1] 0.1 0.2 0.3 0.4 0.5 0.6 0.7 0.8 0.9
```

2.6 Operations on vectors

Binary arithmetics, relational and logical operations that use vectors are performed in an element-wise fashion, i.e. element by element. Here is a sample of the many possibilities that element-wise operations entail.

```
# Grid of length 11
# Create two different vectors of a same length
x <- c(1,2,3,4,5)
y <- rep(2,times=5)

# Powers x^y
x

## [1] 1 2 3 4 5
```

```
y

## [1] 2 2 2 2 2

x^y

## [1]  1  4  9 16 25

# The same operation can be done when the exponent
# is just one number (it is recycled)
x^2

## [1]  1  4  9 16 25

# Other elemet-wise operations
x + y

## [1] 3 4 5 6 7

x - y

## [1] -1  0  1  2  3

x*y

## [1]  2  4  6  8 10

x/y

## [1] 0.5 1.0 1.5 2.0 2.5

# Comparison
z <- x < y
z
```

```
## [1]  TRUE FALSE FALSE FALSE FALSE
```

```
w <- x == y
w
```

```
## [1] FALSE  TRUE FALSE FALSE FALSE
```

```
# Elementwise logical operations
z & w
```

```
## [1] FALSE FALSE FALSE FALSE FALSE
```

```
z | w
```

```
## [1]  TRUE  TRUE FALSE FALSE FALSE
```

Probably the most important fact to remember when using binary operations on vectors is *recycling*. As seen in the examples, it is important that the two vectors involved in the element-wise operation have the same length. If this is not the case, R does not stop execution, but recycles the shortest vector so as to have the same length as the other one, simply adding as many components from component 1, as are needed. This mechanism is illustrated here.

```
# Vector y is shorter than vector x
x <- 1:6
y <- 1:4

# Vector y is recycled into c(1,2,3,4,1,2)
x
```

```
## [1] 1 2 3 4 5 6
```

```
y
```

```
## [1] 1 2 3 4
```

```
x+y
```

```
## Warning in x + y:  longer object length is not a multiple of
shorter object length

## [1] 2 4 6 8 6 8
```

The warning generated automatically from R is meant to make the user aware that the two vectors used do not have the same length.

Most of the functions available in R can be used with vectors too. A function applied on a vector of length n will yield another vector of length n whose element i is the result of applying the function to element i of the original vector. It goes without saying that this is a very useful feature of the language because it makes it possible to carry out the same task in parallel.

```
# Grid of radians values between 0 and 2*pi
x <- seq(0,2*pi,length=13)
x
```

```
##  [1] 0.0000000 0.5235988 1.0471976 1.5707963 2.0943951
##  [6] 2.6179939 3.1415927 3.6651914 4.1887902 4.7123890
## [11] 5.2359878 5.7595865 6.2831853
```

```
# Sine function calculated for all 13 points
# of the grid
y <- sin(x)
y
```

```
##  [1]  0.000000e+00  5.000000e-01  8.660254e-01  1.000000e+00
##  [5]  8.660254e-01  5.000000e-01  1.224606e-16 -5.000000e-01
##  [9] -8.660254e-01 -1.000000e+00 -8.660254e-01 -5.000000e-01
## [13] -2.449213e-16
```

```
# Cosine
z <- cos(x)

# Sin^2 + cos^2 = 1
y^2+z^2
```

```
##  [1] 1 1 1 1 1 1 1 1 1 1 1 1 1
```

2.7 Exercises on R vectors

Exercise 01

Generate and display:

1. A vector of values $-2, -1, 0, 1, 2, 3, 4, 5, 6$.
2. A vector of values $10, 9, 8, 7, 6$.
3. A vector of values $0, 0.5, 1, 1.5, 2, 2.5, 3$, starting from the vector with values $0, 1, 2, 3, 4, 5, 6$.

Exercise 02

Generate a regular grid between $-\pi$ and $+\pi$ of 100 points. Find out the length Δx between two contiguous points of the grid.

Exercise 03

Create the following pattern,

$$1 \quad 1 \quad 1 \quad 2 \quad 2 \quad 3$$

using function `rep`.

Exercise 04

Create the numeric pattern,

$$1 \quad 2 \quad 3 \quad 2 \quad 2 \quad 1 \quad 2 \quad 3 \quad 2 \quad 2$$

using function `rep`.

Exercise 05

Create the numeric pattern,

$$1 \quad 1 \quad 1 \quad 2 \quad 2 \quad 3 \quad 1 \quad 1 \quad 1 \quad 2 \quad 2 \quad 3 \quad 1 \quad 1 \quad 1 \quad 2 \quad 2 \quad 3$$

using function `rep`.

Exercise 06

Create a vector `x` of length 30 using the following expression:

```
set.seed(123)
x <- sample(seq(0,1,length=101),size=30,replace=TRUE)
```

Print the value of the 9th, 18th and 27th component of the vector `x`. The `set.seed` function fixes the generation of the pseudo-random numbers so that the above code will always output the same numbers.

Exercise 07

Consider the vector x with the following components:

$$0 \quad 1 \quad 2 \quad 3 \quad 4 \quad 5 \quad 6 \quad 7 \quad 8 \quad 9$$

Select and print only the components of x which contain even numbers. Then print the other components.

Exercise 08

Using a regular grid x of 361 values in the range $[0,\ 2\pi]$, calculate the corresponding values of the function

$$2\sin(x) - \cos(x),$$

and store them in a vector called y. Find the indices of x corresponding to 0, π and 2π, and verify that the values of y at these positions are -1, 1 and -1.

Exercise 09

Consider the two vectors x and y of different lengths:

$$x: \quad 1 \quad 2 \quad 3 \quad 4 \quad 5, \qquad y: \quad 2 \quad 4$$

What do you expect the result of `x + y` to be? Justify your answer and verify its correctness with R.

2.8 Documentation

As with all programming languages, R cannot be fully learnt in one go. In general users learn a very small subset of the R most common commands, those employed to carry out the tasks useful to them. More commands are met and assimilated as one delves deeper and deeper into the syntax. Each unknown command must to be explored to understand its input, output and underlying algorithms. Documentation for any command is available under the tab *Help* in the bottom-right window of RStudio. The same documentation can also be called with the command `help (command-name)` or the equivalent command `?command-name`. Each documentation includes many lines of explanation and examples with code that can actually be copied and pasted into the console. The examples can also be generated directly in the console via the command `example(command-name)`.

Other types of documentation are available abundantly on the internet in the form of demonstrations, tutorials, discussion lists, YouTube videos, etc.

2.9 Graphics commands

R includes a large variety of graphics commands. Indeed, one of the main attractions of the platform is the accuracy and beauty of the graphics produced. This is made possible by several functions available in the default installation of the platform and many more available in software components (the so-called *packages*) that can be added to the main platform at any time after initial installation. In this section we will explore the most basic and commonly-used functions to produce certain types of graphics. More functions will be explored later in the chapter. A separate section is devoted to an important graphics package called `ggplot2` [23] which takes the production of graphics to a very high level.

The most widespread type of graph used is the plot representing the relation between two variables. It is produced via the function `plot`. At the easiest level `plot` takes in two numeric arguments, appropriately called `x,y`, that need to be same-length vectors.

```
# Parabola between -1 and 1 (x and y have same length)
x <- seq(-1,1,length=100)
y <- x^2

# With default options, the plot is made of open circles
plot(x,y)
```

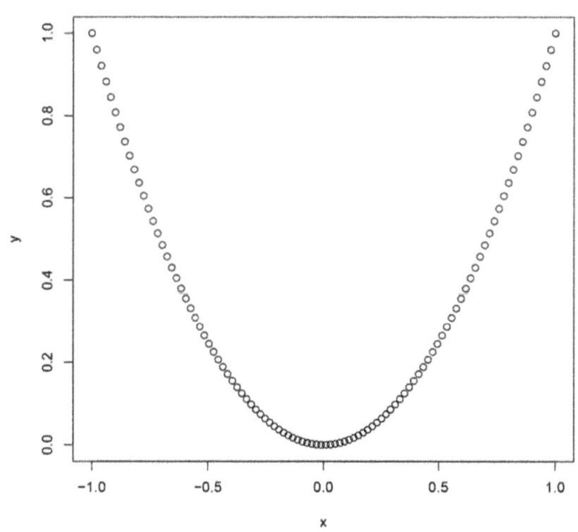


```
# The type of plot provides major changes
plot(x,y,type="l") # Lines
```

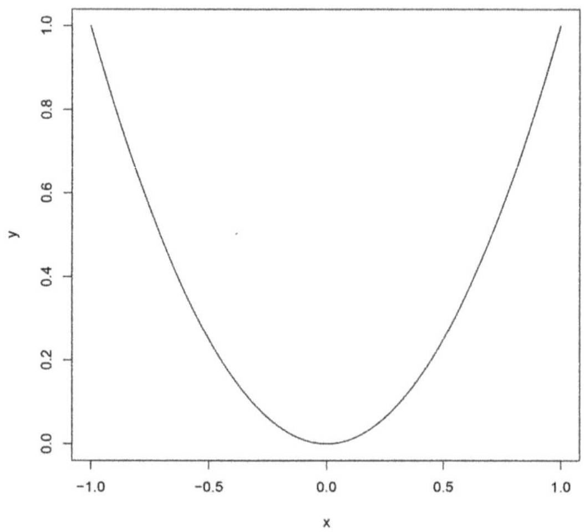


```
# The plotted symbol can be changed into other symbols
plot(x[1:50],y[1:50],pch=16) # Full circles
```

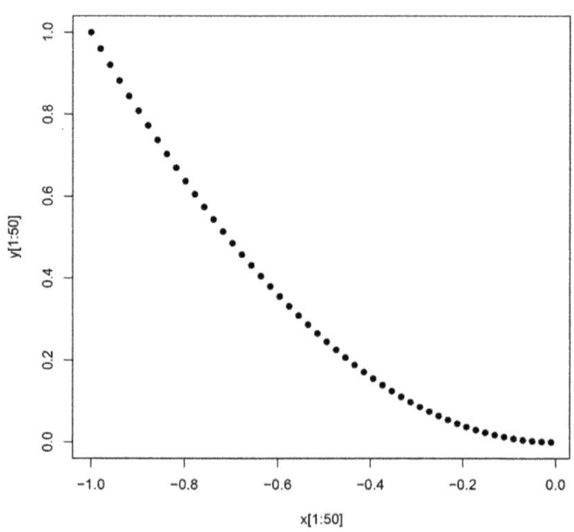


```
plot(x[51:100],y[51:100],pch=3)  # Crosses
```

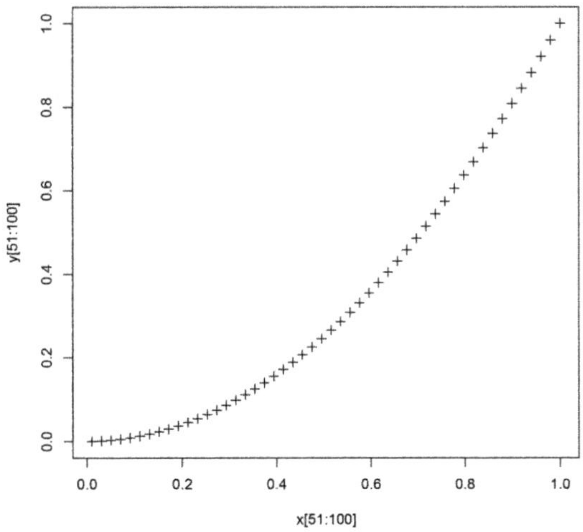


```
# Sine between 0 and 2*pi
x <- seq(0,2*pi,length=20)
y <- sin(x)

# Both coloured symbols and lines
plot(x,y,type="b",col="green")
```

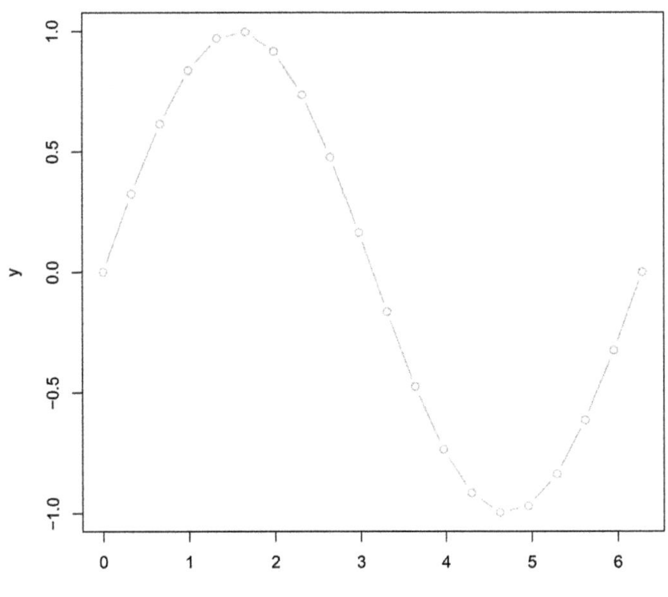

From the simple plots just shown, it is clear that their look and complexity depends on the options used within the command `plot`. Some of them are listed in table 2.3. Multiple plots on the same graph can be achieved with the command `points` as many times as needed, after having used `plot` the first time. A legend and more *decorations* can be added to the same graph with the command `legend` and variants related to plotting. The following snippet shows some of the improvements that can be added to make a graph more appealing.

```
# Sine and cosines between 0 and 2*pi
x <- seq(0,2*pi,length=100)
y <- sin(x)
z <- cos(x)

# Both vurves below have width 2
# Initial plot (for sine) with title and axes labels
plot(x,y,type="l",xlab="x",ylab="Sine and Cosine",
lwd=2,main="Trigonometric Functions")



# Add cosine as a red curve with command points
points(x,z,type="l",col=2,lwd=2)

# Where sine=cosine
x1 <- pi/4
y1 <- sqrt(2)/2
x2 <- 5*pi/4
y2 <- -sqrt(2)/2
points(x1,y1,pch=3,cex=3,col=4,lwd=2)
points(x2,y2,pch=3,cex=3,col=4,lwd=2)

# Add legend
legend(x=pi,y=0.9,
legend=c("Sine","Cosine","Sine = Cosine"),
col=c(1,2,4),
pch=c(-1,-1,3),
lty=c(1,1,-1),
lwd=2)
```

Table 2.3. Some valid options for the `plot` graphics command.

Option	Description	Some values	
Type	1-Character string giving the type of plot desired	`"p"`	Points
		`"l"`	Lines
		`"b"`	Both
		`"c"`	Empty points joined by lines
		`"o"`	Overplotted points and lines
		`"s","S"`	Stair steps
		`"h"`	Histogram-like vertical lines
		`"n"`	No points or lines
col	The colors for lines and points	1	or `"black"`
		2	or `"red"`
		3	or `"green"`
		4	or `"blue"`
		5	or `"cyan"`
		6	or `"magenta"`
		7	or `"yellow"`
		8	or `"grey"`
pch	Plotting characters or symbols	1	Empty circle
		2	Empty upward triangle
		3	Cross
		4	X
		5	Empty diamond
		8	Star
		15	Full box
		16	Full circle
		17	Full upward triangle
		18	Full diamond
lty	Line types	1	Solid
		2	Dashed
		3	Dotted
		4	Dot-dash
		5	Long-dash
		6	Two-dash
xlim	The x limits (x1, x2) of the plot		
ylim	The y limits (y1, y2) of the plot		
main	Main title for the plot		
sub	Sub title for the plot		
xlab	Label for the x-axis, defaults to a description of `x`		
ylab	label for the y-axis, defaults to a description of `y`		
lwd	line width (1,2,3,...)		

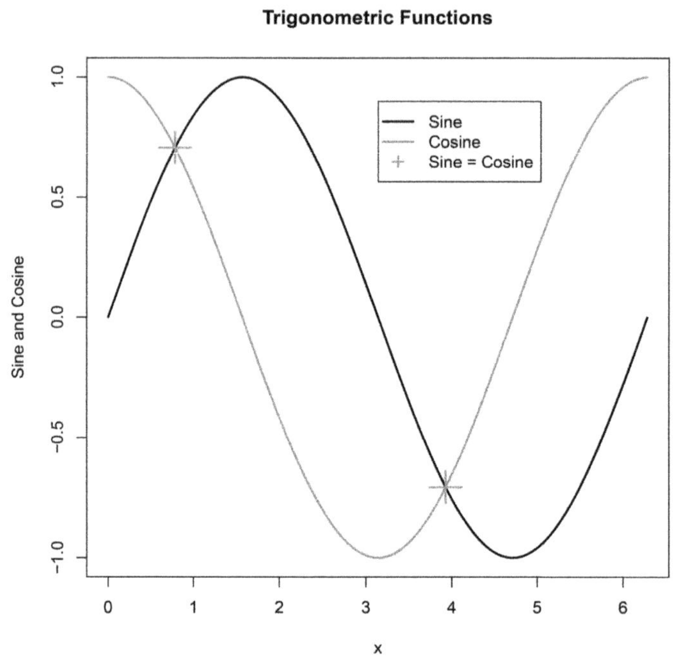

It is perhaps important to deepen the way in which a legend is constructed. For each element of the plot added (in this case one black curve, one red curve and two blue crosses), one text in the parameter `legend` is included (`"Sine","Cosine","Sine=Cosine"`). Each one of these is then coloured with the parameter `col` and it is assigned the corresponding line type or symbol. Here it is useful to observe that a -1 means 'no symbol' or 'no line' assigned.

It is also possible to plot functions whose analytic form is known, using the command `curve`. This command includes the range of the curve plotted. More curves can be included in the same graph, turning the default `add=FALSE` option into `add=TRUE`.

```
# Sine between 0 and 3*pi
curve(sin(x),from=0,to=3*pi,lwd=2)

# Add cosine between pi/4 and 5*pi/4
curve(cos(x),from=pi/4,to=5*pi/4,
col=2,lwd=2,add=TRUE)
```

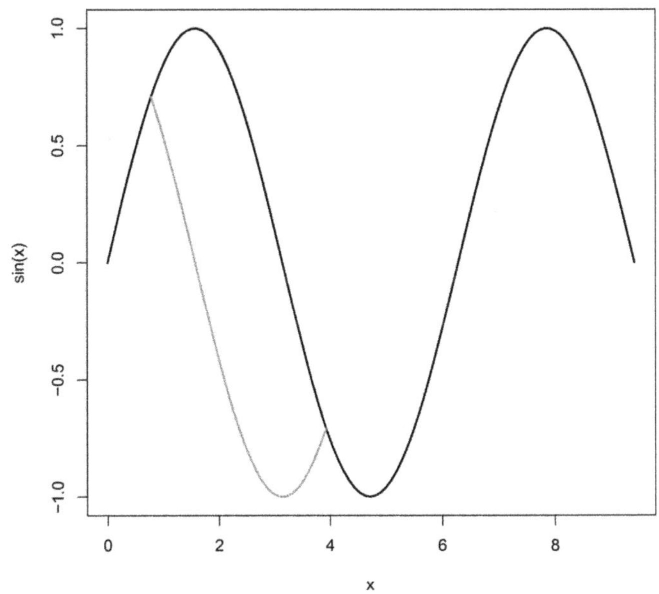

2.10 Exercises on simple graphics

Exercise 10

Using the function `plot`, draw an empty square with vertices at (0, 0), (0, 1), (1, 1) and (1, 0), in black. Draw also the two diagonals in red.

Repeat the same pattern two times, using function `rect` first, and `polygon` second. In both cases, the diagonals can be drawn using the function `segments`.

Make sure to use inline help and/or information on the internet for all functions needed.

Exercise 11

Plot the function $f(x) = |x| + x^2$ in the interval $x \in [-2, 3]$, on two separate plots using, respectively, `plot` and `curve`. For the second plot, use a green, dashed line.

Exercise 12

In the interval $x \in [-\pi, \pi]$, draw the following three experimental points,

x	$y = f(x)$
$-\pi$	-1
0	0
π	1

as black upward triangles, and the following three curves:

a. $f(x) = \sin(x)$, in red
b. $f(x) = x^3 + (1/\pi + \pi^2)x$, in green
c. $f(x) = x/\pi$, in blue.

Make sure that the second curve is a line twice as thick as the other two curves.

2.11 R functions

The R commands are, in fact, better known as *R functions*. In R all commands correspond to functions. A *function* is a chunk of code that takes in some input and returns, when called, some output. The functional structure of any command can be revealed simply by typing the command without brackets in the console. This is what we obtain, for example, with the command `ls`.

```
ls

## function (name, pos = -1L, envir = as.environment(pos),
all.names = FALSE,
##     pattern, sorted = TRUE)
## {
##     if (!missing(name)) {
##         pos <- tryCatch(name, error = function(e) e)
##         if (inherits(pos, "error")) {
##             name <- substitute(name)
##             if (!is.character(name))
##                 name <- deparse(name)
##             warning(gettextf("%s converted to character string",
##                 sQuote(name)), domain = NA)
##             pos <- name
##         }
##     }
##     all.names <- .Internal(ls(envir, all.names, sorted))
##     if (!missing(pattern)) {
##         if ((ll <- length(grep("[", pattern, fixed = TRUE))) &&
##             ll != length(grep("]", pattern, fixed = TRUE))) {
##             if (pattern == "[") {
##                 pattern <- "\\["
##                 warning("replaced regular expression pattern
##                  '[' by'\\\\['")
##             }
##             else if (length(grep("[^\\\\]\\[<-", pattern))) {
##                 pattern <- sub("\\[<-", "\\\\\\[<-", pattern)
##                 warning("replaced '[<-' by '\\\\[<-' in regular
expression pattern")
##             }
##         }
##         grep(pattern, all.names, value = TRUE)
##     }
##     else all.names
## }
## <bytecode: 0x0000000013c95918>
## <environment: namespace:base>
```

When using **R** we can thus look into the source code associated with any function. The important elements of a function are:

(a) Function name, e.g. `ls`.
(b) Function arguments, i.e. a list of objects taken in by the function as input for the internal code. For example some of the arguments of the `ls` function are `name`, `pos`, etc.
(c) Opening and closing braces, { }, between which all internal code is strictly to be located.

A simple example will show how easy is to create a function named arbitrarily `silly_function`. It takes in two numbers and returns a vector of length 3. The first component is the sum of the input numbers, the second component their difference and the third component their product.

```
# Definition of function. All code is between braces
silly_function <- function(a,b) {
# Define a vector with three zeros
v <- rep(0,3)

# First component is the sum
v[1] <- a+b

# Second component is the difference
v[2] <- a-b

# Third component is the product
v[3] <- a*b

# Output: vector v
return(v)
}

# Now let's make use of `silly_function'
silly_function(2,3)
```

```
## [1]  5 -1  6
```

```
w <- silly_function(0,-2)
w
```

```
## [1] -2  2  0
```

An important feature of a function is the possibility to omit input arguments when calling them, because they have been assigned default values. This is, in fact, a powerful tool in the hands of the programmer as it allows certain parameters to have suggested values preserving, at the same time, the freedom to change their pre-assigned values. For instance, in the case of the function just defined, one could add a third input argument, a logical parameter, which causes the third component of the output always to be zero, when the logical parameter is equal to `TRUE` (the `if` construct present in the amended code will be explained later). The default value assigned to the third parameter is `FALSE`, indicating that in general the third component of the output is not forced to be zero.

```
# Definition of function. All code is between braces
silly_function <- function(a,b,Z=FALSE) {
# Define a vector with three zeros
v <- rep(0,3)

# First component is the sum
v[1] <- a+b

# Second component is the difference
v[2] <- a-b

# Third component is the product only
# if Z=FALSE
if (!Z) v[3] <- a*b

# Output: vector v
return(v)
}

# Now let's make use of `silly_function'
silly_function(2,3,TRUE)
```

```
## [1]  5 -1  0
```

```
w <- silly_function(1,-2)
w
```

```
## [1] -1  3 -2
```

2.12 Exercises on R functions

Exercise 13

Write a function with input `n` that outputs the first n integers as a vector.

Exercise 14

Modify the previous function (remember to change the name in order to preserve the previous function) to create a vector between two integer numbers.

Exercise 15

Write a function `spc` to sum the sine and cosine of any angle `x` and a function `smc` that calculates the difference between the sine and cosine of any angle `x`. Finally, write a function that calculates the ratio

$$T(x) = \frac{\sin(x) + \cos(x)}{\sin(x) - \cos(x)},$$

for any given angle. Verify that the result is identical to the one returned by the function

$$\frac{\tan(x) + 1}{\tan(x) - 1}.$$

Finally, use the input `x <- c(0,pi/6,pi/4,pi/3,pi/2)` in the last function. What do you think is happening?

2.13 Objects, data types and data structures

Everything in R is an *object*. Each object can be characterised by the type of data it is associated with. Such *data types* belong to one of the following six categories, *logical*, *numeric*, *integer*, *complex*, *character*, *raw*, also summarised in table 2.4. A data type can be ascertained using `typeof`. When using this command we can think of the value returned as to a property of the specific R object. Thus, we can have vectors of differing lengths all presenting the same data type.

Table 2.4. R data types.

Data type	Description	Example
Logical	Boolean values	`TRUE,FALSE`
Numeric	Integer and real numbers	`2,-2,3.14`
Integer	Numbers explicitly integers	`13L,2L`
Complex	Complex numbers with a real and imaginary part	`0+1i,1+0i,2+3i`
Character	Single characters or strings of characters	`"a","A","Hello!"`
Raw	Raw bytes representation	`48 65 6c 6c 6f 21` is the hexadecimal representation of 'Hello!'

```
# Create a logical, a numeric, an integer and a complex object
w <- 2 == 3
x <- pi/2
y <- 10L
z <- c(0+1i,-1+0i)

# Create a character and a raw object
A <- c("Hello","world")
B <- as.raw(132)
B # Printed in hexadecimal form: 8*16 + 4*1
```

```
## [1] 84
```

```
# What data type? - Irrespective of whether
# or not they are vectors
typeof(w)
```

```
## [1] "logical"
```

```
typeof(x)
```

```
## [1] "double"
```

```
typeof(y)
```

```
## [1] "integer"
```

```
typeof(z)
```

```
## [1] "complex"
```

```
typeof(A)
```

```
## [1] "character"
```

```
typeof(B)
```

```
## [1] "raw"
```

In fact, a vector is just one of the possible *data structures* available in R. There are five base data structures: the *vector* (also known as atomic vector), the *matrix*, the *array*, the *list* and the *data frame*. They can be organised in a table once their dimensionality (1d, 2d and *n*d) and their homogeneity are taken into account:

	Homogeneous	Heterogeneous
1d	Vector	List
2d	Matrix	Data Frame
nd	Array	

So far we have only practiced with vectors. In the next sections we will introduce and practice with the other base data structures. The best way to learn about data type and structure of an object is through the command `str`.

```
# Simple vectors of numbers and letters
x <- seq(0,1,length=5)
y <- LETTERS[1:6]

# str returns the type and the length. No mention
# of data structure means it's a vector
str(x)
```

```
## num [1:5] 0 0.25 0.5 0.75 1
```

```
str(y)
```

```
## chr [1:6] "A" "B" "C" "D" "E" "F"
```

```
# Generate an identity matrix of size 4. str
# returns no name for the data structure, but the
# two sets of numbers reveal it's a matrix
M <- diag(4)
M
```

```
##      [,1] [,2] [,3] [,4]
## [1,]    1    0    0    0
## [2,]    0    1    0    0
## [3,]    0    0    1    0
## [4,]    0    0    0    1
```

```
str(M)
```

```
## num [1:4, 1:4] 1 0 0 0 0 1 0 0 0 0 ...
```

```
# List of datasets built in R. str returns information
# that this is a list and then provides names, etc.
# This output has been here truncated to save space.
ltmp <- data()
str(ltmp)
```

```
## List of 4
##  $ title  : chr "Data sets"
##  $ header : NULL
...
...
```

```
# One of the many data frames available in R, mtcars.
# str returns the structure name, data frame, and other
# information (here shortened to save space).
str(mtcars)
```

```
## 'data.frame': 32 obs. of  11 variables:
##  $ mpg : num  21 21 22.8 21.4 18.7 18.1 14.3 24.4 22.8 19.2 ...
##  $ cyl : num  6 6 4 6 8 6 8 4 4 6 ...
...
...
```

Besides `typeof` and `str`, two more commands can be used to explore R objects. `length` returns the number of components of the object. For a vector, `length` returns the vector size, for a matrix it returns the number of elements in the matrix (in the example above, `M` has $4 \times 4 = 16$ elements), for a list it returns the number of elements (or fields) of the list and for a data frame it returns the number of its variables. These are the lengths of the five objects in the code displayed above.

```
length(x)
```

```
## [1] 5
```

```
length(y)
```

```
## [1] 6
```

```
length(M)
```

```
## [1] 16
```

```
length(ltmp)
```

```
## [1] 4
```

```
length(mtcars)
```

```
## [1] 11
```

The command `attributes` returns the so-called *metadata*, arbitrarily associated with the object in order to make it different from other objects. Attributes can be thought of as a named list (see later for the list data structure). While the command `attributes` displays all attributes of a given object, the command `attr` displays information on a specific attribute.

```
# x and y are vectors without attributes
attributes(x)
```

```
## NULL
```

```
attributes(y)
```

```
## NULL
```

```
# The only attribute of a matrix is its dimension
attributes(M)
```

```
## $dim
## [1] 4 4
```

```
# A list has two attributes
attributes(ltmp)
```

```
## $names
## [1] "title"   "header"  "results" "footer"
##
## $class
## [1] "packageIQR"
```

```
# Each attribute can be listed separately with attr()
attr(M,"dim")
```

```
## [1] 4 4
```

```
attr(ltmp,"names")
```

```
## [1] "title"   "header"  "results" "footer"
```

```
attr(ltmp,"class")
```

```
## [1] "packageIQR"
```

Clearly the attributes of an object are meant to add various types of information to an object without substantially changing the structure of that object. They are a very useful feature of R. R objects can also be tested for their specific type and structure using the `is.` series of commands, as shown here.

```
# The objects used here have been created previously for
# the demonstrations of section 2.10

# The type of x and M is numeric
is.numeric(x)
```

```
## [1] TRUE
```

```
is.numeric(M)
```

```
## [1] TRUE
```

```
# The type of w is logical
is.logical(w)
```

```
## [1] TRUE
```

```
# The type of y is character
is.character(y)
```

```
## [1] TRUE
```

```
# w,x,y are all vectors - is.atomic is the correct test
is.atomic(w)
```

```
## [1] TRUE
```

```
is.atomic(x)
```

```
## [1] TRUE
```

```
is.atomic(y)
```

```
## [1] TRUE
```

```
# M has "matrix" structure
is.matrix(M)
```

```
## [1] TRUE
```

Many more `is.` commands are obviously available to test the type and structure of the many R objects.

2.14 Matrices and arrays

Matrices are very important and ubiquitous mathematical quantities in physics. Arrays are not as important, but can be associated with mathematical objects like tensors. It is, thus, important to know how these two different objects are implemented in R. We can think of matrices as containers of numbers, characters or other homogeneous data type with dimension 2. Arrays can have two or more dimensions; thus, technically, an array is also a matrix. In fact, both matrices and arrays are vectors with a `dim` (dimension) attribute, which is 2 for matrices and 2 or higher integers for arrays. There are specific commands to create matrices and arrays. They are, respectively, `matrix` and `array`. It is important to know that the individual components of a matrix are automatically filled by columns, unless explicitly forbidden to do so. Arrays follow a similar filling pattern.

```
# A 3 X 6 matrix and a 2 X 3 X 3 array contain 18 elements
M <- matrix(LETTERS[1:18],nrow=3)
M
```

```
##      [,1] [,2] [,3] [,4] [,5] [,6]
## [1,] "A"  "D"  "G"  "J"  "M"  "P"
## [2,] "B"  "E"  "H"  "K"  "N"  "Q"
## [3,] "C"  "F"  "I"  "L"  "O"  "R"
```

```
A <- array(1:18,c(2,3,3))
A
```

```
## , , 1
##
##      [,1] [,2] [,3]
## [1,]    1    3    5
## [2,]    2    4    6
##
## , , 2
##
##      [,1] [,2] [,3]
## [1,]    7    9   11
## [2,]    8   10   12
##
## , , 3
##
##      [,1] [,2] [,3]
## [1,]   13   15   17
## [2,]   14   16   18
```

```
# The only attribute is dimension
attributes(M)
```

```
## $dim
## [1] 3 6
```

```
dim(M)
```

```
## [1] 3 6
```

```
attributes(A)
```

```
## $dim
## [1] 2 3 3
```

```
dim(A)
```

```
## [1] 2 3 3
```

```
# Filling by row is possible
Mrow <- matrix(LETTERS[1:18],nrow=3,byrow=TRUE)
Mrow
```

```
##      [,1] [,2] [,3] [,4] [,5] [,6]
## [1,] "A"  "B"  "C"  "D"  "E"  "F"
## [2,] "G"  "H"  "I"  "J"  "K"  "L"
## [3,] "M"  "N"  "O"  "P"  "Q"  "R"
```

To change the filling order of an array is not as straightforward as for a matrix. The most convenient way to arrange it is by first creating an empty array and then filling it using the appropriate indexing.

```
# Create an empty array first.
A <- array(NA,c(2,3,3))
A
```

```
## , , 1
##
##      [,1] [,2] [,3]
## [1,]   NA   NA   NA
## [2,]   NA   NA   NA
##
## , , 2
```

```
##
##      [,1] [,2] [,3]
## [1,]   NA   NA   NA
## [2,]   NA   NA   NA
##
## , , 3
##
##      [,1] [,2] [,3]
## [1,]   NA   NA   NA
## [2,]   NA   NA   NA
```

```
# Now fill numbers 1 to 18 in columns
A[1,1,] <- 1:3
A[2,1,] <- 4:6
A[1,2,] <- 7:9
A[2,2,] <- 10:12
A[1,3,] <- 13:15
A[2,3,] <- 16:18
A
```

```
## , , 1
##
##      [,1] [,2] [,3]
## [1,]    1    7   13
## [2,]    4   10   16
##
## , , 2
##
##      [,1] [,2] [,3]
## [1,]    2    8   14
## [2,]    5   11   17
##
## , , 3
##
##      [,1] [,2] [,3]
## [1,]    3    9   15
## [2,]    6   12   18
```

```
# Display slice corresponding to the first slot with index 1
A[1,,]
```

```
##      [,1] [,2] [,3]
## [1,]    1    2    3
## [2,]    7    8    9
## [3,]   13   14   15
```

In relation to matrix, R includes the most common matrix operations, transpose (`t()`), inverse (`solve`), multiplication (`%*%`) and some types of matrices with given properties like the identity matrix, the diagonal matrix and the triangular matrix.

```
# Identity matrix of dimension 4
I4 <- diag(4)
I4
```

```
##      [,1] [,2] [,3] [,4]
## [1,]    1    0    0    0
## [2,]    0    1    0    0
## [3,]    0    0    1    0
## [4,]    0    0    0    1
```

```
# A diagonal matrix of dimension 3
D <- diag(1:3)
D
```

```
##      [,1] [,2] [,3]
## [1,]    1    0    0
## [2,]    0    2    0
## [3,]    0    0    3
```

```
# Generic 4 X 4 matrix
M <- matrix(1:16,ncol=4)
M
```

```
##      [,1] [,2] [,3] [,4]
## [1,]    1    5    9   13
## [2,]    2    6   10   14
## [3,]    3    7   11   15
## [4,]    4    8   12   16
```

```
# The matrix product with the identity returns
# the same matrix
M%*%I4

##      [,1] [,2] [,3] [,4]
## [1,]    1    5    9   13
## [2,]    2    6   10   14
## [3,]    3    7   11   15
## [4,]    4    8   12   16

# This is an orthogonal matrix
O <- matrix(c(sqrt(3)/2,-1/2,0,1/2,sqrt(3)/2,0,0,0,1),
nrow=3,byrow=TRUE)
O

##           [,1]       [,2] [,3]
## [1,] 0.8660254 -0.5000000    0
## [2,] 0.5000000  0.8660254    0
## [3,] 0.0000000  0.0000000    1

# The matrix product with its transpose is the identity
O%*%t(O)

##      [,1] [,2] [,3]
## [1,]    1    0    0
## [2,]    0    1    0
## [3,]    0    0    1

# Matrix inversion (this is here equal to t(O)
# because O is orthogonal)
Oinv <- solve(O)
Oinv

##            [,1]      [,2] [,3]
## [1,]  0.8660254 0.5000000    0
## [2,] -0.5000000 0.8660254    0
## [3,]  0.0000000 0.0000000    1
```

```
t(O)
```

```
##                [,1]      [,2] [,3]
## [1,]  0.8660254 0.5000000    0
## [2,] -0.5000000 0.8660254    0
## [3,]  0.0000000 0.0000000    1
```

```
# Upper triangular matrix
# The command returns a matrix of TRUE/FALSE
idx <- upper.tri(M)
idx
```

```
##       [,1]  [,2]  [,3]  [,4]
## [1,] FALSE  TRUE  TRUE  TRUE
## [2,] FALSE FALSE  TRUE  TRUE
## [3,] FALSE FALSE FALSE  TRUE
## [4,] FALSE FALSE FALSE FALSE
```

```
# This can be next used to set to 0 the lower part
M[!idx] <- 0
M
```

```
##      [,1] [,2] [,3] [,4]
## [1,]    0    5    9   13
## [2,]    0    0   10   14
## [3,]    0    0    0   15
## [4,]    0    0    0    0
```

To conclude this section it is worth mentioning that specific commands exist to concatenate two or more compatible matrices together. These are a generalisation of `c()`: `cbind` and `rbind`.

```
# Two matrices with a same number of rows, but
# a different number of columns
A1 <- matrix(1:9,ncol=3)
A1
```

```
##      [,1] [,2] [,3]
## [1,]    1    4    7
## [2,]    2    5    8
## [3,]    3    6    9
```

```
A2 <- matrix(rep(0,length=6),ncol=2)
A2
```

```
##      [,1] [,2]
## [1,]    0    0
## [2,]    0    0
## [3,]    0    0
```

```
# Let's create a matrix with more columns
# using cbind
A <- cbind(A1,A2)
A
```

```
##      [,1] [,2] [,3] [,4] [,5]
## [1,]    1    4    7    0    0
## [2,]    2    5    8    0    0
## [3,]    3    6    9    0    0
```

2.15 Lists

A list is similar to a vector in that it is a one-dimensional container with a defined length. But, differently from vectors, lists' elements do not need to be all of the same type or structure. A single list could, for instance, contain numeric, character and logical types at the same time. And some of its elements could be themselves objects with a dimension, like vectors or matrices. Lists are created with the command `list`. As the elements of a list can be of different types (a list is nonhomogeneous), a list will have its own type named quite appropriately *list*.

```
# Create four objects of different types/structure
A <- 1:10
typeof(A)
```

```
## [1] "integer"
```

```
B <- LETTERS[1:5]
typeof(B)
```

```
## [1] "character"
```

```
C <- c(TRUE,FALSE,TRUE,TRUE)
typeof(C)
```

```
## [1] "logical"
```

```
D <- diag(c(1,2,3))
typeof(D)
```

```
## [1] "double"
```

```
# Create list
ltmp <- list(A,B,C,D)
str(ltmp)
```

```
## List of 4
##  $ : int [1:10] 1 2 3 4 5 6 7 8 9 10
##  $ : chr [1:5] "A" "B" "C" "D" ...
##  $ : logi [1:4] TRUE FALSE TRUE TRUE
##  $ : num [1:3, 1:3] 1 0 0 0 2 0 0 0 3
```

```
# The type of a list is "list"
typeof(ltmp)
```

```
## [1] "list"
```

The elements of a list can be named, i.e. a name can be assigned to each one of them. Not all elements in a list must be necessarily named. When an element in a list has a name, it can be accessed with the `$` symbol. Alternatively, elements in a list can be accessed, similarly to what happens with vectors, either with the '`[]`' or the '`[[]]`' symbols. They both return the whole content of the specific list element, but in the first case this is forced to be itself a list.

```
# Elements of a list can be named
ltmp <- list(A=A,B=B,C=C,D=D)
str(ltmp)
```

```
## List of 4
##  $ A: int [1:10] 1 2 3 4 5 6 7 8 9 10
##  $ B: chr [1:5] "A" "B" "C" "D" ...
##  $ C: logi [1:4] TRUE FALSE TRUE TRUE
##  $ D: num [1:3, 1:3] 1 0 0 0 2 0 0 0 3
```

```
# A specific element can be accessed using $
str(ltmp$A)
```

```
##  int [1:10] 1 2 3 4 5 6 7 8 9 10
```

```
str(ltmp$D)
```

```
##  num [1:3, 1:3] 1 0 0 0 2 0 0 0 3
```

```
# Elements can be accessed also using [] or [[]]
# The outcomes are objects with different structures
ltmp[1]
```

```
## $A
##  [1]  1  2  3  4  5  6  7  8  9 10
```

```
is.list(ltmp[1])
```

```
## [1] TRUE
```

```
is.atomic(ltmp[1])
```

```
## [1] FALSE
```

```
ltmp[[1]]
```

```
## [1]  1  2  3  4  5  6  7  8  9 10
```

```
is.list(ltmp[[1]])
```

```
## [1] FALSE
```

```
is.atomic(ltmp[[1]])
```

```
## [1] TRUE
```

Needless to say, lists are very powerful and useful data structures because they make it possible to enclose very different data types and data structures within a single object. In fact, it is also possible to have lists whose elements are themselves lists. Finally, named lists are one of the preferred structures to be returned by functions because with them it is possible to return several and different objects within one object.

2.16 Data frames

A *data frame* is the most popular data container in R. In essence a data frame is a list in which all elements can be of different type and structure, but must have the same length. Thus, the type of a data frame is 'list', but it also includes a special attribute, '`class`', enriching this specific list with much more while restricting, at the same time, all its elements to have the same length. There are several built-in data frames in R, most of them available to test functions and other statistical ideas. But arbitrary data frames can be created using the command `data.frame`.

```
# Data frame included in the R installation
str(mtcars)
```

```
## 'data.frame':	32 obs. of  11 variables:
##  $ mpg : num  21 21 22.8 21.4 18.7 18.1 14.3 24.4 22.8 19.2 ...
##  $ cyl : num  6 6 4 6 8 6 8 4 4 6 ...
##  $ disp: num  160 160 108 258 360 ...
##  $ hp  : num  110 110 93 110 175 105 245 62 95 123 ...
##  $ drat: num  3.9 3.9 3.85 3.08 3.15 2.76 3.21 3.69 3.92 3.92 ...
##  $ wt  : num  2.62 2.88 2.32 3.21 3.44 ...
```

```
## $ qsec: num  16.5 17 18.6 19.4 17 ...
## $ vs  : num  0 0 1 1 0 1 0 1 1 1 ...
## $ am  : num  1 1 1 0 0 0 0 0 0 0 ...
## $ gear: num  4 4 4 3 3 3 3 4 4 4 ...
## $ carb: num  4 4 1 1 2 1 4 2 2 4 ...
```

```
# Type and class
typeof(mtcars)
```

```
## [1] "list"
```

```
class(mtcars)
```

```
## [1] "data.frame"
```

```
# Names of columns
names(mtcars)
```

```
##  [1] "mpg"  "cyl"  "disp" "hp"   "drat" "wt"   "qsec" "vs"
##  [9] "am"   "gear" "carb"
```

```
colnames(mtcars)
```

```
##  [1] "mpg"  "cyl"  "disp" "hp"   "drat" "wt"   "qsec" "vs"
##  [9] "am"   "gear" "carb"
```

```
# Names of observations (rows)
rownames(mtcars)
```

```
##  [1] "Mazda RX4"           "Mazda RX4 Wag"
##  [3] "Datsun 710"          "Hornet 4 Drive"
##  [5] "Hornet Sportabout"   "Valiant"
##  [7] "Duster 360"          "Merc 240D"
##  [9] "Merc 230"            "Merc 280"
## [11] "Merc 280C"           "Merc 450SE"
## [13] "Merc 450SL"          "Merc 450SLC"
```

```
## [15] "Cadillac Fleetwood"  "Lincoln Continental"
## [17] "Chrysler Imperial"   "Fiat 128"
## [19] "Honda Civic"         "Toyota Corolla"
## [21] "Toyota Corona"       "Dodge Challenger"
## [23] "AMC Javelin"         "Camaro Z28"
## [25] "Pontiac Firebird"    "Fiat X1-9"
## [27] "Porsche 914-2"       "Lotus Europa"
## [29] "Ford Pantera L"      "Ferrari Dino"
## [31] "Maserati Bora"       "Volvo 142E"
```

```
# Length corresponds to ncol
length(mtcars)
```

```
## [1] 11
```

```
ncol(mtcars)
```

```
## [1] 11
```

```
# Number of observations (same for all columns)
nrow(mtcars)
```

```
## [1] 32
```

```
length(mtcars$mpg)
```

```
## [1] 32
```

```
length(mtcars$hp)
```

```
## [1] 32
```

```
length(mtcars$am)
```

```
## [1] 32
```

```
# Create a new data frame without assigning names.
# Names are created automatically.
newD <- data.frame(1:10,LETTERS[1:10],rep(TRUE,length=10))
names(newD)
```

```
## [1] "X1.10"                          "LETTERS.1.10."
## [3] "rep.TRUE..length...10."
```

```
rownames(newD)
```

```
##  [1] "1"  "2"  "3"  "4"  "5"  "6"  "7"  "8"  "9"  "10"
```

```
# Create a new data frame with names
newD <- data.frame(A=1:10,B=LETTERS[1:10],
C=rep(TRUE,length=10))
names(newD)
```

```
## [1] "A" "B" "C"
```

```
rownames(newD)
```

```
##  [1] "1"  "2"  "3"  "4"  "5"  "6"  "7"  "8"  "9"  "10"
```

```
rownames(newD) <- c(letters[1:10])
rownames(newD)
```

```
##  [1] "a" "b" "c" "d" "e" "f" "g" "h" "i" "j"
```

One of the possible elements of a data frame is the *factor*. This is in essence an integer vector with two attributes added, *class* and *levels*. The class enriches the integer type so that the class of a factor object is 'factor'. The levels indicate the unique different elements in the factor object. An example can make things clearer.

```
# Data frame "iris", included in the R installation
str(iris)

## 'data.frame': 150 obs. of  5 variables:
##  $ Sepal.Length: num  5.1 4.9 4.7 4.6 5 5.4 4.6 5 4.4 4.9 ...
##  $ Sepal.Width : num  3.5 3 3.2 3.1 3.6 3.9 3.4 3.4 2.9 3.1 ...
##  $ Petal.Length: num  1.4 1.4 1.3 1.5 1.4 1.7 1.4 1.5 1.4 1.5 ...
##  $ Petal.Width : num  0.2 0.2 0.2 0.2 0.2 0.4 0.3 0.2 0.2 0.1 ...
##  $ Species     : Factor w/ 3 levels "setosa","versicolor",..:
##                                          1 1 1 1 1 1 1 ...

# Attributes of "Species"
attributes(iris$Species)

## $levels
## [1] "setosa"     "versicolor" "virginica"
##
## $class
## [1] "factor"

typeof(iris$Species)

## [1] "integer"

class(iris$Species)

## [1] "factor"

# The full factor vector (it only displays 3 values)
iris$Species

##   [1] setosa     setosa     setosa     setosa     setosa
##   [6] setosa     setosa     setosa     setosa     setosa
##  [11] setosa     setosa     setosa     setosa     setosa
##  [16] setosa     setosa     setosa     setosa     setosa
##  [21] setosa     setosa     setosa     setosa     setosa
##  [26] setosa     setosa     setosa     setosa     setosa
##  [31] setosa     setosa     setosa     setosa     setosa
##  [36] setosa     setosa     setosa     setosa     setosa
##  [41] setosa     setosa     setosa     setosa     setosa
##  [46] setosa     setosa     setosa     setosa     setosa
##  [51] versicolor versicolor versicolor versicolor versicolor
##  [56] versicolor versicolor versicolor versicolor versicolor
##  [61] versicolor versicolor versicolor versicolor versicolor
##  [66] versicolor versicolor versicolor versicolor versicolor
##  [71] versicolor versicolor versicolor versicolor versicolor
```

```
##  [76] versicolor versicolor versicolor versicolor versicolor
##  [81] versicolor versicolor versicolor versicolor versicolor
##  [86] versicolor versicolor versicolor versicolor versicolor
##  [91] versicolor versicolor versicolor versicolor versicolor
##  [96] versicolor versicolor versicolor versicolor versicolor
## [101] virginica  virginica  virginica  virginica  virginica
## [106] virginica  virginica  virginica  virginica  virginica
## [111] virginica  virginica  virginica  virginica  virginica
## [116] virginica  virginica  virginica  virginica  virginica
## [121] virginica  virginica  virginica  virginica  virginica
## [126] virginica  virginica  virginica  virginica  virginica
## [131] virginica  virginica  virginica  virginica  virginica
## [136] virginica  virginica  virginica  virginica  virginica
## [141] virginica  virginica  virginica  virginica  virginica
## [146] virginica  virginica  virginica  virginica  virginica
## Levels: setosa versicolor virginica
```

Factors can be very useful when data frames are used for various operations and statistical calculations. But when new data frames are created, the user might want to avoid certain columns to be transformed into factors, which is the default behaviour. To avoid the default behaviour the option '`stringsAsFactors = FALSE`' needs to be used.

```
# Create a new data frame using default mode
newD <- data.frame(A=1:5,B=LETTERS[1:5],
C=rep(TRUE,length=5))
str(newD)

## 'data.frame': 5 obs. of  3 variables:
##  $ A: int  1 2 3 4 5
##  $ B: Factor w/ 5 levels "A","B","C","D",..: 1 2 3 4 5
##  $ C: logi  TRUE TRUE TRUE TRUE TRUE

class(newD$B)

## [1] "factor"

# Create a new data frame and avoid default to factor
newD <- data.frame(A=1:5,B=LETTERS[1:5],
C=rep(TRUE,length=5),stringsAsFactors=FALSE)
str(newD)

## 'data.frame': 5 obs. of  3 variables:
##  $ A: int  1 2 3 4 5
##  $ B: chr  "A" "B" "C" "D" ...
##  $ C: logi  TRUE TRUE TRUE TRUE TRUE

class(newD$B)

## [1] "character"
```

Besides its attributes, which are very important for the central role played in statistical calculations, a data frame can be essentially treated as a matrix, even though the addition of attributes represents a heavier load on computer memory. When large-size data frames are handled, it might be worth temporarily turning data frames into matrices before calculations are performed. The transformation can be done using the command `as.matrix`.

```
# Data frame previously created
newD

##   A B    C
## 1 1 A TRUE
## 2 2 B TRUE
## 3 3 C TRUE
## 4 4 D TRUE
## 5 5 E TRUE

attributes(newD)

## $names
## [1] "A" "B" "C"
##
## $class
## [1] "data.frame"
##
## $row.names
## [1] 1 2 3 4 5

# Turn the data frame into a matrix
M <- as.matrix(newD)
M

##      A   B   C
## [1,] "1" "A" "TRUE"
## [2,] "2" "B" "TRUE"
## [3,] "3" "C" "TRUE"
## [4,] "4" "D" "TRUE"
## [5,] "5" "E" "TRUE"

attributes(M)

## $dim
## [1] 5 3
##
## $dimnames
## $dimnames[[1]]
## NULL
##
## $dimnames[[2]]
## [1] "A" "B" "C"
```

It is worth remarking that one of the attributes of the matrix produced is `dimnames`, which is a list of length 2. The first element is a character vector with the data frame row names and the second element a character vector with the data frame column names. The transformation process tries to retain as much information as possible. Some of it, though, is lost as a matrix, differently from a data frame, needs to have the same-type elements. In the example just seen, all data are turned into characters, even the numeric ones.

2.17 Data aggregation

In data analysis, in general, the goal is not just to inspect individual data points, but to summarise patterns across groups. This process is known as *data aggregation*. For instance, one might want to compute the average income per region, the total rainfall per month, or the number of observations in each experimental condition. Aggregation is one of the core tasks in exploratory data analysis and often a necessary step before applying statistical models.

Since data in R are typically organised in data frames, it is important to learn how to summarise them in flexible and efficient ways. Several base R functions allow this, depending on whether one is working with vectors or entire data frames. The most general of these is `aggregate`, which will be discussed first. We will also explain how to use `tapply` and `by`.

2.17.1 Aggregation with `aggregate`

The function `aggregate` is one of the most versatile tools in base R for grouped data summaries. It works on data frames and allows the user to compute summary statistics (such as the mean, sum, or standard deviation) of one or more variables, grouped according to one or more factors. Its typical usage follows a formula interface:

```
aggregate(response ~ group, data = df, FUN = summary_function)
```

For example, to compute the average miles per gallon (mpg) for each cylinder class (cyl) in the built-in mtcars dataset, we type:

```
aggregate(mpg ~ cyl, data = mtcars, FUN = mean)
```

```
##   cyl      mpg
## 1   4 26.66364
## 2   6 19.74286
## 3   8 15.10000
```

This returns a data frame with one row per cylinder group, showing the corresponding mean of mpg. Multiple grouping variables can be used by extending the formula:

```
aggregate(mpg ~ cyl + gear, data = mtcars, FUN = mean)
```

```
##   cyl gear    mpg
## 1   4    3 21.500
## 2   6    3 19.750
## 3   8    3 15.050
## 4   4    4 26.925
## 5   6    4 19.750
## 6   4    5 28.200
## 7   6    5 19.700
## 8   8    5 15.400
```

This computes the average mpg for each combination of cyl and gear. Likewise, multiple response variables can be handled by combining them with `cbind`:

```
aggregate(cbind(mpg, hp) ~ cyl, data = mtcars, FUN = mean)
```

```
##   cyl      mpg        hp
## 1   4 26.66364  82.63636
## 2   6 19.74286 122.28571
## 3   8 15.10000 209.21429
```

This returns the group-wise means of both mpg and hp, grouped by cyl. The output of `aggregate` is always a data frame, which makes it convenient for further inspection or plotting. It is worth noting that character vectors are treated as grouping factors, and that missing values in the data may need to be handled explicitly by passing `na.rm = TRUE` to the summary function:

```
aggregate(mpg ~ cyl, data = mtcars,
FUN = function(x) mean(x, na.rm = TRUE))
```

```
##   cyl      mpg
## 1   4 26.66364
## 2   6 19.74286
## 3   8 15.10000
```

The aggregate function is a powerful way to produce grouped summaries with minimal code and will often serve as the default option when working within base R.

2.17.2 Aggregation with tapply

While aggregate works on data frames, the function tapply provides a simpler mechanism for computing group summaries when working with individual vectors. It is particularly useful when one wants to summarise a numeric vector according to the levels of a factor. The general form is:

```
tapply(X, INDEX, FUN)
```

- x is the vector to be summarised.
- INDEX is a factor (or a list of factors) defining the groups.
- FUN is the summary function to apply to each group.

A basic example, again using the mtcars dataset, computes the mean mpg for each level of the factor cyl:

```
tapply(mtcars$mpg, mtcars$cyl, mean)
```

```
##        4        6        8
## 26.66364 19.74286 15.10000
```

The result is a named vector, with one element per group. The column names, in this case, are the values of the factor cyl, that is 4,6,8. If multiple grouping variables are used, INDEX can be given as a list:

```
tapply(mtcars$mpg, list(mtcars$cyl, mtcars$gear), mean)
```

```
##       3      4    5
## 4 21.50 26.925 28.2
## 6 19.75 19.750 19.7
## 8 15.05     NA 15.4
```

```
# No data for combination cyl = 8 and gear = 4
subset(mtcars, cyl == 8 & gear == 4)
```

```
##  [1] mpg  cyl  disp hp   drat wt   qsec vs   am   gear carb
## <0 rows> (or 0-length row.names)
```

This produces a matrix-like output, showing the group-wise means for each combination of cylinder count and number of gears (the factor names for `gear` are along the columns, while names for `cyl` are along the rows). The `NA` value present in the data comes from data not existing for the combination with `cyl = 8` and `gear = 4`. As with `aggregate`, missing values must be handled explicitly in the function passed to `tapply`:

```
tapply(mtcars$mpg, mtcars$cyl, function(v) mean(v, na.rm = TRUE))
```

```
##        4        6        8
## 26.66364 19.74286 15.10000
```

The `NA` is there for the same reason as before. But the means have not changed, meaning that there were no `NA`s in this dataset.

Although `tapply` is not as flexible as `aggregate` for working with full data frames, it is extremely concise and efficient for vector-level aggregation, making it a popular tool for quick summaries.

2.17.3 Aggregation with `by`

The function `by` applies a given function to subsets of a data frame, where the subsets are defined by one or more grouping factors. It is a convenient way to run summary functions on parts of a data frame without manually splitting it. The general form is:

```
by(data,INDICES,FUN)
```

Let us try `by` on the usual dataframe `mtcars`:

```
ltmp <- by(mtcars[,c("mpg","hp")],mtcars$cyl,colMeans)
print(class(ltmp))
```

```
## [1] "by"
```

```
print(length(ltmp))
```

```
## [1] 3
```

```
print(ltmp)
```

```
## mtcars$cyl: 4
##      mpg       hp
## 26.66364 82.63636
## -------------------------------------------------------
## mtcars$cyl: 6
##       mpg        hp
##  19.74286 122.28571
## -------------------------------------------------------
## mtcars$cyl: 8
##      mpg       hp
##  15.1000 209.2143
```

```
print(ltmp[[1]])
```

```
##      mpg       hp
## 26.66364 82.63636
```

The object returned is essentially a list, with a special attribute (see documentation). Each field of the list returned includes the data aggregated. The aggregating function in the above example is `colMeans`, which returns the mean of each column. Multiple grouping variables can be used by passing `INDICES` as a list:

```
ltmp <- by(mtcars[,c("mpg","hp")],list(mtcars$cyl,mtcars$gear),
colMeans)
print(length(ltmp))
```

```
## [1] 9
```

```
print(ltmp[1:3])
```

```
## [[1]]
##  mpg   hp
## 21.5 97.0
##
## [[2]]
##    mpg     hp
##  19.75 107.50
##
## [[3]]
##      mpg       hp
##  15.0500 194.1667
```

This produces summaries for each combination of cylinder count and gear count. Since by passes each subset directly to the function, handling missing values depends entirely on that function's arguments. For example, to safely ignore missing values when computing means:

```
by(mtcars[,c("mpg","hp")],mtcars$cyl,
function(df) colMeans(df,na.rm=TRUE))
```

```
## mtcars$cyl: 4
##      mpg       hp
## 26.66364 82.63636
## ------------------------------------------------------------
## mtcars$cyl: 6
##       mpg        hp
##  19.74286 122.28571
## ------------------------------------------------------------
## mtcars$cyl: 8
##      mpg       hp
##  15.1000 209.2143
```

Compared to `aggregate` and `tapply`, the main difference is in the output format: `by` returns results in a structured list rather than a flat data frame or vector. This makes it useful for more complex summaries that don't reduce to a single number per group.

2.18 Exercises on R objects and data handling

Exercise 16

a. Create the following objects in R:

```
a <- 42
b <- "42"
c <- TRUE
d <- charToRaw("Hello")
e <- as.raw(c(0x48, 0x65, 0x6C, 0x6C, 0x6F))
```

b. Use `typeof` to determine the storage type of each object. What is the difference between `a` and `b`? What is the difference between `d` and `e`? For the raw objects `d` and `e`: convert them back to a character string using `rawToChar`.
c. Use `typeof` on a data frame and on one of its columns. Why are the results different? What does this tell you about how R stores data?

Exercise 17

Type in the following (exact) syntax in a console:

```
G <- list(mtcars,diag(5),solve(diag(5)),LETTERS[1:10])
```

a. What is the `typeof` for `G` and `G[[1]]`? Explain the difference.
b. What data structure have `G[[2]]` and `G[[3]]`? Print the objects to verify this is true.
c. What is the data structure and data type of `G[[4]]`?

Exercise 18

Select a built-in data frame different from `mtcars` and transform it into a matrix. Is it possible straight away? Or, do you need to take some decisions on the variables contained, first?

Exercise 19

Create an $2 \times 3 \times 4$ array and fill it naturally (i.e. following the built-in order of your machine) with the first 24 integers.

Exercise 20

Generate an Hermitian matrix. Recall: an *Hermitian matrix*, H, obeys the property

$$H^\dagger = H,$$

where $\dagger$ indicates a complex-conjugate transpose of the original matrix. This property means that H has real numbers along the diagonal, while for the other components h_{ij} (in general complex numbers), we have:

$$h_{ij}^* = h_{ji}.$$

Exercise 21

A quick way of generating all combinations of finite factors is with the command `expand.grid`. Use the help facility in R to understand how this command works. Then create a data frame for all possible combinations of:

1. `colour`: values `"g","r","b"`;
2. `size`: values `"xs","s","m","l","xl"`;
3. `gender`: value `"m","f"`.

Finally, create a list of the three vectors `colour`, `size`, and `gender`.

Exercise 22

Consider the dataset `penguins`. A `help(penguins)` reveals the following description:

> Data on adult penguins covering three species found on three islands in the Palmer Archipelago, Antarctica, including their size (flipper length, body mass, bill dimensions), and sex.
>
> Use `aggregate` to find the average of the `penguins`' flipper length, body mass, and bill dimension, aggregating data by `island`, `species`, and `sex`. Where can you find the penguins with highest body mass index? And are they male or female?

Exercise 23

Carry out the same aggregation of the previous exercise, this time using function `by`.

2.19 Structure of an R program

From the very start of this chapter, you have been programming: writing a sequence of instructions for the computer to execute, using the syntax provided by R. Whether you were calculating a number, transforming a dataset, or producing a plot, the underlying process was always the same: break the task into logical steps and tell R exactly what to do, in the right order.

An R program is more than a list of commands. It is an organised combination of components: data structures to hold information, functions to perform tasks, and control elements to decide what happens and when it happens. Understanding how these elements fit together is the key to writing meaningful and useful code.

In this section, we will first outline the basic components of an R program. Most of them have already been explained and demonstrated earlier. Then we will show how to organise them into functions. Finally, we will explore the control structures that guide the flow of execution. We will conclude with a complete example that brings all these pieces together. No exercises will be included for this section as programming is the task that has been performed from the beginning of the book and that will be continuously performed till its end: plenty to exercise with.

2.19.1 Basic components of an R program

Some of the units of an R program were already encountered earlier in this chapter. They form the raw material from which more elaborate programs are constructed.

- **Expressions and statements**. An expression is any valid piece of R code that produces a value, such as a calculation, a variable reference, or a function call. A statement is a complete instruction, often consisting of an assignment or a function call whose result is not stored. Both concepts have been used repeatedly since the beginning of the chapter.
- **Assignments**. Variables are created and updated through assignment, which stores the result of an expression into a named object. The most used assignment is `<-` as it characterises R as a separate programming language.
- **Comments and documentation**. Comments, introduced by the # symbol, are ignored by R but are essential for making code clear and maintainable. Their importance and use have already been demonstrated in the many examples written so far.
- **Data structures**. The principal ways of storing and organising data in R—vectors, lists, matrices, data frames, and factors—have all been described in detail earlier in the chapter. They occur frequently in R programs as inputs, intermediate results, or outputs, and form a fundamental part of program design.
- **Functions**. Functions group together a set of instructions under a single name, allowing them to be reused with different inputs. They help organise code into logical units, make programs easier to read, and reduce repetition. The definition of a function, the handling of arguments, and the return of results are key aspects of their use. We have seen the creation and use of simple functions earlier in the chapter.

- **Control blocks**. Control blocks determine the flow of execution in a program: whether certain code is run, repeated, or skipped. They include conditional statements and loops. As they have not yet been treated in this chapter, a dedicated section follows to explain their use and provide examples.

2.19.2 Control blocks

Control blocks are programming structures that determine the order in which instructions are executed. They allow a program to make decisions, repeat actions, or skip parts of the code depending on specific conditions.

Conditional execution

Conditional execution lets the program choose what to do based on whether a logical test is `TRUE` or `FALSE`. The simplest form is `if`:

```
x <- 5
if (x > 0) {
print("x is positive")
}
```

```
## [1] "x is positive"
```

It is interesting to observe that printing, in this case, was prompted by the insertion of the `if` block. An `if ... else` block provides an alternative when the condition is not met:

```
x <- -3
if (x > 0) {
print("x is positive")
} else {
print("x is not positive")
}
```

```
## [1] "x is not positive"
```

For element-wise operations on vectors, the vectorised `ifelse()` function is useful:

```
v <- c(3,-1,0)
ifelse(v > 0,"positive","non-positive")
```

```
## [1] "positive"     "non-positive" "non-positive"
```

Loops
Loops repeat a block of code multiple times. A `for` loop iterates over a sequence:

```
for (i in 1:3) {
print(paste("Iteration",i))
}
```

```
## [1] "Iteration 1"
## [1] "Iteration 2"
## [1] "Iteration 3"
```

A `while` loop continues as long as a condition is `TRUE`:

```
count <- 1
while (count <= 3) {
print(paste("Count is",count))
count <- count+1
}
```

```
## [1] "Count is 1"
## [1] "Count is 2"
## [1] "Count is 3"
```

A `repeat` loop runs indefinitely until stopped with `break`:

```
x <- 1
repeat {
if (x > 3) break
print(x)
x <- x+1
}
```

```
## [1] 1
## [1] 2
## [1] 3
```

You can skip the rest of the current iteration with `next`:

```
for (i in 1:5) {
if (i == 3) next
print(i)
}
```

```
## [1] 1
## [1] 2
## [1] 4
## [1] 5
```

2.19.3 Avoiding loops

While loops are useful, they can become slow and verbose in R when working with large datasets. Many operations can be expressed more clearly and efficiently using the apply family of functions, which repeat an operation over elements of an object without explicit looping.

`apply` Applies a function to the rows or columns of a matrix or data frame. The second argument specifies the margin: `1` for rows, `2` for columns.

```
m <- matrix(1:9, nrow = 3)
apply(m, 1, sum)  # sum of each row
```

```
## [1] 12 15 18
```

```
apply(m, 2, mean) # mean of each column
```

```
## [1] 2 5 8
```

`lapply` Applies a function to each element of a list (or vector, coerced to a list) and returns a list of results.

```
lst <- list(a = 1:3, b = 4:6)
lapply(lst, sum)
```

```
## $a
## [1] 6
##
## $b
## [1] 15
```

sapply A simplified version of lapply that tries to return a vector or matrix instead of a list when possible.

```
sapply(lst, sum)
```

```
## a b
## 6 15
```

In many cases, `sapply` produces more compact output, but `lapply` is safer when consistent output format is not guaranteed. These functions avoid the need to write explicit for loops, often making the code shorter, faster, and clearer, particularly when applied to tabular or list-like data.

2.20 R packages

R is not a monolithic block of code. It is, rather, built from a series of independently-created modules called *packages*. The various functions and objects found in any R installation come, in fact, from specific packages. The most immediate way to look at what packages are available when running R is `via` the commans `search`.

```
# What packages are loaded in the current session?
search()
```

```
## [1] ".GlobalEnv"          "package:knitr"
## [3] "package:stats"       "package:graphics"
## [5] "package:grDevices" "package:utils"
## [7] "package:datasets"    "package:methods"
## [9] "Autoloads"           "package:base"
```

Let's ignore for now ".GlobalEnv" and "Autoloads" which are connected to the important R concept of *environment*. In the current R session, the packages loaded in memory (i.e. those whose functionality is ready to use) are stats, grDevices, datasets, knitr, graphics, utils, methods and base. Any function or object available in one of these packages is ready to use. Most of the time we can query from which package a specific function comes using the command `environment`. Some functions do not return a valid answer (they return `NULL`) because they are not directly coded in the R programming language, but are coded in other programming languages (more on this later).

```
# From which package do these functions come?
environment(ls)

## <environment: namespace:base>

environment(plot)

## <environment: namespace:graphics>

environment(rnorm)

## <environment: namespace:stats>

environment(sum)

## NULL
```

Clearly, `ls` belongs to package *base*, `plot` to package *graphics* and `rnorm` to package *stats*. The command `sum`, on the other hand, does not return a package and this means that it has been coded in a programming language different from R, but it is made available in R.

The greatest majority of the available R packages are not loaded in any given R session. In order to load a package in the current session, one has to use the command `library`. In the following example we load the package *MASS* which contains several interesting sets of data as data frames. Such a list can be extracted using the command `data`.

```
# Initially the MASS package is not loaded
search()
```

```
##  [1] ".GlobalEnv"        "package:knitr"
##  [3] "package:stats"     "package:graphics"
##  [5] "package:grDevices" "package:utils"
##  [7] "package:datasets"  "package:methods"
##  [9] "Autoloads"         "package:base"
```

```
# Thus, MASS' dataset 'michelson' is not found
michelson
```

```
## Error in eval(expr, envir, enclos):  object 'michelson' not found
```

```
# MASS is loaded
library(MASS)
search()
```

```
##  [1] ".GlobalEnv"        "package:MASS"
##  [3] "package:knitr"     "package:stats"
##  [5] "package:graphics"  "package:grDevices"
##  [7] "package:utils"     "package:datasets"
##  [9] "package:methods"   "Autoloads"
## [11] "package:base"
```

```
# Among the many datasets available we find 'michelson' as
# the 59th dataset of MASS
ltmp <- data()
ltmp$results[59, ]
```

```
##                                Package
##                                 "MASS"
##                                LibPath
## "C:/Program Files/R/R-3.6.3/library"
##                                   Item
##                            "michelson"
##                                  Title
##    "Michelson's Speed of Light Data"
```

```
michelson[1:10, ]
```

```
##    Speed Run Expt
## 1    850   1    1
## 2    740   2    1
## 3    900   3    1
## 4   1070   4    1
## 5    930   5    1
## 6    850   6    1
## 7    950   7    1
## 8    980   8    1
## 9    980   9    1
## 10   880  10    1
```

When R is started there is usually a very small number of packages automatically loaded in the workspace. Many more packages are available for immediate loading, but still they form a small part of the entire wealth of available ones. These can become part of the local (personal) R installation `via` a process known as *package installation*. The main repository with an updated list of packages is the CRAN repository [24]. As we speak, CRAN lists the amazingly huge number of 15806 packages! It is obviously difficult (if not impossible) to search for packages in such a large pool. It is much easier to use the 'Task Views' section of CRAN [25] which lists packages within the related field of study or according to applications.

2.21 The tidyverse

The *tidyverse* is a collection of R packages designed for data science, built around a shared philosophy and consistent syntax. A good reference to start learning about the tidyverse is the book *R for Data Science* [26].

All packages included in the tidyverse follow the principles of tidy data: each variable is a column, each observation is a row, and each type of observation forms its own table. While the tidyverse is not strictly necessary for computational physics (base R can handle all numerical, statistical, and visualisation needs) it offers a bridge to the world of data science. This alignment opens new avenues for interdisciplinary work, especially in areas such as large-scale data analysis, reproducible research, and effective communication of results. For computational physicists working in data-rich environments, the tidyverse can be a powerful ally.

The most relevant tidyverse packages for computational physics are:

- `tibble`, a modern reimagining of data frames with stable column names, consistent types, compact printing, and no accidental type conversions;
- `readr`, a fast and consistent tool for importing and exporting rectangular data, ideal for reading experiment logs or simulation output;

- `dplyr`, a *Grammar of Data*. It renders data manipulation with concise verbs like `filter`, `select`, `mutate`, `summarise`, and `group_by`. It will be describe in detail later on;
- `tidyr`, for reshaping and tidying datasets with tools such as `pivot_longer` and `pivot_wider`, useful for converting between wide simulation arrays and long, analysis-ready tables;
- `purrr`, which brings functional programming tools like `map` and `map_dfr` for applying computations across many parameters, seeds, or datasets without explicit loops;
- `stringr`, which offers a consistent, readable interface for string handling, useful when parsing file names, processing logs, or handling text-based metadata;
- `ggplot2`, a *Grammar of Graphics*. It provides a powerful and elegant approach to creating publication-quality visualisations and will be discussed in detail later on.

The current packages in the tidyverse can be installed with the simple command:

```
install.packages("tidyverse")
```

While in this book there is no room to delve into the many aspects and the full usefulness of the tidyverse, in the next sections we will explore `dplyr` and `ggplot2` more closely. A systematic and detailed study of the tidyverse can be initiated with reference [26].

2.22 `dplyr` and the Grammar of Data

In the tidyverse, `dplyr` is the main tool for manipulating and transforming data. It is built around a small set of intuitive functions called *verbs* that each performs one clear operation on a dataset. Together, these verbs form what is often called the *Grammar of Data*: a set of basic rules and operations that can be combined to express almost any data transformation. At the heart of this grammar are:

- *tibbles*. Modern data frames that store the data in a predictable, readable format.
- *verbs*. Functions like `select`, `filter`, `mutate`, `summarise`, and `arrange` that each do one job well.
- *pipes*. Operators (|>) that link verbs together into a clear, top-to-bottom workflow[1].

[1] Earlier versions of the tidyverse commonly used the pipe operator `%>%`, provided by the `magrittr` package. Starting from R 4.1, a native pipe operator |> has been introduced into base R. Both behave similarly for most tidyverse workflows, though |> is slightly simpler and does not require extra packages, while `%>%` still offers some advanced features (such as flexible placement of the input using the dot symbol).

- *grouping*. The ability to split data into subsets with group_by so that verbs operate within each subset independently.

Because each step produces a new tibble and leaves the original untouched, dplyr makes it easy to experiment with different transformations, compare results, and ensure a reproducible workflow. In other words, this grammar of data offers a concise and expressive way to handle data, from simple column selections to multi-step analyses involving filtering, transformation, and summarisation.

To demonstrate the workings of dplyr, let us adopt one of the data frames already used previously, mtcars. The first action needed is to turn this data frame into a *tibble* object, so that it can obey stricter rules and be available for all the tidyverse functions.

```
# Load the tidyverse
library(tidyverse)

# Transform mtcars data frame into a tibble data frame
mtcars_tbl <- as_tibble(mtcars)

# Different dumps
str(mtcars)
```

```
## 'data.frame':  32 obs. of  11 variables:
##  $ mpg : num  21 21 22.8 21.4 18.7 18.1 14.3 24.4 22.8 19.2 ...
##  $ cyl : num  6 6 4 6 8 6 8 4 4 6 ...
##  $ disp: num  160 160 108 258 360 ...
##  $ hp  : num  110 110 93 110 175 105 245 62 95 123 ...
##  $ drat: num  3.9 3.9 3.85 3.08 3.15 2.76 3.21 3.69 3.92 3.92 ...
##  $ wt  : num  2.62 2.88 2.32 3.21 3.44 ...
##  $ qsec: num  16.5 17 18.6 19.4 17 ...
##  $ vs  : num  0 0 1 1 0 1 0 1 1 1 ...
##  $ am  : num  1 1 1 0 0 0 0 0 0 0 ...
##  $ gear: num  4 4 4 3 3 3 3 4 4 4 ...
##  $ carb: num  4 4 1 1 2 1 4 2 2 4 ...
```

```
str(mtcars_tbl)
```

```
## tibble [32 x 11] (S3: tbl_df/tbl/data.frame)
##  $ mpg : num [1:32] 21 21 22.8 21.4 18.7 18.1 14.3 24.4 22.8 19.2 ...
##  $ cyl : num [1:32] 6 6 4 6 8 6 8 4 4 6 ...
##  $ disp: num [1:32] 160 160 108 258 360 ...
##  $ hp  : num [1:32] 110 110 93 110 175 105 245 62 95 123 ...
##  $ drat: num [1:32] 3.9 3.9 3.85 3.08 3.15 2.76 3.21 3.69 3.92 3.92 ...
##  $ wt  : num [1:32] 2.62 2.88 2.32 3.21 3.44 ...
##  $ qsec: num [1:32] 16.5 17 18.6 19.4 17 ...
##  $ vs  : num [1:32] 0 0 1 1 0 1 0 1 1 1 ...
##  $ am  : num [1:32] 1 1 1 0 0 0 0 0 0 0 ...
##  $ gear: num [1:32] 4 4 4 3 3 3 3 4 4 4 ...
##  $ carb: num [1:32] 4 4 1 1 2 1 4 2 2 4 ...
```

As we can see clearly from the above output, the actual data have not changed, but the structure around them is different. The best way to realise it is through some additional operations, using grouping and one of the verbs, the processes being connected through *pipes*. We want to group the result using the factor `cyl` and then ask for summaries of `mpg`, by levels of `cyl`.

```
# Grouping by cyl
mtcars_tbl |> group_by(cyl)
```

```
## # A tibble: 32 x 11
## # Groups:   cyl [3]
##      mpg   cyl  disp    hp  drat    wt  qsec    vs    am  gear  carb
##    <dbl> <dbl> <dbl> <dbl> <dbl> <dbl> <dbl> <dbl> <dbl> <dbl> <dbl>
##  1  21       6  160    110  3.9   2.62  16.5     0     1     4     4
##  2  21       6  160    110  3.9   2.88  17.0     0     1     4     4
##  3  22.8     4  108     93  3.85  2.32  18.6     1     1     4     1
##  4  21.4     6  258    110  3.08  3.22  19.4     1     0     3     1
##  5  18.7     8  360    175  3.15  3.44  17.0     0     0     3     2
##  6  18.1     6  225    105  2.76  3.46  20.2     1     0     3     1
##  7  14.3     8  360    245  3.21  3.57  15.8     0     0     3     4
##  8  24.4     4  147.    62  3.69  3.19  20       1     0     4     2
##  9  22.8     4  141.    95  3.92  3.15  22.9     1     0     4     2
## 10  19.2     6  168.   123  3.92  3.44  18.3     1     0     4     4
## # i 22 more rows
```

What you see is still a 32-row tibble (one row per car). The console indicates that the data are grouped (you can see something like `Groups: cyl [3]`), but the rows are not collapsed. Repeated values of `cyl` (4, 6, 8) are expected here because `group_by` does not summarise anything, it only tags the data so that the next operation knows how to work within each cylinder group. The operation targeted is `summarise`:

```
mtcars_tbl |> group_by(cyl) |> summarise(mpg_mean=mean(mpg))
```

```
## # A tibble: 3 x 2
##     cyl mpg_mean
##   <dbl>    <dbl>
## 1     4     26.7
## 2     6     19.7
## 3     8     15.1
```

Now we can see that a new tibble is created with only two columns, `cyl` and the mean of `mpg` by levels of `cyl`. We have done nothing different from what was done using `aggregate`, simply the way of doing it has changed that is, we ask tibbles to be transformed into other tibbles following some instructions (*verbs* and *grouping*).

Intermediate results in the pipeline can be stored into R objects (typically tibbles), as normally done in other R contexts:

```
# Store intermediate result
ave_mpg_by_cyl <-
mtcars_tbl |> group_by(cyl) |> summarise(mpg_mean=mean(mpg))

# What's inside?
print(ave_mpg_by_cyl)
```

```
## # A tibble: 3 x 2
##     cyl mpg_mean
##   <dbl>    <dbl>
## 1     4     26.7
## 2     6     19.7
## 3     8     15.1
```

In fact, pipelines can be short to check results, or articulated if intermediate tibbles are needed for further future processing or if they constitute important findings to be stored and explained. For example, we can create an intermediate tibble where the operation of grouping by `cyl` is performed, and later on various summaries are calculated as needed, as shown here:

```
# Use group-by only once
cars_by_cyl <- mtcars_tbl |> group_by(cyl)

# ... and summaries repeatedly, as needed
ave_mpg <-
cars_by_cyl |> summarise(mpg_mean=mean(mpg))
print(ave_mpg)

## # A tibble: 3 x 2
##     cyl mpg_mean
##   <dbl>    <dbl>
## 1     4     26.7
## 2     6     19.7
## 3     8     15.1

ave_hp  <-
cars_by_cyl |> summarise(hp_mean=mean(hp))
print(ave_hp)

## # A tibble: 3 x 2
##     cyl hp_mean
##   <dbl>   <dbl>
## 1     4    82.6
## 2     6   122.
## 3     8   209.
```

An example of a long, articulated pipeline, the starting tibble, `mtcars_tbl`, will be here processed through a sequence of transformations that include selection, filtering, mutation and summaries, all linked by the pipe operator. The seven steps of the process planned are:

1. Only some columns of interest are retained: `mpg,cyl,hp,wt,am,gear` (many more columns are included in the starting tibble).
2. Cars with fewer than four gears are excluded, leaving only those with four or five.
3. The `cyl` column is converted into a factor, reflecting that it is categorical, and the column `am` is relabelled so that 0 and 1 appear as `auto` and `manual`.
4. Additional columns are created with `mutate`:
   - `hp_kw`: engine power in kilowatts (conversion from horsepower).
   - `wt_kg`: car weight in kilograms (conversion from thousands of pounds).
   - `pwr_density`: the ratio of power to mass, a useful measure of performance.

5. The data are grouped by cylinder count (`cyl`) and transmission type (`am`), so that each combination is treated as a separate subset.
6. For each group, we calculate the average fuel efficiency (named `mpg_mean`), the median power density (named `pwr_median`), and the number of cars in the group (`n`).
7. The summary table is finally ordered by cylinder count and transmission type for clarity.

```
mtcars_tbl |>
select(mpg,cyl,hp,wt,am,gear) |>
filter(gear >= 4) |>
mutate(
cyl = as.factor(cyl),
am = factor(am,labels=c("auto","manual"))
) |>
mutate(
hp_kw = hp*0.7457,
wt_kg = wt*453.592,
pwr_density = hp_kw/wt_kg
) |>
group_by(cyl,am) |>
summarise(
mpg_mean = mean(mpg),
pwr_median = median(pwr_density),
n = dplyr::n()
) |>
arrange(cyl,am)
```

```
## 'summarise()' has grouped output by 'cyl'.  You can override using the
## '.groups' argument.
```

```
## # A tibble: 5 x 5
## # Groups:   cyl [3]
##   cyl   am     mpg_mean pwr_median     n
##   <fct> <fct>     <dbl>      <dbl> <int>
## 1 4     auto       23.6     0.0408     2
## 2 4     manual     28.1     0.0613     8
## 3 6     auto       18.5     0.0588     2
## 4 6     manual     20.6     0.0690     3
## 5 8     manual     15.4     0.146      2
```

The tibble produced shows five groups because the combination `cyl=8`, `am=auto` does not contain data points. For each combination, the requested summary is correctly displayed. It is worth noting that the number of data points in a combination can be calculated with the function `dplyr::n`. The structure of the input is also organised so that each step is arranged in a new row, with a pipe ending it. Furthermore, the `mutate` and `summarise` verbs are organised as blocks with changes between open and closed parentheses, as this makes the changes stand clear.

It is also worth explaining the printed message that `dplyr` has included in the output, warning about the current grouping. The message included by `dplyr` is simply telling us what has happened to the grouping structure after `summarise`. By default, `summarise(` drops only the last grouping variable. Since we grouped by both `cyl` and am, the summary result is still internally grouped by `cyl`. That is why the message says that the output remains grouped, and it reminds us that we can override this behaviour with the `.groups` argument. If we add `. groups = "drop"`, the resulting tibble is returned ungrouped. This makes no difference to the summary values themselves (the numbers are identical) but it avoids carrying along hidden grouping into later steps, where it could produce unexpected results.

```
mtcars_tbl |>
select(mpg,cyl,hp,wt,am,gear) |>
filter(gear >= 4) |>
mutate(
cyl = as.factor(cyl),
am = factor(am,labels=c("auto","manual"))
) |>
mutate(
hp_kw = hp*0.7457,
wt_kg = wt*453.592,
pwr_density = hp_kw/wt_kg
) |>
group_by(cyl,am) |>
summarise(
mpg_mean = mean(mpg),
pwr_median = median(pwr_density),
n = dplyr::n(),
.groups = "drop"
) |>
arrange(cyl,am)
```

```
## # A tibble: 5 x 5
##    cyl   am     mpg_mean pwr_median     n
##    <fct> <fct>     <dbl>      <dbl> <int>
## 1 4      auto       23.6     0.0408     2
## 2 4     manual     28.1     0.0613      8
## 3 6     auto       18.5     0.0588      2
## 4 6     manual     20.6     0.0690      3
## 5 8     manual     15.4     0.146       2
```

The printed output now shows a plain tibble, without any grouped structure. This is often clearer to display, and safer when the tibble is in a pipeline because the inherent group structure might trigger unwanted results.

To conclude this short and demonstrative section on `dplyr`, its strength lies not only in its concise verbs but in the way these can be chained into clear, reproducible pipelines. The same grammar scales from small examples like `mtcars` to large, real-world datasets, and it provides a natural bridge to other tidyverse tools, such as `tidyr` for reshaping and `ggplot2` for visualisation.

2.23 `ggplot2` and the Grammar of Graphics

2.23.1 Why a Grammar of Graphics?

When we think of plots, we often imagine distinct 'types' such as scatterplots, bar charts, or histograms. In reality, all of these graphical forms can be built from a small number of reusable components. The central idea behind the `ggplot2` package is the *Grammar of Graphics*, which provides a declarative way to describe plots in terms of these components.

The grammar is based on the following schematic pipeline:

$$\text{data} \longrightarrow \text{aesthetics} \longrightarrow \text{geometries} \longrightarrow \text{statistics} \longrightarrow \text{scales} \longrightarrow \text{facets} \longrightarrow \text{theme}.$$

Each part plays a specific role. **Data** are supplied as a tidy table. Here 'tidy' refers to the structure of the dataset: each column corresponds to a variable, each row to an observation, and each entry to a single value. In practice, most tidyverse functions (including `ggplot2`) return tibbles, so tidy data are usually stored as tibbles. **Aesthetic** mappings describe how variables are linked to visual attributes such as position, colour, or size. **Geometries** specify how to draw the data (points, lines, bars). **Statistical** transformations summarise or modify the data before plotting. **Scales** control how variables are mapped to axes or colours. **Facets** allow splitting into subplots. Finally, **themes** govern the final appearance.

The great advantage of this approach is its composability. By changing only one element of the specification (for example, replacing points with hexagonal bins), the overall plot can be transformed without rewriting the rest of the code. This makes `ggplot2` an extremely powerful and flexible tool for both exploratory and presentation graphics.

2.23.2 The basic building blocks

After having introduced the overall logic of the Grammar of Graphics, describing how plots can be seen as a sequence of components, let us turn to the details, beginning with the first three elements of the grammar: data, aesthetic mappings, and geometries. Together, these form the minimum specification required to produce a plot.

- **Data**. Every plot starts from a tidy dataset (tibble), where each variable is a column and each observation is a row.
- **Aesthetics**. The `aes` function is used to connect variables in the data to graphical properties such as the position on the axes, the colour of a point, or

the size of a symbol. An *aesthetic mapping* means that the appearance of the graphical element will change when the underlying data change.

- **Geometries**. Geometries, provided by functions with names of the form `geom_*`, specify the type of object drawn: points, lines, bars, and so on. The same dataset can give rise to very different plots simply by changing the geometry.

As a first example, consider the built-in dataset `mtcars`, which records fuel consumption and performance characteristics for a selection of car models. The following code produces a scatter plot of weight against miles per gallon:

```
ggplot(mtcars,aes(x=wt,y=mpg)) +
geom_point()
```

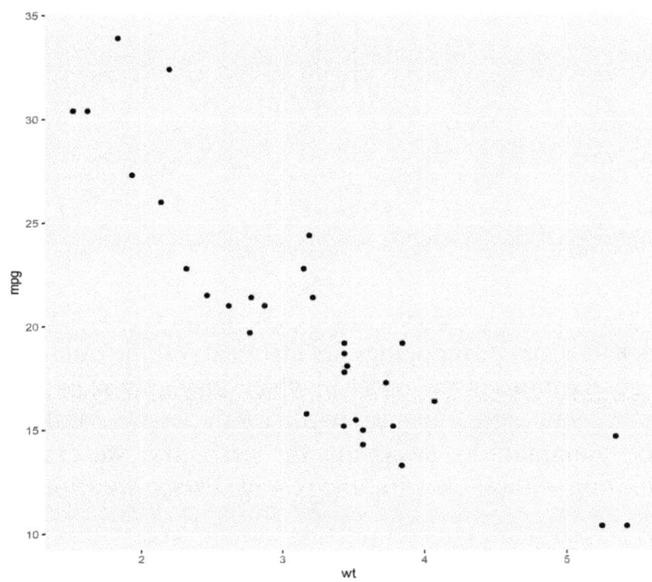

This minimal specification contains all three essential ingredients: a dataset (`mtcars`), aesthetic mappings (weight to the x-axis and fuel economy to the y-axis), and a geometry (`geom_point`). The result is the straightforward scatter plot, shown in the above figure.

A striking feature of the grammar is that changing only the geometry transforms the entire appearance of the plot, while all other elements remain unchanged. For instance, if we replace `geom_point` with `geom_line`, the plot changes:

```
ggplot(mtcars,aes(x=wt,y=mpg)) +
geom_line()
```

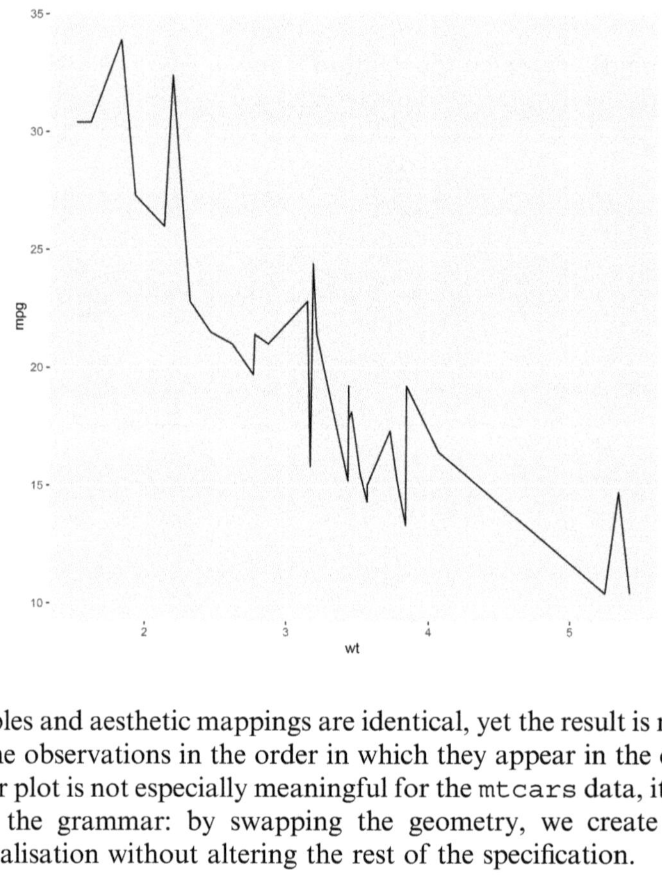

The variables and aesthetic mappings are identical, yet the result is now a line plot connecting the observations in the order in which they appear in the dataset. While this particular plot is not especially meaningful for the `mtcars` data, it illustrates the flexibility of the grammar: by swapping the geometry, we create a completely different visualisation without altering the rest of the specification.

2.23.3 Mappings versus settings

An important distinction in `ggplot2` is between *mappings* and *settings*. Both affect the appearance of the plot, but they operate in different ways.

- Mappings. A mapping links a variable in the dataset to a visual property of the plot. For example, we can colour points by the number of cylinders using `aes(colour=cyl)`:

```
ggplot(mtcars,aes(x=wt,y=mpg,colour=factor(cyl))) +
geom_point()
```

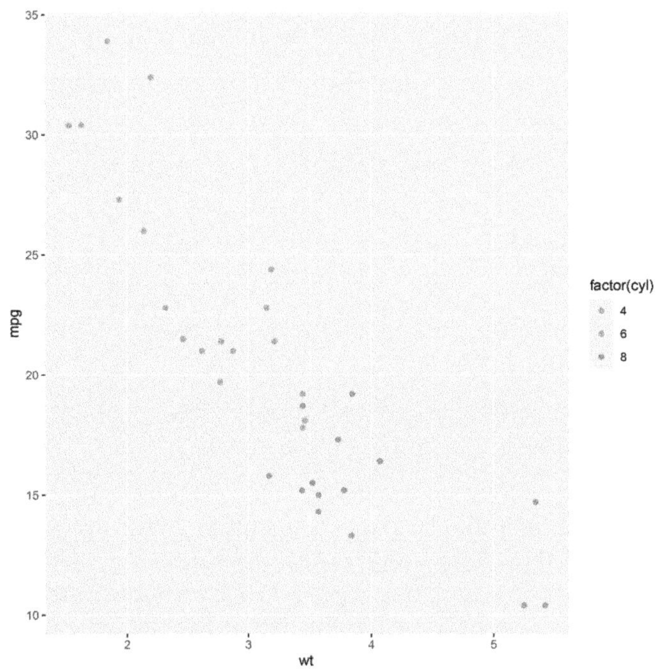

Here the colour of each point depends on the corresponding value of `cyl` in the data.

- Settings. A setting, on the other hand, fixes a visual property to a constant value, independent of the data. This is done by specifying the property outside `aes`. For example, to make all points blue we can write:

```
ggplot(mtcars,aes(x=wt,y=mpg)) +
geom_point(colour="blue")
```

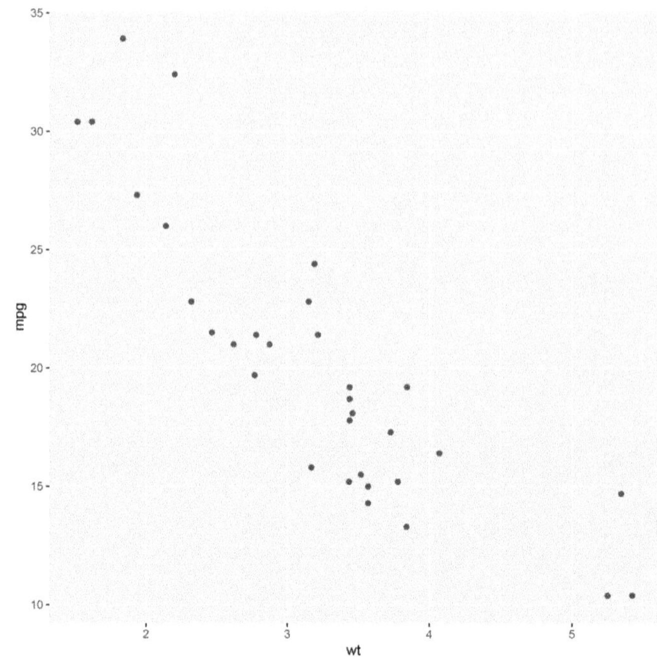

In this case the colour does not vary with the data: every point is drawn in the same style. An interesting aspect revealing the autonomy in taking decision of `ggplot2` is that a legend has been added automatically to the plot when colours changed, while this is not the case, because not necessary, when the colour does not change.

To summarise, the difference between mappings and settings is central to the grammar of graphics. Mappings introduce variation according to the dataset, while settings apply a uniform appearance across all graphical elements.

2.23.4 Layers: adding information

Plots in `ggplot2` are built by adding layers. Each layer can introduce a new geometry, a statistical transformation, or a change in the way the data are displayed. Layers are added with the + operator, which allows us to build up a plot step by step. Let us look at two ways of adding layers, as *multiple geometries* and as *different mappings per layer*.

- **Multiple geometries**. It is common to combine more than one geometry in the same plot. For instance, we can display points together with a smooth curve fitted to the data:

```
ggplot(mtcars,aes(x=wt,y=mpg)) +
geom_point() +
geom_smooth()
```

```
## 'geom_smooth()' using method = 'loess' and formula = 'y ~ x'
```

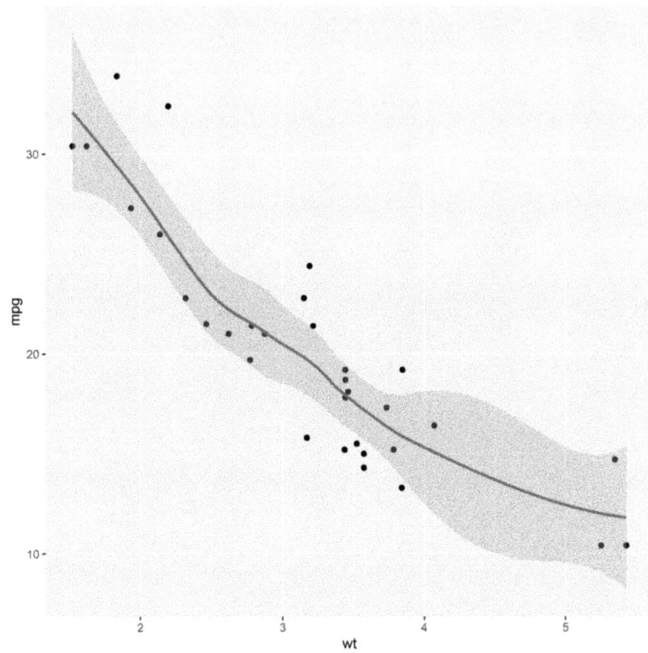

The scatterplot shows the raw data, while the smooth line gives a summary of the overall trend. The message printed out automatically by `ggplot2` explains how and using which parameters, the smooth line has been created (by the function `loess`).

- **Different mappings per layer**. Mappings defined in the initial `ggplot` call are inherited by all subsequent layers. However, each layer may also have its own mappings. For example, we can colour the points by the number of cylinders, but leave the smooth curve uncoloured:

```
ggplot(mtcars,aes(x=wt,y=mpg)) +
geom_point(aes(colour=factor(cyl))) +
geom_smooth()
```

```
## 'geom_smooth()' using method = 'loess' and formula = 'y ~ x'
```

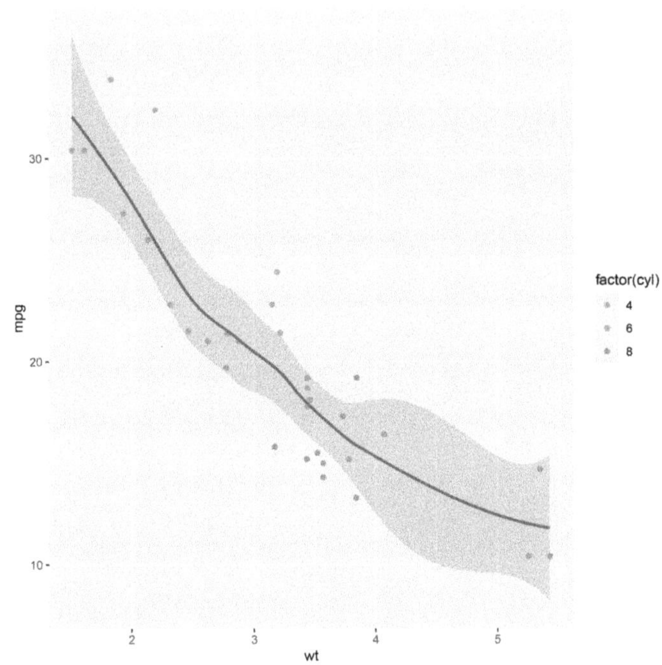

This layering principle is really interesting and powerful: raw data and summaries can be combined, annotations can be added, and alternative representations can coexist in the same figure.

2.23.5 Scales and transformations

Scales control how values in the dataset are mapped to positions, colours, sizes, or other visual properties of the plot. By default, `ggplot2` chooses scales automatically, but these can be modified to highlight features of the data or to change the way the plot is interpreted.

For example, in relation to the transformations of one or both axes, a common adjustment is to apply a logarithmic scale, which can make multiplicative relationships appear more linear. For example, plotting engine displacement against horsepower on a log scale yields:

```
ggplot(mtcars,aes(x=disp,y=hp)) +
geom_point() +
scale_x_log10()
```

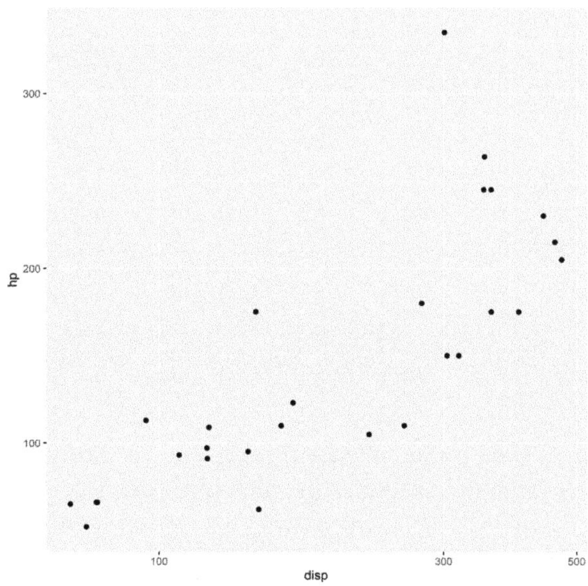

The scale function modifies only the representation of the axis but the raw data remain unchanged.

Another example relates to scales applied to colours. For continuous variables it is often useful to choose a perceptually uniform colour scale. The `viridis` package provides such scales, and they integrate with `ggplot2`:

```
ggplot(mtcars,aes(x=wt,y=mpg,colour=hp)) +
geom_point() +
scale_colour_viridis_c()
```

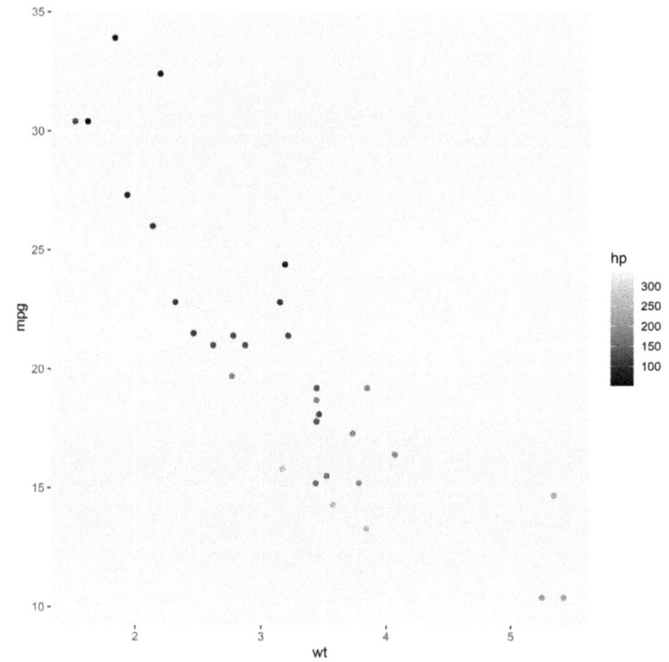

Here the colour of each point encodes the horsepower of the car, with the colour gradient chosen to be both aesthetically pleasing and accessible to colour-blind readers.

To summarise, we could say that scales are the link between data values and their graphical representation. They allow us to change units, apply transformations, and select how continuous or categorical variables are visually displayed.

2.23.6 Facets

Faceting is a convenient way to split a dataset into subsets and display each subset in its own panel. This makes it possible to compare patterns across different categories side by side, while keeping the same axes and scales. The function `facet_wrap` creates a separate panel for each level of a categorical variable, arranging the panels in a wrapping sequence:

```
ggplot(mtcars,aes(x=wt,y=mpg)) +
geom_point() +
facet_wrap(~ cyl)
```

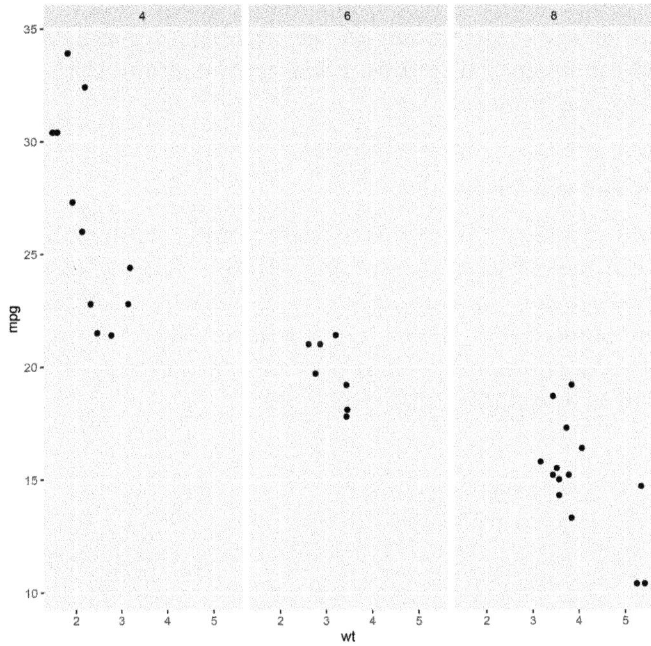

This produces one scatterplot for each value of `cyl`, making it easy to compare cars with four, six, and eight cylinders. The function `facet_grid` arranges panels in a grid defined by two categorical variables:

```
ggplot(mtcars,aes(x=wt,y=mpg)) +
geom_point() +
facet_grid(gear ~ cyl)
```

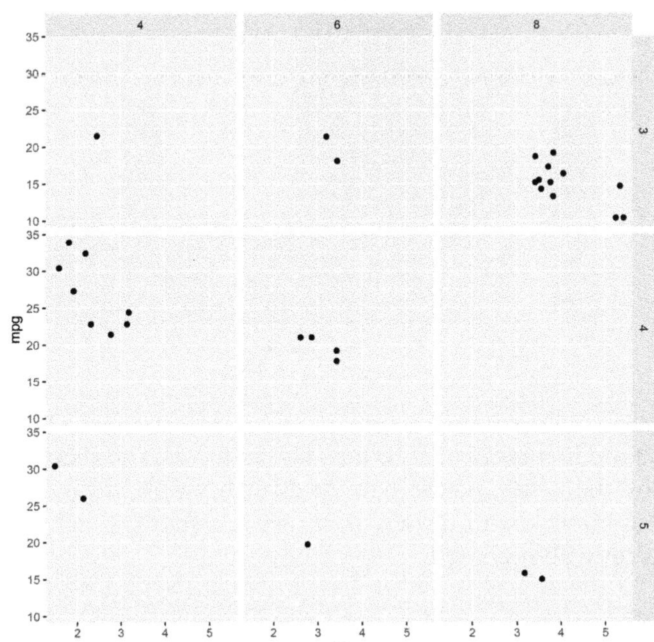

In this case, the rows correspond to the number of gears, while the columns correspond to the number of cylinders. Each panel contains the subset of cars with the corresponding combination.

2.23.7 Statistics and summaries

Every geometry in `ggplot2` is associated with a statistical transformation, called a *stat*. For instance, `geom_point` uses the identity statistic (no transformation), `geom_bar` uses counting, and `geom_smooth` computes a fitted curve. In many cases, the default statistic works well, but it can be useful to control this explicitly. An example is the function `geom_histogram`, which displays the distribution of a continuous variable by default using `stat_bin`:

```
ggplot(mtcars,aes(x=mpg)) +
geom_histogram(binwidth=2)
```

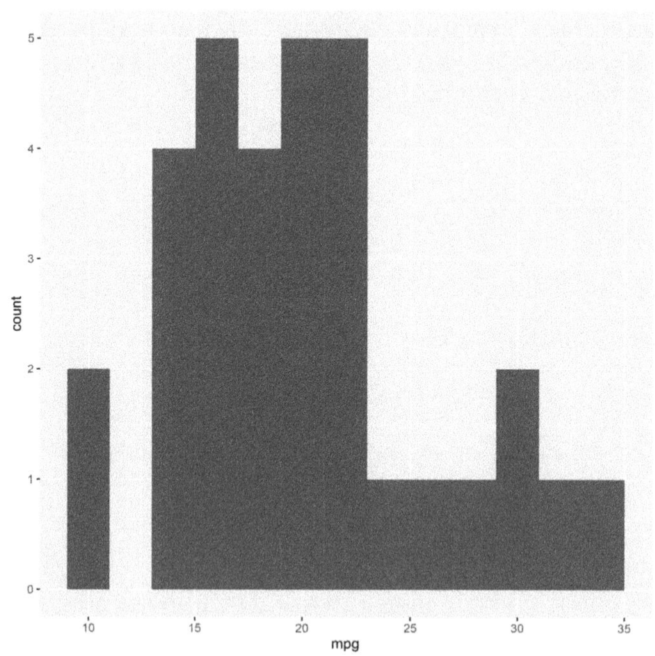

Here the data are first grouped into bins of width two, and then the count per bin is represented by the height of the bar. Another example is the function `stat_summary`, which allows us to display arbitrary summaries, such as the mean and standard deviation within groups:

```
ggplot(mtcars,aes(x=factor(cyl),y=mpg)) +
stat_summary(fun=mean,geom="point") +
stat_summary(fun.data=mean_se,geom="errorbar",width=0.3)
```

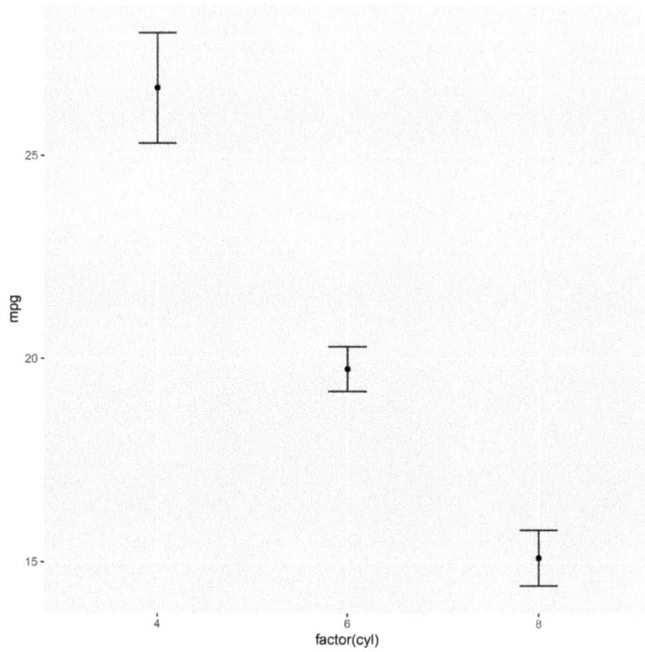

In this example, each cylinder category is represented by the average fuel efficiency, with error bars showing one standard error above and below the mean.

In practice, one often has a choice: either allow the geometry to apply its default statistical transformation, or compute the summary directly with `dplyr` and then display the result using a simple geometry such as `geom_point` or `geom_col`. Both approaches are valid, and the choice depends on whether one wishes the calculation to be visible in the plotting code itself, or performed in a separate data-processing step.

2.23.8 Themes, labels, and annotations

Once the main structure of a plot is in place, it is often desirable to improve its readability and appearance. This can be achieved by adjusting labels, choosing a theme, and adding annotations.

- **Labels**. The function `labs` is used to modify axis labels, the title, and the legend title:

```
ggplot(mtcars,aes(x=wt,y=mpg,colour=factor(cyl))) +
geom_point() +
labs(
x = "Weight (1000 lbs)",
y = "Miles per gallon",
colour = "Cylinders",
title = "Fuel efficiency by car weight and engine size"
)
```

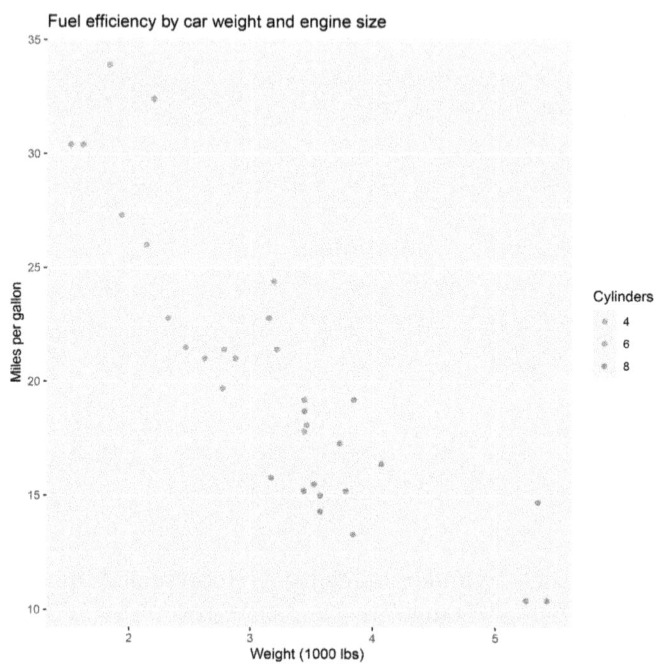

- **Themes**. Themes control the overall look of the plot: background, grid lines, and text styles. Several ready-made options are available, such as `theme_minimal`, `theme_classic`, or `theme_dark`. A same plot can have quite a different overall appearance. For example, a minimal theme produces a very simple-looking plot:

```
ggplot(mtcars,aes(x=wt,y=mpg)) +
geom_point() +
theme_minimal()
```

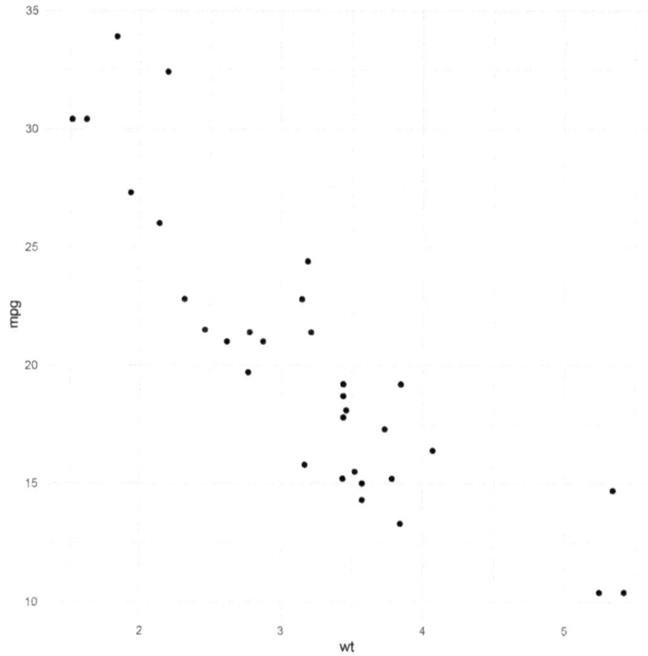

On the other hand, a dark theme produces a plot appearing very different from the simple-looking one, even if the representation is equivalent:

```
ggplot(mtcars,aes(x=wt,y=mpg)) +
geom_point() +
theme_dark()
```

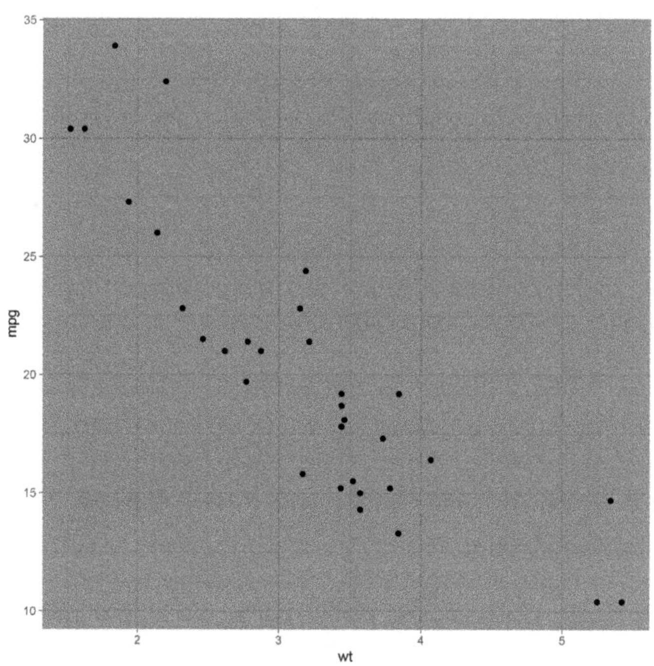

It goes without saying that themes are especially useful to adapt figures to the style of a report or publication.

- **Annotations**. Individual features of the data can be highlighted using the `annotate` function. For example, we can label the car with the highest fuel efficiency:

```
# Identify cars
best <- mtcars[which.max(mtcars$mpg), ]

# Now do the plot
ggplot(mtcars,aes(x=wt,y=mpg)) +
geom_point() +
annotate("text",x=best$wt,y=best$mpg+1,
label=rownames(best))
```

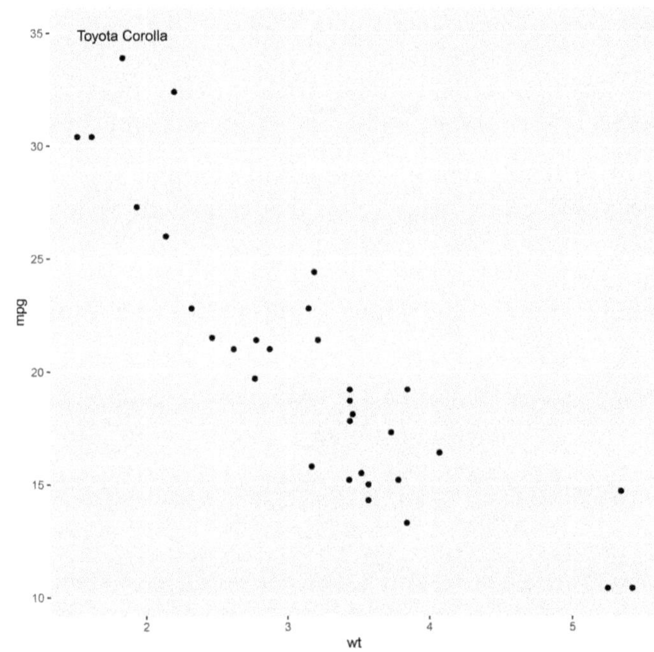

A few readers will have found this last example very interesting as it is always tricky to identify and label one or just a few points in a scatter plot.

As it is easy to see, labels, themes, and annotations provide the finishing touches that turn a draft plot into a polished figure, ready for presentations.

2.23.9 Putting it all together: the Hertzsprung–Russell diagram

We conclude this introduction to `ggplot2` with a more substantial example that combines all the elements discussed so far. The example is drawn from astrophysics and shows how stellar data can be displayed in the classic Hertzsprung–Russell (H–R) diagram (see for instance reference [27], or [28]).

For this example we download the data from the *HYG stellar catalogue* [29], version HYG 4.2, which contains basic information about several hundred thousand stars, including their absolute magnitude (`absmag`) and colour index (`ci`). These two variables are the essential ingredients for constructing the H–R diagram. The preparation step in R looks like this (there might be variants to this code, depending on the format in which data are stored):

```
# This URL might change for other versions of the catalogue
hyg_url <-
"https://www.astronexus.com/downloads/catalogs/hygdata_v42.csv.gz"

# Load the data into a tibble (use a function from readr)
hyg <- read_csv(hyg_url)
```

```
## Rows:  119626 Columns:  37
## -- Column specification --
## Delimiter:  ","
## chr  (8):  gl, bf, proper, spect, bayer, con, base, var
## dbl (29):  id, hip, hd, hr, ra, dec, dist, pmra, pmdec, rv, mag,
## absmag,...
##
## i Use 'spec()' to retrieve the full column specification for this
## data.
## i Specify the column types or set 'show_col_types = FALSE' to quiet
## this message.
```

```
# This object is a tibble
print(class(hyg))
```

```
## [1] "spec_tbl_df" "tbl_df"      "tbl"         "data.frame"
```

```
# The tibble has 37 columns and 119626 rows.
# Show first four lines and 5 columns of this tibble
print(hyg[1:4,1:5])
```

```
## # A tibble: 4 x 5
##        id   hip     hd    hr gl
##     <dbl> <dbl>  <dbl> <dbl> <chr>
## 1       0    NA     NA    NA <NA>
## 2       1     1 224700    NA <NA>
## 3       2     2 224690    NA <NA>
## 4       3     3 224699    NA <NA>
```

The HYG catalogue contains many thousands of stars, but not all records are suitable for plotting an H–R diagram. Some entries have missing or nonsensical values for the absolute magnitude (`absmag`) or the colour index (`ci`), while others lie outside the ranges where these quantities are physically meaningful. Before any plotting, it is therefore important to perform a basic cleaning step. In practice, we begin by converting the relevant columns to numeric form and discarding stars that fall outside reasonable ranges: absolute magnitude between -10 and 15, and colour index between -0.5 and 3.5. These thresholds retain the bulk of normal stars while removing extreme or corrupted entries. The cleaned dataset will then provide a reliable foundation for constructing the H–R diagram.

```
hyg_clean <- hyg |>
mutate(
# Ensure numeric type (some CSVs read these as character)
absmag = as.numeric(absmag),
ci = as.numeric(ci)
) |>
# Keep physically reasonable ranges for an H--R diagram
filter(
!is.na(absmag),
!is.na(ci),
between(absmag, -10, 15),
between(ci, -0.5, 3.5)
)
```

This step removes rows with missing or corrupted values and restricts the data to sensible magnitude and colour ranges, producing a stable basis for the plots that follow.

The simplest version of the Hertzsprung–Russell diagram is a scatter plot of absolute magnitude against colour index. While informative, such a plot shows all stars with the same symbol, ignoring the fact that stars themselves have different colours that are closely related to their physical properties. To capture this additional dimension, we enrich the dataset with new columns that allow us to represent stellar colour directly in the plot. Two methods are used.

1. The HYG catalogue includes a field giving the spectral type of each star, such as ‘G2V’ or ‘K5III’. By extracting the first letter of this string we obtain the broad spectral class (O, B, A, F, G, K, M). This classification corresponds to surface temperature and therefore to colour: O and B stars are blue, F and G stars are white or yellow, while K and M stars are orange or red.
2. Another approach is to compute an approximate effective temperature from the colour index. A standard formula, introduced by F J Ballesteros in 2012 [30], gives

$$T_{\text{eff}} \approx 4600\left(\frac{1}{0.92\,(B - V) + 1.7} + \frac{1}{0.92\,(B - V) + 0.62}\right), \tag{2.1}$$

where $(B - V)$ is the symbol used by Ballesteros to indicate the colour index. This approximation makes it possible to assign a continuous colour scale to the stars, ranging from blue for high temperatures to red for low temperatures.

Both methods provide a way of making the H–R diagram colourful, either by discrete categories (spectral class) or by a continuous physical variable (temperature). In the next step we implement these enrichments in the dataset. Starting from the cleaned table `hyg_clean`, we extract the broad spectral class (O, B, A, F, G, K, M) and compute an approximate effective temperature from the colour index $(B - V)$ using relation (2.1).

```
hyg_enriched <- hyg_clean |>
mutate(
# Broad spectral class from the first letter of 'spect'
# (e.g. "G2V" -> "G")
spect_class = substr(spect,1,1),
spect_class = ifelse(
spect_class %in% c("O","B","A","F","G","K","M"),
spect_class, NA
),
# Treat as an ordered factor for nicer legends
spect_class = factor(spect_class,
levels=c("O","B","A","F","G","K","M")),

# Approximate effective temperature from B-V (ci)
# [Ballesteros, 2011]
# T_eff = 4600 * (1/(0.92*(B-V)+1.7) + 1/(0.92*(B-V)+0.62))
teff = 4600*(1/(0.92*ci+1.7)+1/(0.92*ci+0.62))
)
```

The new columns `spect_class` (categorical) and `teff` (continuous, in Kelvin) allow us to colour the H–R diagram either by broad spectral type or by physical temperature. After their creation, though, we need to remove rows that might have become unusable for plotting, following the operations carried out by the `mutate` verb. This avoids warnings about dropped rows and ensures that each plotting tibble contains only complete, in-range values for the variables actually mapped in the figure.

```
# For first method
hyg_plot_class <- hyg_enriched |>
filter(
!is.na(ci),
!is.na(absmag),
!is.na(spect_class)
)

# For second method
hyg_plot_teff <- hyg_enriched |>
filter(
!is.na(ci),
!is.na(absmag),
is.finite(teff),
teff > 1500,
teff < 50000
)
```

Let us now look at how to produce the coloured plot with the first method. The method is essentially a mapping (inside aes) so each point inherits its colour from the spect_class value. The magnitude axis is also reversed to follow astronomical convention (brighter stars at the top). Because the HYG sample is dense, we use small points and partial transparency to reduce overplotting. Finally, we supply a manual palette that follows the canonical O→M hue progression (blue → red) and s

```
# HR diagram coloured by spectral class
ggplot(hyg_plot_class,aes(x=ci,y=absmag,colour=spect_class)) +
geom_point(alpha=0.35,size=0.6) +
scale_y_reverse() +
scale_colour_manual(
values = c(
O = "midnightblue",
B = "royalblue",
A = "deepskyblue",
F = "lightskyblue",
G = "gold",
K = "orange",
M = "red3"
),
na.translate = FALSE,
name = "Spectral class"
) +
labs(
x = "Colour index (B--V proxy 'ci')",
y = "Absolute magnitude",
title = "Hertzsprung--Russell diagram (coloured by spectral class)"
) +
theme_minimal()
```

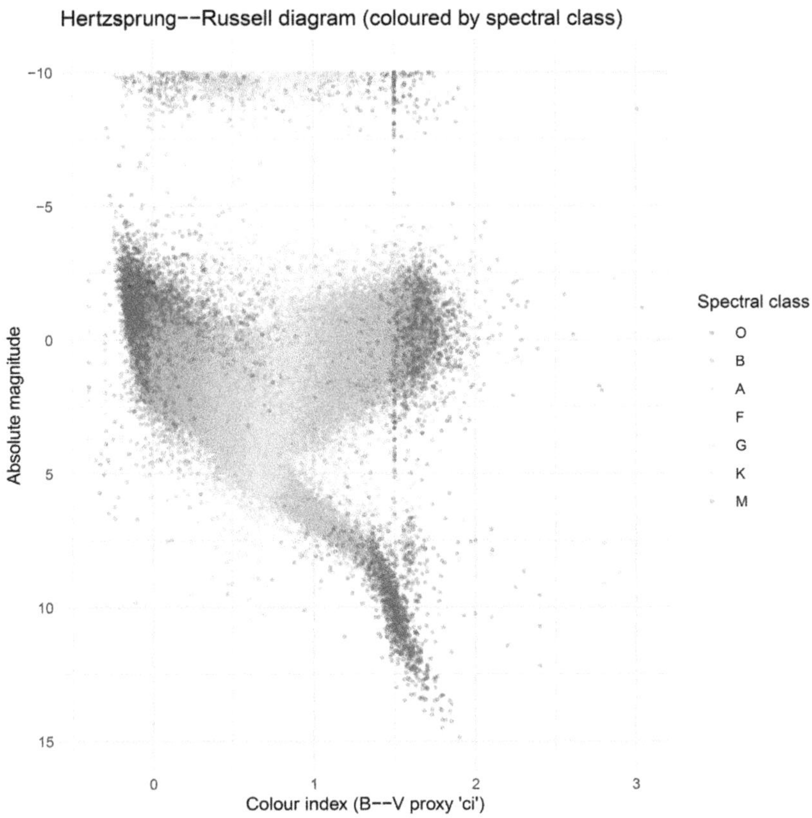

A few remarks mght be useful, at this stage.

- **Mapping versus setting**. `colour = spect_class` is a mapping (inside `aes`), so the legend is generated automatically and reflects data values. The specific hues come from `scale_colour_manual`.
- **Axis convention**. `scale_y_reverse()` puts bright (low magnitude) at the top.
- **Overplotting control**. `alpha = 0.35` and `size = 0.6` keep dense regions readable; one can adjust to taste depending on how many rows survive the filters used.
- **Missing/other types**. `na.translate = FALSE` hides an 'NA' legend key for stars without a recognised class. If the catalogue includes rare classes (e.g. L/T), one will have to either map them to a colour or set them to `NA` before plotting.

For the second method, we map the stellar effective temperature to colour. As the temperature decreases from left to right in the H–R plane (hot, blue stars at low `ci`;

cool, red stars at high `ci`), a continuous colour scale conveys physics directly. We again reverse the magnitude axis to follow astronomical convention.

```
ggplot(hyg_plot_teff,aes(x=ci,y=absmag,colour=teff)) +
geom_point(alpha=0.35,size=0.6) +
scale_y_reverse() +
scale_colour_gradientn(
colours = c("darkblue","deepskyblue","white",
"gold","orange","red"),
name = expression(T[eff]~"(K)")
) +
labs(
x = "Colour index (B--V proxy 'ci')",
y = "Absolute magnitude",
title = "Hertzsprung--Russell diagram (continuous colour by T[eff])"
) +
theme_minimal()
```

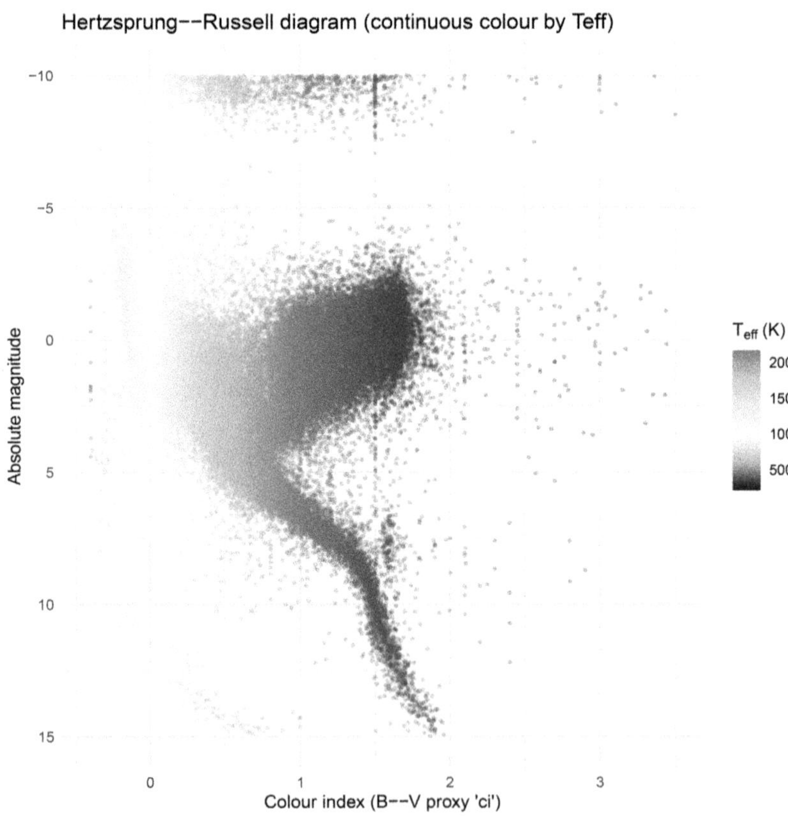

Also for this second piece of code, a few remarks are useful:

- **Aesthetic mapping**. We map `colour = teff`*inside*`aes`, so the legend is generated from the data and the scale controls its appearance.
- **Axis convention**. `scale_y_reverse()` puts bright (low magnitude) stars at the top.
- **Perceptual order**. The gradient runs blue → red to match the physical hot → cool progression. For print-friendly options, one could replace the gradient with a perceptually uniform palette.
- **Point density**. Small size and `alpha = 0.35` mitigate overplotting. On very dense catalogues, `geom_bin2d` or `geom_hex()` are good alternatives.
- **Temperature bounds**. The earlier `hyg_plot_teff` filter ($1500 \leqslant T_{\text{eff}} \leqslant 50\,000$ K) avoids implausible or ill-conditioned colour mappings and prevents 'removed row' warnings at draw time.

The plots and procedures demonstrated in this section are necessarily more elaborate than the simple examples encountered earlier, and they draw on many of the computational tools provided by the tidyverse and `ggplot2`. The reader should not be discouraged if the logic feels dense on a first encounter. A careful re-reading will reveal the key ideas and, in particular, the interplay between data, aesthetics, geometries, scales, facets, statistics, and themes that together form the Grammar of Graphics as implemented in `ggplot2`.

2.24 The `comphy` package

To support the material presented in part II of this book, the R package `comphy` (short for **com**putational **phy**sics) was created. The package collects a wide range of functions implementing the algorithms introduced in the text, covering numerical differentiation, integration, the solution of ordinary differential equations, and other topics central to computational physics.

It is important to stress that `comphy` is not intended as a repository of the most efficient or state-of-the-art routines available in R. Excellent, highly optimised libraries already exist for professional use. Instead, the goal of `comphy` is pedagogical: to make the essential algorithmic mechanisms transparent to the student. Each function has been written with clarity and simplicity in mind, closely reflecting the algorithms as they were derived in the chapters of part II. This allows students to bridge the gap between theoretical derivation and practical implementation, and to see how fundamental computational techniques can be translated into working code.

By experimenting with `comphy`, students can test the behaviour of different methods, explore their accuracy and limitations, and gain an intuitive understanding of computational approaches to physics. Once the fundamental principles are mastered in this way, more advanced or specialised packages can be explored with confidence.

The list of functions included in `comphy` at the time this book was completed, is included in appendix H.

2.25 A suggested approach to a machine's rounding

We would like to conclude this chapter on R with some considerations related to rounding off errors, given what is explained in chapter 1. R uses double precision numbers most of the time and problematic or even noticeable round off errors are rare. But, as we know, they cannot be completely eliminated and one must be always vigilant to their possible appearance, when setting up simple or less-simple calculations. It would be time-consuming and complicated to study such errors systematically, and in this introductory book it is suggested that an *ad hoc* approach is adopted to investigate the errors thoroughly. What counts when the investigation is carried out, is knowledge of the machine's precision and the ability to print out or display values with the correct number of digits.

2.25.1 Numeric characteristics of a computer

How numbers are stored in memory, in any implementation of R, depends on the computer on which R is installed. This information is specific and must be found out individually prior to starting calculations or writing code. A very useful internal variable that helps greatly in this respect is `.Machine`, which stores, among other things:

- `double.xmin` The smallest number representable in normal floating point form. It matters when numbers are extremely close to zero.
- `double.xmax` The largest number representable in normal floating point form.
- `double.eps` If represented with ϵ, this is the smallest floating point number for which $1 + \epsilon \neq 1$. It is therefore equivalent to the *machine epsilon*, ϵ_{mach} of chapter 1. This number is in general much larger than `double.xmin`. Numbers smaller than ϵ are subject to the precision's problems connected to round off errors, so that $1 + r = 1$, if $0 \leqslant r < \epsilon$.[2]
- `double.neg.eps` If represented with ϵ^-, this is the smallest (positive) floating point number for which $1 - \epsilon^- \neq 1$. It has characteristics similar to those of ϵ.
- `integer.max` The largest integer (number stored in memory as an integer).

These quantities come in handy when certain calculations face the machine's limits. In such circumstances, `if` conditions, `while` loops and the like, could fail to function properly, without exact knowledge of these limits.

The following code shows how one can decide which number makes a difference between 1 and $1 + f\epsilon$, where f is a fraction of 1. Given that usually intervals like

[2] It is important to note that `double.eps` is not the smallest representable positive number, but rather the distance between 1 and the next larger representable number. This quantity measures the *relative precision* of floating point arithmetic: near 1, consecutive floating point numbers are separated by roughly $\epsilon_{\text{mach}} \approx 10^{-16}$. More generally, the gap between representable numbers grows in proportion to the magnitude of the number, so around x the spacing is of order $\epsilon\,|x|$. By contrast, `double.xmin` (about 10^{-308}) gives the absolute smallest positive normalized number that can be represented, relevant only when values approach zero and underflow occurs.

$[-0.5, 0.5)$ are those used for rounding, it is suspected that f might be related to 0.5. But, and this is the spirit with which the student is encouraged to proceed, it doesn't cost much to test the idea with simple calculations.

```
# Value of eps
eps <- .Machine$double.eps
print(eps)

## [1] 2.220446e-16

# 1 + eps is different from 1
print(1+eps == 1)

## [1] FALSE

# Now let's use a fraction of eps, smaller than eps
new_eps <- 0.99*eps
print(1+new_eps == 1)

## [1] FALSE

# What's the fraction of eps from which 1
# and 1+new_eps are not anymore different?
# Use a loop
for (i in 10:0) {
fr <- i/10
ans <- 1+fr*eps == 1
sfr <- sprintf("%5.1f ",fr)
smtp <- paste(c(sfr,ans))
print(smtp)
}
```

```
## [1] "  1.0 " "FALSE"
## [1] "  0.9 " "FALSE"
## [1] "  0.8 " "FALSE"
## [1] "  0.7 " "FALSE"
## [1] "  0.6 " "FALSE"
## [1] "  0.5 " "TRUE"
## [1] "  0.4 " "TRUE"
## [1] "  0.3 " "TRUE"
## [1] "  0.2 " "TRUE"
## [1] "  0.1 " "TRUE"
## [1] "  0.0 " "TRUE"
```

```
# It seems that fractional values greater than 0.5
# yields a quantity which "matters".
# Some more, finer tests
for (i in -5:5) {
fr <- 0.5+i/100
ans <- 1+fr*eps == 1
sfr <- sprintf("%5.2f ",fr)
smtp <- paste(c(sfr,ans))
print(smtp)
}
```

```
## [1] " 0.45 " "TRUE"
## [1] " 0.46 " "TRUE"
## [1] " 0.47 " "TRUE"
## [1] " 0.48 " "TRUE"
## [1] " 0.49 " "TRUE"
## [1] " 0.50 " "TRUE"
## [1] " 0.51 " "FALSE"
## [1] " 0.52 " "FALSE"
## [1] " 0.53 " "FALSE"
## [1] " 0.54 " "FALSE"
## [1] " 0.55 " "FALSE"
```

We can also check whether the smallest and largest normal floating point numbers, `double.xmin` and `double.xmax`, respectively, have value equal to their estimate in equation (1.1). For the smallest value we should have

$$2^{-1023+1} = 2^{-1022} = 2.225\ 07 \cdots \times 10^{-308},$$

and for the largest value

$$(2 - 2^{-1023})2^{1023+1} = 1.797\ 69 \cdots \times 10^{308},$$

because for double precision (normally used by R), $m = 1023$, $e_{\min} = -1023$ and $e_{\max} = 1024$. These values can be verified straightaway, using more digits for the precision.

```
# Increase precision
backup_opts <- options()
options(digits=12)

# Smallest normal floating point
print(.Machine$double.xmin)
```

```
## [1] 2.22507385851e-308
```

```
# Largest normal floating point
print(.Machine$double.xmax)
```

```
## [1] 1.79769313486e+308
```

```
# Back to default options
options(backup_opts)
```

Undesired results coming from simple operations can also arise in conjunction with large numbers, like large integers, for example. It should be, first of all, clear that actual integers are restricted to a given range. The `integer.max` variable of `.Machine` reveals this range.

```
# Largest positive integer
xL <- .Machine$integer.max
print(typeof(xL))
```

```
## [1] "integer"
```

```
print(xL)
```

```
## [1] 2147483647
```

```
# Largest negative integer
```

```
xM <- -xL
print(typeof(xM))
```

```
## [1] "integer"
```

```
print(xM)
```

```
## [1] -2147483647
```

```
# Integers greater than xL or smaller than xM are converted
# to double precision floating point numbers
y <- xL+1
print(y)
```

```
## [1] 2147483648
```

```
print(typeof(y))
```

```
## [1] "double"
```

```
y <- xL-1
print(y)
```

```
## [1] 2147483646
```

```
print(typeof(y))
```

```
## [1] "double"
```

Integers larger than `integer.max` do not exist and must be converted to a double precision number. As there are only zeros after the floating point, R displays the number as if it were an integer, but care must be adopted for subsequent calculations. The limitation in range can cause problems when addition of subtraction of large numbers is involved. Consider for instance an addition in which one of the two integers is much larger than the other. The result is not what is expected, the smaller number is ignored, because of round off errors. Even the 52 bits needed to store the large integer transformed into a double precision number, are not enough to register the smaller integer.

```
# Two large numbers (floats). One much bigger
x <- 1e+20
y <- 1e+40
print(typeof(c(x,y)))

## [1] "double"

# Their sum is totally inaccurate
z <- x+y
print(typeof(z))

## [1] "double"

print(z)

## [1] 1e+40
```

One should be aware of risks, like the one shown above, associated with the handling of large numbers. There exist various solutions to handle these situations in R. A well-known solution is provided by the package `gmp`, which is an adaptation from the GNU Multiple Precision Arithmetic Library in the C programming language [31]. A new class, `bigz` (a *Big Integer*), is introduced by `gmp` that signals R to treat any `bigz` object as a special object to which enhanced precision is provided in various ways.

```
# Import gmp library
library(gmp)

##
## Attaching package:  'gmp'
## The following objects are masked from 'package:base':
##
##     %*%, apply, crossprod, matrix, tcrossprod

# Define large numbers in a different way
x <- as.bigz(10)^20
y <- as.bigz(10)^40
print(class(c(x,y)))

## [1] "bigz"

print(typeof(c(x,y)))

## [1] "raw"

# This time the smallest number is recorded
z <- x+y
print(z)

## Big Integer ('bigz') :
## [1] 10000000000000000000100000000000000000000
```

Details on all the features included with gmp can be found in the documentation accompanying the package.

2.25.2 Printing or displaying significant digits

When dealing with the empirical estimation of round off errors, in fact, of all types of computational errors, it is important to be aware of the ways in which numbers are rounded with the functions `round`, `floor`, `ceiling`, `trunc` and `signif`. Details on each of these functions can be read in the related help pages. Here we provide a short description.

Functions with one input. These are `ceiling`, `floor` and `trunc`. They take in one number and return an integer.

- `ceiling(x)` Return the closest integer to x, in the direction away from 0, if $x > 0$, and towards 0, if $x < 0$.
- `floor(x)` Return the closest integer to x, in the direction towards from 0, if $x > 0$, and away from 0, if $x < 0$.
- `trunc(x)` Return the closest integer to x, always in the direction towards 0.

```
# Let us use a positive and negative number for the demo
xpos <- 1.23456
xneg <- -1.23456

# ceiling, floor, trunc
print(ceiling(xpos))
```

```
## [1] 2
```

```
print(ceiling(xneg))
```

```
## [1] -1
```

```
print(floor(xpos))
```

```
## [1] 1
```

```
print(floor(xneg))
```

```
## [1] -2
```

```
print(trunc(xpos))
```

```
## [1] 1
```

```
print(trunc(xneg))
```

```
## [1] -1
```

Functions with two inputs. These are `round` and `signif`. They take in two numbers and return a fraction (in decimals).

- `round(x,digits=0)` Rounds x to the closest number of decimals (after the point), indicated by `digits`.
- `signif(x,digits=0)` Rounds x to the closest number of digits (all digits, before and after the point), indicated by `digits`.

```
# Use the same positive and negative numbers of previous chunk

# round to 0 (default), 1 and 4 decimals
print(c(round(xpos),round(xneg)))

## [1]  1 -1

print(c(round(xpos,1),round(xneg,1)))

## [1]  1.2 -1.2

print(c(round(xpos,4),round(xneg,4)))

## [1]  1.2346 -1.2346

# Round to 1 and 3 significant digits
print(c(signif(xpos,1),signif(xneg,1)))

## [1]  1 -1

print(c(signif(xpos,3),signif(xneg,3)))

## [1]  1.23 -1.23
```

The only issue with both `round` and `signif` involves numbers with a 5 deciding the rounding. These functions always select the closest *even* number. This can create confusion, but it is a good choice, statistically speaking, as it makes the likelihood of systematic errors smaller.

```
# Create two numbers (with 5 as last digit)
x1 <- 1.55
x2 <- 1.45
# The first is rounded up and the second rounded down
# This is called "rounding to even"
print(round(x1,1))

## [1] 1.6

print(round(x2,1))

## [1] 1.4

# Also signif shows this behaviour
print(signif(x1,2))

## [1] 1.6

print(signif(x2,2))

## [1] 1.4
```

The 'round to even' rule can raise serious issues when exact results are expected. For this reason it is important to know that the functions `round` and `signif` apply such a rule.

One should also notice that R has built in rules to display or print numbers, which are connected to the current number of digits, imposed by the function `options`. By default, the number of digits is 7. This explains the following output.

```
# The following number has 13 significant digits
x <- 1.234567890123

# Default options allows 7 digits
print(x)

## [1] 1.234568

# Change digits in options
```

```
bkup <- options()
options(digits=9)
print(x)

## [1] 1.23456789

# Change again. But the 0 is not displayed because not significant
options(digits=10)
print(x)

## [1] 1.23456789

options(digits=11)
print(x)

## [1] 1.2345678901
```

What happens when the number of digits is changed before using, for example, `signif`? In general `options` wins over `signif`, but that, in addition, depends on the binary representation of the given number. Consider the following code.

```
# 0.1 is not representable as a binary
# with a finite number of digits
x <- 0.1

# With default digits=7
print(x)

## [1] 0.1

# Increase to 15, 16, 17
# The 17th significant digit is different from 0
# and so it is displayed
options(digits=15)
print(x)

## [1] 0.1
```

```
options(digits=16)
print(x)

## [1] 0.1

options(digits=17)
print(x)

## [1] 0.10000000000000001

# Back to default
options(bkup)
```

signif does not seem to change this behaviour.

```
# Still using 0.1
print(signif(x,digits=5))

## [1] 0.1

# Maybe this is unexpected
options(digits=17)
print(signif(x,digits=5))

## [1] 0.10000000000000001
```

The take-home message is that when tinkering with the system's parameters (like the number of digits displayed), one should always test and check behaviours ahead of the following applications.

2.25.3 Testing round off errors empirically

We have seen how R represents very small and very large numbers, and how rounding rules can lead to unexpected outcomes: for example, when ϵ is smaller than machine precision, the expression $1 + \epsilon$ simply evaluates to 1. This illustrates how subtle rounding effects can appear in even the simplest of computations. However, when numerical expressions involve many successive operations on small and large numbers, a precise theoretical analysis of error propagation becomes exceedingly difficult, and in many cases not feasible at all. Instead, a practical approach is to test empirically the stability and reliability of the expressions one intends to use. This involves experimenting with alternative formulations, introducing small perturbations, and comparing results. Three examples will clarify this point.

Example 2.1 *Consider the difference*

$$\sqrt{x+1} - \sqrt{x},$$

for very large values of x. *Algebraically, this expression is equal to*

$$\frac{1}{\sqrt{x+1} + \sqrt{x}},$$

so the two formulas should always give the same result. In floating-point arithmetic, however, their behaviour is very different. Let us consider, for example, a large value for x, *that is,* $x = 10^{12}$, *so that* $\sqrt{x} = 10^6$. *We need to figure out a way to estimate* $\sqrt{x+1}$ *numerically, to compare it with computer results. We could, for instance, use a Taylor expansion of* $\sqrt{x+1}$ *around* x:

$$\sqrt{x+1} \approx \sqrt{x} + \frac{1}{2\sqrt{x}} - \frac{1}{8x^{3/2}} + \cdots$$

$$\Downarrow$$

$$\sqrt{10^{12}+1} \approx 10^6 + 5 \times 10^{-7} - 1.25 \times 10^{-19} + \cdots$$

The two square roots differ only by about 5×10^{-7}, *while their size is around* 10^6. *Subtracting two nearly equal numbers of size one million to extract a difference in the millionth's decimal place inevitably leads to loss of precision: most significant digits cancel out. In R this scenario can be demonstrated fairly easily:*

```
x <- 1e12
expr1 <- sqrt(x+1)-sqrt(x)
expr2 <- 1/(sqrt(x+1)+sqrt(x))
print(expr1)

## [1] 5.000038e-07

print(expr2)

## [1] 5e-07
```

The variable `expr1` *is inaccurate due to rounding off cancellation. The true value is about* $5.000\,000\,0000 \times 10^{-7}$. *The second variable,* `expr2`*, was formulated to avoid subtraction of nearly equal terms and delivers the correct value within machine precision.*

This example illustrates a general strategy: when an expression involves subtraction of almost equal numbers, it is worth seeking an equivalent formulation that does not suffer from such cancellation.

Example 2.2 *This is another interesting example, still involving a difference between two close numbers, that shows how algebraic manipulations are most of the time needed to stabilise numerical results. For small* x*, the direct formula*

$$f(x) = \frac{1 - \cos(x)}{x^2}$$

is numerically fragile. Since

$$\cos(x) = 1 - \frac{x^2}{2} + \frac{x^4}{24} - \cdots,$$

we have

$$1 - \cos(x) = \frac{x^2}{2} - \frac{x^4}{24} + \frac{x^6}{720} - \cdots \quad \Rightarrow f(x) = \frac{1}{2} - \frac{x^2}{24} + \frac{x^4}{720} - \cdots$$

Thus, $1 - \cos(x)$ *is of order* x^2*, because its Taylor expansion starts with the term* $\frac{x^2}{2}$. *This means that for small* x *the quantity* $1 - \cos(x)$ *is very small compared to 1. When we compute it directly as* $1 - \cos(x)$*, we are subtracting two numbers that are almost equal: 1 and* $\cos(x)$. *In floating-point arithmetic, as seen, such a subtraction is dangerous because most of the leading digits cancel out, leaving only a few digits of the difference.*

To understand how serious this is, recall that evaluating $\cos(x)$ *in floating point comes with a tiny absolute error, say about* $\pm\epsilon$. *When* x *is not too small, this error is negligible. But when we form* $1 - \cos(x)$*, the 'true' result is of size* $\sim x^2$*. The relative error in this result is therefore roughly the absolute error of* $\cos(x)$*, which is* ϵ*, divided by the true size, which is* x^2*. Hence the relative error in* $1 - \cos(x)$ *scales like*

$$\frac{\epsilon}{x^2}.$$

Now, double precision arithmetic has $\epsilon \approx 10^{-16}$*. So if* $|x|$ *is about* 10^{-8}*, we find* $\epsilon/x^2 \approx 1$*. At this point the relative error in* $1 - \cos(x)$ *is of order one, which means we may lose all significant digits in the result. That is why once x is smaller than about* $\sqrt{\epsilon} \approx 1.5 \times 10^{-8}$*, the direct formula becomes numerically unreliable: the subtraction discards the information we are actually trying to compute. A more stable formulation comes from the identity*

$$1 - \cos(x) = 2\sin^2(x/2).$$

The initial expression, $f(x)$*, can now be re-written as*

$$f(x) = \frac{2\sin^2(x/2)}{x^2} = \frac{1}{2}\left(\frac{\sin(x/2)}{x/2}\right)^2,$$

which avoids subtracting nearly equal numbers. The function $\sin(y)/y$ *is well-behaved near* $y = 0$ *(it tends to 1), so the expression maintains accuracy for very small* x*. In R we can demonstrate this behaviour comparing* $f(x)$ *with the improved version obtained above and with a simple Taylor approximation retaining just the first two terms of the expansion:*

```
# Unstable and stable expressions
f_direct  <- function(x) (1-cos(x))/x^2
f_stable  <- function(x) 0.5*(sin(x/2)/(x/2))^2
f_taylor <- function(x) 0.5-x^2/24 # two-term Taylor

# Try on smaller and smaller numbers
xs <- c(1e-3, 1e-6, 1e-9, 1e-12)
cbind(x = xs,
direct = f_direct(xs),
stable = f_stable(xs),
taylor = f_taylor(xs))
```

```
##           x    direct stable  taylor
## [1,] 1e-03 0.5000000    0.5     0.5
## [2,] 1e-06 0.5000445    0.5     0.5
## [3,] 1e-09 0.0000000    0.5     0.5
## [4,] 1e-12 0.0000000    0.5     0.5
```

What is happening?

- *For moderately small x* (e.g. 10^{-3}, 10^{-6}), *all three agree to many digits.*
- *For very small x* (e.g. 10^{-9}, 10^{-12}), `f_direct` *degrades due to cancellation, while* `f_stable` *still matches the series* $1/2 - x^2/24$ *to machine precision.*

The rule of thumb in this second example, again, is to prefer algebraic rearrangements that avoid subtracting nearly equal quantities. When in doubt, one should validate empirically by (i) comparing equivalent formulations (here, direct versus stable), and (ii) checking against a trusted local approximation (here, the Taylor series) on a range of small x.

Example 2.3 *Consider the* logistic–sigmoid *written as*

$$\sigma(x) = \frac{\mathrm{e}^x}{1 + \mathrm{e}^x}.$$

Algebraically this is fine, but numerically it is hazardous. In double precision, e^x *overflows to* $+\infty$ *for large x (in R, around* $x \approx 709$*), so* $\sigma(1000)$ *computed this way becomes* $\infty/(1 + \infty)$*, which evaluates to* `NaN` *or may collapse to 1 only by accident of how the operations are ordered. At the other end, for very negative x (e.g.* $x = -1000$*),* e^x *underflows to* 0*, and* $\sigma(x) = 0/(1 + 0) = 0$*. While that value is correct, the formula is numerically brittle across ranges of x.*

A stable algebraic rearrangement avoids these issues:

$$\sigma(x) = \frac{1}{1 + \mathrm{e}^{-x}}.$$

For large positive x, e^{-x} *is tiny and no overflow occurs. For large negative x,* e^{-x} *is huge and may overflow, so we instead use a piecewise version that is stable on both sides:*

$$\sigma(x) = \begin{cases} \dfrac{1}{1 + \mathrm{e}^{-x}}, & x \geqslant 0, \\ \dfrac{\mathrm{e}^x}{1 + \mathrm{e}^x}, & x < 0. \end{cases}$$

Each branch keeps the exponential applied to a quantity that does not blow up. An appropriate R code to demonstrate this issue follows:

```
# unstable version
sigma_unstable <- function(x) exp(x) / (1 + exp(x))

# Stable version
sigma_stable <- function(x) {
out <- x
i1 <- (x >= 0)
out[i1] <- 1/(1+exp(-x[i1]))                # safe when x >= 0
out[!i1] <- exp(x[!i1])/(1+exp(x[!i1])) # safe when x < 0
out
}

# Test
xs <- c(-1000,-50,0,50,1000)
cbind(x = xs,
unstable = sigma_unstable(xs),
stable = sigma_stable(xs))
```

```
##          x    unstable      stable
## [1,] -1000 0.00000e+00 0.00000e+00
## [2,]   -50 1.92875e-22 1.92875e-22
## [3,]     0 5.00000e-01 5.00000e-01
## [4,]    50 1.00000e+00 1.00000e+00
## [5,]  1000         NaN 1.00000e+00
```

What is happening?

- *For* $x = 1000$, `sigma_unstable` *attempts* exp(1000) *and overflows;* `sigma_stable` *returns 1 accurately.*
- *For* $x = -1000$, *both return* 0, *but the stable version does so without relying on underflow.*
- *For moderate* x, *both agree.*

A general lesson to learn here is that when exponentials appear, one should think about the range of inputs. We should prefer algebraic forms that keep exponentials of large-magnitude arguments out of the computation (e.g. $\sigma(x) = 1/(1 + e^{-x})$*), and consider branch-wise definitions that are explicitly stable across regions.*

The three examples above show that numerical problems do not only arise from obvious mistakes but also from subtle interactions between floating-point representation and algebraic structure. Catastrophic cancellation, overflow, and loss of precision can all occur in expressions that look perfectly innocent on paper. The lesson for working computational physicists is then twofold:

- Always question whether the algebraic form about to be implemented is numerically reliable.

- Test expressions empirically with a range of input values before relying on them. Simple checks often reveal instabilities that theory alone may not predict.

By keeping these principles in mind and by reformulating expressions when needed, one can avoid many hidden traps of round–off error and build numerical algorithms that are both accurate and robust.

References

[1] Robinson D 2017 The Impressive Growth of R https://stackoverflow.blog/2017/10/10/impressive-growth-r/ (Accessed: 2020-04-23)
[2] The rise of R programming language and its usefulness in data science 2019 https://blog.eduonix.com/software-development/rise-r-programming-language-usefulness-data-science/ (Accessed: 2020-04-23)
[3] Team Dataflair 2019 Uncover the R applications https://data-flair.training/blogs/r-applications/ (Accessed: 2020-04-23)
[4] Woodie A 2019 Is Python strangling R to death? https://www.datanami.com/2019/08/15/is-python-strangling-r-to-death/ (Accessed: 2020-04-23)
[5] Tung L 2019 R vs Python https://www.zdnet.com/article/r-vs-python-rs-out-of-top-20-programming-languages-despite-boom-in-statistical-jobs/ (Accessed: 2020-04-23)
[6] Braun W J and Murdoch D J 2008 *A First Course in Statistical Programming with* R (Cambridge University Press)
[7] Gardener M 2012 *Beginning* R (John Wiley & Sons)
[8] Maindonald J and Braun W J 2006 *Data Analysis and Graphics using* R (Cambridge University Press)
[9] Owen W J 2010 The R guide https://cran.r-project.org/doc/contrib/Owen-TheRGuide.pdf
[10] Adler J 2010 *R in a Nutshell* (O'Reilly & Associates)
[11] Albert J and Rizzo M 2012 *R by example* (Springer)
[12] Crawley M J 2012 *The R Book* (John Wiley & Sons)
[13] Verzani J 2005 *Using R for Introductory Statistics* (Taylor and Francis)
[14] Matloff N 2011 *The Art of R Programming* (No Starch Press)
[15] Davies T M 2016 *The Book of* R (No Starch Press)
[16] The R Foundation The R project for statistical computing https://www.r-project.org/ (Accessed: 2020-04-28)
[17] The RStudio Team RStudio https://rstudio.com/ (Accessed: 2020-04-28)
[18] Faria J C Tinn-R https://sourceforge.net/projects/tinn-r/ (Accessed: 2020-04-28)
[19] Maechler M *et al* 2008 EMACS speaks statistics http://ess.r-project.org/index.php?Section=about, 2008 (Accessed: 2020-04-28)
[20] Friedrichsmeier T *et al* RKWard https://rkward.kde.org/index.html (Accessed: 2020-04-28)
[21] Eclipse Foundation Eclipse StatET: Tooling for the R language https://projects.eclipse.org/projects/science.statet (Accessed: 2020-04-28)
[22] Fox J Rcommander (Rcmdr) https://www.rcommander.com/ (Accessed: 2020-04-28)
[23] Wickham H 2016 *ggplot2: Elegant Graphics for Data Analysis* (Springer)
[24] The R Foundation The comprehensive R archive network https://cran.r-project.org (Accessed: 2020-06-19)
[25] The R Foundation CRAN task views https://cran.r-project.org (Accessed: 2020-06-19)

[26] Wickham H, Rundel M C and Grolemund G 2023 *R for Data Science: Import, Tidy, Transform, Visualize, and Model Data* 2nd edn (O'Reilly & Associates)
[27] Clayton D D 1983 *Principles of Stellar Evolution and Nucleosynthesis* (University of Chicago Press)
[28] Carroll B W and Ostlie D A 2017 *An Introduction to Modern Astrophysics* 2nd edn (Cambridge University Press)
[29] Nash D 2019 Hyg stellar database (v3.0) *Dataset Merging the Hipparcos, Yale Bright Star, and Gliese Catalogs* Available at https://www.astronexus.com/hyg and mirrored on GitHub/Cod berg.
[30] Ballesteros F J 2012 New insights into black bodies *Europhys. Lett.* **97** 34008
[31] GNU 2020 The GNU multiple precision arithmetic library https://gmplib.org/ (Accessed: 2022-09-24)

Part II

Core computational physics

The second part of the book focuses on the core numerical techniques that underpin much of computational physics. Each chapter develops specialised algorithms and shows how these can be implemented efficiently in R.

The emphasis here is on both the theoretical foundation of the methods and their practical implementation. Code is written from scratch whenever possible, and encapsulated into functions that form the basis of the `comphy` package, developed alongside the text.

Throughout this part, the reader learns how to analyse algorithmic performance, estimate errors, and choose appropriate numerical strategies for different types of problems. Numerous worked examples and exercises are included to reinforce understanding and encourage hands-on engagement.

By the end of this part, the reader will have acquired both a conceptual understanding of the most important numerical algorithms in physics and the skills to implement them in R from first principles.

IOP Publishing

Computational Physics with R

James Foadi

Chapter 3

Interpolation

3.1 Introduction

Very simply, *interpolation* is the determination of any non-tabulated value of a function. Consider, for instance, a function $f(x)$ for which the correct analytic values at the specific points $x = 0.2, 0.4, 0.8$ are displayed in table 3.1. What are the values of $f(x)$ at a different set of points, say $x = 0.3$ and $x = 0.6$? Unless more information is provided, these will never be known with certainty, especially when the analytic expression of $f(x)$ is unknown. An appropriate interpolation technique can, thus, provide the unknown values and estimates of the errors, whether the analytic form of $f(x)$ is available or not.

Interpolation was once a necessity as only tabulated values were available (e.g. trigonometric or logarithmic table). In modern days interpolation is not needed when the analytic form of $f(x)$ is known because computers can rapidly calculate numeric values of very many functions with high precision. However, it remains important to understand and learn interpolation techniques because the analytic form of $f(x)$ can be too complicated or unavailable in certain instances and because interpolation ideas are the basis for other important numerical methods.

The techniques covered in this section are:

- Linear interpolation.
- Lagrangian interpolation, Neville–Aitken method.
- Divided differences.
- Cubic splines.

Table 3.1. Correct values for $f(x)$.

x	0.2	0.4	0.8
$f(x)$	0.402 113	2.389 096	−2.365 639

Other methods are available and more in-depth descriptions can be found in modern, specialised literature.

3.2 Linear interpolation

Consider table 3.1. Values between $x = 0.2$ and $x = 0.4$, or between $x = 0.4$ and $x = 0.8$ are not known and have to be guessed. The easiest way to obtain such a guess is to use *linear interpolation*. This technique consists in defining non-tabulated values with the points of the segments joining each consecutive pair of tabulated values (see figure 3.1). Suppose a linear interpolation has to be carried out between the two points $(x_1, f_1 \equiv f(x_1))$ and $(x_2, f_2 \equiv f(x_2))$. The straight line passing through these two points is described by the equation,

$$\frac{y - f_1}{f_2 - f_1} = \frac{x - x_1}{x_2 - x_1}$$

Given that $y = f(x)$, the above equation means that the linear interpolation $f_{\text{int}}(x)$ of $f(x)$ is given by the following expression,

$$f_{\text{int}}(x) = f_1 + \frac{x - x_1}{x_2 - x_1}(f_2 - f_1) \tag{3.1}$$

For example, the value of the function tabulated in table 3.1 at $x = 0.3$ is linearly interpolated by,

$$f_{\text{int}}(0.3) \approx 0.402113 + \frac{0.3 - 0.2}{0.4 - 0.2}(2.389096 - 0.402113) = 1.395604$$

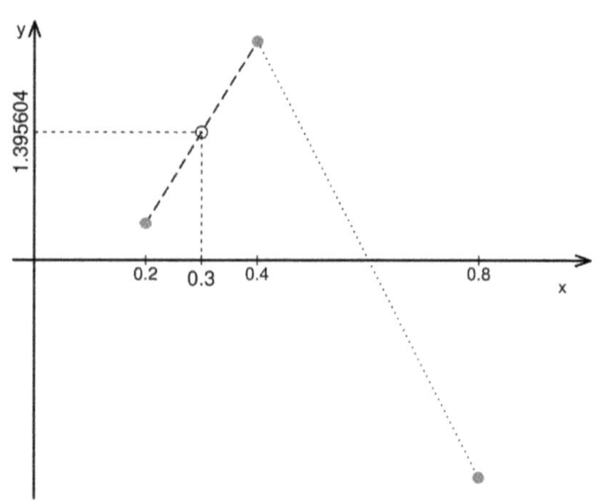

Figure 3.1. Linear interpolation. All values corresponding to x between 0.2 and 0.4 are defined by the points on the dashed segment, while all values corresponding to x between 0.4 and 0.8 are defined by the points on the dotted segment. For example, the linearly-interpolated value for $x = 0.3$ is 1.395 604, as highlighted in the picture (see main text).

The interpolated value is, in general, different from the correct value of the function, unless the function is itself a linear function. Therefore, the linear interpolation comes with an associated error, $\Delta f_{\text{int}}(x)$:

$$f(x) = f_{\text{int}}(x) + \Delta f_{\text{int}}(x) \tag{3.2}$$

When the analytic form of $f(x)$ is known, the analytic expression for the error,

$$\Delta f_{\text{int}}(x) \equiv f(x) - f_{\text{int}}(x), \tag{3.3}$$

is provided by the following formula,

$$\Delta f_{\text{int}}(x) = \frac{f''(\xi)}{2}(x - x_1)(x - x_2), \tag{3.4}$$

where ξ is a value between x_1 and x_2, depending on the specific value of x at which the error needs to be calculated. Formula (3.4) is proven in appendix A.1. From a practical point of view this expression is mostly used to estimate the largest possible error, as the precise value of ξ is not available.

Example 3.1 *Suppose* $f(x) = \sin(x)$ *and that tabulated values for* $x_1 = 0.2$ *and* $x_2 = 0.5$ *are, respectively,* $f_1 = 0.198\,669$ *and* $f_2 = 0.479\,426$. *Let us calculate the linearly interpolated value of sine for* $x = 0.3$ *and evaluate the largest interpolation error committed. Using the interpolation formula (3.1) we have,*

$$f_{\text{int}}(0.3) = f_1 + \frac{0.3 - x_1}{x_2 - x_1}(f_2 - f_1) = 0.198\,669 + \frac{0.3 - 0.2}{0.5 - 0.2}(0.479\,426 - 0.198\,669) = 0.292\,255$$

Therefore, we can use $0.292\,255$ *instead of* $\sin(0.3)$. *The largest possible error committed with such a replacement is calculated using formula (3.4),*

$$\Delta f_{\text{int}}(0.3) = \frac{f''(\xi)}{2}(0.3 - 0.2)(0.3 - 0.5) = \frac{-\sin(\xi)}{2}(0.1)(-0.2)$$

The sine goes from $0.198\,670$ *at* $\xi = 0.2$ *to* $0.479\,426$ at $\xi = 0.5$. *Thus, the largest value for the error is,*

$$\Delta_{\text{max}} = -\frac{0.479\,426}{2}(0.1)(-0.2) = 0.004\,794$$

The exact error is $\sin(0.3) - 0.292\,255 = 0.003\,266$ *smaller, as expected, than the largest estimate,* $0.004\,794$.

Most of the time the analytic form of $f(x)$ is not known and, thus, the expression (3.4) cannot be used to estimate the interpolation error. We will see shortly, though, that it is still possible to calculate a fairly accurate estimate of the error even when the analytic form of $f(x)$ is not known.

3.3 Relevant R code. Function `linpol`

`linpol` takes in the grid values (vector `x`), the tabulated values (vector `f`) and a set of points (vector `x0`) at which the linear interpolation is required. The function returns a vector of interpolated values of the same length as the vector `x0`. The following code snippet finds all linearly interpolated values between a grid of tabulated function points for $f(x) = 2x^2 + (1/2)x^3$. The grid of interpolated points is in this case finer than the grid of tabulated points.

```
# Load functions of package "comphy"
# (only done once for each session)
library(comphy)

# Grid of 11 points
x <- seq(0,1,length.out=11)

# Tabulated function values for f(x)=2*x^2+(1/2)*x^3
f <- 2*x^2+0.5*x^3

# Grid of interpolated points
x0 <- seq(0,1,length=101)

# Linear interpolation
f0 <- linpol(x,f,x0)

# Plot
plot(x,f,pch=16,col=2,xlab=expression(x),ylab=expression(f(x)))
points(x0,f0,type="l")
```

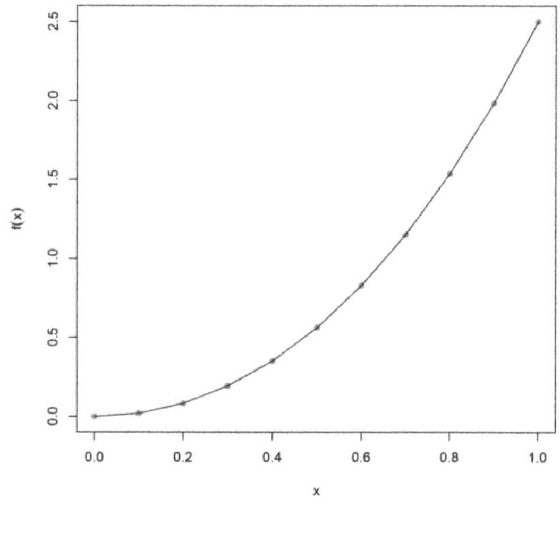

When the interpolated point coincides with a tabulated one, the function returns exactly the tabulated value:

```
# Interpolation for tabulated grid points
# coincides with tabulated values
c(x[1],f[1],x0[1],f0[1])

## [1] 0 0 0 0

c(x[2],f[2],x0[11],f0[11])

## [1] 0.1000 0.0205 0.1000 0.0205
```

3.4 Exercises on linear interpolation

Exercise 01

A function $f(x)$ is known only at the values here tabulated:

x	-2	-1	0	1	2
$f(x)$	3	0	-1	0	3

Use the `linpol` function to calculate the linear interpolation corresponding to the grid $\{x_0 = -2 + 0.1i, \ \ i = 0, 1, \ldots, 60\}$. Plot all values and highlight the known values in red.

Exercise 02

Assume that the function used in exercise 01 is $f(x) = x^2 - 1$. Plot in $x \in [-2, 2]$ the error,

$$\Delta f(x) \equiv f(x) - f_{\text{int}}(x),$$

where f_{int} is the linear interpolation of $f(x)$. Verify that $\Delta f(x)$ satisfies equation (3.4).

Exercise 03

Sample the function

$$f(x) = 2\sin(x) - \cos(2x)$$

at 20 random points in the interval $(0, 2\pi)$. Then find all linear interpolations in the 21 interpolation intervals created in $[0, 2\pi]$. Finally, plot $f(x)$ and the linear interpolation in the same plot.

Exercise 04

Consider the function and linear interpolation found in exercise 03. Write a function called 'which_max' that takes in the correct function values and the related linear interpolations, and returns the value at which the interpolation error

$$\Delta f(x) = f(x) - f_{\text{int}}(x)$$

is the largest. Can you justify the value found, using formula (3.4)?

Exercise 05

A linear interpolation of $f(x) = x^2$ is carried out between $x = 0$ and $x = 1$ using a grid of equally-spaced values, $x_i = (i - 1)d,\ i = 1, \ldots, n + 1,\ d > 0$. What value should be assigned to d in order to keep the interpolation error $|\Delta f(x)| \equiv |f(x) - f_{\text{int}}(x)|$ smaller than an assigned positive number ϵ?

3.5 Lagrangian interpolation

In table 3.1 there are three available values of $f(x)$ that is $f(x_1) \equiv f_1$, $f(x_2) \equiv f_2$, and $f(x_3) \equiv f_3$. Three values make it possible to use a quadratic function for interpolation. The corresponding quadratic curve goes exactly through the three available points and provides between them smooth interpolating arcs. The straightforward procedure to calculate the three parameters describing the quadratic curve, consists in replacing in turn (x_1, f_1), (x_2, f_2), (x_3, f_3) in the equation $f_{\text{int}}(x) = ax^2 + bx + c$ and obtaining a system of three equations in the three unknowns a, b, c. The solution of such a system of linear equations requires additional computing efforts. This is not a problem when the polynomial used to fit the tabulated values is a second degree one, like in the present example, but can add non-trivial computational issues, and run the risk of turning into a severe ill-conditioned problem, when the polynomial degree is higher. For this reason, it is better to resort to the so-called *Lagrange polynomials* which do not require explicit determination of any parameter and provide a systematic expression for any polynomial of degree n, fitting exactly $n + 1$ points. The key ingredient of the Lagrange polynomials, and one that allows passing through the $n + 1$ available points, is the presence of fractions including factors like $(x - x_i)$ or $(x_j - x_i)$. For example, the Lagrange polynomial for the quadratic function has the following form,

$$P_2(x) \equiv \frac{(x - x_2)(x - x_3)}{(x_1 - x_2)(x_1 - x_3)} f_1 + \frac{(x - x_1)(x - x_3)}{(x_2 - x_1)(x_2 - x_3)} f_2 + \frac{(x - x_1)(x - x_2)}{(x_3 - x_1)(x_3 - x_2)} f_3$$

It is easy to verify that $P_2(x_1) = f_1$, $P_2(x_2) = f_2$ and $P_2(x_3) = f_3$ so that the above Lagrange polynomial fits exactly the three points provided.

In what follows, we will deal with the interpolation of $n+1$ points whose coordinates are

$$(x_1, f_1), \quad (x_2, f_2), \quad \dots \quad (x_{n+1}, f_{n+1}),$$

with

$$x_m \equiv x_1 < x_2 < \cdots < x_n < x_{n+1} \equiv x_M.$$

When interpolation is required for $n+1$ points, the general expression is,

$$P_n(x) = \frac{(x-x_2)(x-x_3)\cdots(x-x_{n+1})}{(x_1-x_2)(x_1-x_3)\cdots(x_1-x_{n+1})}f_1 + \frac{(x-x_1)(x-x_3)\cdots(x-x_{n+1})}{(x_2-x_1)(x_2-x_3)\cdots(x_2-x_{n+1})}f_2 + \cdots$$
$$\cdots + \frac{(x-x_1)(x-x_2)\cdots(x-x_n)}{(x_{n+1}-x_1)(x_{n+1}-x_2)\cdots(x_{n+1}-x_n)}f_{n+1} \tag{3.5}$$

The linear interpolation described in section 3.2 can too be expressed as a Lagrange polynomial. More specifically,

$$f_{\text{int}} \equiv P_1(x) = \frac{x-x_2}{x_1-x_2}f_1 + \frac{x-x_1}{x_2-x_1}f_2 = \frac{(x-x_2)f_1 + (x_1-x)f_2}{x_1-x_2}$$

It is straightforward to show that the above expression is equivalent to expression (3.1).

Example 3.2 *Consider the same function and values used in example 3.1, with the added pair* $x_3 = 0.7, f_3 \equiv f(x_3) = 0.644\ 218$. *Let us try and write the expression for the second-degree Lagrange polynomial passing through the three points given, and calculate* Lagrange interpolation *for* $x = 0.3$:

$$\begin{aligned} P_2(x) &= \frac{(x-0.5)(x-0.7)}{(0.2-0.5)(0.2-0.7)}0.198\ 669 + \frac{(x-0.2)(x-0.7)}{(0.5-0.2)(0.5-0.7)}0.479\ 426 \\ &+ \frac{(x-0.2)(x-0.5)}{(0.7-0.2)(0.7-0.5)}0.644\ 218 \end{aligned}$$

or, simplifying,

$$\begin{aligned} P_2(x) &= 0.198\ 669\frac{(x-0.5)(x-0.7)}{0.15} - 0.479\ 426\frac{(x-0.2)(x-0.7)}{0.06} \\ &+ 0.644\ 218\frac{(x-0.2)(x-0.5)}{0.10} \end{aligned}$$

Lagrange interpolation for $x = 0.3$ *is obtained by replacing* $x = 0.3$ *in the above formula. The result is* $P_2(0.3) = 0.296\ 731$. *This value has to be compared with the one obtained with linear interpolation,* $0.292\,255$. *The error committed with Lagrange interpolation is, in this case,* $\sin(0.3) - P_2(0.3) = -0.001\ 211$, *smaller in magnitude than the error* $0.003\,266$ *of linear interpolation.*

A Lagrange polynomial of degree n can also be represented as a linear combination of $n + 1$ so-called *basic Lagrange polynomials* of order k and degree n,

$$L_{n,k}(x) \equiv \frac{(x - x_1)(x - x_2) \cdots (x - x_{k-1})(x - x_{k+1}) \cdots (x - x_{n+1})}{(x_k - x_1)(x_k - x_2) \cdots (x_k - x_{k-1})(x_k - x_{k+1}) \cdots (x_k - x_{n+1})}, \tag{3.6}$$

according to the expansion,

$$P_n(x) = \sum_{i=1}^{n+1} f_i L_{n,i}(x) \qquad \text{where} \qquad f_i = P_n(x_i), \; i = 1, \ldots, n + 1 \tag{3.7}$$

3.5.1 Errors for the Lagrangian interpolation

An analytic expression for the error is possible also when the interpolation is done using Lagrange polynomials. In this case,

$$f(x) = P_n(x) + \Delta P_n(x), \tag{3.8}$$

where,

$$\Delta P_n(x) \equiv f(x) - P_n(x) \tag{3.9}$$

The analytic expression for $\Delta P_n(x)$ is,

$$\Delta P_n(x) = \frac{f^{(n+1)}(\xi)}{(n + 1)!}(x - x_1)(x - x_2) \cdots (x - x_{n+1}), \tag{3.10}$$

where $f^{(n+1)}(x)$ is the $(n + 1)$-th derivative of $f(x)$ and where ξ, which is in the real interval (x_m, x_M), depends on the specific point x at which the error is calculated ($\xi \neq x_1, x_2, \ldots, x_{n+1}$; if ξ is equal to one of the tabulated points, x_i, then $\Delta P_n(x) = 0$, as expected). The calculation of the error with formula (3.10) is possible only if the analytic form of $f(x)$ is known, much the same way as discussed for linear interpolation. A proof to derive expression (3.10) is included in appendix A.2. The magnitude of the error is larger the further away x is from the center of the interval (x_n, x_M). This is explained in appendix A.2. The only exception to this statement is when x coincides with one of the interpolation points because then $x - x_i = 0$ for a value of i, and the whole product becomes zero. Intuitively, this explains why extrapolation is in general not advisable as in such a circumstance x is even more away from the centre of the interpolation interval.

3.5.2 The Neville–Aitken algorithm

This method (and the associated algorithm) was initially developed by Aitken [1, 2] and subsequently modified by Neville in its current form [3]. The main advantage in using the Neville–Aitken (shortened to *N–A*) algorithm is that the lengthy formulas of Lagrange interpolation are replaced by the shorter formulas of linear interpolation.

Example 3.3 *Consider* $f(x) = x^3 - 2x^2 + 3x - 1$, *the correct values of which, at certain points, are shown in table 3.2.*

Table 3.2. Correct values for $f(x) = x^3 - 2x^2 + 3x - 1$.

x	0.1	0.4	0.6	0.8	0.9
$f(x)$	−0.719 000	−0.056 000	0.296 000	0.632 000	0.809 000

The Lagrange interpolation at $x = 0.75$ using all five points is given by the following formula,

$$\begin{aligned}
P_5(0.75) &= \frac{(0.75-0.4)(0.75-0.6)(0.75-0.8)(0.75-0.9)}{(0.1-0.4)(0.1-0.6)(0.1-0.8)(0.1-0.9)}(-0.719\,000)\\
&+ \frac{(0.75-0.1)(0.75-0.6)(0.75-0.8)(0.75-0.9)}{(0.4-0.1)(0.4-0.6)(0.4-0.8)(0.4-0.9)}(-0.056\,000)\\
&+ \frac{(0.75-0.1)(0.75-0.4)(0.75-0.8)(0.75-0.9)}{(0.6-0.1)(0.6-0.4)(0.6-0.8)(0.6-0.9)}(0.296\,000)\\
&+ \frac{(0.75-0.1)(0.75-0.4)(0.75-0.6)(0.75-0.9)}{(0.8-0.1)(0.8-0.4)(0.8-0.6)(0.8-0.9)}(0.632\,000)\\
&+ \frac{(0.75-0.1)(0.75-0.4)(0.75-0.6)(0.75-0.8)}{(0.9-0.1)(0.9-0.4)(0.9-0.6)(0.9-0.8)}(0.809\,000)\\
&= 0.546\,875,
\end{aligned}$$

which clearly shows how lengthy an expression for the Lagrange interpolation can be.

The idea behind the N–A algorithm is to arrive at the $(n + 1)$-points interpolation formula using a series of n successive linear interpolations. While from the point of view of the number of operations involved this might be equivalent to the original Lagrange interpolation, from the coding point of view N–A is much simpler, relying essentially on a single coding line. The algorithm consists of a series of linear interpolations using all $n + 1$ points available, two at a time. Technically speaking, each interpolation can also be an extrapolation, as the interpolating point can lie outside the interpolation interval; the term linear interpolation is still going to be used in the present context. The starting point, also known as *first level* of the N–A algorithm, is represented by the $2(n + 1)$ values, x_i and f_i, where $i = 1, \ldots, n + 1$. These can also be indicated as P_{1i}, where the '1' means 'first level':

$$P_{1i} \equiv f_i \qquad i = 1, \ldots, n + 1. \tag{3.11}$$

Next, n linear interpolations for x are calculated using x_1, x_2, followed by x_2, x_3, etc. The interpolated quantities obtained will be indicated as P_{2i}, $i = 1, \ldots, n$, where the '2' means *second level* of interpolation. The notation is meant to imply that the second level of the N–A algorithm is composed by the n linear interpolations obtained using points x_i and x_{i+1}. In the third level of the N–A algorithm, $n - 1$ linear interpolations are carried out, the first using the two points x_1, x_3 and the second-level values P_{21}, P_{22}, the second using the two points x_2, x_4 and the second-level values P_{22}, P_{23}, and so on. The interpolated values at the third level are indicated as P_{3i}, where $i = 1, \ldots, n - 1$. Each linear interpolation can be calculated using the first-order Lagrange-polynomial form which can be simplified as,

$$P_{ji} \equiv \frac{(x - x_{i+j-1})P_{j-1,i} + (x_i - x)P_{j-1,i+1}}{x_i - x_{i+j-1}} \quad j = 2, \ldots, n+1 \quad i = 1, \ldots, n - j + 2 \quad (3.12)$$

For example, the second-level interpolated value P_{24} is given by the following formula,

$$P_{24} = \frac{(x - x_5)P_{14} + (x_4 - x)P_{15}}{x_4 - x_5},$$

where the two first-level quantities are,

$$P_{14} = f_4, \quad P_{15} = f_5.$$

An interesting consequence of the N–A algorithm is that the last level of interpolation yields the same numerical value as the one obtained using Lagrange interpolation. In fact, the first element (element with $i = 1$) of all levels of the interpolation coincides with the Lagrangian interpolation performed using the first interpolation point x_1, the first two interpolation points x_1, x_2, the first three interpolation points x_1, x_2, x_3, etc.

Example 3.4 *Let us carry out the interpolation required in example 3.3 using five levels of the N–A algorithm and verify that P_{51} coincides with the value* 0.546 875 *obtained with the fifth-order Lagrange interpolation. The form of the various calculations involved is suggested by the following sideways pyramidal scheme:*

$$\begin{array}{ccccc}
P_{11} & & & & \\
 & P_{21} & & & \\
P_{12} & & P_{31} & & \\
 & P_{22} & & P_{41} & \\
P_{13} & & P_{32} & & P_{51} \\
 & P_{23} & & P_{42} & \\
P_{14} & & P_{33} & & \\
 & P_{24} & & & \\
P_{15} & & & &
\end{array}$$

The P_{1i}'s coincide with the original tabulated f_i's, while the other P_{ji}'s are given by formula (3.12). Using the values reported in table 3.2 and carrying out all calculations yield,

```
-0.719000
            0.717500
-0.056000               0.512750
            0.560000                0.546875
 0.296000               0.549500                0.546875
            0.548000                0.546875
 0.632000               0.545750
            0.543500
 0.809000
```

The P_{51} value found with the N–A scheme is exactly what was calculated with the full Lagrange polynomial.

From the example just discussed one can notice that the correct value of the function $f(x) = x^3 - 2x^2 + 3x - 1$ at $x = 0.75$ is 0.546 875, i.e. the result reached already at the fourth level of the N–A algorithm. This is what one would expect in this specific example because the function to interpolate is a third-degree polynomial, exactly matched by a third-order Lagrange polynomial. The fourth level of the N–A algorithm is equivalent to the interpolation with the third-order Lagrange polynomial. The same goes for the fifth-order polynomial. We expect this to be the case also for higher-order polynomials because a third-order polynomial is completely and uniquely determined by knowledge of four of its points. In general, an nth-degree polynomial is completely and uniquely determined by knowledge of $n + 1$ of its points (see appendix A.3). Then knowledge of a number of points higher than $n + 1$ does not improve the accuracy of the interpolating polynomial. This will be noticeable in the hierarchy of values implied by the N–A algorithm, as values will be identical for all columns of level higher than k, where k is the degree of the polynomial, with $k < n$. Things are different when the function to interpolate is not a polynomial and in such an instance the accuracy will increase locally (around the interpolation value) with the number $n + 1$ of known points.

Example 3.5 *Consider the function* $f(x) = \sin^2(x)$ *for* $x \in [0, 9]$ *(in radians). We would like to trace the two curves obtained through interpolation using, in turn, the following two sets of known values:*

First set,

x	0	1.5	3	4.5	6	7.5	9
$f(x)$	0	0.994 996	0.019 915	0.955 565	0.078 073	0.879 844	0.169 842

Second set,

x	0	0.5	1	1.5	2	2.5	3
$f(x)$	0	0.229 849	0.708 073	0.994 996	0.826 822	0.358 169	0.019 915

x	3.5	4	4.5	5	5.5	6	6.5
$f(x)$	0.123 049	0.572 750	0.955 565 13	0.919 536	0.497 787	0.078 073	0.046 277

x	7	7.5	8	8.5	9
$f(x)$	0.431 631	0.879 844	0.978 830	0.637 582	0.169 842

The first set consists of seven correct values of $f(x)$, while the second set includes all points of the first set and 12 additional points, for a total of 19 known values of $f(x)$. It should be clear from the two different interpolating curves that the one traced using more points is closer to the correct curve. The way to proceed is to select a relatively-high number of interpolating points between 0 *and* 9 *and apply the N–A algorithm for each interpolating point. The collection of interpolations found will form one of the two curves to be traced. Here we have selected as interpolating points those belonging to the regular grid of* 51 *points between* 0 *and* 9. *The two curves resulting from the two sets are depicted in figure 3.2, superimposed to the correct analytic curve. The picture clearly demonstrates that a better match to the correct curve can be obtained using a sufficiently-high number of correct function values. It is also clear from the picture that serious errors are generated when the values of the function used for the interpolation are not enough.*

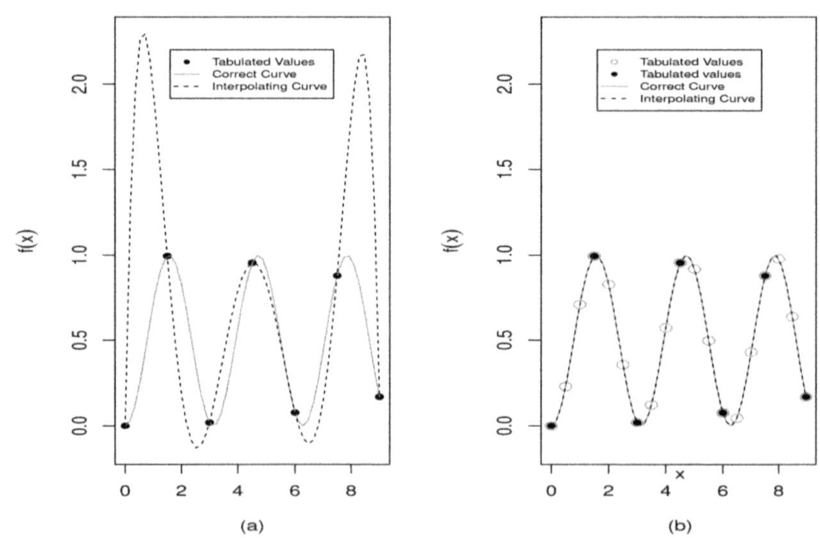

Figure 3.2. Effect of increased sample size in Lagrange interpolation. The function $f(x) = \sin^2(x)$ is plotted as a continuous, red line. The full black circles in (a) and empty red circles in (b) represent the known tabulated values of the function. The curves interpolating $\sin^2(x)$ using seven points, part (a), and 19 points, part (b), are plotted as dashed, black lines. The picture clearly shows that the accuracy of the interpolation improves with the number of known values.

3.6 Relevant R code. Function `nevaitpol`

The function used in `comphy` to implement the N–A algorithm is `nevaitpol`. It takes as input the known values (set of x's and $f(x)$'s) and the interpolating point. It returns a matrix which is, in fact, a triangular matrix as all values in the lower triangular region are set to zero. Column `j` of this matrix has all values related to the level j of the N–A algorithm. The first row has all the possible interpolations achievable with the Lagrangian interpolation.

The following code chunck reproduces figure 3.2. The function to interpolate using N–A is $f(x) = \sin^2(x)$ in the interval [0, 9] (radians). This exercise is meant to stress that interpolation can be a risky operation if the number of known (tabulated) values of the underlying function is too low.

```
# Load library "comphy" in working space
library(comphy)

# Fine grid (51 points) for plotting
# correct and interpolated curves
x0 <- seq(0,9,length.out=51)

# Correct curve (evaluated at x0)
ftrue <- (sin(x0))^2

# First set of tabulated points (too few!)
x1 <- c(0,1.5,3,4.5,6,7.5,9)
f1 <- (sin(x1))^2

# Interpolation using the first set
npts <- length(x1)
f01 <- numeric()
for (i in 1:length(x0)) {
P <- nevaitpol(x1,f1,x0[i])
f01 <- c(f01,P[1,npts]) # The last value in the sideways
}                          # pyramid is the interpolated value

# Second set of tabulated values (enough points!)
x2 <- c(0,0.5,1,1.5,2,2.5,3,3.5,4,4.5,5,5.5,6,6.5,7,7.5,8,8.5,9)
f2 <- (sin(x2))^2

# Interpolation using the second set
npts <- length(x2)
f02 <- numeric()
for (i in 1:length(x0)) {
```

```
P <- nevaitpol(x2,f2,x0[i])
f02 <- c(f02,P[1,npts])
}

# Plot of Figure 3.2 (commented out as the Figure is
# already present in main text)
#m <- min(f1,f2,f01,f02,ftrue)
#M <- max(f1,f2,f01,f02,ftrue)
#par(mfrow=c(1,2))
#plot(x1,f1,pch=16,ylim=c(m,M),xlab="(a)",ylab=expression(f(x)))
#points(x0,ftrue,type="l",col=2)
#points(x0,f01,type="l",lty=2)
#legend(1.6,2.2,legend=c("Tabulated Values","Correct Curve",
#         "Approximated Curve"),col=c(1,2,1),lty=c(0,1,2),
#         pch=c(16,-1,-1),cex=0.7)
#plot(x2,f2,cex=1.3,ylim=c(m,M),xlab="(b)",
#      ylab=expression(f(x)),col=2)
#points(x0,ftrue,type="l",col=2)
#points(x0,f02,type="l",lty=2)
#mtext(side=1,text=expression(x))
#legend(1.6,2.2,legend=c("Tabulated Values","Tabulated values",
#         "Correct Curve","Approximated Curve"),
#         col=c(2,1,2,1),lty=c(0,0,1,2),pch=c(1,16,-1,-1),cex=0.7)
#points(x1,f1,pch=16)
```

3.7 Exercises on Lagrangian interpolation and the Neville–Aitken algorithm

Exercise 06

Find the four basic Lagrangian polynomials $L_{3,i}$, $i = 1, 2, 3, 4$ for the function $f(x) = x^3 - 5x^2 + 3x + 2$ where the four interpolating points are,

$$x_1 = 1,\ x_2 = -1,\ x_3 = 0,\ x_4 = 2$$

Then write the expression for $f(x)$ as a linear expansion in term of the basic Lagrange polynomials. Finally, plot $f(x)$ between $x = -2$ and $x = 3$, with the interpolating points in red.

Exercise 07

Write the analytic form of the Lagrange polynomial of degree 4, $P_4(x)$, that interpolates the five points

$$x_1 = 0,\ x_2 = \pi/6,\ x_3 = \pi/4,\ x_4 = \pi/3,\ x_5 = \pi/2$$

of the function $f(x) = \sin(x)$ in the interval $x \in [0, \pi/2]$. Plot the curves corresponding to both $f(x)$ and $P_4(x)$, and highlight in red the five points of the interpolation.

Exercise 08

Consider the difference between the function of exercise 07, $f(x)$, and its approximation using Lagrange polynomials, $P_4(x)$,

$$\Delta P_4(x) = f(x) - P_4(x)$$

Estimate the largest value of $|\Delta P_4(x)|$, ΔP, in the interval $x \in [0, \pi/2]$.

Exercise 09

Consider the following three known points for a given function $f(x)$:

x	0	0.5	1
$f(x)$	0	0.125	1

Find all interpolated values between 0 and 1 using Lagrangian interpolation ($P_2(x)$) and interpolation with the N–A algorithm, on a grid `x0` of 20 points. Plot both interpolations and the known points to show that they coincide.

Exercise 10

Consider the following six points,

$$x_1 = 0, x_2 = 0.2, x_3 = 0.3, x_4 = 0.6, x_5 = 0.7, x_6 = 1,$$

part of the third degree polynomial,

$$f(x) = x^3 - 5x^2 + 2x + 1.$$

Verify that the values obtained with the N–A algorithm start to be identical at its fourth level. Explain why this is the case.

Exercise 11

Write the code to implement the Lagrange polynomial $P_n(x)$ to interpolate $n + 1$ given values of a function $f(x)$. Apply the code created to find:

- $P_2(1.5)$ when $x_1 = 1,\ x_2 = 2$ and $f_1 \equiv \log(x_1),\ f_2 \equiv \log(x_2)$
- $P_3(1.5)$ when $x_1 = 1,\ x_2 = 2,\ x_3 = 3$ and $f_1 \equiv \log(x_1),\ f_2 \equiv \log(x_2)$, $f_3 \equiv \log(x_3)$
- $P_4(1.5)$ when $x_1 = 1,\ x_2 = 2,\ x_3 = 3,\ x_4 = 4$ and $f_1 \equiv \log(x_1),\ f_2 \equiv \log(x_2)$, $f_3 \equiv \log(x_3),\ f_4 \equiv \log(x_4)$

Verify that the three values found coincide with the values in row 1 of the matrix P found using the N–A algorithm on the last set of interpolation points given.

Exercise 12

Consider the set of known points and values of the function $\log(x)$:

$$x_i = i, \; i = 1{:}\,100, \qquad f_i = log(x_i), \; i = 1{:}\,100$$

Write a program which selects randomly four known points in the set $\{x_i, \; i = 2, 99\}$, includes x_1 and x_{100} as first and last known point and calculates interpolations $P_5(x_i)$ at the locations of the remaining 94 points, using the N–A algorithm. The program should calculate the two following quantities for each random set of interpolations:

$$\Delta P \equiv \langle |P_{1,5} - P_{2,5}| \rangle, \qquad \text{Err} \equiv \left\langle \left| f_i - P_5(x_i) \right| \right\rangle,$$

where $\langle\rangle$ indicates the average corresponding to the 94 remaining points. Plot ΔP versus Err for 1000 simulations, i.e. 1000 random selections of the four known points and 94 remaining points. Repeat the exercise using 18 random points and the first and last sample point as before. What changes do you observe?

3.8 Divided differences

In section 3.5 the analytic form chosen to express Lagrange polynomials was advantageous because it reverts instantly to f_i when $x = x_i$. Thus, a Lagrange polynomial of degree n, $P_n(x)$, is readily written once $n + 1$ known points are made available. While the same polynomial can be written in many different analytic forms, it represents always the same function. For instance, a generic polynomial of degree n, $Q_n(x)$, could be written to highlight the coefficients, c_i, of the powers of x as,

$$Q_n(x) = c_1 + c_2 x + c_3 x^2 + \cdots + c_{n+1} x^n$$

The same polynomial can be written in the following interesting form,

$$Q_n(x) = a_1 + a_2(x - x_1) + a_3(x - x_1)(x - x_2) + \cdots + a_{n+1}(x - x_1)(x - x_2) \cdots (x - x_n), \quad (3.13)$$

which is useful when the values of $Q_n(x)$ are known at the points $x_1, x_2, \ldots, x_{n+1}$. If we call with $f_1, f_2, \ldots, f_{n+1}$ these values, it is possible (although a tedious exercise) to find the exact expression for the coefficients $a_1, a_2, \ldots, a_{n+1}$. Such expression is provided, in the context of interpolation, via the following useful quantities known as *divided differences*. By definition, the first divided difference, $f[x_i, x_j]$, between x_i and x_j is,

$$f[x_i, x_j] \equiv \frac{f_j - f_i}{x_j - x_i} \quad (3.14)$$

The second divided difference $f[x_i, x_j, x_k]$ among x_i, x_j and x_k is defined in terms of first divided differences as,

$$f[x_i, x_j, x_k] \equiv \frac{f[x_j, x_k] - f[x_i, x_j]}{x_k - x_i} \quad (3.15)$$

Divided differences of higher order can be constructed using divided differences of lower order in a similar fashion. In general, the nth order divided difference among $x_i, x_j, x_k, \ldots, x_r, x_s$ is defined as,

$$f[x_i, x_j, x_k, \ldots, x_r, x_s] \equiv \frac{f[x_j, x_k, \ldots, x_s] - f[x_i, x_j, \ldots, x_r]}{x_s - x_i} \tag{3.16}$$

The zeroth-order divided difference $f[x_i]$ can also be defined for completeness. The appropriate form turns out to be

$$f[x_i] \equiv f_i \tag{3.17}$$

As seen in definitions ((3.14), (3.15), (3.16), (3.17)) the symbol $f[\cdots]$ is used as standard notation for a divided difference ([4, 5]). It turns out that the coefficients $a_1, a_2, \ldots, a_{n+1}$ in the polynomial (3.13) coincide exactly with the divided differences:

$$\begin{aligned} Q_n(x) = & f[x_1] \\ & + f[x_1, x_2](x - x_1) \\ & + f[x_1, x_2, x_3](x - x_1)(x - x_2) \\ & + \cdots \\ & + f[x_1, x_2, \ldots, x_{n+1}](x - x_1)(x - x_2) \cdots (x - x_n) \end{aligned} \tag{3.18}$$

As $Q_n(x)$ is an interpolating function, defined to go through the $n + 1$ points (x_1, f_1), $(x_2, f_2), \cdots, (x_{n+1}, f_{n+1})$, it follows that $Q_n(x_1) = f_1, Q_n(x_2) = f_2, \cdots, Q_n(x_{n+1}) = f_{n+1}$. An interesting property of the analytic form (3.18) is that the expression constructed using $n + 1$ points can be considered simply as the expression $Q_{n-1}(x)$ constructed using n of the $n + 1$ points, with the addition of the last one. In other words, the divided differences allow to save computational efforts when it is decided to add more points to the initial set. An example will clarify this important feature of divided differences.

Example 3.6 *A set of four points for a given function $f(x)$ is presented in the following table,*

x	−1	1	2	4
$f(x)$	−6	2	0	14

The four divided differences are highlighted in boldface in the following calculations:

$$f[x_1] = f(-1) \equiv f_1 = \mathbf{-6} \quad \leftarrow$$

$$f[x_1, x_2] = \frac{f_2 - f_1}{x_2 - x_1} = \frac{8}{2} = \mathbf{4} \quad \leftarrow$$

$$f[x_2, x_3] = \frac{f_3 - f_2}{x_3 - x_2} = \frac{-2}{1} = -2$$

$$f[x_3, x_4] = \frac{f_4 - f_3}{x_4 - x_3} = \frac{14}{2} = 7$$

$$f[x_1, x_2, x_3] = \frac{f[x_2, x_3] - f[x_1, x_2]}{x_3 - x_1} = \frac{-6}{3} = \mathbf{-2} \quad \leftarrow$$

$$f[x_2, x_3, x_4] = \frac{f[x_3, x_4] - f[x_2, x_3]}{x_4 - x_2} = \frac{9}{3} = 3$$

$$f[x_1, x_2, x_3, x_4] = \frac{f[x_2, x_3, x_4] - f[x_1, x_2, x_3]}{x_4 - x_1} = \frac{5}{5} = \mathbf{1} \quad \leftarrow$$

Following equation (3.18), the analytic expression for the polynomial is,

$Q_3(x) = -6 + 4(x + 1) - 2(x + 1)(x - 1) + (x + 1)(x - 1)(x - 2) = x^3 - 4x^2 + 3x + 2$

If the following additional points

x	0	3
$f(x)$	2	2

are added to the previous set, there is no need to start the computation of all divided differences again. Only those differences involving the new data will have to be calculated. The new interpolating polynomial is potentially a five degree polynomial, $Q_5(x)$ such that,

$$\begin{aligned} Q_5(x) &= Q_3(x) + f[x_1, x_2, x_3, x_4, x_5](x - x_1)(x - x_2)(x - x_3)(x - x_4) \\ &+ f[x_1, x_2, x_3, x_4, x_5, x_6](x - x_1)(x - x_2)(x - x_3)(x - x_4)(x - x_5) \end{aligned}$$

The following are the necessary divided differences (where, again, the new coefficients of $Q_5(x)$ are highlighted in boldface):

$$f[x_4, x_5] = \frac{f_5 - f_4}{x_5 - x_4} = \frac{-12}{-4} = 3$$

$$f[x_5, x_6] = \frac{f_6 - f_5}{x_6 - x_5} = \frac{0}{3} = 0$$

$$f[x_3, x_4, x_5] = \frac{f[x_4, x_5] - f[x_3, x_4]}{x_5 - x_3} = \frac{-4}{-2} = 2$$

$$f[x_4, x_5, x_6] = \frac{f[x_5, x_6] - f[x_4, x_5]}{x_6 - x_4} = \frac{-3}{-1} = 3$$

$$f[x_2, x_3, x_4, x_5] = \frac{f[x_3, x_4, x_5] - f[x_2, x_3, x_4]}{x_5 - x_2} = \frac{-1}{-1} = 1$$

$$f[x_3, x_4, x_5, x_6] = \frac{f[x_4, x_5, x_6] - f[x_3, x_4, x_5]}{x_6 - x_3} = \frac{1}{1} = 1$$

$$f[x_1, x_2, x_3, x_4, x_5] = \frac{f[x_2, x_3, x_4, x_5] - f[x_1, x_2, x_3, x_4]}{x_5 - x_1} = \frac{0}{1} = \mathbf{0} \quad \leftarrow$$

$$f[x_2, x_3, x_4, x_5, x_6] = \frac{f[x_3, x_4, x_5, x_6] - f[x_2, x_3, x_4, x_5]}{x_6 - x_2} = \frac{0}{2} = 0$$

$$f[x_1, x_2, x_3, x_4, x_5, x_6] = \frac{f[x_2, x_3, x_4, x_5, x_6] - f[x_1, x_2, x_3, x_4, x_5]}{x_6 - x_1} = \frac{0}{4} = \mathbf{0} \quad \leftarrow$$

It turns out that the new coefficients that should yield x^4 and x^5 are both zero. This means that the interpolating polynomial obtained considering the two additional points coincides, in fact, with $Q_3(x)$. These points belong, thus, to $Q_3(x)$.

The example just illustrated shows that the construction of interpolating polynomials using divided differences is hierarchical. Both the divided differences taking direct part in the expression of the interpolating polynomial and those used only in the overall calculation can be arranged in a sideways pyramidal table similar to the one used in the N–A algorithm. For the divided differences the table has the following look (say, for a six-levels calculation):

$$
\begin{array}{llllll}
f[x_1] & & & & & \\
 & f[x_1, x_2] & & & & \\
f[x_2] & & f[x_1, x_2, x_3] & & & \\
 & f[x_2, x_3] & & f[x_1, x_2, x_3, x_4] & & \\
f[x_3] & & f[x_2, x_3, x_4] & & f[x_1, x_2, x_3, x_4, x_5] & \\
 & f[x_3, x_4] & & f[x_2, x_3, x_4, x_5] & & f[x_1, x_2, x_3, x_4, x_5, x_6] \\
f[x_4] & & f[x_3, x_4, x_5] & & f[x_2, x_3, x_4, x_5, x_6] & \\
 & f[x_4, x_5] & & f[x_3, x_4, x_5, x_6] & & \\
f[x_5] & & f[x_4, x_5, x_6] & & & \\
 & f[x_5, x_6] & & & & \\
f[x_6] & & & & &
\end{array}
$$

When the interpolating polynomial has a degree k lower than the number $n + 1$ of available points, then the columns with index greater than $k + 1$ will be filled with zeros, while in the $(k + 1)$-th column all divided differences will be the same. In example 3.6, for instance, the interpolating polynomial had degree $k = 3$; for this reason, all divided differences in the $3 + 1 = 4$-th column are the same (they have value 1), and the divided differences in the fifth and sixth columns are zero. With functions different from polynomials, things develop differently. In general, the columns of divided differences will never be exactly zero, but will become smaller and smaller in absolute value. Beyond a given column, say $k + 1$, the divided differences will be so small that not including them in the polynomial (which will then be a polynomial of degree k) does not cause the approximation to be appreciably different from the correct function. Several criteria can be used to decide when the divided differences are small enough to be neglected. Typically, they will make use of the average of their absolute values.

To conclude this section, it is important to remember that a divided difference of order n is invariant to the ordering of the x_i's:

$$
f[x_1, \ldots, x_{n+1}] = f[\pi(x_1, \ldots, x_{n+1})], \tag{3.19}
$$

where $\pi()$ indicates any permutation of the $n + 1 x_i$'s.

Example 3.7 *This example is treated in detail in section 3.10. Twenty values equally-spaced between* 0 *and* 10 *form a grid on which the function* $f(x) = \sin(x)$ *is tabulated. The absolute values of all divided differences corresponding to column* 8 *of the table is* $112\,389 \times 10^{-9}$. *For columns* 12 *and* 13, *instead, we have, respectively,* 16×10^{-9} and 1×10^{-9}. *One could use an* ad hoc *criterion in which columns corresponding to values smaller than* 1×10^{-9} *are excluded. This means, in the specific case under scrutiny, that a polynomial that makes use only of* 12 *points (i.e. a polynomial of degree* 11*) is reasonably close to the correct function. The same might not be true of polynomials of degree lower than* 11, *like the interpolating polynomial of degree* 7, *passing through* 8 *of the tabulated points. It is important to stress that only a polynomial of degree* 19 *passes exactly through the* 20 *tabulated points; polynomials of degree lower than* 19 *might pass close to the tabulated points, but not exactly through all of them. In fact, a polynomial of degree k will need a selection of* $k + 1$ *of the* 20 *available points to be*

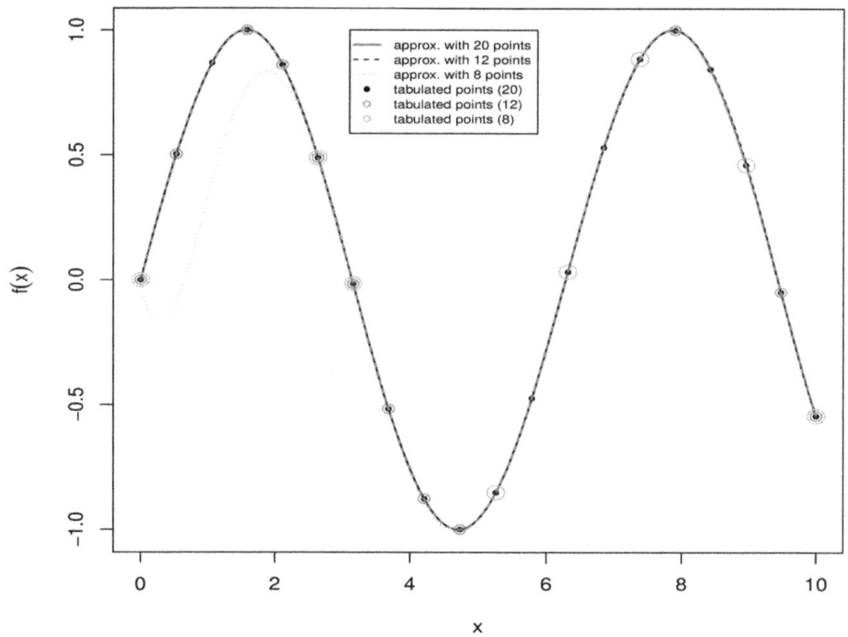

Figure 3.3. Interpolating polynomials passing through 20 points (red line), 12 points (black, dashed line) and 8 points (green, dotted line). The full set of 20 tabulated points is represented by black full circles, while the selected sets for the 12-points and 8-points cases are indicated by red and green empty circles, respectively. Clearly, 8 points do not lead to a good approximation, while 12 points yield a very good interpolating curve, passing very close to all the 20 tabulated points.

uniquely determined. For such interpolations making use of a smaller number of points it is in general convenient to include the two points at the extreme of the interpolation interval, so to avoid the issues connected with extrapolation. The interpolating polynomials passing through 8, 12 *and all* 20 *points are shown in figure 3.3.*

3.8.1 Error estimation using divided differences. The next-term rule

When interpolating using divided differences, the error coincides with the one given for Lagrange interpolation (equation (3.10)) because the polynomial is the same (uniqueness of polynomials passing through a same set of points). With divided differences, though, it is possible to estimate the error also when the analytic expression of $f(x)$ is not known. In appendix A.4 the following result is proved:

$$\frac{f^{(n)}(\xi)}{n!} = f[x_1, x_2, \ldots, x_{n+1}] \tag{3.20}$$

To estimate the derivative in (3.10) one needs to add an additional point, different from x, and calculate the new divided difference. The addition of the new point can then give us an estimate of the interpolation error. This practical idea is commonly known as *next-term rule*: **the error ΔP_n is approximately the value of the next term added to P_n**. An example can help to clarify this rule.

Example 3.8 *Consider the function* $f(x) = x^3$ *in the interval* $[0, 1]$ *and the two tabulated points* $x_1 = 0$ *and* $x_2 = 1$. *With knowledge of only two points, one uses linear interpolation. The polynomial is,*

$$P_1(x) = f_1 + f[x_1, x_2](x - x_1) = 0 + \frac{f_2 - f_1}{x - x_1} = \frac{1 - 0}{1 - 0}x = x$$

The exact error is, therefore,

$$\Delta P_1(x) = f(x) - P_1(x) = x^3 - x = x(x^2 - 1)$$

The exact expression of the error $\Delta P_1(x)$, *according to the formula for Lagrangian interpolation, is*

$$\Delta P_1(x) = \frac{f''(\xi)}{2}x(x - 1),$$

where $\xi \in (0, 1)$. *As the function* f *is known, we can calculate the minimum and maximum of* $f''(\xi)$ in $(0, 1)$:

$$f''(\xi) = 6\xi \Rightarrow f''(0) = 0, f''(1) = 6.$$

Therefore, the lower and upper bounds for the error estimation are (between 0 *and* 1, $3x(x - 1)$ *is always less than or equal to* 0*):*

$$\frac{f''(\xi)}{2}x(x - 1) \Rightarrow 3x(x - 1), 0.$$

The error, and lower and upper bounds for all values of x *between* 0 *and* 1 *are shown in figure 3.4. In the figure it is evident that the error,* $x(x^2 - 1)$, *is always within its lower and upper bound. If the function were not known analytically, we could estimate the error with the next-term rule only if the function is known at at least another point. If, for instance, the new point* $x_3 = 1.1$ *is added to the previous points, the divided difference is*

$$f[0, 1, 1.1] = 2.1,$$

and the estimated error between 0 *and* 1 *is*

$$2.1x(x - 1).$$

If the new point is taken inside the interpolating interval, say $x_3 = 0.6$, *the divided difference is*

$$f[0, 1, 1.1] = 1.6,$$

and the estimated error between 0 *and* 1 *is*

$$1.6x(x - 1).$$

One can first of all check in figure 3.4 that both error estimates are reasonably close to the analytic error, but also that the estimate with added point inside the interpolating interval is more accurate than the one with added point outside it.

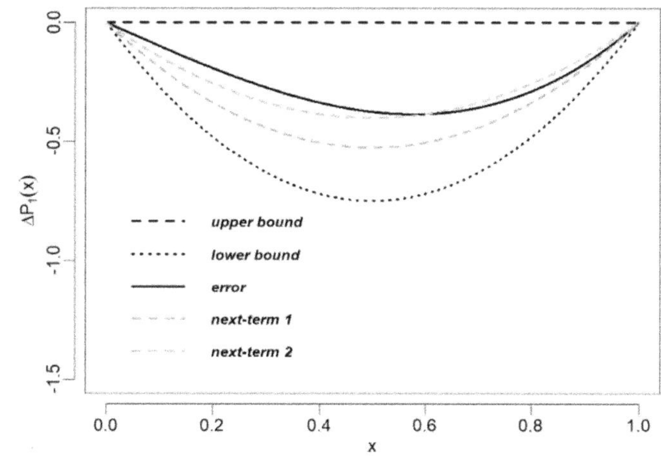

Figure 3.4. Estimation of the error when $f(x) = x^3$ is interpolated using divided differences. The polynomial of order 1, $P_1(x) = x$ yields an error $\Delta P_1(x) = x(x^2 - 1)$, represented by the full line. As the function to interpolate is known, upper (0) and lower ($3x(x - 1)$) bounds for the error can be calculated. In the plot it is evident that the error is always contained between its lower and upper bounds. The remaining two curves are estimates of the error found using the next-term rule (see text).

Example 3.9 *The function* $f(x) = x^2 e^{-x/2}$ *is tabulated at* $x_1 = 0, f_1 = 0$, $x_2 = 1, f_2 = 0.606\ 53$, $x_3 = 2, f_3 = 1.471\ 52$ and $x_4 = 3, f_4 = 2.008\ 17$. *How can we estimate the error with the next-term rule at* $x = 1.5$*?*

If we use all the four values, there is nothing left to calculate the next term. We can, therefore, select x_1, x_2 *and* x_3 *to calculate the interpolation, and use* x_4 *to estimate the error. It should be better, though, to use* x_1, x_2 *and* x_4 *for the interpolation and* x_3 *for the estimate, as* $x_3 = 2$ *is closer to* $x = 1.5$ *than* $x_4 = 3$*. The same should be verified if the omitted point were* x_1*.*

Let us then calculate divided differences for x_1, x_2 *and* x_4*:*

$$f[x_1] = 0, \quad f[x_1, x_2] = 0.606\ 53, \quad f[x_1, x_2, x_4] = 0.031\ 43.$$

The interpolation at $x = 1.5$ *is thus,*

$$P_2(1.5) = f[0] + f[0,\ 1](1.5) + f[0,\ 1,\ 3](1.5)(1.5 - 1) = 0.933\ 37.$$

The correct value with six significant digits is

$$f(1.5) = (1.\ 5^2)e^{-1.5/2} = 1.062\ 83,$$

and the error turns out to be

$$\Delta P_2(1.5) = 1.062\ 83 - 0.933\ 37 = 0.129\ 46.$$

An estimate of this error can be found with the next-term rule. We simply need to add $x_3 = 2$ *to the previous collection of points. The divided differences are:*

$$f[x_1] = 0, f[x_1, x_2] = 0.606\ 53, f[x_1, x_2, x_4] = 0.031\ 43, f[x_1, x_2, x_4, x_3] = -0.097\ 80.$$

The next-term estimate, considering that the next term is

$$f[x_1, x_2, x_4, x_2]x(x-1)(x-3) = -0.097\,80x(x-1)(x-3),$$

turns out to be

$$\Delta P_2(1.5) \approx -0.097\,80(1.5)(1.5-1)(1.5-3) = 0.110\,03.$$

The estimated value 0.110 03 *is relatively close to the correct value* 0.129 46*, as expected.*

If the interpolation had been carried out using x_1, x_2 *and* x_3*, with the next-term rule applied adding* x_4*, the interpolation would have been*

$$P_2(1.5) = 1.006\,72,$$

with a true error equal to

$$\Delta P_2(1.5) = 1.062\,83 - 1.006\,72 = 0.056\,11,$$

and a next-term estimate equal to

$$\Delta P_2(1.5) \approx -0.097\,80(1.5)(1.5-1)(1.5-2) = 0.036\,68$$

(remember that the divided difference of order n is invariant to the ordering of the x_i*'s). It is straightforward to calculate the relative error for the two cases, and observe that the estimate of the error is more accurate in the first case.*

3.9 Relevant R code. Functions `divdif`, `polydivdif` and `decidepoly_n`

In `comphy`, calculations for the divided differences are carried out by the function `divdif`. Input for this function is the points, $(x_i, f(x_i) \equiv f_i)$; the output is a triangular matrix P in which the bottom triangular part is filled with zeros and the top triangular part contains the divided differences.

```
# Load library "comphy" in working space
library(comphy)

# The function whose values have to be
# interpolated is f(x)=x^4-3*x^3+2*x^2+1
```

```
# 5 known points
x <- c(-2,-1,0,1,2)
f <- x^4-3*x^3+2*x^2+x+1

# Put them in a matrix, just to display them
print(matrix(c(x,f),ncol=2))
```

```
##      [,1] [,2]
## [1,]   -2   47
## [2,]   -1    6
## [3,]    0    1
## [4,]    1    2
## [5,]    2    3
```

```
# Divided differences
P <- divdif(x,f)
print(P)
```

```
##      [,1] [,2] [,3] [,4] [,5]
## [1,]   47  -41   18   -5    1
## [2,]    6   -5    3   -1    0
## [3,]    1    1    0    0    0
## [4,]    2    1    0    0    0
## [5,]    3    0    0    0    0
```

```
# Let's add a sixth known point, part of the
# same curve. The new divided differences will
# be zero as only five points are needed to
# define a polynomial of degree 4
xnew <- c(x,3)
fnew <- c(f,3^4-3*3^3+2*3^2+3+1)

# Display the new set of points
print(matrix(c(xnew,fnew),ncol=2))
```

```
##      [,1] [,2]
## [1,]   -2   47
## [2,]   -1    6
## [3,]    0    1
## [4,]    1    2
```

```
## [5,]    2    3
## [6,]    3   22
```

```
# Divided differences again
Pnew <- divdif(xnew,fnew)

# The new column has only zeros
print(Pnew)
```

```
##      [,1] [,2] [,3] [,4] [,5] [,6]
## [1,]   47  -41   18   -5    1    0
## [2,]    6   -5    3   -1    1    0
## [3,]    1    1    0    3    0    0
## [4,]    2    1    9    0    0    0
## [5,]    3   19    0    0    0    0
## [6,]   22    0    0    0    0    0
```

It is worth stressing that the order of the known points can be changed. The divided differences will change accordingly, with the exception of the highest-order divided difference, so that the interpolating polynomial will remain unchanged.

```
# Shuffle the previous set of known points
idx <- sample(1:6,size=6,replace=FALSE)
print(idx)
```

```
## [1] 1 5 2 3 4 6
```

```
xnew <- xnew[idx]
fnew <- fnew[idx]

# Re-do divided differences
Pnew <- divdif(xnew,fnew)

# The matrix is different, but the last
# column is still zero
print(Pnew)
```

```
##      [,1] [,2] [,3] [,4] [,5] [,6]
## [1,]   47  -11   10   -4    1    0
## [2,]    3   -1    2   -1    1    0
## [3,]    6   -5    3    0    0    0
## [4,]    1    1    3    0    0    0
## [5,]    2   10    0    0    0    0
## [6,]   22    0    0    0    0    0
```

The interpolating polynomial in `comphy` is implemented in the function `polydivdif`. The input consists of an interpolation grid, `x0`, and the known points, `x, f`. There exists also the option of deciding how many known points, `np`, to use for the interpolation, in addition to the two points at the beginning and end of the interpolating interval. By default the function uses all known points, but `np` can be equal to or greater than 2, as the points at the beginning and the end of the interval have to be always included, due to the meaning of interpolation.

```
# Let's use the known points x,f generated earlier

# x0 is a very fine grid
x0 <- seq(-2,2,length.out=1000)

# Interpolating polynomial (np unchanged)
lDD <- polydivdif(x0,x,f)

# Interpolated value
f0 <- lDD$f0

# Plot with kown points
plot(x,f,pch=16,cex=1.5,xlab="x",ylab="f(x)")
points(x0,f0,type="l",col=2)
```

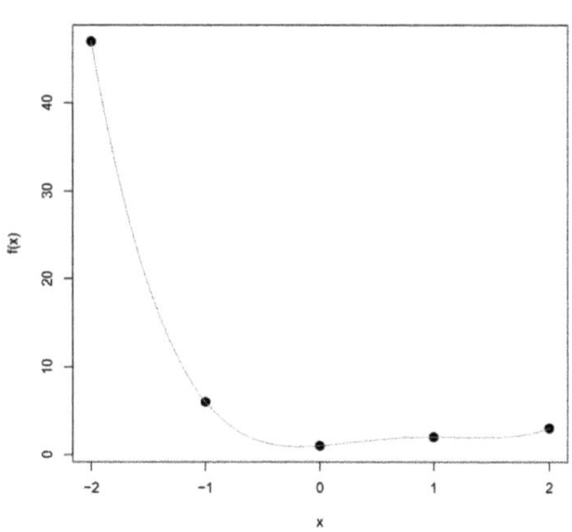

```
# Now use only 3 out of 5 points: first, last and
# one selected randomly
set.seed(7634)
lDD <- polydivdif(x0,x,f,np=3)
f0new <- lDD$f0

# The plot reveals which point was selected randomly
plot(x0,f0new,type="l",xlab="x",ylab="f(x)",col=3)
points(x0,f0,type="l",col=2)
points(x,f,pch=16,cex=1.5,col=1)
```

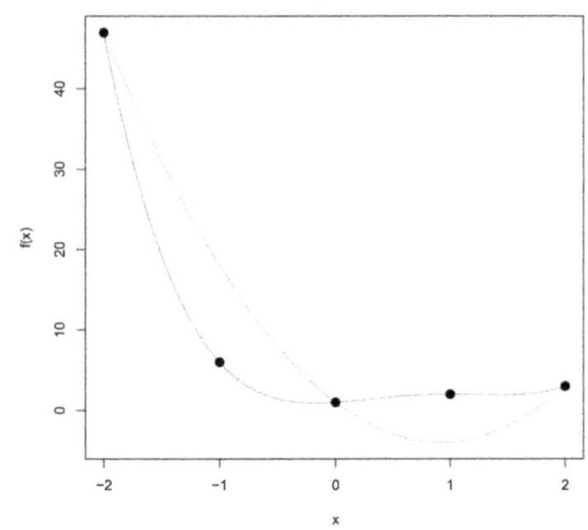

As mentioned at the end of section 3.8, the divided differences tend to become smaller and smaller at each level of interpolation. When the function to interpolate is a polynomial of degree k, all divided differences from level $k + 2$ and above will be zero. If the function to interpolate is not a polynomial, such values will become smaller and smaller until, at a given $k + 2$ they will be smaller than a given threshold. If this threshold is very small, then the polynomial of degree k interpolates the function very well. Both scenarios have been implemented in the `comphy` function, `decidepoly_n`, which takes the known points and returns the degree k of the best-interpolating polynomial, considered as the one whose divided differences average falls below a given threshold `thr` (default is 10^{-6}), at all columns with index $k + 1$ and above.

In the code presented below, two guesses for best-matching polynomial are attempted. The first is on an actual polynomial of degree 5, where 30 known points are provided. The second is on a trigonometric polynomial where, again, 30 known points are provided.

```
#
# First case: Polynomial of degree 5 - x^5-4*x^2+1
#

# Known points
x <- seq(-pi,2*pi,length.out=30)
f1 <- x^5-4*x^2+1

# Degree of the best-interpolating polynomial
n1 <- decidepoly_n(x,f1)
print(n1)


## [1] 5


# Second case: trigonometric polynomial - 2*sin(0.5*x)-cos(x)
#                                          +4*sin(x)
# Known data
f2 <- 2*sin(0.5*x)-cos(x)+4*sin(x)

# Degree of the best interpolating polynomial
n2 <- decidepoly_n(x,f2)
print(n2)


## [1] 20


# How good is the match? - Visually ...
# Use only 20 points
set.seed(8775)
x0 <- seq(-pi,2*pi,length.out=1000)
lDD <- polydivdif(x0,x,f2,np=n2)
f0 <- lDD$f0
x <- lDD$x
f <- lDD$f
curve(2*sin(0.5*x)-cos(x)+4*sin(x),from=-pi,to=2*pi,
xlab="x",ylab="f(x)")
points(x,f,pch=16)
points(x0,f0,type="l",col=2)
```

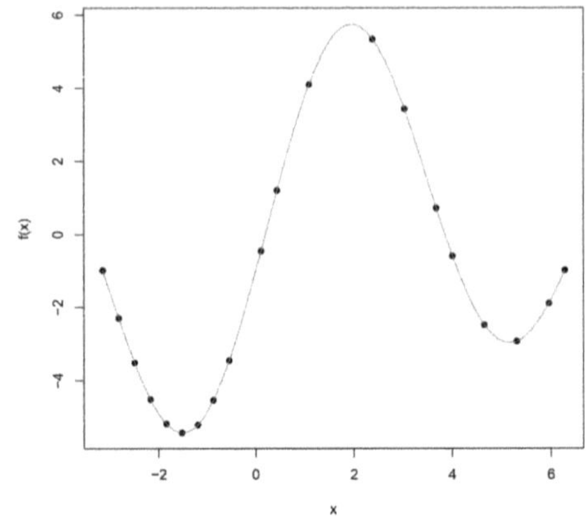

In the first case no more than 6 points were needed as the polynomial had degree 5. The function returned the correct guess, `n1=5`. In the second case the maximum value could have been `n2=30` but, in fact, the function returned `n2=20`, thus indicating that with the threshold `thr`$=10^{-6}$ a polynomial of degree 20 matched the trigonometric polynomial closely, in the given interval $[-\pi, 2\pi]$. The plot shows indeed that the correct curve and its polynomial interpolation overlap very closely. To carry out the interpolation with `plydivdif`, we have only used 20 points (option `np=20`), as indicated by the degree's guess. The 20 randomly selected points are those shown in the plot.

3.10 Exercises on divided differences

Exercise 13

Find the divided differences for the points tabulated.

x	1.000	2.500	3.000	4.000	4.500
$f(x)$	0.000	1.833	2.197	2.773	3.008

What is the value for $x = 1.5$ using all five points? And using only the first four points?

Exercise 14

Find the coefficients $a_1, \ldots, a_6$ of the function,

$$f(x) = a_1 + a_2(x+1) + a_3(x+1)(x-1) + a_4(x+1)(x-1)(x-2)+$$
$$a_5(x+1)(x-1)(x-2)(x-4) + a_6(x+1)(x-1)(x-2)(x-4)(x-5),$$

equal to the following polynomial,

$$P_5(x) = x^5 - 2x^4 - x^3 + 3x^2 - 6$$

Exercise 15

The function,

$$f(x) = \cos(x/2) - \sin(x)$$

in the interval $[-2\pi, 2\pi]$ can be interpolated by a four degrees polynomial, $Q_4(x)$, passing through the points x_1, x_2, x_3, x_4, x_6 where,

$$x_1 = -2\pi,\ x_2 = -\pi,\ x_3 = 0,\ x_4 = \pi/2,\ x_6 = 2\pi$$

Using knowledge of $f(x)$ at $x_5 = \pi$, compute the error,

$$\Delta Q_4(x) \equiv f(x) - Q_4(x)$$

at $x = (3/2)\pi$, using the next-term rule.

Exercise 16

Considering the expression (3.18) for the interpolation using divided differences, prove that the following recurring equation is correct:

$$Q_n(x) = Q_{n-1}(x) + \frac{f_{n+1} - Q_{n-1}(x_{n+1})}{(x_{n+1} - x_1) \cdots (x_{n+1} - x_n)}(x - x_1) \cdots (x - x_n)$$

Exercise 17

Consider the following known points of a function $f(x)$:

x	−2	−1	0	1	3	2
$f(x)$	−160	−1	10	17	835	116

In a regular grid containing 1000 points between −2 and 3, calculate the fourth-degree interpolating polynomial $Q_4(x)$ using the first five values tabulated and the divided differences. Next, calculate the fifth-degree interpolating poynomial $Q_5(x)$ using the last tabulated value and the formula introduced in the previous exercise. Finally, compute $Q_5(x)$ via divided differences using all six tabulated points. Plot the three curves found and verify that the last two curves coincide.

3.11 Cubic splines

In this chapter, polynomials have been used to interpolate a function between known points. Other functions could have been used, of course. The reason why the theory of polynomial interpolation has been treated in detail is connected to the *Weierstrass approximation theorem* [6]:

Suppose $f(x)$ is a continuous real-valued function defined on the real interval $[a, b]$. For every $\epsilon > 0$, there exists a polynomial $P(x)$ such that for all x in $[a, b]$, we have $|f(x) - P(x)| < \epsilon$.

The above theorem is meant to describe the extent of a polynomial's oscillations around the given function it is interpolating. Figure 3.5 renders the idea. In the interval $[a, b]$, it is always possible to find a polynomial $P(x)$ whose values are (everywhere in $[a, b]$) contained inside the strip bounded by the curves $f(x) + \epsilon$ and $f(x) - \epsilon$. This situation is known as *uniform approximation* because it is possible to determine an error ϵ which is valid throughout the interval $[a, b]$. So when using polynomials we are assured we can be as close to the function $f(x)$ as we need to be. On the other end, we know that a polynomial of degree n can have up to $n - 1$ maxima or minima, those features we have referred previously as polynomial oscillations. Such oscillations can result in large errors. This tends to happen especially when the function $f(x)$ is mostly flat and just a few of its points are known. An example is provided by the Gaussian in figure 3.6, where 3, 5, 9 and 11 points are known. The Gaussian function is mostly flat in the interval $[-5, 5]$. This feature does not help a correct interpolation with polynomials as they tend to present large oscillations, away from the correct function. On the other hand, a polynomial does not oscillate too much if limited to small intervals of x and, in fact, it has the desirable properties of being continuous and 'smooth', i.e. with continuous derivatives. For this reason, a new type of interpolation curve has been created to maintain the desirable characteristics of a polynomial curve, while eliminating its tendency to undergo large oscillations. The curve is known as *cubic spline* and it is a 'collage' of several third-degree polynomials.

In fact, a cubic spline is a piecewise curve whose components are cubic polynomials each one going through at least one known point of the function,

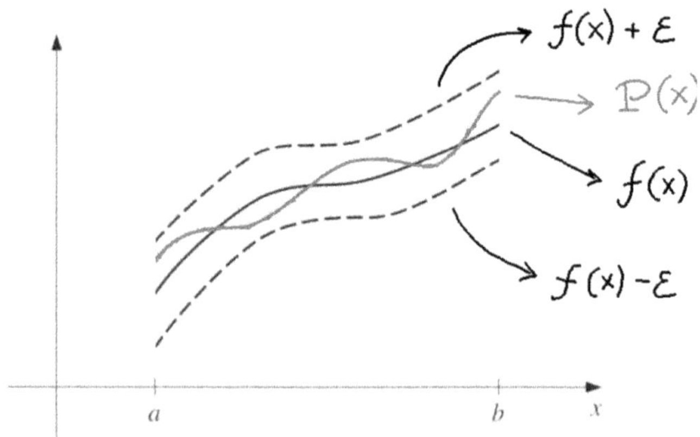

Figure 3.5. Illustration of Weierstrass approximation theorem. Given a function $f(x)$ in an interval $[a, b]$, a polynomial (in red) can always be found whose oscillations are contained within the two functions $f(x) + \epsilon$ and $f(x) - \epsilon$.

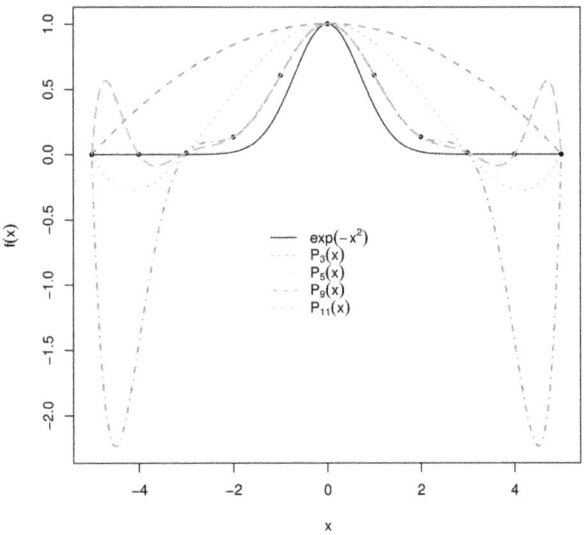

Figure 3.6. Interpolation of a Gaussian function using 3, 5, 9 and 11 known points. Most of the Gaussian function is flat so that polynomial interpolation leads to unwanted large oscillations.

and with parameters calculated so to present continuity, jointly with their first and second derivatives, at all known points. Such conditions yield a class of curves fitting the data exactly, while being very smooth and avoiding unwanted oscillations. Let us look at how to build cubic splines in detail. The goal is to pass exactly through the $n + 1$ known points of the function $f(x)$,

$$(x_1, f_1),\ (x_2, f_2),\ \ldots,\ (x_{n+1}, f_{n+1}),$$

where $f_i \equiv f(x_i)$. Let us call the curved segment between (x_k, f_k) and (x_{k+1}, f_{k+1}) as $q_k(x)$ and the curved segment between (x_{k+1}, f_{k+1}) and (x_{k+2}, f_{k+2}) as $q_{k+1}(x)$ (see figure 3.7). These two functions are the type of cubic polynomial patches composing the whole spline. The polynomial $q_k(x)$ is so defined:

$$q_k(x) \equiv a_{k,1}(x - x_k)^3 + a_{k,2}(x - x_k)^2 + a_{k,3}(x - x_k) + a_{k,4} \tag{3.21}$$

The curved segment is obviously completely determined once the four coefficients $\{a_{k,i},\ i = 1, 2, 3, 4\}$ are calculated. The conditions mentioned above and defining the spline can be translated, in the case of $q_k(x)$ and $q_{k+1}(x)$, in the following set of equations:

$$\begin{aligned} q_k(x_k) &= f_k \\ q_{k+1}(x_{k+1}) &= f_{k+1} \\ q_{k+1}(x_{k+2}) &= f_{k+2} \\ q_k(x_{k+1}) &= q_{k+1}(x_{k+1}) \\ q'_k(x_{k+1}) &= q'_{k+1}(x_{k+1}) \\ q''_k(x_{k+1}) &= q''_{k+1}(x_{k+1}) \end{aligned} \tag{3.22}$$

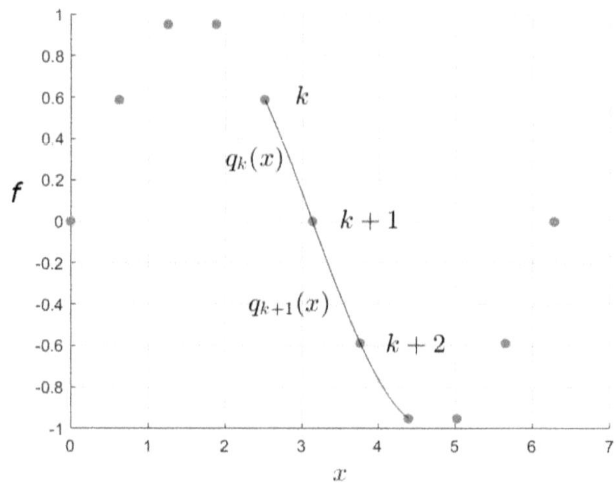

Figure 3.7. Cubic splines components at point k and $k + 1$.

The first two equations simply state that the two components go through the k-th and $(k + 1)$-th point, respectively. The third equation states that q_{k+1} goes through the $(k + 2)$-th point. The fourth equation is the continuity equation for the function values at the point joining the two curved segments, while the fifth and sixth equations relate to the continuity of first and second derivatives. We have mentioned how these last two conditions make the spline a smooth interpolating curve. To summarize, when three points, $k, k + 1, k + 2$ are used to build two cubic polynomial segments, 2×4 coefficients are needed, but only six equations are provided. Similarly, it is relatively straightforward to verify that with $n + 1$ data points given, only $4n - 2$ equations are available for the determination of $4n$ coefficients. For this reason, two additional conditions have to be added in relation to the first and last point, for example fixing the slope (first derivative) of the first and last cubic polynomial segment at a prefixed value (often 0). Once the $4n$ coefficients have been determined, all the n polynomial segments forming the cubic spline can be drawn, each one within its strict interval of validity.

The determination of the $4n$ coefficients $\{a_{k,i},\ k = 1, \ldots, n,\ i = 1, 2, 3, 4\}$ goes through the solution of a system of linear equations. We will not describe the system or its solution in depth, although one of the exercises in section 3.13 asks to find the solution for a cubic spline passing through four points. Rather, the `comphy` function dealing with cubic splines interpolation will be used to illustrate the methodology, and particular care will be devoted to the different types of additional conditions that allow the determination of the cubic splines coefficients.

3.12 Relevant R code. Function `spline`

The function used here to implement cubic splines is called `spline` and is not part of the `comphy` package, but it is found among the base packages installed with R. The function takes in the two vectors containing the known points' coordinates, $\{x_i, f_i,\ i = 1, \ldots, n + 1\}$ and the fine grid at which points interpolation is calculated. The type of cubic spline used will depend on the conditions used to determine the end segments (see section 3.11). In the `spline` function these conditions are known as *methods.* The default method is indicated with `"fmm"` and it stands for the Forsythe, Malcolm and Moler method [7]. The condition that allows one to determine the first and last spline (curved segment) according to this method is that these segments coincide with the cubic polynomials calculated using the first four points and the last four points. Another method often implemented is the `"natural"` method. The first and last curved segments are here determined so to have the second derivatives at points x_1 and x_{n+1} equal to zero. This means that approaching x_1 and x_{n+1}, the curved segments will straighten up, becoming close to straight segments (for which the curvature, i.e. the second derivative, is zero). Another method, used occasionally, is the `"periodic"` method which is used when the function to interpolate is periodic. This implies that $f(x_1)$ must be equal to $f(x_{n+1})$. `spline` includes other methods to determine the end segments, and these can be looked up using `?spline`.

Let us demonstrate this function with a four-degrees polynomial, $f(x) = x^4 - 2x^3 + x^2 + 1$. The known points are not equally spaced. The default method, used here, is `"fmm"`.

```
# Load library "comphy" in working space
library(comphy)

# The function whose values have to be
# interpolated is f(x)=x^4-3*x^3+2*x^2+1

# 4 known points
x <- c(-5,-3,0,5)
f <- x^4-2*x^3+x^2+1

# Fine grid
x0 <- seq(-5,5,length.out=1000)

# Cubic spline with default method, "fmm"
lCS <- spline(x,f,xout=x0)

# Plot correct curve and splines
frange <- range(lCS$y)
curve(x^4-2*x^3+x^2+1,from=-5,to=5,lwd=2,
xlab="x",ylab="f(x)",ylim=frange)
points(x,f,pch=16)
points(lCS$x,lCS$y,type="l",lty=2,col=2,lwd=2)
```

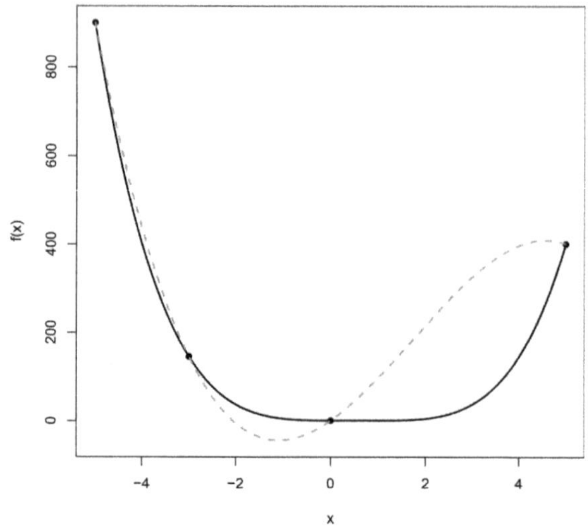

The interpolation is not very good for the last segment because the method "fmm" yields just one third-degree polynomial both when considering the first and the last segment. This polynomial should be the same polynomial obtained with the divided differences when the four known points are provided for interpolation.

```
# Overlap previous cubic spline with polynomial
# from divided differences
lDD <- polydivdif(x0,x,f)

# Overlap plot
frange <- range(lCS$y,lDD$f0)
plot(lCS$x,lCS$y,type="l",ylim=frange,
lwd=2,xlab="x",ylab="f(x)")
points(x0,lDD$f0,type="l",col=2,lty=2,lwd=2)
```

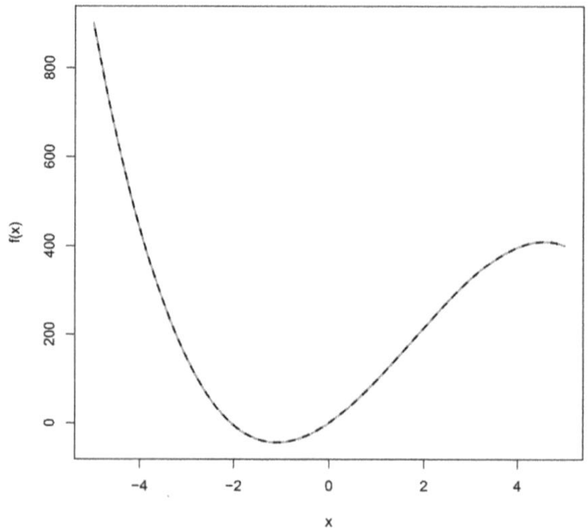

The situation changes slightly when the condition imposed is the "natural" method. Now we should see that the first and last segment approximate straight lines.

```
# Cubic spline with method "natural"
lCSn <- spline(x,f,xout=x0,method="natural")

# Recall previous plots
frange <- range(lCS$y,lCSn$y)
curve(x^4-2*x^3+x^2+1,from=-5,to=5,lwd=2,
xlab="x",ylab="f(x)",ylim=frange)
points(x,f,pch=16)
points(lCS$x,lCS$y,type="l",lty=2,col=2,lwd=2)
points(lCSn$x,lCSn$y,type="l",lty=3,col=3,lwd=3)
```

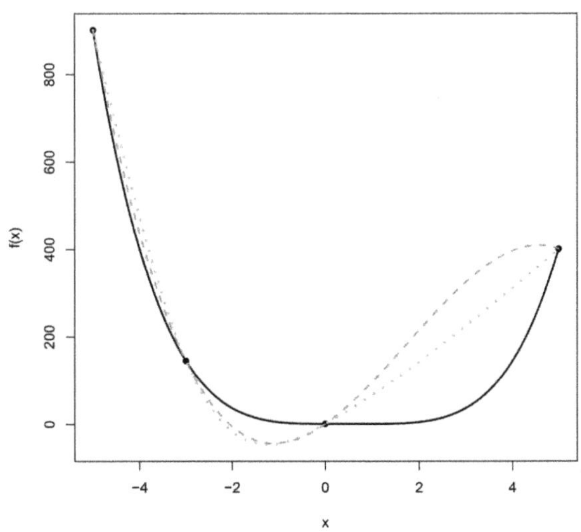

When the parameter xout is not provided, the function can still build up cubic splines between the two optional parameters xmin and xmax (default min(x) and max(x), respectively); in this case the interpolation grid will be generated automatically with a number of points that can be decided with the parameter n (default equal to three times the number of known points).

```
# Max and min. Regular grid with number of
# points generated by defaut
lCS <- spline(x,f)

# By default the number of grid points is 3 times
# the number of known points
print(length(x))
```

```
## [1] 4
```

```
print(length(lCS$x))
```

```
## [1] 12
```

```
# The number of grid points can also be increased
lCS <- spline(x,f,n=200)
print(length(lCS$x))
```

```
## [1] 200
```

3.13 Exercises on cubic splines

Exercise 18

Interpolate, using a grid of 200 points in the interval $x \in [-4, 4]$, the Gaussian function,

$$f(x) = \exp\left(-\frac{x^2}{10}\right),$$

where the known points are,

$$x_1 = -4,\ x_2 = -2,\ x_3 = -1,\ x_4 = -1/2,\ x_5 = -1/4$$
$$x_6 = 0,\ x_7 = 1/4,\ x_8 = 1/2,\ x_9 = 1,\ x_{10} = 2,\ x_{11} = 4,$$

using cubic splines with the Forsythe, Malcolm and Moler method for end segments. Compare graphically the cubic splines with a polynomial fit using divided differences.

Exercise 19

Given the same known x points of exercise 18, apply them to function,

$$f(x) = \exp(-x^2/0.1)$$

which has a narrower peak than the Gaussian of exercise 17. Carry out the same interpolation and comparison done in exercise 17, on this new set of points.

Exercise 20

Using the `comphy` function `polydivdif`, demonstrate visually that the first and last segment of a cubic spline used to interpolate,

$$x_i = -\pi + (i - 1)\pi/4, \quad i = 1, \ldots, 9$$
$$f_i = \sin(x_i) + 2\cos(3x_i), \quad i = 1, \ldots, 9$$

with the Forsythe, Malcolm and Moler method, belong to the cubic polynomials passing respectively through the first and last four known points. You can use a regular fine grid containing 1000 points for interpolation.

References

[1] Aitken A C 1929 A general formula of polynomial interpolation *Proc. Edinb. Math. Soc.* **1** 199–203

[2] Aitken A C 1931 On interpolation by iteration of proportional parts, without the use of differences *Proc. Edinb. Math. Soc.* **3** 56–7

[3] Neville E H 1934 Iterative interpolation *J. Indian Math. Soc.* **20** 87–120

[4] Divided Differences, Wikipedia https://en.wikipedia.org/wiki/Divided_differences (Accessed: 2019-09-03)

[5] Burden R L, Faires D J and Burden A M 2016 *Numerical Analysis* (Cengage Learning)

[6] Kreyszig E 2006 *Advanced Engineering Mathematics* (John Wiley & Sons)

[7] Malcolm M A, Forsythe G E and Moler C B 1977 *Computer Methods for Mathematical Computations* (John Wiley & Sons)

IOP Publishing

Computational Physics with R

James Foadi

Chapter 4

Computation using matrices

Matrix operations appear in so many parts of physics and underlie so many numerical methods that it is really difficult, if not impossible, to omit them from a course in computational physics. For this reason, it has been decided to introduce and deepen this topic early in the book, in order to build some of the numerical foundations used in later chapters.

Matrices are also key to the solution of systems of algebraic linear equations. Therefore this chapter also deals with the numerical solution of such systems.

The reader is assumed to have studied matrices and matrix algebra in foundational courses. Most of the material needed as background is synthetically collected in appendix B. A good (albeit old) reference for physicists is *Matrix Theory for Physicists* by J Heading [1].

4.1 The matrix form of a system of linear equations

A generic system of n linear equations in n unknowns,

$$\begin{cases} a_{11}x_1 + a_{12}x_2 + \cdots + a_{1n}x_n & = b_1 \\ a_{21}x_1 + a_{22}x_2 + \cdots + a_{2n}x_n & = b_2 \\ \cdots & = \cdots \\ a_{n1}x_1 + a_{n2}x_2 + \cdots + a_{nn}x_n & = b_n \end{cases} \tag{4.1}$$

can be expressed in matrix form as,

$$A\mathbf{x} = \mathbf{b} \tag{4.2}$$

if,

$$A = \begin{pmatrix} a_{11} & a_{12} & \cdots & a_{1n} \\ a_{21} & a_{22} & \cdots & a_{2n} \\ \cdots & \cdots & \cdots & \cdots \\ a_{n1} & a_{n2} & \cdots & a_{nn} \end{pmatrix}, \quad \mathbf{x} = \begin{pmatrix} x_1 \\ x_2 \\ \cdots \\ x_n \end{pmatrix}, \quad \mathbf{b} = \begin{pmatrix} b_1 \\ b_2 \\ \cdots \\ b_n \end{pmatrix}$$

doi:10.1088/978-0-7503-2632-2ch4

This matrix form is, in principle, very useful because, provided A is an invertible matrix, it yields a unique solution to the system:

$$\mathbf{x} = A^{-1}\mathbf{b} \tag{4.3}$$

Most of the calculations involved in the above solution go in the determination of the matrix inverse. Inverting a matrix is a very time-consuming operation. For this reason, much of the efforts in finding the solution of a linear system have gone to researching methods to avoid matrix inversion.

4.2 The importance of triangular matrices

An *upper-triangular* matrix is a square matrix with all elements below the diagonal, equal to zero; a *lower-triangular* matrix has all the elements above the diagonal, equal to zero. Systems of linear equations associated with upper-triangular or lower-triangular matrices can be solved very quickly, as demonstrated in the following example.

Example 4.1 *The solution of the system,*

$$\begin{cases} 2x_1 & = & 4 \\ x_1 - 3x_2 & = & 11 \\ 3x_1 + x_2 + x_3 & = & 4 \\ -x_1 + x_2 + 5x_3 + x_4 & = & 6 \end{cases}$$

can be found very quickly by substitution, starting with the value of x_1 in the first equation. This value, once replaced in the second equation, provides an equation with the only unknown x_2. This second unknown can then be found easily and used, with x_1 in the third equation, which will contain the only unknown x_3. Once this is found, x_1, x_2, x_3 are replaced in the last equation so that x_4 is readily calculated. The series of operations just described appears below:

$$2x_1 = 4 \Rightarrow x_1 = 2$$

$$\Downarrow$$

$$x_1 - 3x_2 = 11 \Rightarrow 2 - 3x_2 = 11 \Rightarrow x_2 = -3$$

$$\Downarrow$$

$$3x_1 + x_2 + x_3 = 4 \Rightarrow 3(2) - 3 + x_3 = 4 \Rightarrow x_3 = 1$$

$$\Downarrow$$

$$-x_1 + x_2 + 5x_3 + x_4 = 6 \Rightarrow -2 - 3 + 5(1) + x_4 = 6 \Rightarrow x_4 = 6$$

The solution of the system is,

$$x_1 = 2,\ x_2 = -3,\ x_3 = 1,\ x_4 = 6$$

At each step, the numerical value of one unknown is found. This is made possible by the particular form of the system that contains only one unknown in the first equation, one additional unknown in the second equation, and so on. This system of equations corresponds to the lower-triangular matrix,

$$A = \begin{pmatrix} 1 & 0 & 0 & 0 \\ 1 & -3 & 0 & 0 \\ 3 & 1 & 1 & 0 \\ -1 & 1 & 5 & 1 \end{pmatrix}$$

A system of linear equations associated with an upper-triangular matrix would be solved performing the same series of substitutions used before, with the difference that one starts this time from the last equation.

Triangular matrices are therefore associated with easy-to-solve systems of linear equations. Thus, a group of methods used to solve these systems attempts at transforming the original matrix associated with the (non triangular) system into a triangular matrix, as will be explained in the next section.

4.3 The Gaussian elimination method

This section will introduce one of the best-known methods to turn a linear system into a triangular one. Let us consider a specific example with three equations and three unknowns:

$$\begin{cases} 2x_1 - x_2 + 3x_3 &= 9 \\ x_1 + x_2 + 4x_3 &= 15 \\ 3x_1 + 2x_2 + x_3 &= 10 \end{cases}$$

We would like to turn the corresponding coefficients,

$$A = \begin{pmatrix} 2 & -1 & 3 \\ 1 & 1 & 4 \\ 3 & 2 & 1 \end{pmatrix}$$

into an upper-triangular matrix, U, which will therefore be associated with an equivalent system of equations. The type of operations that can be applied to an equation in order not to change its solution are addition, subtraction, multiplication and division, both by a constant or by two equivalent quantities. Such operations will change the equations' coefficients and the numeric values on the right hand sides. The goal here is to change them so that some will be zero, resulting in a triangular system. The system's coefficients and the numeric values on the right hand sides can be arranged into a so-called *augmented matrix*, A', which can be constructed from A by adding the vector column $\mathbf{b}$ of equation (4.2):

$$A' = \begin{pmatrix} 2 & -1 & 3 & 9 \\ 1 & 1 & 4 & 15 \\ 3 & 2 & 1 & 10 \end{pmatrix}$$

We can then start changing the augmented matrix's values using the so-called *row operations*. Purpose of a row operation is to turn the element immediately below the diagonal of A to zero. For instance, the following row operation involves row one, R_1, and row two, R_2,

$$R_2 \quad \rightarrow -\frac{1}{2}R_1 + R_2$$

This operation means that all the elements of the first row (R_1) are multiplied by $-1/2$ and added to the second row (R_2). The elements of the first row are not changed, but the elements of the second row are (this is why the symbol $R_2 \quad \rightarrow$ is used). The transformed augmented matrix is

$$\begin{pmatrix} 2 & -1 & 3 & 9 \\ 0 & 3/2 & 5/2 & 21/2 \\ 3 & 2 & 1 & 10 \end{pmatrix}.$$

The element in R_2 immediately under the diagonal has now been transformed into a zero. Next, we would like to turn the element in row 3 and column 1 to zero. For this purpose, it suffices to change R_3 with the first row multiplied by $-3/2$ and added to the third row. In symbols,

$$R_3 \quad \rightarrow -\frac{3}{2}R_1 + R_3$$

This second row operation turns the augmented matrix to,

$$\begin{pmatrix} 2 & -1 & 3 & 9 \\ 0 & 3/2 & 5/2 & 21/2 \\ 0 & 7/2 & -7/2 & -7/2 \end{pmatrix}$$

Now, only the first element of the first column is not a zero. One can continue to apply these to transform the initial matrix into an upper-triangular matrix. In fact, one can turn the element in row 3 and column 2 to zero with only an additional row operation, which is given by

$$R_3 \quad \rightarrow -\frac{7}{3}R_2 + R_3$$

and the augmented matrix is, accordingly, transformed as

$$\begin{pmatrix} 2 & -1 & 3 & 9 \\ 0 & 3/2 & 5/2 & 21/2 \\ 0 & 0 & -28/3 & -28 \end{pmatrix}$$

The first 3×3 part of the matrix is now an upper-triangular matrix. This means that the series of operations needed to achieve this result, are finished. The resulting augmented matrix can now be switched back into its corresponding triangular system of equations:

$$\begin{cases} 2x_1 - x_2 + 3x_3 &= 9 \\ (3/2)x_2 + (5/2)x_3 &= 21/2 \\ -(28/3)x_3 &= -28 \end{cases}$$

Solutions can be fast calculated from this system which yields $x_3 = 3$, $x_2 = 2$ and $x_1 = 1$. The procedure just illustrated is known as *Gaussian elimination*. This method finds the unique solution of a system of n linear equations in n unknowns, if one exists. The key ingredient of Gaussian elimination is the transformation of matrix A in equation (4.1) into an upper-triangular matrix.

The method also applies to instances with no solution of infinite solutions. If the process of triangularisation leads to some of the rows being entirely filled with zeros, then the number of valid equations is less than n and either there cannot be a unique solution, or the equations lead to inconsistent results.

Example 4.2 *The following system of form* $A\mathbf{x} = \mathbf{b}$,

$$\begin{cases} 2x_1 + 3x_2 - x_3 &= 10 \\ -x_1 - (3/2)x_2 + (1/2)x_3 &= -5 \\ 4x_1 + 6x_2 - 2x_3 &= 15 \end{cases}$$

has no solution. To see this, let us carry out row operations to transform the system into its upper triangular form. A first round of row operations is,

$$R_2 \to (1/2)R_1 + R_2$$

$$R_3 \to -2R_1 + R_3$$

The resulting transformed augmented matrix, $(A'|\mathbf{b}')$, *is*

$$\begin{pmatrix} 2 & 3 & -1 & 10 \\ 0 & 0 & 0 & 0 \\ 0 & 0 & 0 & -5 \end{pmatrix}$$

The appearance of rows completely made of zeros in A' *signals the possibility that the system has no solutions, or an infinite number of solutions. The reason for this is evident when considering that a row of zeros in matrix* A' *is equivalent to the equality,*

$$0 = b_i' \qquad i = 1, 2$$

where b_i' *is the* i component of the transformed vector $\mathbf{b}'$ *in the system* $A'\mathbf{x} = \mathbf{b}'$. *When* $b_i' = 0$, *the corresponding equation is valid for any set of values of* x_1, x_2, x_3*; on the other hand, if* $b_i' \neq 0$, *the corresponding equation is never valid. Thus, the system above has no solution because one of the rows with zeros has* $b_i' = -5 \neq 0$.

Each group of row operations turning all values below the diagonal to zero can be represented by a lower triangular matrix L_i (with i a positive integer). For example, the first group of row operations transforming,

$$A' = \begin{pmatrix} 2 & -1 & 3 & 9 \\ 1 & 1 & 4 & 15 \\ 3 & 2 & 1 & 10 \end{pmatrix}$$

into

$$A'' = \begin{pmatrix} 2 & -1 & 3 & 9 \\ 0 & 3/2 & 5/2 & 21/2 \\ 0 & 7/2 & -7/2 & -7/2 \end{pmatrix}$$

is equivalent to the matrix product,

$$A'' = L_1 A',$$

where

$$L_1 = \begin{pmatrix} 1 & 0 & 0 \\ -1/2 & 1 & 0 \\ -3/2 & 0 & 1 \end{pmatrix}$$

Similarly, the final reduced matrix,

$$A''' = \begin{pmatrix} 2 & -1 & 3 & 9 \\ 0 & 3/2 & 5/2 & 21/2 \\ 0 & 0 & -28/3 & -28 \end{pmatrix}$$

can be obtained as the product,

$$A''' = L_2 A'',$$

where

$$L_2 = \begin{pmatrix} 1 & 0 & 0 \\ 0 & 1 & 0 \\ 0 & -7/3 & 1 \end{pmatrix}$$

It is straightforward to check that the product of two lower triangular matrices is still a lower triangular matrix. Therefore, the two sets of row operations (three row operations in total) needed to turn A' into A''' are equivalent to the product of the initial augmented matrix by a lower triangular matrix, L:

$$A''' = L_2 L_1 A' \equiv LA'$$

This is the case for any number of row operations that turn the augmented matrix A' into its final reduced form, A_r:

$$A_r = L\ A' \tag{4.4}$$

This result will turn out to be useful for the LU decomposition of a square matrix, which will be illustrated later.

4.4 Relevant R code. Function `gauss_elim`

The function available in `comphy` to find the solution of a linear system is `gauss_elim`. We begin by testing the function on a simple system with three equations and three unknowns:

$$\begin{cases} 3x_1 + x_2 + x_3 &= 6 \\ x_1 - x_2 + 2x_3 &= 4 \\ -x_1 + x_2 + x_3 &= 2 \end{cases}$$

The solution is $x_1 = 1,\ \ x_2 = 1,\ \ x_3 = 2$, as it can be verified by inspection. The input for `gauss_elim` is the augmented matrix,

$$\begin{pmatrix} 3 & 1 & 1 & 6 \\ 1 & -1 & 2 & 4 \\ -1 & 1 & 1 & 2 \end{pmatrix},$$

in which the first 3×3 block is formed by the unknowns' coefficients and the last column by the constants. The function returns a vector with the solution.

```
# Load library "comphy" in working space
library(comphy)

# Prepare the augmented matrix
M <- matrix(c(3,1,-1,1,-1,1,1,2,1,6,4,2),ncol=4)

# Solution via Gaussian elimination
x <- gauss_elim(M)
print(x) # x[1]=x_1, x[2]=x_2, x[3]=x_3

## [1] 1 1 2
```

The same function returns a warning message and no solution in those cases where there is no solution or when there is an infinite number of solutions. An example of the first scenario was given above and it is repeated here:

$$\begin{cases} 2x_1 + 3x_2 - x_3 &= 10 \\ -x_1 - (3/2)x_2 + (1/2)x_3 &= -5 \\ 4x_1 + 6x_2 - 2x_3 &= 15 \end{cases}$$

```
# Prepare the augmented matrix
M <- matrix(c(2,-1,4,3,-3/2,6,-1,1/2,-2,10,-5,15),ncol=4)

# Solution via Gaussian elimination
# A "no-solution" message is printed and x is NULL
x <- gauss_elim(M)

## This system has no solution or infinite solutions.

print(x) # x[1]=x_1, x[2]=x_2, x[3]=x_3

## NULL
```

For an example with an infinite number of solutions we can use,

$$\begin{cases} 2x_1 + x_2 - 3x_3 &= 0 \\ 4x_1 + 2x_2 - 6x_3 &= 0 \\ x_1 - x_2 + x_3 &= 0 \end{cases}$$

```
# Prepare the augmented matrix
M <- matrix(c(2,4,1,1,2,-1,-3,-6,1,0,0,0),ncol=4)

# Solution via Gaussian elimination
# An "infinite solutions" message is printed and x is NULL
x <- gauss_elim(M)

## This system has no solution or infinite solutions.

print(x) # x[1]=x_1, x[2]=x_2, x[3]=x_3

## NULL
```

The function prints out the same message for both *no solutions* and *infinite solutions* case as there is no numerical solution in both cases.

The key engine of gauss_elim is the reduction of the augmented matrix. This is performed by the comphy function transform_upper. An important part of the transformation to upper triangular is the potential swapping of rows at each set of row operations. The reason goes as follows. In a row operation, a division occurs between a matrix number and a number on the diagonal. If the number on the diagonal is very small or zero, then the resulting division yields a very large number or infinity, both of which have the potential to swamp the transformed matrix. A way to avoid the swamping is to switch rows so to have decreasing values on the diagonal, from the largest to the smallest. If one or more values on the diagonal is zero, they will be shifted at the bottom of the diagonal. Small numbers, similarly, will appear at the bottom as well and will have less effect on the matrix components. The lines of code responsible for the rows swapping in transform_upper are reproduced here:

```
# .........
# missing lines of code
# .........
# Main loop
for (i in 1:(n-1)) {
  # Swap rows to have largest values on diagonal
idx <- which(abs(M[i:n,i]) == max(abs(M[i:n,i])))
idx <- idx[length(idx)] # If more than one, pick the last
idx <- i+idx-1
N <- M[i,]
M[i,] <- M[idx,]
M[idx,] <- N

# .........
# missing lines of code
# .........
```

Other options to speed up calculations and to further reduce rounding-off errors are possible. They have not been implemented in gauss_elim to keep this function conceptually simple, but they are present in other R functions (see later).

4.5 LU decomposition

As seen in section 4.3, equation (4.4), the set of row operations reducing the augmented matrix can be represented by a lower triangular matrix, which we label as Q. Clearly, the same matrix reduces the matrix of coefficients, A, into an upper triangular matrix, U:

$$Q\,A = U$$

Given the very peculiar form of a lower triangular matrix, it is not too difficult to see that the inverse of a lower triangular matrix is also a lower triangular matrix. Therefore, the previous equation leads to the following result for square matrices,

$$A = L\ U \tag{4.5}$$

where L is the inverse of Q, $L = Q^{-1}$. A square matrix can be decomposed into the product of a lower triangular and an upper triangular matrix. The decomposition is known as *LU decomposition.* A simple example can illustrate the result. Consider the 3×3 matrix A:

$$\begin{pmatrix} 4 & -2 & 1 \\ -3 & -1 & 4 \\ 1 & -1 & 3 \end{pmatrix}$$

The first two row operations filling the first column with two zeros are

$$R_2 \to (3/4)R_1 + R_2$$

$$R_3 \to -(1/4)R_1 + R_3$$

After these row operations the matrix becomes,

$$\begin{pmatrix} 4 & -2 & 1 \\ 0 & -5/2 & 19/4 \\ 0 & -1/2 & 11/4 \end{pmatrix}$$

The next (and last) row operation needed for this case is

$$R_3 \to -(1/5)R_2 + R_3$$

After this operation, the matrix assumes the following upper triangular form:

$$U = \begin{pmatrix} 4 & -2 & 1 \\ 0 & -5/2 & 19/4 \\ 0 & 0 & 9/5 \end{pmatrix}$$

A lower triangular matrix, L, with ones on the diagonal and the negative of the coefficients of the row operations in its lower section can be uniquely constructed. In the above case, this matrix corresponds to:

$$L = \begin{pmatrix} 1 & 0 & 0 \\ -3/4 & 1 & 0 \\ 1/4 & 1/5 & 1 \end{pmatrix}$$

It is straightforward to verify that the product LU yields A.

The LU decomposition of a matrix is not unique. In the examples treated so far, the lower triangular matrix had ones on the diagonal. This feature comes from the direct translation of row operations into square matrices L_i with ones on their diagonal. However, in other LU decompositions, the upper triangular matrix is that with ones on its diagonal. The Gaussian elimination method contains implicitly the calculation of an LU decomposition because it is based on row operations. But the row-swapping part in the algorithm scrambles the order of the rows. Thus, an LU product derived from Gaussian elimination does not necessarily return the original matrix A; quite often it yields a matrix equivalent to A with some of its rows swapped. In fact, it is more correct to say that a matrix A has not always an LU decomposition, but a permuted matrix, PA (where P is the square matrix containing ones and zeros and that determines the rows permutation), has. Therefore, formula (4.5) should be replaced by the following formula, covering all scenarios:

$$P\ A = L\ U \tag{4.6}$$

An LU decomposition is, in fact, still possible when a matrix is singular and has rank k, as long as the leading principal minor of order k is not zero[1]. If this is not the case, the matrix has no LU decomposition. Therefore, not all square matrices have an LU decomposition.

There are essentially two methods to calculate an LU decomposition, the *Crout method* and the *Doolittle method.*

4.5.1 The Crout method

The upper triangular matrix has all ones on the diagonal. We can force this form on U, while leaving a generic form for L, and calculate the deriving equations that are necessary to determine both sets of numbers. More specifically, if l_{ij} are the components of L, u_{ij} the components of U and a_{ij} the components of the matrix A to be decomposed, then the n^2 unknown components l_{ik}, u_{ij} which are also different from the zeros of the triangular matrices and from the ones along the diagonal of U, can be exactly determined using the known n^2 components a_{ij} of A. The conditions are formed using the LU product (4.5). Let's see how by using an example with a small n, say $n = 4$. Then, the matrix product becomes:

$$\begin{pmatrix} l_{11} & 0 & 0 & 0 \\ l_{21} & l_{22} & 0 & 0 \\ l_{31} & l_{32} & l_{33} & 0 \\ l_{41} & l_{42} & l_{43} & l_{44} \end{pmatrix} \begin{pmatrix} 1 & u_{12} & u_{13} & u_{14} \\ 0 & 1 & u_{23} & u_{24} \\ 0 & 0 & 1 & u_{34} \\ 0 & 0 & 0 & 1 \end{pmatrix} = \begin{pmatrix} a_{11} & a_{12} & a_{13} & a_{14} \\ a_{21} & a_{22} & a_{23} & a_{24} \\ a_{31} & a_{32} & a_{33} & a_{34} \\ a_{41} & a_{42} & a_{43} & a_{44} \end{pmatrix}$$

The 10 components l_{ij} of L and 6 components u_{ij} of U can now be calculated using the 16 relations implicit in the above matrix equation. From an algorithmical point of view, one starts with the product with the first column of U which yields,

$$l_{11} = a_{11},\ l_{21} = a_{21},\ l_{31} = a_{31},\ l_{41} = a_{41}$$

[1] A *leading principal minor* of order k is defined as the determinant of the matrix obtained by deleting the last n–k rows and columns of the original matrix.

Next, the product with the first row of L yields,

$$u_{12}l_{11} = a_{12},\ u_{13}l_{11} = a_{13},\ u_{14}l_{11} = a_{14},$$

from which the three unknown u_{ij} (three and not four because $u_{11} = 1$) can be derived. Then, we proceed by alternating between calculating a column of L and a row of U. For example, the three unknown non-zero elements in the second column of L are obtained with the equations:

$$u_{12}l_{21} + l_{22} = a_{22},\ u_{12}l_{31} + l_{32} = a_{32},\ u_{12}l_{41} + l_{42} = a_{42},$$

which give,

$$l_{22} = a_{22} - l_{21}u_{12},\ l_{32} = a_{32} - l_{31}u_{12},\ l_{42} = a_{42} - l_{41}u_{12}$$

It is important to note that the components of U and L in the equation above have been already computed. If that were not the case, it would not be possible to proceed with the calculation. In other words, the second column of L can be calculated once its first column and u_{12} are known. We can then proceed by calculating the last two unknown components of the second row of U (because the first component is zero and the second is one), simply by considering the second row of L

$$l_{21}u_{12} + l_{22} = a_{22},\ l_{21}u_{13} + l_{22}u_{23} = a_{23},\ l_{21}u_{14} + l_{22}u_{24} = a_{24}$$

The first relation had already been used to determine l_{22}, so we are only interested in the last two relations, from which we find

$$u_{23} = \frac{a_{23} - l_{21}u_{13}}{l_{22}},\ u_{24} = \frac{a_{24} - l_{21}u_{14}}{l_{22}}$$

In general, there will be two sets of relations from which the l_{ij} and u_{ij} components can be calculated. These are

$$l_{ij} = a_{ij} - \sum_{k=1}^{j-1} l_{ik}u_{kj}, \quad \text{with} \quad i \geqslant j,\ i = 2, 3, \ldots, n \tag{4.7}$$

and

$$u_{ij} = \left(a_{ij} - \sum_{k=1}^{i-1} l_{ik}u_{kj}\right)/l_{ii}, \quad \text{with} \quad j > i,\ j = 2, 3, \ldots, n \tag{4.8}$$

The sums with index k in the above formulas are not implemented as loops in the R code because the products are carried out in parallel as `L[i,] %*% U[,j]` and the resulting values are added together with the function `sum`. The code, contained in the function `LUdeco`, is displayed here.

```
# .........
# missing lines of code
# .........

# Initial matrices (and first n + n -1 components)
L <- matrix(rep(0,times=n*n),ncol=n)
L[,1] <- A[,1]
U <- diag(n)
U[1,2:n] <- A[1,2:n]/A[1,1]

# Remaining components
for (j in 2:n) {

# First set of relations for L
for (i in j:n) {
sss <- sum(L[i,1:(j-1)]*U[1:(j-1),j])
L[i,j] <- A[i,j]-sss
}

# Second set of relations for U
# Here i refers to the columns of U
if (j < n) {
for (i in (j+1):n) {
sss <- sum(L[j,1:(i-1)]*U[1:(i-1),i])
U[j,i] <- (A[j,i]-sss)/L[j,j]
}
}
}

# .........
# missing lines of code
# .........
```

In the code displayed, it is important not to be confused by the swap of indices, i, j, for the second set of relations, those to calculate the components of U. A condition to limit j had also to be added in order to avoid j becomeing larger than n.

4.5.2 The Doolittle method

This method aims at having all ones along the diagonal of L, rather than U. Thus, in order to form the equations needed to derive the unknown l_{ij} and u_{ij}, the same procedure highlighted for the Crout method can be followed, taking care to calculate each time a row of U first, followed by a column of L. The code for this procedure is thus very similar to the one used for the Crout method, with appropriate swappings of the indices.

```
# .........
# missing lines of code
# .........

# Initial matrices (and first n + n -1 components)
U <- matrix(rep(0,times=n*n),ncol=n)
U[1,] <- A[1,]
L <- diag(n)
L[2:n,1] <- A[2:n,1]/A[1,1]

# Remaining components
for (j in 2:n) {

# First set of relations for U
# Here i refers to the columns of U
for (i in j:n) {
sss <- sum(L[j,1:(j-1)]*U[1:(j-1),i])
U[j,i] <- A[j,i]-sss
}

# Second set of relations for L
if (j < n) {
for (i in (j+1):n) {
sss <- sum(L[i,1:(i-1)]*U[1:(i-1),j])
L[i,j] <- (A[i,j]-sss)/U[j,j]
}
}
}

# .........
# missing lines of code
# .........
```

4.6 Tridiagonal systems

There are cases in which most of the coefficients in the linear system are zero. In such instances, it is more convenient to find the solution using methods which are more appropriate to the specific structure of the system. One of such cases is when the system has *tridiagonal form*, where there are at most two or three non-zero coefficients per equation. The general form of a tridiagonal system is

$$\begin{cases} a_{11}x_1 + a_{12}x_2 & & = b_1 \\ a_{21}x_1 + a_{22}x_2 + a_{23}x_3 & & = b_2 \\ \qquad \cdots \qquad \cdots & \cdots & \cdots \\ & a_{n-1,n}x_{n-1} + a_{nn}x_n & = b_n \end{cases} \tag{4.9}$$

The coefficients matrix derived from the above system is full of zeros:

$$A = \begin{pmatrix} a_{11} & a_{12} & 0 & \dots & 0 & 0 \\ a_{21} & a_{22} & a_{23} & \dots & 0 & 0 \\ & \dots & & \dots & & \\ 0 & 0 & 0 & & a_{n,n-1} & a_{nn} \end{pmatrix}$$

It is one of the simplest examples of a so-called *sparse matrix*, a matrix in which the greatest majority of elements are zero. The matrix A is called tridiagonal since, besides the elements along the main diagonal, only the elements on the two diagonals adjacent to the main one are different from zero.

To solve a tridiagonal system, we essentially reduce A to an upper triangular matrix. Given the massive presence of zeros, such a reduction turns out to be less operation-intensive than that of a standard matrix. To start with, we can re-write the augmented matrix (A and the vector of constants, $\mathbf{b}$) using the $4n$ elements of the four vectors,

$$\boldsymbol{\alpha} = \begin{pmatrix} 0 \\ \alpha_2 \\ \alpha_3 \\ \dots \\ \alpha_n \end{pmatrix}, \quad \boldsymbol{\beta} = \begin{pmatrix} \beta_1 \\ \beta_2 \\ \beta_3 \\ \dots \\ \beta_n \end{pmatrix}, \quad \boldsymbol{\gamma} = \begin{pmatrix} \gamma_1 \\ \gamma_2 \\ \gamma_3 \\ \dots \\ 0 \end{pmatrix}, \quad \mathbf{b} = \begin{pmatrix} b_1 \\ b_2 \\ b_3 \\ \dots \\ b_n \end{pmatrix}$$

The augmented matrix can thus be written as,

$$\begin{pmatrix} \beta_1 & \gamma_1 & 0 & \dots & 0 & 0 & b_1 \\ \alpha_2 & \beta_2 & \beta_3 & \dots & 0 & 0 & b_2 \\ & \dots & & \dots & & & \\ 0 & 0 & 0 & & \alpha_n & \beta_n & b_n \end{pmatrix}$$

It is immediate to see that no component of $\boldsymbol{\alpha}$ is present in row 1 and no component of $\boldsymbol{\gamma}$ is present in the last row; this is why the first component of $\boldsymbol{\alpha}$ and the last component of $\boldsymbol{\gamma}$ have both been fixed to 0. After the transformation to an upper triangular matrix, we expect all elements of $\boldsymbol{\alpha}$ to be 0. Let's see how this can be achieved using the usual row operations. What we will need to calculate, essentially, are three new vectors, a vector $\boldsymbol{\beta}'$ replacing $\boldsymbol{\beta}$, a vector $\boldsymbol{\gamma}'$ replacing $\boldsymbol{\gamma}$ and a vector $\mathbf{b}'$ replacing $\mathbf{b}$. The first row is initially left unchanged:

$$\beta_1' = \beta_1, \quad \gamma_1' = \gamma_1, \quad b_1' = b_1$$

The second row is transformed (in order to have 0 as the first element) as follows:

$$\beta_2' = \beta_2 - \frac{\alpha_2}{\beta_1'}\gamma_1', \quad \gamma_2' = \gamma_2, \quad b_2' = b_2 - \frac{\alpha_2}{\beta_1'}b_1'$$

All remaining rows are transformed similarly. The transformations for row i are:

$$\begin{aligned} \beta_{i+1}' &= \beta_{i+1} - \frac{\alpha_{i+1}}{\beta_i'}\gamma_i', \quad \gamma_{i+1}' = \gamma_{i+1}, \\ b_{i+1}' &= b_{i+1} - \frac{\alpha_{i+1}}{\beta_i'}b_i', \qquad i = 1, \dots n-1 \end{aligned} \tag{4.10}$$

Eventually, after this first series of row operations, all components of $\boldsymbol{\alpha}$ will be zero and the matrix will be reduced to an upper triangular matrix. It is also worth noting that the components of γ remain unchanged. At this point, the reduced matrix can be used, with back-substitution, to find the solution. The first unknown to be found is x_n:

$$x_n = \frac{b'_n}{\beta'_n}$$

The remaining unknowns are found through the recurring formula,

$$x_i = \frac{b'_i - \gamma'_i x_{i+1}}{\beta'_i}, \qquad i = n - 1,\ldots,1 \tag{4.11}$$

Although the solution has been here illustrated *via* two series of algebraic equations, what was really done was an LU decomposition of matrix A and a re-casting of the original system into two separate systems (see exercise 03 in section 4.8). The R code to implement the above algorithm, known as *Thomas algorithm*, consists of two simple loops, reproduced below.

```
# .........
# missing lines of code
# .........

# Main algorithm
for (i in 1:(m-1)) {
ss <- alpha[i+1]/beta[i]
beta[i+1] <- beta[i+1]-ss*gamma[i]
b[i+1] <- b[i+1]-ss*b[i]
}

# .........
# missing lines of code
# .........

# Solution by back-substitution
x <- rep(0,length=m)
x[m] <- b[m]/beta[m]
for (i in (m-1):1) {
x[i] <- (b[i]-gamma[i]*x[i+1])/beta[i]
}

# .........
# missing lines of code
# .........
```

The fast solution for a tridiagonal system is used frequently in connection with the numerical solution of differential equations.

4.7 Relevant R code. Functions LUdeco and solve_tridiag

The function `LUdeco` takes in a square matrix of size n and one of two possible character strings, `"crout"` or `"doolittle"`, which fixes the method used to perform the decomposition. The output consists of a named list where the first component, `L`, is the lower triangular matrix, the second component, `U`, is the upper triangular matrix and the third component, `ord`, is an integer vector of length n containing the row permutations needed to make the decomposition possible. When such a decomposition is not possible, the function returns a `NULL` and prints out a message about the singularity of the matrix. A few examples follow.

```
# Load library "comphy" in working space
library(comphy)

# First example. A non-singular 4 X 4 matrix
# generated with random integers
set.seed(1298)
itmp <- sample(-10:10,size=16,replace=TRUE)
A <- matrix(itmp,ncol=4)
print(A)
```

```
##      [,1] [,2] [,3] [,4]
## [1,]    0   -5    0    1
## [2,]   -6    9   -8    6
## [3,]    1    5    5    7
## [4,]    0    9   -8    6
```

```
# LU decomposition using Crout (default)
ltmp <- LUdeco(A)
print(ltmp$L) # Lower triangular
```

```
##      [,1] [,2]      [,3]     [,4]
## [1,]   -6  0.0  0.000000 0.000000
## [2,]    0  9.0  0.000000 0.000000
## [3,]    1  6.5  9.444444 0.000000
## [4,]    0 -5.0 -4.444444 6.058824
```

```
print(ltmp$U) # Upper triangular
```

```
##      [,1] [,2]       [,3]       [,4]
## [1,]    1 -1.5  1.3333333 -1.0000000
## [2,]    0  1.0 -0.8888889  0.6666667
## [3,]    0  0.0  1.0000000  0.3882353
## [4,]    0  0.0  0.0000000  1.0000000
```

```
# A permutation of the original matrix A
ltmp$L %*% ltmp$U
```

```
##      [,1] [,2] [,3] [,4]
## [1,]   -6    9   -8    6
```

```
## [2,]    0    9   -8    6
## [3,]    1    5    5    7
## [4,]    0   -5    0    1
```

```
print(ltmp$ord) # Permuted rows of original matrix
```

```
## [1] 2 4 3 1
```

```
# Same matrix, decomposed using Doolittle
ltmp <- LUdeco(A,"doolittle")
print(ltmp$L) # Lower triangular
```

```
##            [,1]       [,2]       [,3] [,4]
## [1,]  1.0000000  0.0000000  0.0000000    0
## [2,]  0.0000000  1.0000000  0.0000000    0
## [3,] -0.1666667  0.7222222  1.0000000    0
## [4,]  0.0000000 -0.5555556 -0.4705882    1
```

```
print(ltmp$U) # Upper triangular
```

```
##      [,1] [,2]      [,3]     [,4]
## [1,]   -6    9 -8.000000 6.000000
## [2,]    0    9 -8.000000 6.000000
## [3,]    0    0  9.444444 3.666667
## [4,]    0    0  0.000000 6.058824
```

```
# A permutation of the original matrix A
ltmp$L %*% ltmp$U
```

```
##      [,1] [,2] [,3] [,4]
## [1,]   -6    9   -8    6
## [2,]    0    9   -8    6
## [3,]    1    5    5    7
## [4,]    0   -5    0    1
```

```
print(ltmp$ord) # Permuted rows of original matrix
```

```
## [1] 2 4 3 1

# LU decomposition of a singular matrix
A <- matrix(c(1,0,1,0,0,1,0,0,0,0,1,0,1,0,0,0),ncol=4)
print(A)

##      [,1] [,2] [,3] [,4]
## [1,]    1    0    0    1
## [2,]    0    1    0    0
## [3,]    1    0    1    0
## [4,]    0    0    0    0

print(det(A)) # The matrix is singular

## [1] 0

ltmp <- LUdeco(A)
print(ltmp$L) # Lower triangular

##      [,1] [,2] [,3] [,4]
## [1,]    1    0    0    0
## [2,]    0    1    0    0
## [3,]    1    0    1    0
## [4,]    0    0    0    0

print(ltmp$U) # Upper triangular

##      [,1] [,2] [,3] [,4]
## [1,]    1    0    0    1
## [2,]    0    1    0    0
## [3,]    0    0    1   -1
## [4,]    0    0    0    1

# The product LU reproduces A without row permutation
ltmp$L %*% ltmp$U
```

```
##      [,1] [,2] [,3] [,4]
## [1,]    1    0    0    1
## [2,]    0    1    0    0
## [3,]    1    0    1    0
## [4,]    0    0    0    0

print(ltmp$ord) # no permutation

## [1] 1 2 3 4

# A singular matrix with no LU decomposition
A <- matrix(c(1,0,0,1,0,0,0,0,0,0,1,0,1,0,0,1),ncol=4)
print(A)

##      [,1] [,2] [,3] [,4]
## [1,]    1    0    0    1
## [2,]    0    0    0    0
## [3,]    0    0    1    0
## [4,]    1    0    0    1

print(det(A)) # The matrix is singular

## [1] 0

# LU decomposition not possible
ltmp <- LUdeco(A)

## The input matrix is singular and it does not have
##  an LU decomposition.

print(ltmp)

## NULL
```

The function `solve_tridiag` takes in the augmented matrix, $M = (A|\mathbf{b})$, and returns the system's solution as components of a vector of length n. This function is optimised (and only valid) for tridiagonal systems; if used with non-tridiagonal systems it returns a wrong solution. `solve_tridiag` is used in a similar fashion to `gauss_elim`.

```
# Augmented matrix for the given system
M = matrix(c( 1,-8, 0, 0,-15,
2,-2,-7, 0,-23,
0, 7, 3,-6, -1,
0, 0, 8,-7, -4),ncol=5,byrow=TRUE)

# Solution for a tridiagonal system
x1 <- solve_tridiag(M)

# The same solution with generic Gaussian elimination
x2 <- gauss_elim(M)

# The two solutions are the same
print(x1)
```

```
## [1] 1 2 3 4
```

```
print(x2)
```

```
## [1] 1 2 3 4
```

4.8 Exercises on systems of linear equations

Exercise 01

Find the solution of the following system of five equations and five unknowns

$$\begin{cases} 2x_1 + x_2 - 3x_3 - 5x_4 + x_5 &= -4 \\ 7x_1 - x_2 - 5x_5 &= 3 \\ -6x_1 + 8x_2 - 2x_4 - 4x_5 &= -2 \\ -3x_1 + 8x_2 + x_3 + 8x_4 + 6x_5 &= 0 \\ 5x_1 + 2x_2 + 7x_3 + 8x_4 - 2x_5 &= -2 \end{cases}$$

using the function `gauss_elim`. Verify the solution found by substituting it in the original system (use R code).

Exercise 02

Use `gauss_elim` to verify that the system,

$$\begin{cases} 3x_1 + 2x_2 - x_3 - 4x_4 &= 10 \\ x_1 - x_2 + 3x_3 - x_4 &= -4 \\ 2x_1 + x_2 - 3x_3 &= 16 \\ -x_2 + 8x_3 - 5x_4 &= 3 \end{cases}$$

has either no solution, or an infinite number of solutions. Next, use the function `transform_upper` and quantitative reasoning to prove that the system has, in fact, no solution.

Exercise 03

The LU decomposition of a square matrix A can be used to solve an algebraic system of linear equations. Starting from the system in matrix form,

$$A\mathbf{x} = \mathbf{b}$$

and using the LU decomposition of A we get

$$L\ U\ \mathbf{x} = \mathbf{b}$$

If a new set of unknowns, $\mathbf{y}$, defined as

$$\mathbf{y} = U\mathbf{x}$$

is introduced, then the original system becomes a lower triangular system

$$L\mathbf{y} = \mathbf{b}$$

which can be quickly solved using back-substitution. Eventually, the solution $\mathbf{x}$ can be found quickly by solving the upper triangular system

$$U\mathbf{x} = \mathbf{y}$$

as the previously unknown $\mathbf{y}$ now has a numeric value.

Write a program that makes use of the LU decomposition and of the considerations written above, to solve the system presented in exercise 01. Compare the solution found with the solution of exercise 01.

Exercise 04

Computer time to solve a linear system is very short with modern processors. It is normally less than a second when the system size includes 100 or less equations. As it is very tedious and time-consuming to fill matrices of size 100 or more, we can test solution time of the `gauss_elim` function using matrices with elements generated randomly. Generate a random matrix of size $n = 100$, using the sampling of integers between -5 and $+5$ and using a fixed random seed for generation, in order to compare your results with those of the solution presented here. Use the seed 7821 for matrix A and the seed 7659 for the constant vector, $\mathbf{b}$. Use the R function `Sys.time` to find out the execution time. Try your procedure with various values of n, say $n = 100$, 500, 1000.

Exercise 05

Write a function that returns coefficients and constants for a tridiagonal system $A\mathbf{x} = \mathbf{b}$. A is a square tridiagonal matrix of size n and $\mathbf{b}$ a vector of length n. The non-zero elements of A and the elements of $\mathbf{b}$ have to be sampled randomly (with repetition) among the numbers

$$-5, -4, -3, -2, -1, 1, 2, 3, 4, 5.$$

Use the integer seed 1243 for the random sampling of matrix A and the integer seed 8731 for the random sampling of vector $\mathbf{b}$.

Using $n = 100, 500, 1000$, compare the execution time to find the solution using the functions `gauss_elim` and `solve_tridiag`. As the last function is supposed to exploit the special structure of a tridiagonal system, execution times for it are expected to be shorter than for `gauss_elim`.

4.9 Determinants

The calculation of determinants is computationally intensive as it consists of several multiplications/divisions and additions/subtractions. This number of operations can be notably reduced if the original matrix is transformed into a triangular matrix. In such a case, the determinant can be simply calculated as the product of the elements on the diagonal. An example with a generic matrix of size 4 can demonstrate effectively why this is the case with, for example, upper triangular matrices. The matrix is

$$A = \begin{pmatrix} a_{11} & a_{12} & a_{13} & a_{14} \\ 0 & a_{22} & a_{23} & a_{24} \\ 0 & 0 & a_{33} & a_{34} \\ 0 & 0 & 0 & a_{44} \end{pmatrix}$$

and its determinant can be for example computed using the expansion by minors via the first column:

$$|A| = a_{11} \begin{vmatrix} a_{22} & a_{23} & a_{24} \\ 0 & a_{33} & a_{34} \\ 0 & 0 & a_{44} \end{vmatrix}$$

Then we can keep expanding according to the first column of each new minor, because the first element of such minors is the only one to be different from zero:

$$|A| = a_{11}a_{22} \begin{vmatrix} a_{33} & a_{34} \\ 0 & a_{44} \end{vmatrix} = a_{11}a_{22}a_{33}a_{44}$$

As said earlier, it turns out that the determinant is equivalent to the product of the elements on the diagonal of the triangular matrix.

The onus is essentially on transforming the original matrix into an upper triangular matrix using, for example, Gaussian elimination. As this can involve a permutation of the matrix's rows, the permutation's parity will have to be considered when computing the determinant. The product of the elements on the diagonal will have to be multiplied by -1 if the permutation has odd parity.

4.10 Relevant R code. Functions `condet` and `oddity`

This function essentially carries out the same row operations of Gaussian elimination in order to transform the input matrix into an upper triangular matrix. As previously explained, the determinant can at this point be calculated as the product of the elements along the diagonal. The only caveat is to keep track of all row permutations because the final result will be multiplied by $+1$ or -1 depending on whether the parity of the permutation is even or odd. This step is carried out by the function `oddity`. The demonstration code pasted here is compared with the default R code (function `det`).

```
# Load library "comphy" in working space
library(comphy)

# The determinant of an identity matrix is 1
A <- diag(20)
print(condet(A))

## [1] 1

print(det(A)) # Comparison with built in function

## [1] 1

# Matrix with random elements
set.seed(9887)
A <- matrix(rnorm(100),ncol=10)
print(condet(A))

## [1] -349.462

print(det(A)) # Comparison with built in function
```

```
## [1] -349.462
```

```
# Large size matrices to compare execution time
# (small numbers to avoid large values of the determinant)
set.seed(4781)
A <- matrix(sample(c(0,0.35),size=10000,replace=TRUE),ncol=100)
st <- Sys.time()
D1 <- condet(A)
et <- Sys.time()
print(paste("Time for condet determinant: ",et-st))
```

```
## [1] "Time for condet determinant:  0.0260000228881836"
```

```
st <- Sys.time()
D2 <- det(A)
et <- Sys.time()
print(paste("Time for det determinant: ",et-st))
```

```
## [1] "Time for det determinant:  0.0019989013671875"
```

```
print(D1)
```

```
## [1] -2415.983
```

```
print(D2)
```

```
## [1] -2415.983
```

The execution time is much reduced (10 times) for the built-in function `det` because this makes use of a faster matrix decomposition. An example of the workings of `oddity` is shown below. Given, for instance, the vector containing the first 10 integers without any permutation, the parity is $+1$. But if we swap, say, 2 with 8, there is one permutation and the parity is -1.

```
# Original set of integers from 1 to 10
v <- 1:10

# No permutation. Parity is +1
print(oddity(v))

## [1] 1

# One swap (2 with 8). Parity is -1
v[2] <- 8
v[8] <- 2
print(v)

##  [1]  1  8  3  4  5  6  7  2  9 10

print(oddity(v))

## [1] -1
```

4.11 Built-in functions for matrix operations

Up to this point we have made an effort to build our own functions to handle matrix calculations, in order to highlight the numerical problems involved with the specific algorithms. However, the wide variety of calculations used in linear algebra forms a well-established research field in which several and varied functions have been created. R includes built-in functions that deal with matrix operations and linear algebra. Furthermore, there are additional packages dealing with specific classes of matrices, like the sparse matrices, and one should always try and search for the appropriate function and package for the given application.

In the rest of this chapter we will explore additional topics related to matrices and linear algebra, this time using the functions built in the base platform and/or those available in other packages.

4.12 Matrix inverse with `solve`

The R built-in function to solve the system of linear equations,

$$A\mathbf{x} = \mathbf{b}$$

is `solve`. Its input consists of the A matrix and the $\mathbf{b}$ coefficients, separately, rather than the augmented matrix that we have used in `gauss_elim`. The function works using LU decomposition. `solve` is also used to calculate a matrix inverse. This happens if the **b** input is omitted, as the function then interprets it as the identity matrix, and therefore, the **x** solution must necessarily be the inverse, A^{-1}. In the following lines, the same example of section 4.7 is used to demonstrate `solve`.

```
# Load library "comphy" in working space
library(comphy)

M = matrix(c( 1,-8, 0, 0,-15,
2,-2,-7, 0,-23,
0, 7, 3,-6, -1,
0, 0, 8,-7, -4),ncol=5,byrow=TRUE)

# Solution for a tridiagonal system
x1 <- solve_tridiag(M)

# The same solution with generic Gaussian elimination
x2 <- gauss_elim(M)

# Solution using solve (A and b separate)
x3 <- solve(M[,1:4],M[,5])

# The three solutions are the same
print(x1)

## [1] 1 2 3 4

print(x2)

## [1] 1 2 3 4

print(x3)
```

```
## [1] 1 2 3 4

# Inverse of a simple (diagonal) matrix
A <- diag(2,nrow=4,ncol=4)
print(A)

##      [,1] [,2] [,3] [,4]
## [1,]    2    0    0    0
## [2,]    0    2    0    0
## [3,]    0    0    2    0
## [4,]    0    0    0    2

Ainv <- solve(A)
print(Ainv)

##      [,1] [,2] [,3] [,4]
## [1,]  0.5  0.0  0.0  0.0
## [2,]  0.0  0.5  0.0  0.0
## [3,]  0.0  0.0  0.5  0.0
## [4,]  0.0  0.0  0.0  0.5

# More complicated matrix
set.seed(7753)
A <- matrix(sample(-5:5,size=64,replace=TRUE),ncol=8)
Ainv <- solve(A)

# Product should give the identity matrix, where its diagonal
# elements are all ones and off-diagonal are zero
Id <- A %*% Ainv
 print(diag(Id))

 ## [1] 1 1 1 1 1 1 1 1

 OffDiag <- (A %*% Ainv) - Id
  print(sum(abs(OffDiag)))

  ## [1] 0
```

4.13 Cholesky decomposition with `chol`

The *Cholesky decomposition* applies to symmetric and positive-definite matrices. If A is a symmetric and positive-definite matrix (see appendix B), it can be decomposed as

$$A = L\ L^T, \tag{4.12}$$

where L is a lower triangular matrix. L^T is an upper triangular matrix and therefore the Cholesky decomposition is a special case of the LU decomposition. The Cholesky decomposition applies also to the case of hermitian matrices, where L^T is replaced by the transpose conjugate (*adjoint*), $L^\dagger$:

$$A = L\ L^\dagger, \tag{4.13}$$

Since one requires A being symmetric and positive-definite, the algorithm for decomposition is faster than LU decomposition of a generic matrix. The primary choice for a function to perform the Cholesky decomposition in R is `chol`. Unfortunately, `chol` does not work at present with hermitian matrices. It is also important to know that the function does not check for symmetry; this means that it is possible to provide a non-symmetric matrix as input and the function could still return some output, not necessarily possessing any meaning. A last comment concerns the matrix returned by `chol`. Rather than a lower triangular matrix, it returns an upper triangular one. Therefore `chol` produces L^T rather than L. These are a couple of examples.

```
# Load library "comphy" in working space
library(comphy)

A <- matrix(c( 1, 1, 1, 1,
1, 2, 3, 4,
1, 3, 6,10,
1, 4,10,20),ncol=4,byrow=TRUE)

# Cholesky decomposition
L <- chol(A) # L is here an upper, rather than lower triangular
print(t(L))
```

```
##      [,1] [,2] [,3] [,4]
## [1,]    1    0    0    0
## [2,]    1    1    0    0
## [3,]    1    2    1    0
## [4,]    1    3    3    1
```

```
# The product L^T L should yield A
M <- t(L) %*% L
print(M)

##      [,1] [,2] [,3] [,4]
## [1,]    1    1    1    1
## [2,]    1    2    3    4
## [3,]    1    3    6   10
## [4,]    1    4   10   20

# Presently chol does not work for hermitian matrices
el11 <- 2
el12 <- 1i
el21 <- -1i
el22 <- 2
A <- matrix(c(el11,el21,el12,el22),ncol=2)
print(A)

##      [,1] [,2]
## [1,] 2+0i 0+1i
## [2,] 0-1i 2+0i

L <- chol(A)

## Error in chol.default(A): complex matrices not permitted at present
```

4.14 QR decomposition with `qr`

Consider a real $m \times n$ matrix A with $n \leqslant m$ and rank $k \leqslant n$. Its *QR decomposition* is simply

$$A = Q\ R \tag{4.14}$$

In this expression, Q is an $m \times n$ orthogonal matrix (for which $Q^T\ Q = I$) in which the n columns are n orthonormal vectors and R is an $n \times n$ upper triangular matrix. If $k < n$, the last $n - k$ rows of R will be filled with zeros. It is also possible for A to have $m \leqslant n$ and for k to be less than or equal to m. In this case, Q will be an $m \times m$ orthogonal and square matrix, while R will be a $m \times n$ upper triangular matrix. In this case too, if $k < m$, the last $m - k$ rows of R will be filled with zeros.

The QR decomposition is an important part of several algorithms involved with the calculation of eigenvalues and eigenvectors and the solution of systems of linear equations.

The main function in R responsible for the QR decomposition is `qr`. Its input is simply a matrix A. The output is an object of class *qr*, which is a class created *ad hoc* to extract specific information using certain functions like `qr.Q`, `qr.R`, etc. More information can be found browsing the available documentation. Here we will look at the practical use of this function using two matrices A, one with $m > n$ and the other with $m < n$.

```
# Load library "comphy" in working space
library(comphy)

# Case where m=4 > n=3
A <- matrix(c( 1, 0, 1,
2, 2,-1,
-1, 2, 2,
0, 1,-2),ncol=3,byrow=TRUE)

# qr creates an object of class "qr"
QR <- qr(A)
print(class(QR))

## [1] "qr"

print(names(QR)) # Names of the elements of object QR

## [1] "qr"    "rank"  "qraux" "pivot"

# Matrix A has rank 3. Expect no rows of R to be null
print(QR$rank)

## [1] 3

# Extract matrices Q and R using the appropriate functions
Q <- qr.Q(QR)
print(Q)
```

```
##            [,1]       [,2]        [,3]
## [1,] -0.4082483  0.1154701 -0.53198417
## [2,] -0.8164966 -0.4618802  0.05527108
## [3,]  0.4082483 -0.8082904 -0.42144201
## [4,]  0.0000000 -0.3464102  0.73234185
```

```
R <- qr.R(QR)
print(R) # No rows are null
```

```
##          [,1]       [,2]       [,3]
## [1,] -2.44949 -0.8164966  1.2247449
## [2,]  0.00000 -2.8867513 -0.3464102
## [3,]  0.00000  0.0000000 -2.8948230
```

```
# The column vectors of Q are orthonormal
for (i in 1:3) {
for (j in 1:3) {
print(sum(Q[,i]*Q[,j])) # This is the inner product
}
}
```

```
## [1] 1
## [1] 4.163336e-17
## [1] 1.110223e-16
## [1] 4.163336e-17
## [1] 1
## [1] -4.510281e-17
## [1] 1.110223e-16
## [1] -4.510281e-17
## [1] 1
```

```
# Case where m=3 < n=4
A <- matrix(c(1,0, 1,
2,2,-1,
2,2,-1,
2,0, 2),ncol=4)
QR <- qr(A)
print(QR$rank) # This time the rank is 2. Expect null rows in R
```

```
## [1] 2
```

```
Q <- qr.Q(QR)
print(Q)
```

```
##             [,1]       [,2]       [,3]
## [1,] -0.7071068 -0.5144958 -0.4850713
## [2,]  0.0000000 -0.6859943  0.7276069
## [3,] -0.7071068  0.5144958  0.4850713
```

```
R <- qr.R(QR)
print(R) # Last row is a null row
```

```
##           [,1]       [,2]          [,3]          [,4]
## [1,] -1.414214 -0.7071068 -7.071068e-01 -2.828427e+00
## [2,]  0.000000 -2.9154759 -2.915476e+00  3.231224e-16
## [3,]  0.000000  0.0000000 -8.881784e-16  3.046427e-16
```

```
# But Q still contains orthonormal column vectors
for (i in 1:3) {
for (j in 1:3) {
print(sum(Q[,i]*Q[,j])) # This is the inner product
}
}
```

```
## [1] 1
## [1] -1.110223e-16
## [1] 0
## [1] -1.110223e-16
## [1] 1
## [1] 1.110223e-16
## [1] 0
## [1] 1.110223e-16
## [1] 1
```

4.14.1 Eigenvalues using QR decomposition

Many algorithms available to find the eigenvalues and eigenvectors of a square matrix make use of the QR decomposition within the following procedure. Consider a square matrix, A of size n. This is similar to another square matrix, B of size n, if there exists an invertible matrix S for which,

$$A = S^{-1}B\ S$$

(see appendix B, equation (B.12)). A and B have the same set of eigenvalues. When the matrix S is an orthogonal matrix, Q, then the above similarity transformation becomes,

$$A = Q^T BQ \tag{4.15}$$

Consider now a matrix and its QR decomposition,

$$A = Q_0 \; R_0 \tag{4.16}$$

Using the components found, let us compute a different matrix,

$$A_1 = R_0 \; Q_0 \tag{4.17}$$

Equation (4.17) can be multiplied on the right by matrix Q_0^T and on the left by matrix Q_0. The result is (using the orthogonality of Q_0),

$$Q_0 \; A_1 \; Q_0^T = Q_0 \; R_0$$

or, using equation (4.16),

$$A = Q_0 \; A_1 \; Q_0^T \tag{4.18}$$

The above relation essentially means that A and A_1 are similar (looking at definition (4.15), consider that in the relation above $Q_0 = Q^T$), i.e. they have the same eigenvalues.

A_1 is a new matrix that just happens to be formed out of the components of the QR decomposition of A. This matrix has, in turn, its own QR decomposition with new matrices Q_1, R_1:

$$A_1 = Q_1 \; R_1$$

Using these new component matrices Q_1 and R_1, a new matrix, $A_2 = R_2 \; Q_2$, can be constructed. Matrix A_2 is similar to matrix A_1 (see demonstration on the similarity between A and A_1; this time is $A_1 = Q_1 \; A_2 \; Q_1^T$). But we can also show that A_2 is similar to A. Indeed:

$$A = Q_0 \; A_1 \; Q_0^T \Rightarrow A = Q_0 \; Q_1 \; A_2 \; Q_1^T \; Q_0^T \Rightarrow \quad A = (Q_0 \; Q_1) A_2 (Q_0 \; Q_1)^T$$

So, A_2 has the same eigenvalues as A. But A_2 is computed as the matrix product of an upper triangular matrix with an orthogonal matrix. It is intuitive to realise that this multiplication tends to make A_2 closer to an upper triangular matrix because of the positioning of the zeros. In fact, John Francis, who introduced this method in 1961, proved that the succession of matrices,

$$A_k = R_k \; Q_k, \quad k = 0, 1, \ldots \tag{4.19}$$

tends to an upper triangular matrix. The elements along the diagonal of this limit upper triangular matrix are the eigenvalues of A [2, 3].

It is not too difficult to write a simple program that loops through several iterations including QR decompositions and stops when the elements in the lower triangular part of the matrix obtained fall below a given small threshold (see exercises in section 4.17). When this is the case, the elements along the diagonal are the chosen approximation to the original matrix's eigenvalues. But a program structured in this way will sometimes meet with situations in which the algorithm does not converge, i.e. when the elements in the lower triangular part of the matrix do not get close enough to zero (see exercises in section 4.17). In fact, several improvements both in terms of convergence and speed have been investigated and implemented after Francis' major breakthrough. Modern eigenvalues and eigenvectors search accommodates such changes and are part of the main R functions available for the search. We will not spend time on those algorithms' refinements; the interested and specialist reader can read the light but comprehensive detailed overview by Arbenz [4].

4.15 Eigenvalues and eigenvectors with `eigen`

The main function for the calculation of the eigenvalues and eigenvectors of a matrix in R is `eigen`. It is based on robust, accurate and well established code in the LAPACK library. The function is very straightforward and easy to use. The only input is a square matrix. The output consists of an object of class `eigen`, whose two slots are named `values` and `vectors`. They contain eigenvalues and eigenvectors, respectively. An important feature related to the eigenvectors is that they are defined up to a constant and a comparison between different procedures to calculate eigenvectors of the same matrix will have to take this into account.

```
# Load library "comphy" in working space
library(comphy)

# Small square matrix
set.seed(1754) # Just for reproducibility purpose
A <- matrix(sample(-3:3,size=16,replace=TRUE),ncol=4)
print(A)
```

```
##      [,1] [,2] [,3] [,4]
## [1,]   -3    0    1    2
## [2,]   -1   -1    2    1
## [3,]    1   -3   -2   -3
## [4,]   -2    1    3    0
```

```
# Object of class "eigen"
```

```
lEigen <- eigen(A)
print(class(lEigen))

## [1] "eigen"

# Names of the elements of object QR
print(names(lEigen))

## [1] "values"  "vectors"

# Eigenvalues can be complex numbers
print(lEigen$values)

## [1] -1.115402+3.814215i -1.115402-3.814215i -1.884598+0.497590i
## [4] -1.884598-0.497590i

# First two eigenvectors
print(lEigen$vectors[,1:2])

##                       [,1]                  [,2]
## [1,]  0.1976045+0.1911844i  0.1976045-0.1911844i
## [2,]  0.1067707+0.3589626i  0.1067707-0.3589626i
## [3,] -0.6703138+0.0000000i -0.6703138+0.0000000i
## [4,]  0.1567503+0.5570059i  0.1567503-0.5570059i

# The first Pauli matrix, sigma1, is Hermitian
sigma1 = matrix(c(0,1,1,0),ncol=2)
print(sigma1)

##      [,1] [,2]
## [1,]    0    1
## [2,]    1    0

print(t(sigma1))
```

```
##      [,1] [,2]
## [1,]    0    1
## [2,]    1    0
```

```
# Eigenvalues and eigenvectors of the first Pauli matrix.
# Eigenvalues are real and the eigenvectors orthonormal
lEigen <- eigen(sigma1)
print(lEigen$values)
```

```
## [1]  1 -1
```

```
V <- lEigen$vectors
print(V)
```

```
##           [,1]       [,2]
## [1,] 0.7071068 -0.7071068
## [2,] 0.7071068  0.7071068
```

```
print(t(V) %*% V) # Eigenvectors are orthonormal
```

```
##      [,1] [,2]
## [1,]    1    0
## [2,]    0    1
```

4.16 The singular value decomposition with svd

The *singular value decomposition* of a matrix M (not necessarily square) is a very important algorithmic operation as it permeates a large variety of procedures in the whole of science. It can be considered as a generalisation of *matrix diagonalisation*, where an $n \times n$ square matrix A is decomposed as follows,

$$A = V^{-1}DV, \tag{4.20}$$

where V, D are also $n \times n$ square matrices, and where D is a diagonal matrix with the eigenvalues of A along the diagonal (see also equation (B.12)).

An $m \times n$ real or complex matrix M can always decomposed as follows:

$$M = U\Sigma V^{\dagger} \tag{4.21}$$

In expression (4.21), Σ is a $m \times n$ diagonal and rectangular matrix that looks like, if $m > n$,

$$\Sigma = \begin{pmatrix} \sigma_1 & 0 & 0 & \cdots & 0 & 0 \\ 0 & \sigma_2 & 0 & \cdots & 0 & 0 \\ \cdots & \cdots & \cdots & \cdots & \cdots & \cdots \\ 0 & 0 & 0 & \cdots & 0 & \sigma_n \\ \cdots & \cdots & \cdots & \cdots & 0 & \cdots \\ 0 & 0 & 0 & \cdots & 0 & 0 \end{pmatrix}$$

Matrix U is an $m \times m$ unitary matrix while V is an $n \times n$ unitary matrix. Also, $V^\dagger$ is the adjoint of V (see appendix B.5.1). The elements along the diagonal of Σ are non negative and are called the *singular values* of M, while the expression (4.21) is known as *singular value decomposition* (SVD in short). Formula (4.21) is also called *normal SVD*, to distinguish it from the compact SVD. Indeed, SVD is not defined in a unique way, and different numeric realisations of the three matrices are possible, depending on the order with which the singular values appear along the diagonal and on the scaling adopted for the columns of U and V. It is also possible to formulate the same matrix as a *compact singular value decomposition* if Σ is an $r \times r$ diagonal matrix, where r is the rank of M (it must then be $r \leqslant \min\{m, n\}$). In this case, U is an $m \times r$ matrix, V an $n \times r$ matrix and both satisfy

$$U^\dagger U = I, \qquad V^\dagger V = I, \tag{4.22}$$

with I the identity matrix. Relations (4.22) also apply for the non-compact case as U and V are unitary matrices.

In R the singular value decomposition is achieved using the function `svd`. Input for the function `svd` consists of the $m \times n$ matrix to be decomposed, and of the number of columns expected for U and for V. The last two parameters can be chosen appropriately so to trigger a normal SVD or a compact SVD. The following code demonstrates this feature using known starting component matrices, `U,D,V`.

```
# Load library "comphy" in working space
library(comphy)

# Small rectangular (4 X 3) matrix
# Built starting from a 4 X 4 (U) unitary matrix, a 4 X 3 (D)
# diagonal matrix and a 3 X 3 (V) unitary matrix
U <- matrix(c(0,1,1,1,1,0,-1,1,1,1,0,-1,1,-1,1,0),ncol=4)/sqrt(3)
print(U)
```

```
##            [,1]       [,2]       [,3]       [,4]
## [1,] 0.0000000  0.5773503  0.5773503  0.5773503
## [2,] 0.5773503  0.0000000  0.5773503 -0.5773503
## [3,] 0.5773503 -0.5773503  0.0000000  0.5773503
```

```
## [4,] 0.5773503  0.5773503 -0.5773503  0.0000000
```

```
D <- diag(c(3,2,1),nrow=4,ncol=3)
print(D)
```

```
##      [,1] [,2] [,3]
## [1,]    3    0    0
## [2,]    0    2    0
## [3,]    0    0    1
## [4,]    0    0    0
```

```
V <- matrix(c(0,-1,0,1,0,0,0,0,1),ncol=3)
print(V)
```

```
##      [,1] [,2] [,3]
## [1,]    0    1    0
## [2,]   -1    0    0
## [3,]    0    0    1
```

```
M <- U %*% D %*% t(V)
print(M)
```

```
##           [,1]      [,2]       [,3]
## [1,]  1.154701  0.000000  0.5773503
## [2,]  0.000000 -1.732051  0.5773503
## [3,] -1.154701 -1.732051  0.0000000
## [4,]  1.154701 -1.732051 -0.5773503
```

```
###
### normal SVD
###
# U has 4 columns and V has 3 columns
lSVD <- svd(M,nu=4,nv=3)
class(lSVD) # The returned object is a named list
```

```
## [1] "list"
```

```
names(lSVD)

## [1] "d" "u" "v"

# U and V unitary matrix. Not quite the same signs (due to
# the scaling arbitrariety of the algorithm)
print(lSVD$u)

##               [,1]       [,2]          [,3]          [,4]
## [1,] -8.756053e-17 -0.5773503 -5.773503e-01 -5.773503e-01
## [2,] -5.773503e-01  0.0000000 -5.773503e-01  5.773503e-01
## [3,] -5.773503e-01  0.5773503  1.110223e-16 -5.773503e-01
## [4,] -5.773503e-01 -0.5773503  5.773503e-01 -1.942890e-16

print(lSVD$u %*% t(lSVD$u))

##               [,1]          [,2]          [,3]          [,4]
## [1,]  1.000000e+00  1.665335e-16 -2.220446e-16 -5.436063e-17
## [2,]  1.665335e-16  1.000000e+00  0.000000e+00 -1.676840e-16
## [3,] -2.220446e-16  0.000000e+00  1.000000e+00 -1.567953e-16
## [4,] -5.436063e-17 -1.676840e-16 -1.567953e-16  1.000000e+00

print(lSVD$v)

##      [,1] [,2] [,3]
## [1,]    0   -1    0
## [2,]    1    0    0
## [3,]    0    0   -1

print(lSVD$v %*% t(lSVD$v))

##      [,1] [,2] [,3]
## [1,]    1    0    0
## [2,]    0    1    0
## [3,]    0    0    1
```

```
# But the product returns M
print(lSVD$u %*% diag(lSVD$d,nrow=4,ncol=3) %*% t(lSVD$v))

##          [,1]          [,2]          [,3]
## [1,]  1.154701 -2.626816e-16  5.773503e-01
## [2,]  0.000000 -1.732051e+00  5.773503e-01
## [3,] -1.154701 -1.732051e+00 -1.110223e-16
## [4,]  1.154701 -1.732051e+00 -5.773503e-01

###
### compact SVD
###
# U has 3 columns and V has 3 columns
lSVD <- svd(M,nu=3,nv=3)

# This time U is a 4 X 3 matrix, D is a 3 X 3 matrix
# and V is a 3 X 3 matrix
print(lSVD$u)

##               [,1]       [,2]          [,3]
## [1,] -8.756053e-17 -0.5773503 -5.773503e-01
## [2,] -5.773503e-01  0.0000000 -5.773503e-01
## [3,] -5.773503e-01  0.5773503  1.110223e-16
## [4,] -5.773503e-01 -0.5773503  5.773503e-01

print(lSVD$d)

## [1] 3 2 1

print(lSVD$v)

##      [,1] [,2] [,3]
## [1,]    0   -1    0
## [2,]    1    0    0
## [3,]    0    0   -1

# But the product still returns M
print(lSVD$u %*% diag(lSVD$d,nrow=3,ncol=3) %*% t(lSVD$v))
```

```
##              [,1]          [,2]          [,3]
## [1,]  1.154701 -2.626816e-16  5.773503e-01
## [2,]  0.000000 -1.732051e+00  5.773503e-01
## [3,] -1.154701 -1.732051e+00 -1.110223e-16
## [4,]  1.154701 -1.732051e+00 -5.773503e-01
```

The possibility offered by `svd` to achieve decomposition using a number of columns for U and V smaller than the full number is connected to the intrinsic property of the singular value decomposition to approximate, rather than fully reproduce, the original numeric values of the matrix. Let us consider, for instance, the expression for the normal SVD and assume that $m > n$. Then the matrix can be seen as an expansion in terms of *elemental* $m \times n$ matrices, $\mathbf{u}_i\mathbf{v}_j^T$, between the n columns of U and V:

$$M = (\mathbf{u}_1 \cdots \mathbf{u}_n)\begin{pmatrix} \sigma_1 & \cdots & 0 \\ \cdots & \cdots & \cdots \\ 0 & \cdots & \sigma_n \\ \cdots & \cdots & \cdots \\ 0 & \cdots & 0 \end{pmatrix}\begin{pmatrix} \mathbf{v}_1^T \\ \cdots \\ \mathbf{v}_n^T \end{pmatrix}$$

or,

$$M = \sigma_1\mathbf{u}_1\mathbf{v}_1^T + \sigma_2\mathbf{u}_2\mathbf{v}_2^T + \cdots + \sigma_n\mathbf{u}_n\mathbf{v}_n^T \tag{4.23}$$

Each elemental matrix, $\mathbf{u}_i\mathbf{v}_j^T$, has components resulting from products of factors smaller or equal than 1, because both $\mathbf{u}_i$ and $\mathbf{v}_j$ are unit-length vectors. Therefore, the strength of each contribution to M in the expansion (4.23) is fundamentally given by each singular value, σ_i. Given that all the singular values are ranked from larger to smaller in the standard SVD algorithm (i.e. $\sigma_1 \geqslant \sigma_2 \geqslant \cdots \geqslant \sigma_n$), the contributions corresponding to higher values of the index are less significant than those with lower index. The consequence of such an ordering is that very little is lost in the original matrix M as this is approximated with less elemental contribution from the expansion (4.23), provided that the truncation starts from index n and proceeds with decreasing values of the index. A simple illustration of such an approximation is provided in the following code, in which the 4×3 M matrix just introduced is decomposed with a 4×2 U matrix and a 3×2 V matrix. The *reconstructed* M matrix is only an approximation of the original matrix. Such an approximation, in this specific case, is not particularly good as the discarded singular value, $\sigma_3 = 1$, is not negligible if compared to the other singular values. But there are many situations in which the original matrix (for example an image) has a large size and the elimination of the contribution from the smallest singular values yields an approximating matrix which is not so different from the original one.

```
# Re-calculate U and V using the normal SVD
lSVDfull <- svd(M,nu=4,nv=3)

# Now eliminate some 'elemental' contributions
lSVDpart <- svd(M,nu=2,nv=2)
```

```
# Compare U's and V's
# The result is that some of the columns are missing
print(lSVDfull$u)

##               [,1]       [,2]          [,3]          [,4]
## [1,] -8.756053e-17 -0.5773503 -5.773503e-01 -5.773503e-01
## [2,] -5.773503e-01  0.0000000 -5.773503e-01  5.773503e-01
## [3,] -5.773503e-01  0.5773503  1.110223e-16 -5.773503e-01
## [4,] -5.773503e-01 -0.5773503  5.773503e-01 -1.942890e-16

print(lSVDpart$u)

##               [,1]       [,2]
## [1,] -8.756053e-17 -0.5773503
## [2,] -5.773503e-01  0.0000000
## [3,] -5.773503e-01  0.5773503
## [4,] -5.773503e-01 -0.5773503

print(lSVDfull$v)

##      [,1] [,2] [,3]
## [1,]    0   -1    0
## [2,]    1    0    0
## [3,]    0    0   -1

print(lSVDpart$v)

##      [,1] [,2]
## [1,]    0   -1
## [2,]    1    0
## [3,]    0    0

# The resulting (reconstructed) matrix is an approximation
# of the original matrix

# Original matrix
print(M)
```

```
##            [,1]      [,2]       [,3]
## [1,]  1.154701  0.000000  0.5773503
## [2,]  0.000000 -1.732051  0.5773503
## [3,] -1.154701 -1.732051  0.0000000
## [4,]  1.154701 -1.732051 -0.5773503
```

```
# Approximation
appU <- lSVDpart$u
appD <- diag(lSVDpart$d,nrow=2,ncol=2)
appV <- lSVDpart$v
print(appU %*% appD %*% t(appV))
```

```
##            [,1]          [,2] [,3]
## [1,]  1.154701 -2.626816e-16    0
## [2,]  0.000000 -1.732051e+00    0
## [3,] -1.154701 -1.732051e+00    0
## [4,]  1.154701 -1.732051e+00    0
```

To conclude this section, it is worth mentioning that the singular value decomposition of a matrix M is closely connected to the calculation of eigenvalues and eigenvectors of the matrices $MM^\dagger$ and $M^\dagger M$. For a demonstration, see appendix A, section A.5.

4.17 Exercises on matrix decompositions

Exercise 06

One way to create arbitrary symmetric matrices is to add a generic matrix to its transpose. Prove that the resulting matrix is, indeed, symmetric and create a function called `symmat` that take the matrix size, n, a set of numbers (not necessarily n) as input and returns a symmetric $n \times n$ matrix, with elements derived from the input numbers.

Exercise 07

Apply the Cholesky decomposition to the following symmetric matrix:

$$A = \begin{pmatrix} 2 & 1 & 0 \\ 1 & 2 & 0 \\ 0 & 1 & 1 \end{pmatrix}$$

Calculate the determinant of A using the result of the decomposition and verify that the result is correct with both `condet` and `det`.

Exercise 08

Apply the Cholesky decomposition to the following symmetric matrix:

$$A = \begin{pmatrix} 2 & 1 & -3 \\ 1 & 2 & 0 \\ 0 & 1 & 1 \end{pmatrix}$$

Does the `chol` function return a numeric result? Why?

Exercise09

The QR decomposition is at the foundation of most methods to find the eigenvalues of a matrix. The algorithm is a cycle over the following steps, provided that the matrix whose eigenvalues have to be found is the square matrix of size n, A:

1. Find the QR decomposition of A. This is cycle 1 of the algorithm, and A is called A_0. The QR decomposition yields,

$$A_0 = Q_0 R_0$$

2. Build the new matrix, A_1, for cycle 1,

$$A_1 = R_0 Q_0$$

3. Check that all the elements in the lower triangular part of A_1 are close to 0 (practically below a fixed threshold `zero_cut`). If they are all close to zero the algorithm has terminated successfully and the eigenvalues of A are the elements on the diagonal of A_1. Otherwise, go back to step 1, where in cycle 2 A_1 replaces A_0. In general, in cycle i, A_{i-1} replaces A_{i-2}.
4. If, after a pre-established number of cycles, `nmax`, not all elements of the lower triangular part of A_i, at cycle i, fall below the threshold `zero_cut`, the algorithm has not achieved convergence and it is not possible to find the eigenvalues.

Write a function, called `eigenQR`, that implements the steps above to find the eigenvalues of an input matrix, A. Besides the matrix, the function takes in the threshold, `zero_cut`, basically a small number (default is `1e-6`) and the maximum number of cycles, `nmax` (default is 1000). Apply the function to find the eigenvalues of,

$$A = \begin{pmatrix} -2 & -4 & 2 \\ -2 & 1 & 2 \\ 4 & 2 & 5 \end{pmatrix}$$

Exercise 10

What would be the result of applying `eigenQR` to the following matrix,

$$A = \begin{pmatrix} 0 & 1 \\ 1 & 0 \end{pmatrix}?$$

Try to understand what the issue is in this case, by tracing it back, through inspection of the function, to what is causing it. Verify that the eigenvalues are found with the default R function, `eigen`.

Exercise 11

The requirement for the algorithm of exercise 09 did not include the eigenvectors as part of the output. In fact, obtaining the eigenvectors in general implies using additional algorithmical steps that use matrices in special forms Rather than considering the general case, we will study the special case in which the starting matrix is symmetric. As explained in the theory of the QR algorithm, the similarity transformation between A and the A_k matrix obtained at cycle k of the algorithm is of the form,

$$A = (Q_0\ Q_1\ \cdots\ Q_k)\ A_k\ (Q_0\ Q_1\ \cdots\ Q_k)^T$$

The quantities in parenthesis can be indicated simply with the letter Q, as they are single matrices. The expression of the algorithm after k iteration is thus

$$A = QA_kQ^T$$

In general, A_k is an upper triangular matrix and the columns of Q are not the eigenvectors of A. The only thing that can be stated with accuracy is that the elements on the diagonal of A_k are good approximations of A's eigenvalues. If A is symmetric, though, A_k is a diagonal matrix and the columns of Q are the orthonormal eigenvectors of A. To see this let's consider that for a symmetric matrix $A^T = A$. Therefore, taking the transpose of the expression above we get,

$$A^T = QA_k^TQ^T = A = QA_kQ^T \quad \Rightarrow \quad A_k^T = A_k$$

But A_k is an upper triangular matrix and this can be at the same time symmetric only if the off-diagonal elements are all zero. In conclusion, A_k is a diagonal matrix containing the eigenvalues of A. And, accordingly, the columns of Q will be its ordered eigenvectors.

Modify the algorithm created for exercise 09 so that the matrix of eigenvectors forms part of its output, and apply it to the symmetric matrix, $B = A + A^T$, where A was introduced in that exercise. Find an effective way to show that the eigenvectors found do correspond to the eigenvalues obtained.

Exercise 12

A square matrix A can be *diagonalised* if a diagonal matrix D and another, invertible, square matrix P can be found such that the following relation holds:

$$A = PDP^{-1} \tag{4.24}$$

When this happens, D has the eigenvalues of A along its diagonal, while the columns of P are, in order, the corresponding eigenvectors. This can be seen if the above equation is multiplied, on the left, by P,

$$AP = PD,$$

and if P is re-written in terms of its column vectors, $\mathbf{v}_i$,

$$A(\mathbf{v}_1 \cdots \mathbf{v}_n) = (\mathbf{v}_1 \cdots \mathbf{v}_n)\begin{pmatrix} \lambda_1 & \cdots & 0 \\ \cdots & \cdots & \cdots \\ 0 & \cdots & \lambda_n \end{pmatrix}$$

$$\Downarrow$$

$$(A\mathbf{v}_1 \cdots A\mathbf{v}_n) = (\lambda_1\mathbf{v}_1 \cdots \lambda_n\mathbf{v}_n)$$

Clearly, the expression just derived is a set of n eigenvalue equations for n eigenvectors of A.

Using the function `eigen`, write the diagonal matrix D corresponding to the 4×4 matrix A generated by sampling randomly the integers -1, 0, 1 (use seed 1188). Find also the expression of P and, using matrix multiplication in R, show that $AP = PD = DP$.

Exercise 13

Consider the full set of *Pauli matrices*,

$$\sigma_x = \begin{pmatrix} 0 & 1 \\ 1 & 0 \end{pmatrix}, \quad \sigma_y = \begin{pmatrix} 0 & -\mathrm{i} \\ \mathrm{i} & 0 \end{pmatrix}, \quad \sigma_z = \begin{pmatrix} 1 & 0 \\ 0 & -1 \end{pmatrix}$$

Using the function `eigen`, verify that the eigenvalues of each matrix are $+1$ and -1 and that the corresponding eigenvectors are,

$$\psi_{x+} = \frac{1}{\sqrt{2}}\begin{pmatrix} 1 \\ 1 \end{pmatrix}, \quad \psi_{x-} = \frac{1}{\sqrt{2}}\begin{pmatrix} 1 \\ -1 \end{pmatrix}$$

$$\psi_{y+} = \frac{1}{\sqrt{2}}\begin{pmatrix} 1 \\ \mathrm{i} \end{pmatrix}, \quad \psi_{y-} = \frac{1}{\sqrt{2}}\begin{pmatrix} 1 \\ -\mathrm{i} \end{pmatrix}$$

$$\psi_{z+} = \begin{pmatrix} 1 \\ 0 \end{pmatrix}, \quad \psi_{z-} = \begin{pmatrix} 0 \\ 1 \end{pmatrix}$$

Exercise 14

After having generated a 12×10 random matrix M with the function `sample` and starting from the set $\{-2, -1, 0, 1, 2\}$ and with seed 2673, find its singular values without using the function `svd`. Verified then that the values found are correct by using the function `svd`.

4.18 Iterative methods

The solution of a system of linear equations via Gauss elimination or similar techniques, is often called a *direct method*. This terminology is used in contraposition to techniques based on algorithmic iteration, appropriately called *iterative methods*. In this last section of the chapter, two important iterative methods will be introduced and explained, the *Jacobi method* and the *Gauss–Seidel method*. Iterative methods offer an alternative to direct methods because they do not involve the time and memory-intensive inversion of matrices which, especially when a matrix is large, can be a prohibitive cost to achieving the solution. For this reason, iterative methods are particularly useful to solve linear systems in which the matrix of coefficients is a large *sparse matrix*, i.e. a matrix which is comprised of mostly zeros.

4.18.1 The Jacobi method

The algorithm for the Jacobi method can be explained using matrix notation. The linear system of n equations in n unknowns is written using the $n \times n$ non-singular matrix A, the column vector of coefficients $\mathbf{x}$ and the column vector of known values, $\mathbf{b}$. This is,

$$A\mathbf{x} = \mathbf{b} \tag{4.25}$$

In order for the algorithm to work, i.e. for the iterations to achieve convergence to a finite result, A must be *diagonally dominant*. This means that each element on the diagonal must be, in absolute value, larger than the sum of the absolute value of all the remaining elements on each specific row. In formula, the condition for a matrix to be diagonally dominant is:

$$|a_{ii}| > \sum_{\substack{j=1 \\ j \neq i}}^{n} |a_{ij}|, \qquad i = 1,\ldots,n \tag{4.26}$$

Although the condition just introduced can seem particularly restrictive, in practice there are several applications in which the resulting system turns out to have a diagonally dominant matrix of coefficients.

The matrix A can be decomposed into a sum of three matrices, a lower triangular matrix, L, a diagonal matrix, D, and an upper triangular matrix, U:

$$A = L + D + U \tag{4.27}$$

When equation (4.27) is used in equation (4.25), we obtain:

$$(L + D + U)\mathbf{x} = \mathbf{b} \Rightarrow D\mathbf{x} = \mathbf{b} - (L + U)\mathbf{x} \Rightarrow \mathbf{x} = D^{-1}\mathbf{b} - D^{-1}(L + U)\mathbf{x}$$

Next, D is added and subtracted in the parenthesis:

$$A\mathbf{x} = \mathbf{b} \Rightarrow \mathbf{x} = D^{-1}\mathbf{b} - D^{-1}(L + D + U - D)\mathbf{x},$$

or, using $D^{-1}D = I$ and equation (4.27),

$$A\mathbf{x} = \mathbf{b} \Rightarrow \mathbf{x} = \mathbf{x} + D^{-1}(\mathbf{b} - A\mathbf{x}) \tag{4.28}$$

Equation (4.28) is a typical iterative expression. A new set of unknowns, $\mathbf{x}$, is computed starting from the set of old values, i.e. the values obtained during the previous cycle (still symbolised by $\mathbf{x}$ in the equation), and adding a given quantity that can be indicated as $\Delta\mathbf{x}$. If the new set of values for the unknowns is indicated as $\mathbf{x}^{(i+1)}$ (cycle $i+1$) and the old set is indicated as $\mathbf{x}^{(i)}$ (cycle i), formula (4.28) can be re-written as the *Jacobi algorithm*:

$$\mathbf{x}^{(i+1)} = \mathbf{x}^{(i)} + \Delta\mathbf{x}^{(i)}, \quad \Delta\mathbf{x}^{(i)} = D^{-1}(\mathbf{b} - A\mathbf{x}^{(i)}) \tag{4.29}$$

The method implies a starting set of values for the unknowns. It goes without saying that the closer these values are to the system's solution, the faster the algorithm will converge to the solution. But obviously the solution is not known initially, and generic values, for example a set of zeros, will be adopted to get the algorithm started.

The correction $\Delta\mathbf{x}^{(i)}$ is characterised by the inverse of D. This is, though, a very quick and easy inverse to calculate because D is a diagonal matrix. If d_{jj} are the elements on the diagonal of D, $1/d_{jj}$ are the elements on the diagonal of the diagonal matrix, D^{-1}. Notice that d_{jj} is guaranteed to be different from zero because A is diagonally dominant. An example will render the method more digestible. Consider the following system of three equations in three unknowns:

$$\begin{aligned} 3x_1 - x_2 + x_3 &= 4 \\ x_1 - 4x_2 + x_3 &= -4 \\ 2x_1 + 2x_2 + 7x_3 &= 27 \end{aligned}$$

The system is characterised by the following matrices, useful to determine the Jacobi algorithm:

$$A = \begin{pmatrix} 3 & -1 & 1 \\ 1 & -4 & 1 \\ 2 & 2 & 7 \end{pmatrix}, \mathbf{b} = \begin{pmatrix} 4 \\ -4 \\ 27 \end{pmatrix}, D = \begin{pmatrix} 3 & 0 & 0 \\ 0 & -4 & 0 \\ 0 & 0 & 7 \end{pmatrix}, D^{-1} = \begin{pmatrix} 1/3 & 0 & 0 \\ 0 & -1/4 & 0 \\ 0 & 0 & 1/7 \end{pmatrix}$$

The expression for the increment $\Delta\mathbf{x}^{(i)}$ in components is thus:

$$\begin{aligned} \Delta x_1^{(i)} &= (4 - 3x_1^{(i)} + x_2^{(i)} - x_3^{(i)})/3 \\ \Delta x_2^{(i)} &= (4 + x_1^{(i)} - 4x_2^{(i)} + x_3^{(i)})/4 \\ \Delta x_3^{(i)} &= (27 - 2x_1^{(i)} - 2x_2^{(i)} - 7x_3^{(i)})/7 \end{aligned}$$

If, for instance, at cycle 0 we start with the set,

$$x_1^{(0)} = 0, \quad x_2^{(0)} = 0, \quad x_3^{(0)} = 0,$$

at cycle 1 the updated values will be,

$$x_1^{(1)} = 4/3, \quad x_2^{(1)} = 1, \quad x_3^{(1)} = 27/7$$

At cycle 2, still using the same components of $\Delta\mathbf{x}^{(i)}$, the values become:

$$x_1^{(2)} = 8/21 \approx 0.38, \quad x_2^{(2)} = 193/84 \approx 2.30, \quad x_3^{(2)} = 67/21 \approx 3.19$$

At cycle 8 the three values are,

$$x_1^{(8)} = 1.00, \quad x_2^{(8)} = 2.00, \quad x_3^{(8)} = 3.00$$

which, up to two decimals, coincide with the correct solution of the system.

The incremental quantity $\Delta\mathbf{x}^{(i)}$ in equation (4.29) should become smaller and smaller with increasing cycles, if the algorithm leads to convergence. The quantity can thus be adopted as an indication of how close the numerical approximation is to the final solution. In general one would typically use a relative increase as such a measure. Considering the n components,

$$\Delta x_1^{(i)},\ldots,\Delta x_n^{(i)}$$

of $\Delta\mathbf{x}^{(i)}$, the following quantity,

$$\epsilon \equiv \max_{1\leqslant j\leqslant n}\left\{\frac{|\Delta x_j^{(i)}|}{|x_j^{(i)}|}\right\}$$

could be adopted to stop the iteration. But one of the components of the approximate solution at cycle i could be zero, and the corresponding ratio used to compute a relative increase would become infinity. In both functions `PJacobi` and `GSeidel` (described in the next section) a different criterion is used to halt iterations. If

$$M^{(1)} \equiv \max_{1\leqslant j\leqslant n}\left\{|x_j^{(1)} - x_j^{(0)}|\right\} \tag{4.30}$$

is the largest change in cycle 1, and

$$M^{(i)} \equiv \max_{1\leqslant j\leqslant n}\left\{|x_j^{(i)} - x_j^{(i-1)}|\right\} \tag{4.31}$$

is the largest change in cycle i, the redefined ϵ, measuring the convergence of the algorithm, is:

$$\epsilon \equiv \frac{|M^{(i)} - M^{(1)}|}{M^{(1)}} \tag{4.32}$$

At cycle 1, $\epsilon = 0$. Then it progressively approaches the limiting value 1, as $M^{(i)}$ is supposed to get smaller and smaller. A pre-defined and arbitrary tolerance can then be used to determine when cycling stops. The updated set of values $\mathbf{x}^{(i)}$ will be accordingly adopted as the accepted numerical approximation to the solution. In the example just shown, the value of ϵ at cycle 8 was 0.999 791, so that $1 - \epsilon$ was 0.000 209.

4.18.2 The Gauss–Seidel method

In the Jacobi method, the values are updated all at the same time only after all increments in $\mathbf{x}^{(i)}$ have been calculated. The *Gauss–Seidel method* is a slight variant of the Jacobi method in which each time a component of $\mathbf{x}^{(i)}$ has been calculated, it is immediately used to update the related variable. This means that while the Jacobi method is ideal for applications in parallel computing, the Gauss–Seidel method is better structured for sequential computing.

In this section, the formulas previously expressed in matrix form will be reworked. One starts with the set of n equations in n unknowns:

$$\begin{aligned} a_{11}x_1 + a_{12}x_2 + \cdots + a_{1n}x_n &= b_1 \\ a_{21}x_1 + a_{22}x_2 + \cdots + a_{2n}x_n &= b_2 \\ \cdots &= \cdots \\ a_{n1}x_1 + a_{n2}x_2 + \cdots + a_{nn}x_n &= b_n \end{aligned}$$

Next, the first equation is transformed into $x_1 = f(\{x_j, k \neq 1\})$, the second into $x_2 = f(\{x_j, k \neq 2\})$, etc. The result is a set of n relations of the form:

$$x_i = \frac{b_i}{a_{ii}} - \sum_{\substack{j=1 \\ j\neq i}}^{n} \frac{a_{i,j}}{a_{ii}} x_j, \qquad i = 1,\ldots,n \tag{4.33}$$

This expression forms the core of the Gauss–Seidel algorithm, with the caveat that those values of $\mathbf{x}$ that have already been changed at cycle $i+1$, will be used to change all other values, within the same cycle $i+1$. This means that $x_1^{(i+1)}$ will use $x_2^{(i)},\ldots,x_n^{(i)}$, but $x_2^{(i+1)}$ can use the updated x_1, that is $x_2^{(i+1)}$ will use $x_1^{(i+1)}$, $x_3^{(i)},\ldots,x^{(i)}x_n$. And then $x_3^{(i+1)}$ will use the updated x_1, x_3, etc. The way this algorithm is implemented in R makes use, as shown in the next section, of the *slicing* properties of arrays, in particular of the omission of element i, using $-i$.

4.19 Relevant R code. Functions `PJacobi` and `GSeidel`

Both functions are included in the `comphy` package. Function `PJacobi` exploits the matrix form of the iterative algorithm just described, while `GSeidel` uses a for loop. In both functions, a maximum number of iterations (`nmax`) is allowed (currently `nmax=100000`), but the iterations are halted when the ϵ described in equation (4.32) becomes closer to 1 than a pre-defined amount, named *tolerance* and represented by a variable `tol` with default value 10^{-6}.

The key piece of code that makes possible the implementation of formula (4.33) for Gauss–Seidel in `GSeidel` is the following:

```
.........
.........
.........
for (i in 1:n) {
x[i] <- b[i]/A[i,i] - sum((A[i,-i]/A[i,i])*x[-i])
}
.........
.........
```

In this snippet, the important feature is the use of $-i$ as index of an array in R. When $-i$ is used instead of i in an array, the meaning is that the array will include all the components except the ith one. This is handy with formulas like formula (4.33) where i has to be excluded in the calculation.

Both functions converge to the solution when the matrix of coefficients is diagonally dominant, but there might be instances in which one wants to apply the method also when the matrix of coefficients is not diagonally dominant. In this case a divergence, rather than a convergence, is possible and both functions will stop when the increment gets larger than 10^{100}.

Let us now solve a couple of systems using both functions, with the purpose of demonstrating their use.

```
# Load library "comphy" in working space
library(comphy)

# Small 3 X 3 matrix - Not diagonally dominant
A <- matrix(c(2,8,1,1,3,5,6,2,1),ncol=3)
b <- matrix(c(9,13,7),ncol=1)

# Not forcing solution
x <- PJacobi(A,b)
```

```
## The matrix of coefficients is not diagonally dominant.
## Not attempting solution.
```

```
# Forcing solution (converges after 32 cycles)
x <- PJacobi(A,b,ddominant=FALSE)
```

```
## The matrix of coefficients is not diagonally dominant.
## Attempting solution anyway...
##
```

```
## Number of cycles needed to converge: 22
## Last relative increment: 6.80072923975317e-07

print(x)

## [1] 0.9999996 0.9999997 0.9999997

# Gauss-Seidel converges earlier
x <- GSeidel(A,b,ddominant=FALSE)

## The matrix of coefficients is not diagonally dominant.
## Attempting solution anyway...
##
## Number of cycles needed to converge: 9
## Last relative increment: 7.67678824287188e-07

print(x)

## [1] 1 1 1

# Trying a non diagonally dominant matrix
# Matrix is generated randomly
set.seed(2006)
A <- matrix(runif(25),ncol=5)
b <- matrix(rep(1,times=5),ncol=1)

# Where Jacobi fails, Gauss-Seidel might converge

# Starting solution, null vector
xP <- PJacobi(A,b,ddominant=FALSE)

## The matrix of coefficients is not diagonally dominant.
## Attempting solution anyway...
##
## The increment is getting larger and larger.
## Max value of increment at cycle 269: 2.29660651118428e+100
```

```
# Gauss-Seidel converges
xG <- GSeidel(A,b,ddominant=FALSE)
```

```
## The matrix of coefficients is not diagonally dominant.
## Attempting solution anyway...
##
## Number of cycles needed to converge: 32
## Last relative increment: 6.59038252526223e-07
```

```
# Print values
print(xG)
```

```
## [1] -0.4059801  0.8115036  0.7122835 -0.5637728  0.9443353
```

```
# Is this the solution?
ans <- A %*% matrix(xG,ncol=1) - b
print(ans)  # Apparently so ...
```

```
##                [,1]
## [1,] -5.405590e-07
## [2,] -9.360117e-08
## [3,] -3.065735e-08
## [4,] -2.956979e-07
## [5,]  0.000000e+00
```

4.20 Ill-conditioned systems

As the numerical solution of linear systems consists in general of an approximation to its correct solution, it is desirable for the system not to be affected in an unpredictable way by constraints in the representation of real numbers, like the round-off errors. By this it is meant that a small change in any of the parameters representing the system, for example the components of matrix A or the components of the column vector $\mathbf{b}$, should not be reflected in a large change in the system's solution. An example can clarify this points. Consider the following linear system,

$$A\mathbf{x} = \mathbf{b} \Leftrightarrow \begin{pmatrix} 2 & 3 \\ 1 & 1 \end{pmatrix} \begin{pmatrix} x_1 \\ x_2 \end{pmatrix} = \begin{pmatrix} 7 \\ 3 \end{pmatrix} \Rightarrow \mathbf{x} = \begin{pmatrix} 2 \\ 1 \end{pmatrix}$$

A $\mathbf{b}$ vector with components 7, 3 yields a solution $\mathbf{x}^T = (2 \quad 1)$. Let $\mathbf{b}$ change of a small amount from components 7, 3 to components 6.99, 3.01; how is the solution going to change? We have, in this case,

$$A\mathbf{x} = \mathbf{b} \Leftrightarrow \begin{pmatrix} 2 & 3 \\ 1 & 1 \end{pmatrix} \begin{pmatrix} x_1 \\ x_2 \end{pmatrix} = \begin{pmatrix} 6.99 \\ 3.01 \end{pmatrix} \Rightarrow \mathbf{x} = \begin{pmatrix} 2.04 \\ 0.97 \end{pmatrix}$$

The difference of the new solution from the old solution (here a maximum of 0.04) is of the same order of magnitude as the difference between the old **b** and the new one (here equal to 0.01). The system represented by the above matrix behaves in a desirable way and it is called a *well-conditioned* system.

Consider now a slightly different system:

$$A\mathbf{x} = \mathbf{b} \Leftrightarrow \begin{pmatrix} 2 & 1.99 \\ 1 & 1.00 \end{pmatrix} \begin{pmatrix} x_1 \\ x_2 \end{pmatrix} = \begin{pmatrix} 5.99 \\ 3.00 \end{pmatrix} \Rightarrow \mathbf{x} = \begin{pmatrix} 2 \\ 1 \end{pmatrix}$$

This system has the same solution as the previous one, but A and **b** are slightly different. And now we repeat what was done with the previous system, by changing **b** slightly. This time the solution changes drastically:

$$A\mathbf{x} = \mathbf{b} \Leftrightarrow \begin{pmatrix} 2 & 1.99 \\ 1 & 1.00 \end{pmatrix} \begin{pmatrix} x_1 \\ x_2 \end{pmatrix} = \begin{pmatrix} 6.00 \\ 2.99 \end{pmatrix} \Rightarrow \mathbf{x} = \begin{pmatrix} 4.99 \\ -2.00 \end{pmatrix}$$

This time, a maximum change of 0.1 to the components of **b** has resulted in a change of around 30 times greater to the components of the solution. This is not a desirable feature of a linear system, especially when the solution depends on numerical procedures, subject to uneliminable approximation errors. In this second case, the system is called *ill-conditioned.*

In the present section we will explore why and when a system is ill-conditioned. This will provide us with the essential means to decide when to pay extra attention to the numerical solution of a linear system. Such means rely heavily on the concept of norm of vectors and matrices.

4.20.1 The norm of vectors and matrices

A positive-definite quantity analogous to the norm of vectors in $\mathbb{R}^n$ can be defined for matrices. Such a definition is introduced and demonstrated in appendix B. Several types of norm are possible for matrices (and as we will see shortly for vectors too). Let us focus in this section on the *infinity-norm*, equation (B.20), that is, if A is an $m \times n$ matrix:

$$\|A\|_\infty \equiv \max_{1 \leqslant i \leqslant m} \left\{ \sum_{j=1}^{n} |a_{ij}| \right\}$$

A similar definition is applicable to the norm of a vector, if we consider the vector with n components as an $n \times 1$ matrix. Therefore, for example,

$$\|\mathbf{x}\|_\infty \equiv \max_{1 \leqslant i \leqslant n} \{|x_i|\}$$

And it is immediate to see that the Euclidean norm for vectors is equivalent to the 2-norm for matrices.

Example 4.3 *Consider the two different matrices previously introduced, that is*

$$A_1 = \begin{pmatrix} 2 & 3 \\ 1 & 1 \end{pmatrix}, \quad A_2 = \begin{pmatrix} 2 & 1.99 \\ 1 & 1.00 \end{pmatrix}$$

The infinity-norm of both matrices is:

$$||A_1||_\infty = \max\{2 + 3, 1 + 1\} = 5 \qquad \text{and} \qquad ||A_2||_\infty = \max\{2 + 1.99, 1 + 1\} = 3.99$$

Example 4.4 *The infinity-norm of a vector of length n,*

$$\mathbf{x} = \begin{pmatrix} x_1 \\ x_2 \\ \cdots \\ x_n \end{pmatrix}$$

according to the definition given above is

$$||\mathbf{x}||_\infty = \max\{|x_1|,\ldots,|x_n|\}$$

The same result can be obtained if $\mathbf{x}$ *is transformed into an* $n \times n$ *diagonal matrix with the n components along the diagonal,*

$$||X||_\infty \equiv ||\mathbf{x}||_\infty \qquad \text{where} \qquad X = \begin{pmatrix} x_1 & 0 & \cdots & 0 \\ 0 & x_2 & \cdots & 0 \\ \cdots & \cdots & \cdots & \cdots \\ 0 & 0 & \cdots & x_n \end{pmatrix}$$

4.20.2 Ill-conditioned systems and matrix norm

Consider the linear system $A\mathbf{x} = \mathbf{b}$ and the related system where $\mathbf{b}$ has been slightly changed into a new vector $\mathbf{b}'$ and where, accordingly, the solution vector is too slightly changed into a $\mathbf{x}'$. We are interested to know how the difference $\Delta\mathbf{x} = \mathbf{x} - \mathbf{x}'$ is influenced by the difference $\Delta\mathbf{b} = \mathbf{b} - \mathbf{b}'$. In appendix A, equation (A.16), it is shown that

$$\frac{1}{||A^{-1}||\,||A||}\frac{||\Delta\mathbf{b}||}{||\mathbf{b}||} \leqslant \frac{||\Delta\mathbf{x}||}{||\mathbf{x}||} \leqslant ||A||\,||A^{-1}||\frac{||\Delta\mathbf{b}||}{||\mathbf{b}||}, \tag{4.34}$$

where the norms can be one of the types described in appendix B. The quantity $||A||\,||A^{-1}||$ is called *condition number* and it is indicated with cond:

$$\text{cond}(A) = ||A||\,||A^{-1}|| \tag{4.35}$$

Looking at the inequality (4.34), if the condition number of the matrix of coefficient of an equation is 1, then the relative error $||\Delta\mathbf{x}||/||\mathbf{x}||$ of the solution is exactly equal to the relative error, $||\Delta\mathbf{x}||/||\mathbf{b}||$, of $\mathbf{b}$. In this scenario, a small round-off error in $\mathbf{b}$ has very little effect on the final solution. But if the condition number of A is large, $\text{cond}(A) \gg 1$, then the relative error can be, according to inequality (4.34), much smaller or much larger than the relative error on $\mathbf{b}$. This is the sort of situation occurring in ill-conditioned systems. We can therefore associate ill-conditioning to matrices of coefficients with a large condition number.

Example 4.5 *Consider the well-conditioned system studied at the beginning of this section and the related system obtained changing* $\mathbf{b}^T = (7 \quad 3)$ *with* $\mathbf{b}'^T = (6.993.01)$. *In this example we have, using the infinity-norm,*

$$\|A\| = 5, \qquad \|A^{-1}\| = 4 \Rightarrow \text{cond}(A) = 20$$

The condition number can then be used to calculate within which interval the relative solution error, $\|\Delta\mathbf{x}\|/\|\mathbf{x}\|$, *will be, once the relative change in* $\mathbf{b}$, $\|\Delta\mathbf{b}\|/\|\mathbf{b}\|$, *is known. We have indeed,*

$$\Delta\mathbf{b} = \begin{pmatrix} -0.01 \\ 0.01 \end{pmatrix} \quad \Rightarrow \quad \|\Delta\mathbf{b}\| = 0.01$$

and

$$\mathbf{b} = \begin{pmatrix} 7 \\ 3 \end{pmatrix} \Rightarrow \|\mathbf{b}\| = 7$$

According to inequality (4.34) we therefore expect the following variability for the relative error of the solution:

$$\frac{1}{20}\frac{0.01}{7} \leqslant \frac{\|\Delta\mathbf{x}\|}{\|\mathbf{x}\|} \leqslant 20\frac{0.01}{7} \quad \Rightarrow 0.000\,07 \leqslant \frac{\|\Delta\mathbf{x}\|}{\|\mathbf{x}\|} \leqslant 0.028\,57$$

The relative error in $\mathbf{x}$ *can in fact be calculated and it turns out to be,*

$$\frac{\|\Delta\mathbf{x}\|}{\|\mathbf{x}\|} = \frac{0.04}{2} = 0.02$$

So this relative error, 0.02, *is indeed within the interval* [0.000 07, 0.028 57] *provided by formula (4.34). It is, furthermore relatively small because the condition number is* 20, *a relatively small number.*

Let us now consider the ill-conditioned case where,

$$\|A\| = 3.99, \qquad \|A^{-1}\| = 300 \Rightarrow \text{cond}(A) = 1197$$

This time the condition number is relatively high and we expect the interval within which the relative solution error is estimated to be much larger than before. Even starting with small relative right hand side differences,

$$\frac{\|\Delta\mathbf{b}\|}{\|\mathbf{b}\|} = \frac{0.01}{5.99} = 0.001\,67,$$

the high condition number will make the interval within which the relative solution error lies, quite spacious:

$$0.000\ 00 \leqslant \frac{||\Delta \mathbf{x}||}{||\mathbf{x}||} \leqslant 1.998\ 33$$

The relative solution error is indeed, this time,

$$\frac{||\Delta \mathbf{x}||}{||\mathbf{x}||} = \frac{3}{2} = 1.5,$$

within the range of the inequality. All calculations presented here will be part of one of the exercises in section (4.21).

4.21 Exercises on iterative methods and ill conditioning

Exercise 15

Create a 20×20 real random matrix, A, using numbers generated from a normal distribution with mean 2 and standard deviation 0.5, and making sure to start the random generation with integer seed 3367 (fill the matrix column by column). Next, create a 20 real column vector, $\mathbf{b}$, filled with a random sample with repetitions from the set $\{-1, 1\}$, and making sure to start the sampling with integer seed 4189.

(a) Attempt to solve the linear system $A\mathbf{x} = \mathbf{b}$ using the functions `PJacobi` and `GSeidel` in the `comphy` package. Notice that the matrix A is not diagonally dominant.

(b) Turn A into a diagonally dominant matrix by adding the same positive integer to all the elements along its diagonal. What is the smallest positive integer that makes A diagonally dominant?

(c) With the diagonally dominant matrix found in part b, find the numerical solution of the system, using both `PJacobi` and `GSeidel`.

Exercise 16

To study the convergence of both the Jacobi and the Gauss–Seidel algorithm, the norm of the approximate solutions, $\mathbf{x}^{(i)}$, can be measured and plotted with respect to the iteration number. This is one of the many ways to study convergence, initially at a qualitative level.

In this exercise you should try and modify `PJacobi` to observe convergence qualitatively. Call the function you have modified, `MPJacobi`. Its input is the same as the one for `PJacobi`, but its output, besides the solution `x`, is the norm of all approximations from cycle 1 to the last cycle. Then plot the norms against the iteration number and verify that the values converge to a finite value. Try calculating the solution using different starting solutions and comment qualitatively on the character of the convergence.

Use the following non diagonally dominant matrix A and column vector $\mathbf{b}$

$$A = \begin{pmatrix} 3 & 4 & -1 & 1 & 5 \\ -1 & 1 & 0 & 4 & -1 \\ 1 & 2 & 3 & 5 & -1 \\ 1 & 4 & 0 & -1 & 0 \\ -1 & 2 & 0 & 3 & 4 \end{pmatrix}, \quad \mathbf{b} = \begin{pmatrix} 1 \\ 1 \\ 1 \\ 1 \\ 1 \end{pmatrix}$$

for all calculations.

Exercise 17

Carry out the calculations presented in chapter 4 to introduce ill-conditioning. In the text we have two matrices. The first,

$$A_1 = \begin{pmatrix} 2 & 3 \\ 1 & 1 \end{pmatrix},$$

is well-conditioned, the second,

$$A_2 = \begin{pmatrix} 2 & 1.99 \\ 1 & 1.00 \end{pmatrix},$$

is ill-conditioned. They have been used to solve the two systems,

$$A_1\mathbf{x} = \mathbf{b}_1, \qquad A_2\mathbf{x} = \mathbf{b}_2,$$

where

$$\mathbf{b}_1 = \begin{pmatrix} 7 \\ 3 \end{pmatrix}, \qquad \mathbf{b}_2 = \begin{pmatrix} 5.99 \\ 3 \end{pmatrix}.$$

Conditioning of the two systems is manifested when $\mathbf{b}_1$ and $\mathbf{b}_2$ are *perturbed*, i.e. slightly changed into the two vectors,

$$\mathbf{b}_1' = \begin{pmatrix} 6.99 \\ 3.01 \end{pmatrix}, \qquad \mathbf{b}_2' = \begin{pmatrix} 6.00 \\ 2.99 \end{pmatrix}.$$

These can be re-written as,

$$\mathbf{b}_1' = \mathbf{b}_1 - \Delta\mathbf{b}_1, \qquad \mathbf{b}_2' = \mathbf{b}_2 - \Delta\mathbf{b}_2,$$

with

$$\Delta\mathbf{b}_1 = \begin{pmatrix} -0.01 \\ 0.01 \end{pmatrix}, \qquad \Delta\mathbf{b}_2 = \begin{pmatrix} 0.01 \\ -0.01 \end{pmatrix}$$

Using the `norm` function in R, calculate the following quantities,

$$\|A_1\|, \ \|A_2\|, \ \|A_1^{-1}\|, \ \|A_2^{-1}\|,$$

$$\|\mathbf{b}_1\|, \ \|\mathbf{b}_2\|,$$

$$\|\Delta \mathbf{b}_1\|,\ \|\Delta \mathbf{b}_2\|.$$

Calculate also the solutions of the linear systems with the original and perturbed right hand sides, using the `solve` R function. You should verify that the relative error, $\|\Delta \mathbf{x}\|$, satisfies the inequality provided in the text:

$$\frac{1}{\|A^{-1}\|\|A\|}\frac{\|\Delta \mathbf{b}\|}{\|\mathbf{b}\|} \leqslant \frac{\|\Delta \mathbf{x}\|}{\|\mathbf{x}\|} \leqslant \|A\|\|A^{-1}\|\frac{\|\Delta \mathbf{b}\|}{\|\mathbf{b}\|}.$$

For all calculations use both the infinity-norm (yielding the numbers in the text) and the Frobenius norm.

Exercise 18

Most of the time, values of $\|\Delta \mathbf{x}\|/\|\mathbf{x}\|$, where $\mathbf{x}$ is the solution of a linear system, are far from their upper limit. More specifically, when in the linear system,

$$A\mathbf{x} = \mathbf{b}$$

the right hand side is perturbed, the solution is changed so that the relative error acquires values in the following interval,

$$\frac{1}{\|A^{-1}\|\|A\|}\frac{\|\Delta \mathbf{b}\|}{\|\mathbf{b}\|} \leqslant \frac{\|\Delta \mathbf{x}\|}{\|\mathbf{x}\|} \leqslant \|A\|\|A^{-1}\|\frac{\|\Delta \mathbf{b}\|}{\|\mathbf{b}\|},$$

where $\|\Delta \mathbf{b}\|/\|\mathbf{b}\|$ measures the relative change in the right hand side of the linear system. Very often, $\|\Delta \mathbf{x}\|/\|\mathbf{x}\|$ is far from the upper limit of the inequality, $\|A\|\|A^{-1}\|\|\Delta \mathbf{b}\|/\|\mathbf{b}\|$, so that the system behaves as well-conditioned, even if the matrix condition number is high. In fact, both the right hand side of the system and its change, $\Delta \mathbf{b}$, are important quantities when the stability of the solution is contemplated.

It is possible to investigate the range of values of $\|\Delta \mathbf{x}\|/\|\mathbf{x}\|$, once A, $\mathbf{b}$ and $\Delta \mathbf{b}$ are given. In this exercise you are required to prove, using the Frobenius norm, that $\|\Delta \mathbf{x}\|/\|\mathbf{x}\|$ never reaches its upper limit, if A is a 2×2 matrix.

Exercise 19

The function `illcond_sample` in `comphy` is a *toy* function that creates examples of linear systems that manifest ill-conditioning in a dramatic way. The highest theoretical limit for the solution relative error is often unreachable (see exercise 18), but evident effects of ill-conditioning can be seen also without the relative error to reach such a limit. It is possible to find high values of the relative error, given the matrix A, if sampling of $\mathbf{b}$ and $\Delta \mathbf{b}$ is carried out in a statistically-meaningful way, i.e. with a high number of randomly generated values. This is what is achieved by `illcond_sample`.

1. Study the documentation and code of `illcond_sample` and try to understand how the function works.
2. The matrix,

$$A = \begin{pmatrix} 2 & 1.99 \\ 1 & 1.00 \end{pmatrix}$$

was used in the text to demonstrate the effects of ill-conditioning. With,

$$\mathbf{b} = \begin{pmatrix} 5.99 \\ 3.00 \end{pmatrix}, \ \mathbf{b}' = \begin{pmatrix} 6.00 \\ 2.99 \end{pmatrix}$$

$$\Downarrow$$

$$\Delta\mathbf{b} \equiv \mathbf{b} - \mathbf{b}' = \begin{pmatrix} -0.01 \\ 0.01 \end{pmatrix},$$

the relative error using the Frobenius norm turns out to be, for the above values of $\mathbf{b}$ and $\Delta\mathbf{b}$,

$$\frac{||\Delta\mathbf{x}||}{||\mathbf{x}||} = 1.894\ 21$$

The upper limit for this case is,

$$||A^{-1}||\,||A||\frac{||\Delta\mathbf{b}||}{||\mathbf{b}||} = 2.102\ 58$$

Use `illcond_sample` to try and find a relative solution error greater than the one presented here. You will observe that it is relatively easy to reach a case with a relative solution error higher than the 1.894 21 presented here, but this in general corresponds to an upper limit much higher than 2.102 58. We can empirically observe that for the specific matrix presented, the ratio between the relative solution error and its upper limit struggles to reach the value $1.894\ 21/2.102\ 58 = 0.900\ 90$ of the case treated here.

Note that some of the numbers here are different from those in the main text. This is because in the main text we made use of the infinity-norm, while in this section we make use of the Frobenius norm.

References

[1] Heading J 1958 *Matrix Theory for Physicists* (Longmans Green)

[2] Francis J G F 1961 The QR transformation, a unitary analogue to the lr transformation–part 1 *Comput. J.* **4** 265–71

[3] Francis J G F 1962 The QR transformation–part 2 *Comput. J.* **4** 332–45

[4] Arbenz P 2018 The QR Algorithm http://people.inf.ethz.ch/arbenz/ewp/Lnotes/chapter4.pdf (Accessed: 2021-10-13)

IOP Publishing

Computational Physics with R

James Foadi

Chapter 5

Data fitting

In the chapter on interpolation, we discussed the local approximation of a function $f(x)$ with a polynomial in a finite interval. A set of known points of the function was given,

$$(x_1, f_1), (x_2, f_2), \ldots, (x_{n+1}, f_{n+1}),$$

where $f_i \equiv f(x_i)$, and the polynomial passing through all the $n+1$ points was determined with one of the techniques explained. The polynomial calculated replaces the function $f(x)$ at all points of the closed interval

$$[\min\{x_i\}, \max\{x_i\}], \quad i = 1, \ldots, n+1$$

With interpolation, the focus is on the most appropriate function passing exactly through all data (points) provided. In this chapter the focus is different. The function is postulated as having a specific analytic and parametric form (mostly we will deal with polynomials), and the goal is to find the parameters that make this function transit as close as possible to all the points provided. The function $f(x)$ thus obtained will not, in general, pass through all points, and for this reason it is going to be more appropriate to use the term *curve fitting* rather than *interpolation*. Another term used often in statistics and data analysis to indicate the same concept is *regression*. Curve fitting a technique frequently used to guess a specific physical law emerging from data of a given experiment. It is therefore of primary importance in physics.

In this chapter, the *least squares* procedure to accomplish curve fitting will be introduced and explained. The first of the possible and frequently used analytic forms to approximate $f(x)$ will be the linear form as this is the simplest, yet it contains all the important ingredients of the method. The least squares method applied to linear functions is often called *linear regression*.

doi:10.1088/978-0-7503-2632-2ch5

5.1 Least squares for a straight line

Consider a group of n points of coordinates (x_i, y_i), $i = 1,\ldots,n$ which are, in general, not aligned on a single straight line (figure 5.1). For this reason, a legitimate question is whether it is possible to find a criterion according to which a unique straight line goes through the given 'cloud of points'. Such a criterion is known as *least squares* criterion:

A straight line, $y = f(x) = mx + q$, passing through a cloud of points (x_i, y_i), $i = 1,\ldots,n$ can be uniquely determined with the requirement that the sum of squared differences,

$$S \equiv \sum_{i=1}^{n} \epsilon_i^2, \quad \epsilon_i \equiv y_i - f(x_i), \quad i = 1,\ldots,n \tag{5.1}$$

reaches the minimum possible value.

This sum is a function of the two parameters m and q because the difference $\epsilon_i \equiv y_i - f(x_i)$, called *residual*, is a function of the two parameters:

$$\epsilon_i = y_i - f(x_i) = y_i - (mx_i + q) = y_i - mx_i - q \tag{5.2}$$

The function to minimise, i.e. the sum of the squared residuals, is thus defined as follows:

$$S(m, q) = \sum_{i=1}^{n} (y_i - mx_i - q)^2 \tag{5.3}$$

The values of the parameters that minimise S can be found by setting the two partial derivatives to zero:

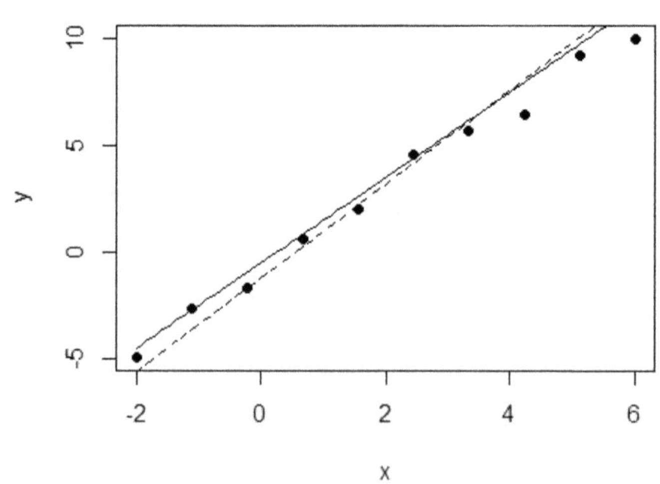

Figure 5.1. The n points in the picture are not aligned on a straight line. Straight lines passing through such a *cloud* of points can be determined using various criteria. Two such straight lines are included in the picture.

$$\partial S/\partial m = 0 = \sum_{i=1}^{n} 2(y_i - mx_i - q)(-x_i)$$
$$\partial S/\partial q = 0 = \sum_{i=1}^{n} 2(y_i - mx_i - q)(-1)$$

Once the equations are divided by −2 and the related summations expanded (considering also that each sum is over n elements), the result are the so-called *least squares normal equations*,

$$\begin{aligned} m\sum_{i=1}^{n} x_i^2 + q\sum_{i=1}^{n} x_i &= \sum_{i=1}^{n} x_i y_i \\ m\sum_{i=1}^{n} x_i + qn &= \sum_{i=1}^{n} y_i \end{aligned} \tag{5.4}$$

The solution of this linear system yields the value of the two parameters and, therefore, leads to the required straight line.

Example 5.1 *The coordinates of all points reproduced in figure 5.1 are listed in the following table*:

x	−2.00	−1.11	−0.22	0.67	1.56	2.44	3.33	4.22	5.11	6.00
y	−4.93	−2.64	−1.68	0.58	2.01	4.56	5.68	6.46	9.21	9.99

The key quantities needed for the estimation of q and m in equation (5.4) can be calculated directly using the (x_i, y_i) values provided:

$$\sum_{i=1}^{10} x_i^2 = 105.185, \sum_{i=1}^{10} x_i = 20.000, \quad \sum_{i=1}^{10} x_i y_i = 181.035, \sum_{i=1}^{10} y_i = 29.253$$

where we have used the fact that the fitting is done using 10 *points*, $n = 10$. *The system derived to estimate m and q and the solutions are, in this case*:

$$\begin{cases} 105.185m + 20.000q &= 181.035 \\ 20.000m + 20q &= 29.253 \end{cases} \Rightarrow m = 1.880, q = -0.834$$

5.2 Multilinear least squares

When linear fitting is required for functions of two or more variables, the principle to derive the normal equations (least squares) does not change, but the equations derived will be different from those in equation (5.4) because more independent variables are involved. For instance, a linear fitting in 3D will consist of a function of two variables, x_1, x_2, that can be written as,

$$f(x_1, x_2) = a_1x_1 + a_2x_2 + a_3,$$

where the parameters a_1, a_2, a_3 are to be determined following the least squares criterion. In general, the linear fitting in an $(m+1)$th dimensional space consists of the following linear function of m variables, $x_1,\ldots,x_m$ and including $m+1$ parameters, $a_1,\ldots,a_{m+1}$:

$$f(x_1,\ldots,x_m) = a_1x_1 + \cdots + a_mx_m + a_{m+1} \tag{5.5}$$

To set up search for the normal (least squares) equations, it is necessary to add a second index to the variables x_i; the first index, running from 1 to n, takes into account the number of points used for the fitting, while the second index, running from 1 to $m+1$, takes into account the number of variables of the model. Each point then is represented by $m+1$ coordinates, m reserved for the model's variables, $x_1,\ldots,x_m$, and one reserved for the approximate value of the function, y. Thus, a point in the $(m+1)$th dimensional space can be indicated as,

$$P_i \equiv (x_{i1}, x_{i2},\ldots,x_{im}, y_i) \tag{5.6}$$

The residuals and the sum of squared residuals will be, accordingly:

$$\epsilon_i(a_1,\ldots,a_m, a_{m+1}) \equiv y_i - a_1x_{i1} - \cdots - a_mx_{im} - a_{m+1} \tag{5.7}$$

$$S(a_1,\ldots,a_m, a_{m+1}) = \sum_{i=1}^{n}\epsilon_i^2(a_1,\ldots,a_m, a_{m+1}) \tag{5.8}$$

The $m+1$ normal equations, derived by setting to zero the $m+1$ partial derivatives of $S(a_1,\ldots,a_m, a_{m+1})$, will be this time (see appendix A.7):

$$\left\{\begin{array}{l}
\left(\sum_{i=1}^{n}x_{i1}^2\right)a_1 + \left(\sum_{i=1}^{n}x_{i1}x_{i2}\right)a_2 + \cdots + \left(\sum_{i=1}^{n}x_{i1}x_{im}\right)a_m + \left(\sum_{i=1}^{n}x_{i1}\right)a_{m+1} = \left(\sum_{i=1}^{n}x_{i1}y_i\right)\\
\left(\sum_{i=1}^{n}x_{i2}x_{i1}\right)a_1 + \left(\sum_{i=1}^{n}x_{i2}^2\right)a_2 + \cdots + \left(\sum_{i=1}^{n}x_{i2}x_{im}\right)a_m + \left(\sum_{i=1}^{n}x_{i2}\right)a_{m+1} = \left(\sum_{i=1}^{n}x_{i2}y_i\right)\\
\qquad\cdots\quad\cdots\\
\left(\sum_{i=1}^{n}x_{im}x_{i1}\right)a_1 + \left(\sum_{i=1}^{n}x_{im}x_{i2}\right)a_2 + \cdots + \left(\sum_{i=1}^{n}x_{im}^2\right)a_m + \left(\sum_{i=1}^{n}x_{im}\right)a_{m+1} = \left(\sum_{i=1}^{n}x_{im}y_i\right)\\
\left(\sum_{i=1}^{n}x_{i1}\right)a_1 + \left(\sum_{i=1}^{n}x_{i2}\right)a_2 + \cdots + \left(\sum_{i=1}^{n}x_{im}\right)a_m + na_{m+1} = \left(\sum_{i=1}^{n}y_i\right)
\end{array}\right. \tag{5.9}$$

It is possible to show that this system has a unique solution in most cases and that it always has at least one solution (see for example [1]).

Example 5.2 *The five points in the 3D space (figure 5.2),*

x_1	−1.49	1.42	4.17	−1.72	2.62
x_2	0.56	2.27	3.47	−0.73	−0.62
y	28.51	36.03	43.49	26.34	37.18

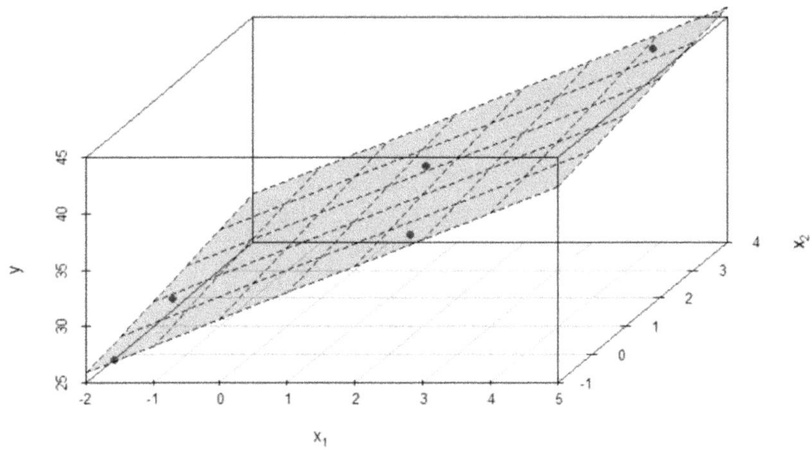

Figure 5.2. Least squares plane fitting the five data points listed in example 5.2. The five points are illustrated as five blue dots.

can be fitted by a plane whose equation is,

$$f(x_1, x_2) = a_1x_1 + a_2x_2 + a_3$$

The quantities in equation (5.9) needed to find the three parameters a_1, a_2, a_3 are:

- $\sum_{i=1}^{5} x_{i1}^2 = 31.476$
- $\sum_{i=1}^{5} x_{i1}x_{i2} = 16.453$
- $\sum_{i=1}^{5} x_{i1} = 5.000$
- $\sum_{i=1}^{5} x_{i1}y_i = 242.202$
- $\sum_{i=1}^{5} x_{i2}^2 = 18.377$
- $\sum_{i=1}^{5} x_{i2} = 4.946$
- $\sum_{i=1}^{5} x_{i2}y_i = 848.424$
- $\sum_{i=1}^{5} y_i = 171.554$

The system of normal equations is thus:

$$\begin{cases} 31.476a_1 & + & 16.453a_2 & + & 5.001a_3 & = & 242.202 \\ 16.453a_1 & + & 18.377a_2 & + & 4.946a_3 & = & 848.424 \\ 5.001a_1 & + & 4.946a_2 & + & 5a_3 & = & 171.554 \end{cases}$$

The solution of the above system is:

$$a_1 = 2.37, \quad a_2 = 0.68, \quad a_3 = 31.27$$

This means that the plane,

$$y = 2.37x_1 + 0.68x_2 + 31.27$$

is the linear function fitting the five data points provided, according to the least squares target.

5.3 The matrix form of linear least squares

The system (5.9) of normal equations can be expressed in matrix form as,

$$F\mathbf{a} = \mathbf{g}, \quad \mathbf{a} = \begin{pmatrix} a_1 \\ a_2 \\ \cdots \\ a_m \\ a_{m+1} \end{pmatrix}, \quad \mathbf{g} = \begin{pmatrix} \sum_{i=1}^{n} x_{i1} y_i \\ \sum_{i=1}^{n} x_{i2} y_i \\ \cdots \\ \sum_{i=1}^{n} x_{im} y_i \\ \sum_{i=1}^{n} y_i \end{pmatrix} \tag{5.10}$$

The matrix F is a symmetric matrix (see appendix B) and has the following form:

$$F = \begin{pmatrix} \sum_{i=1}^{n} x_{i1}^2 & \sum_{i=1}^{n} x_{i1} x_{i2} & \cdots & \sum_{i=1}^{n} x_{i1} x_{im} & \sum_{i=1}^{n} x_{i1} \\ \sum_{i=1}^{n} x_{i2} x_{i1} & \sum_{i=1}^{n} x_{i2}^2 & \cdots & \sum_{i=1}^{n} x_{i2} x_{im} & \sum_{i=1}^{n} x_{i2} \\ & & \cdots & & \\ \sum_{i=1}^{n} x_{im} x_{i1} & \sum_{i=1}^{n} x_{im} x_{i2} & \cdots & \sum_{i=1}^{n} x_{im}^2 & \sum_{i=1}^{n} x_{im} \\ \sum_{i=1}^{n} x_{i1} & \sum_{i=1}^{n} x_{i2} & \cdots & \sum_{i=1}^{n} x_{im} & n \end{pmatrix} \tag{5.11}$$

The parameters of the linear least squares procedure can therefore be calculated with any of the R functions to solve linear systems of equations. Furthermore, the symmetry of F is indicative of a decomposition that turns out to be very useful for the building up of the normal equations, without having to remember its coefficients.

A symmetric matrix can be decomposed as a product of another matrix and its transpose. Indeed, given any matrix A, the product $A^T A$ is symmetric:

$$F = A^T A \Rightarrow F^T = (A^T A)^T = A^T (A^T)^T = A^T A = F$$

The matrix A needed to form the F in equation (5.11) is an $n \times (m + 1)$ matrix where the first m columns contain the components of the n data points, while the last column has all its components equal to 1. To be more specific, if $\mathbf{x}_j$ is the column vector whose components are the jth coordinates of all the n data points, and $\mathbf{1}$ is the column vector whose components are all equal to 1, then:

$$A = \begin{pmatrix} \mathbf{x}_1 & \mathbf{x}_2 & \cdots & \mathbf{x}_m & \mathbf{1} \end{pmatrix} = \begin{pmatrix} x_{11} & x_{12} & \cdots & x_{1m} & 1 \\ x_{21} & x_{22} & \cdots & x_{2m} & 1 \\ & & \cdots & & \\ x_{n1} & x_{n2} & \cdots & x_{nm} & 1 \end{pmatrix} \tag{5.12}$$

By direct calculation it is straightforward to check that $A^T A$ is actually equal to the matrix F given in equation (5.11). It is, in fact, possible to define matrix A with the 1's placed as the first column, rather than the last; this corresponds to an expression of the multilinear model equal to:

$$f(x_1,\ldots,x_m) = a_0 + a_1 x_1 + \cdots + a_m x_m \tag{5.13}$$

in which, differently from the expression (5.5), the free coefficient is a_0, rather than a_{m+1}. Indeed, the column of 1's can be also placed anywhere in the middle of A, if a specific position for the free coefficient is adopted. It is also possible to find that A is defined as its transpose. The final treatment of the multilinear least squares in matrix form in this case is equivalent to the one described here, but one will have to be careful of the initial swap of rows with columns.

Column $\mathbf{g}$ in equation (5.10), too, can be formulated starting from the data, in a way similar to how F was formulated *via* A, starting from the data points' coordinates. It is straightforward to verify that,

$$\mathbf{g} = A^T \mathbf{y}, \qquad \mathbf{y} = \begin{pmatrix} y_1 \\ y_2 \\ \cdots \\ y_n \end{pmatrix} \tag{5.14}$$

Using both equations (5.10) and (5.14), the following result is readily obtained:

$$F\mathbf{a} = \mathbf{g} \quad \Leftrightarrow \quad A^T A\mathbf{a} = A^T \mathbf{y} \Rightarrow A\mathbf{a} = \mathbf{y}$$

The significance of this result is that the normal equations for least squares are equivalent to an overdetermined system of n equations in m unknowns, where $n \geqslant m$:

$$F\mathbf{a} = \mathbf{g} \Leftrightarrow A\mathbf{a} = \mathbf{y} \tag{5.15}$$

From an intuitive point of view, the equivalence (5.15) conveys the idea that it is possible to think of a set of n points as belonging to the hyperplane (5.5), even if they do not satisfy its analytic expression, in the least squares sense as they satisfy the normal equations associated with the plane.

Example 5.3 *Let us write down matrix A and vector column $\mathbf{y}$ of expressions (5.10) and (5.14), when the data points of the least squares exercise are those of example 5.2. In the example there are 5 data points, each point with coordinates $x - 1$ and x_2. A will therefore have 5 rows and 3 columns, because the last column is made of 1's. The vector $\mathbf{y}$, more simply, is exactly equal to the last row of the table in example 5.2. More explicitly*

$$A = \begin{pmatrix} 1.49 & 0.56 & 1 \\ 1.42 & 2.27 & 1 \\ 4.17 & 3.47 & 1 \\ -1.72 & -0.73 & 1 \\ 2.62 & -0.62 & 1 \end{pmatrix}, \qquad \mathbf{y} = \begin{pmatrix} 28.51 \\ 36.03 \\ 43.49 \\ 26.34 \\ 37.18 \end{pmatrix}$$

It is then straightforward to verify that $A^T A$ and $A^T \mathbf{y}$ reproduce the coefficients of the system in the same example.

5.4 Relevant R code. Functions `solveLS` and `solve`

The function responsible to carry out multilinear least squares fitting in `comphy` is called `solveLS`. The function's engine is the `solve` function in the `base` package. This is the reference R function to solve linear equation systems, as mentioned in chapter 4. `solveLS` essentially builds the F matrix and the $\mathbf{g}$ column vector so that `solve` can find the unique solution of $F\mathbf{a} = \mathbf{g}$, if that exists, otherwise `solveLS` returns `NULL` printing, at the same time, a message stating that the solution is not unique. In the following demonstration `solveLS` is used to carry out least squares fitting in three different situations, when the points are not on the fitting plane, when they are exactly on the fitting plane (thus returning a sum of squared residuals equal to zero), and when they are aligned on a segment so that the fitting plane is not unique. The plane for the first two scenarios has equation,

$$y = 2x_1 - x_2 + 3,$$

while in the third scenario the points are chosen in the straight line described by the relations,

$$x_1 + x_2 = 1, \qquad y = 3(x_1 + x_2) - 2$$

There is an infinite number of planes passing through that line so that the solution to the least squares problem in this case is not unique.

```
# Load comphy package
library(comphy)

# Create data points for case 2 - on the plane (x_1,x_2,y)
p1 <- c(0,1,2)
p2 <- c(1,0,5)
p3 <- c(1,1,4)
p4 <- c(0,2,1)
p5 <- c(2,0,7)
x2 <- matrix(c(p1,p2,p3,p4,p5),ncol=3,byrow=TRUE)

# Data points for case 1 - randomly away from plane
x1 <- x2
x1[,3] <- x1[,3]+rnorm(5,mean=0.5,sd=0.1)

# Create data points for case 3 - on a straight line
p1 <- c(0,1,1)
p2 <- c(1,0,1)
p3 <- c(2,-1,1)
p4 <- c(3,-2,1)
p5 <- c(-1,2,1)
x3 <- matrix(c(p1,p2,p3,p4,p5),ncol=3,byrow=TRUE)

# Coefficients for case 1
a1 <- solveLS(x1)
```

```
## Sum of squared residuals: 0.007199
```

```
print(a1)

## [1]  2.1395761 -0.9169707  3.3705736

# Coefficients for case 2
a2 <- solveLS(x2)

## Sum of squared residuals: 0.000000

print(a2)

## [1]  2 -1  3

# Coefficients for case 3
a3 <- solveLS(x3)

## There are infinite solutions to this least squares fitting.

print(a3)

## NULL
```

The slight difference between the parameters of case 1 and case 2 is compatible with the fact that the data points for case 1 are not on the plane, but away from it of a small, random, distance.

It is also important to notice that `solve` stops execution for A and $\mathbf{y}$ of an overdetermined system because it can only be used when A is a square matrix.

5.5 Polynomial least squares

It can be surprising to learn that linear regression can be applied to fit data points with polynomials, even though a polynomial is in general not a linear function. The reason is that the linearity is not meant to be in the variable, x, of the function, but in the parameters of the model. An example will help to see why. Consider the second-degree polynomial,

$$P_2(x) = a_1x^2 + a_2x + a_3$$

It can be considered a multilinear model with two variables x_1, x_2 where,

$$x_1 \equiv x^2, \qquad x_2 = x$$

Then the least squares procedure can be applied to the second-degree polynomial exactly in the same way that is applied to a linear model with two variables. The same idea can be extended to polynomials of degree m, once the following substitutions are considered:

$$x_1 = x^m, \quad x_2 = x^{m-1},\ldots,x_m = x \tag{5.16}$$

From what was just stated, it seems that polynomial regression does not require extra procedures, compared with multilinear regression. But there is a caveat in the use of polynomial regression because the matrix associated with the normal equations becomes severely ill conditioned (see chapter 4) with high values of the polynomial's degree. The reason is connected to the so-called *multicollinearity* of the independent variables, which means that the correlation between two or more x_i is close to 1. Due to the expression for the estimate of the parameters a_i, which include the summation of products like x_ix_j, high correlation means that these numbers can be high and lead to expressions potentially becoming infinite. An example of multicollinearity is treated in one of the exercises in section 5.11. For additional information see, for instance, reference [2][1]. This is not a problem in real applications because the modelling is done mostly using polynomials of third or fourth degree, maximum.

Example 5.4 *The velocity of a rain drop is measured at regular time intervals for the first* 120 *seconds, and it is reported in the following table.*

t (s)	v (m s^{-1})
0	0.00
12	99.06
24	187.10
36	267.55
48	313.78
60	331.46
72	375.50
84	407.76
96	431.45
108	418.20
120	455.31

[1] A simple way to look at the problem is when, say, two of the m independent variables are linearly dependent (perfect collinearity). In this case, the matrix A (see equation (5.12)) has rank less than m, so that F,the matrix of the normal equations, cannot be inverted, it is singular.

To fit a quadratic and cubic polynomial to these data, we need to convert them into data for a two-variables and three-variables model.

The quadratic model consists of the following relations:

$$y = a_1x_1 + a_2x_2 + a_3, \qquad x_1 = t^2, x_2 = t, y = v$$

The original table is, accordingly, changed into the following one.

x_1	x_2	y
0	0	0.00
144	12	99.06
576	24	187.10
1296	36	267.55
2304	48	313.78
3600	60	331.46
5184	72	375.50
7056	84	407.76
9216	96	431.45
11 664	108	418.20
14 400	120	455.31

The parameters found using least squares on this model are:

$$a_1 = -0.034, \quad a_2 = 7.639, \quad a_3 = 14.063$$

The sum of squared residual for fitting the quadric polynomial turns out to be 2182.530.

For the cubic model we have:

$$y = a_1x_1 + a_2x_2 + a_3x_3 + a_4, \qquad x_1 = t^3, x_2 = t^2, x_3 = t, y = v$$

so that the original table is changed into the following one.

x_1	x_2	x_3	y
0	0	0	0.00
1728	144	12	99.06
13 824	576	24	187.10
46 656	1296	36	267.55
110 592	2304	48	313.78
216 000	3600	60	331.46
373 248	5184	72	375.50
592 704	7056	84	407.76
884 736	9216	96	431.45
1 259 712	11 664	108	418.20
1 728 000	14 400	120	455.31

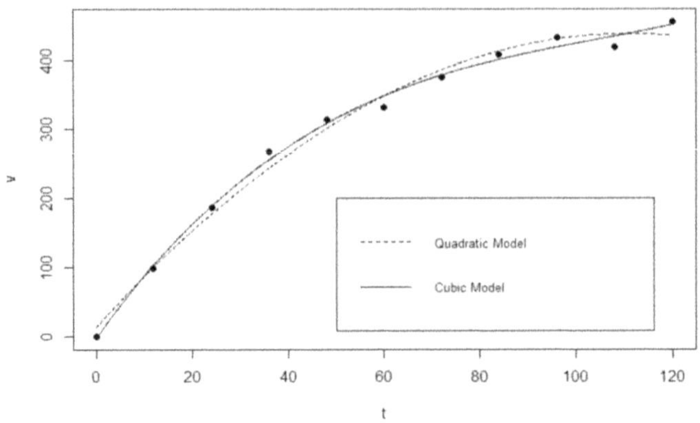

Figure 5.3. Quadratic and cubic polynomial least squares fitting for the data points of example 5.4.

The parameters found using least squares on this model are:

$$a_1 = 0.000\,26,\ a_2 = -0.080\,57,\ a_3 = 9.750\,24,\ a_4 = -1.880\,35$$

while the sum of squared residual for this polynomial fit turns out to be 970.941.

A couple of observations are in place. In both models, the cubic and quadratic leading terms are not particularly high. This means that fitting using a linear function explains most of the data variability. Second, the sum of squared residuals, which essentially describes how close the curve found is to the data points, is smaller for the cubic than the quadratic model. This is also reflected in the closer proximity of the cubic curve to the data points in the plot of figure 5.3. Finally, the magnitude of the numbers involved in the calculation is high and this can be a cause of ill-conditioning because big numbers (or very small ones, if the powers affect numbers with absolute value between 0 and 1) run the risk of swamping several intermediate results.

5.6 What degree polynomial? Underfitting and overfitting

Fitting a curve to data points does not amount, in general, to finding the correct model that produces such data. One should be happy to find a model that justifies as much as possible the variability of the available data points. Sometimes the model is known because it corresponds to a specific physical law. In this special instance, data fitting has the purpose of determining the parameters present in the analytic expression of the physical law. The researcher is more often left to guess the model, without knowing which one is closest to the most appropriate analytic form. When the number of data points is $n + 1$ and the model used for the investigations is a polynomial, it should be intuitively clear that the closer its degree is to n the better is the fit, if this is measured by the sum of squared residuals. Eventually, when the degree is n, the polynomial goes exactly through all the $n + 1$ data points (see section 3.8), and the sum of squared residuals is exactly zero. Increasing the polynomial's degree even further causes no difference to the sum of squared residuals, which will remain zero. In this case the model is said to be *overfitting* the data; this means that the fitting function is trying to model the random errors of the data (noise), besides the main analytic form that justifies data variability. At the opposite spectrum of complexity is the *underfitting* of the data, where the model does not account even for the main data variability. When polynomial regression is involved, both overfitting and underfitting can be associated, respectively, with a too high or too low polynomials' degree.

From what was just explained, it seems that data overfitting cannot be determined using the sum of squared residuals because a value of zero for such quantity is the best possible value, and there are infinitely many polynomials to yield such a value. Similarly, how can the sum of squared residuals indicate a polynomial that is underfitting the data? We might encounter the same value of this quantity for two different sets of data and yet in one case the model could be appropriate while in the other the model could underfit the data. As the sum of squared residuals is going to be mentioned several times in what follows, let us indicate it with the abbreviation SSE (*sum of squared errors*) from now on.

To build up a justification for the criterion that will eventually be adopted to select the appropriate degree for the polynomial fit, let us consider the data used in example 5.4. If these are fitted with polynomials of increasing degree, from 0 (a constant) to, say, 8, and if the related SSE is calculated for each regression, a plot of SSE versus m, the polynomials' degree, is possible (see figure 5.4).

The plot reveals an interesting feature, when the increasing polynomial's degree is interpreted as the attempt of the model to account for data variability. In the figure we can see that when one goes from a constant (polynomial with degree 0) to a straight line (polynomial with degree 1), data variability is quantitatively described in greater detail because the SSE drops dramatically from 218 025 to 23 287. A further, and noticeable, improvement is obtained adopting the quadratic model, as the SSE drops to 2182. It is not immediately clear from the plot whether the cubic model still helps in accounting for data variability, but the value of SSE still drops considerably and lands at 971. Noticeable decrease in SSE could be the criterion we

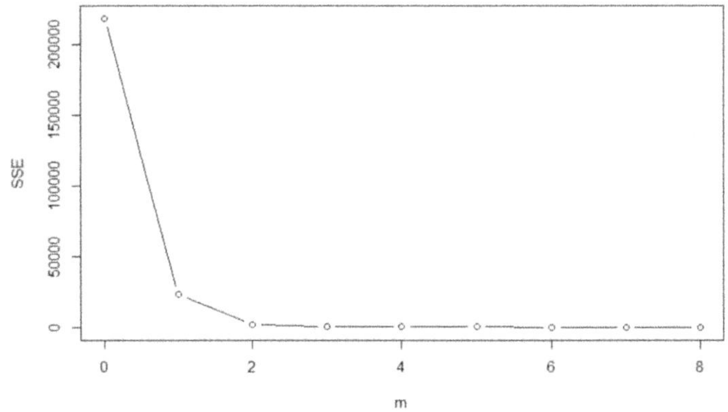

Figure 5.4. Sum of squared residuals (SSE) versus degree of fitting polynomials, for regressions of data in example 5.4 with polynomials of degree 0 to 8. There is an abrupt change going from $m = 0$ to $m = 2$, and then the change is less evident.

were looking for to decide what is the most appropriate degree to adopt for a polynomial regression. Statistically too it makes sense. In statistics, the simplest model, yet the one that accounts for most of the data variability, is the preferred choice (Occam's razor). Starting with the simplest model, and until the drop is significant, one should increase the degree and stop when the fall in SSE is no longer noticeable. In fact, it turns out that the SSE can be replaced by a more sensible and related quantity,

$$\sigma_e^2 = \frac{\sum_{i=1}^{n} \epsilon_i^2}{n - m - 1}, \tag{5.17}$$

which is the variance of the regression's residuals because $n - m - 1$ represents the degrees of freedom in its calculation (essentially and in short, $m + 1$ of the n data points are used in the regression to calculate the $m + 1$ parameters of the model). Quite often, the square root of σ_e^2, i.e. σ_e, is known as the *standard error of the fit*. An intuitive understanding can be provided when thinking to σ_e^2 in terms of the ratio between SSE and $n - m - 1$. When the degree of the model is increased, both the numerator and denominator of such a fraction will decrease and the new value of σ_e^2 will be the result of two effects. The first is the tendency to overfitting (corresponding to the reduction of SSE), while the second is the increased cost of the model because the number of parameters (and the related number of calculations) is increased (corresponding to the reduction of $n - m - 1$). Thus, thinking to σ_e^2 as variance of the residuals, large drops correspond to models properly accounting for data variability, while small drops basically indicates an increase in model complexity not matched for an equally significant reduction in the variance of the residuals. This is an indication that the new model is likely to be modelling the noise rather than the signal. In using σ_e^2 to decide on best polynomial's degree it should be clear that the decision is aided by the behaviour of a numerical quantity, but that it is ultimately subject to a subjective decision.

5.7 Relevant R code. Functions `which_poly` and `polysolveLS`

These functions are both in the `comphy` package. The first function suggests a degree for polynomial regression, based on the methodology developed in section 5.6. Here a plot of σ_e^2 versus m, the polynomial's degree, is displayed and both quantities are also returned as a data frame. The second function carries out the actual regression, returning the estimate of the model's parameters.

Let us demonstrate `which_poly` using a simple example of a cubic polynomial:

$$y = \frac{1}{3}x^3 - \frac{1}{2}x^2 - 2x + 1$$

The data consist of 30 points simulated as random errors around this curve, distributed according to the normal distribution with mean 0 and standard deviation 0.6.

```
# Load comphy package
library(comphy)

# Generate 30 data points for a cubic
# y = x^3/3 - x^2/2 -2x + 1
#
set.seed(4213)
x <- seq(-2,3,length=30)
eps <- rnorm(30,mean=0,sd=0.6)
y <- x^3/3-x^2/2-2*x+1+eps
pts <- data.frame(x=x,y=y)

# Check which polynomial's degree is appropriate
# It turns out that m=3 is the right choice
ddd <- which_poly(pts)
```

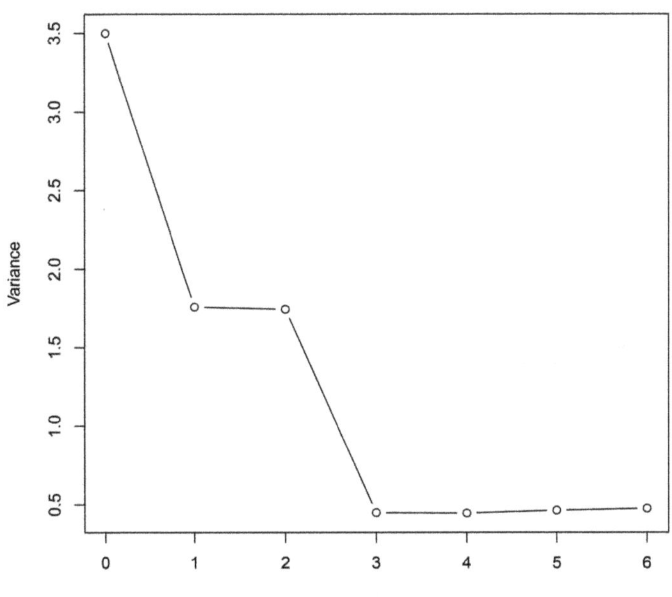

In the plot, σ_e^2 drops two times and then stabilises after $m = 3$. This is a clear indication that the cubic polynomial is the appropriate model that takes most data variability into account, without unnecessary parameters modelling the noise. In the next code snippet, the regression is performed with $m = 3$ and the data displayed with the regression curve (full line) and with both curves for the linear and quadratic regression (dashed lines) for comparison.

```
# Polynomial regression with m=3
ltmp <- polysolveLS(pts,3)
a <- ltmp$a
print(a)
```

```
## [1]  0.4185437 -0.7619175 -2.1118885  1.3856494
```

```
# Values for chosen regression model (m=3)
xx <- seq(-2,3,length=1000)
yy <- a[1]*xx^3+a[2]*xx^2+a[3]*xx+a[4]

# Add regression curves for m=1 and m=2 (just to compare)
ltmp <- polysolveLS(pts,1)
a <- ltmp$a
yy1 <- a[1]*xx+a[2]
ltmp <- polysolveLS(pts,2)
a <- ltmp$a
yy2 <- a[1]*xx^2+a[2]*xx+a[3]

# Plot
plot(pts,pch=16,xlab="x",ylab="y")
points(xx,yy,type="l")
points(xx,yy1,type="l",lty=3)
points(xx,yy2,type="l",lty=3)
```

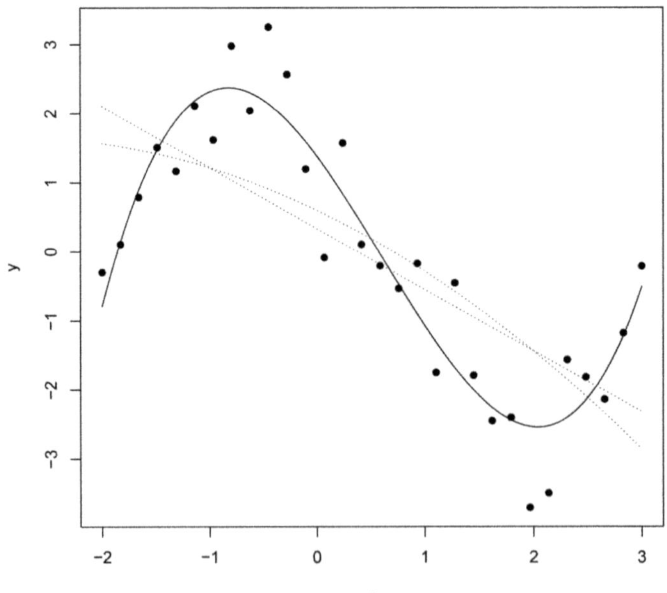

Two observations are in place. First, the regression coefficients, 0.419, −0.762, −2.112, 1.386, are not exactly those of the model, 0.333, −0.500, −2.000, 1.000, but are not far from them. The difference is obviously caused by the data points not positioned exactly on the correct cubic curve. Second, both the linear and quadratic regressions are close to each other; this justifies the close value of σ_e^2 found previously for both regressions.

5.8 Nonlinear fitting using linear least squares

The technique of multilinear regression can be used also for nonlinear fitting because, as shown in the case of polynomial regression, the necessary condition is for the model to be linear in the parameters, rather than the variables. Thus, it is possible to perform regressions consisting of linear combinations of nonlinear expressions of the variables like, for instance,

$$f(x_1, x_2) = x^2 + 2xy + y^2 \quad \text{or} \quad f(x_1, x_2, x_3) = 2\sin(x_1) - \log(x_2) + 3x_1x_3$$

The models to use, for the above functions, are:

$$f(x_1, x_2) = a_1x^2 + a_2xy + a_3y^2 \quad \text{and} \quad f(x_1, x_2, x_3) = a_1\sin(x_1) + a_2\log(x_2) + a_3x_1x_3$$

The possibilities to use linear regression to model nonlinear functions are thus endless, once the intermediate nonlinear expressions are known. This turns out not to be a rare occurrence in physics, where particular forms of the laws behind specific observations are in general known or guessed.

Example 5.5 *Consider the motion of a projectile shot from a cannon tilted at an angle* $\alpha = 60°$ *with respect to the ground and with a starting speed equal to* 200 m s^{-1}. *It is relatively straightforward to work out the projectile's trajectory, which is described by the following equation (see for example [3]):*

$$y = -\frac{g}{2v^2\cos^2(\alpha)}x^2 + \tan(\alpha)x,$$

where the motion happens in the vertical, (x, y), *plane,* $g = 9.81$ m s^{-2} *is the constant gravitational acceleration and where the air friction has been neglected. As the constant term is missing, it is clear from the equation for the trajectory that the projectile is shot from the origin,* (0, 0). *When the numeric values for* g, v, α *are replaced in the formula, the following expression is obtained,*

$$y = a_1x^2 + a_2x + a_3 \qquad \text{where } a_1 = -0.000\,49,\ a_2 = 1.732\,05,\ a_3 = 0$$

The measurement of the projectile at 10 *different time values is reported in the following table.*

x	y
1067.81	1299.60
1132.54	1320.63
1648.46	1484.93
1767.54	1538.48
1913.10	1536.40
2006.30	1480.06
2020.40	1480.32
2149.52	1472.84
2257.35	1407.19
2411.80	1330.35

From the data we can try to recover the trajectory of the projectile and subsequently work out when it is at its highest point above the ground. As the trajectory has a quadratic form, we can attempt polynomial regression with degree $m = 2$. The result is:

$$a_1 = -0.000\,46,\ a_2 = 1.634\,21,\ a_3 = 68.593, \qquad \text{SSE} = 2648.215$$

The first two coefficients are not very different from those calculated theoretically, but the constant, a_3, is definitely not zero, the reason being that the least squares algorithm tries to minimise SSE and this seems to be achieved by letting a_3 drift away from 0.

As the initial position of the projectile is a known background fact and as this corresponds to the quadratic used as model to have $a_3 = 0$, one can decide to attempt the regression without a_3 in the model, i.e. with the model,

$$y = a_1x^2 + a_2x$$

which is still a linear model of the two expressions x^2 and x. The data for this new linear model are listed in the following table.

$x_1 = x^2$	$x_2 = x$	$y = y$
1 140 218	1067.81	1299.60
1 282 647	1132.54	1320.63
2 717 420	1648.46	1484.93
3 124 198	1767.54	1538.48
3 659 952	1913.10	1536.40
4 025 240	2006.30	1480.06
4 082 016	2020.40	1480.32
4 620 436	2149.52	1472.84
5 095 629	2257.35	1407.19
5 816 779	2411.80	1330.35

The regression, this time, yields:

$$a_1 = -0.000\,48,\ a_2 = 1.717\,40,\ a_3 = 0, \qquad \text{SSE} = 2820.327$$

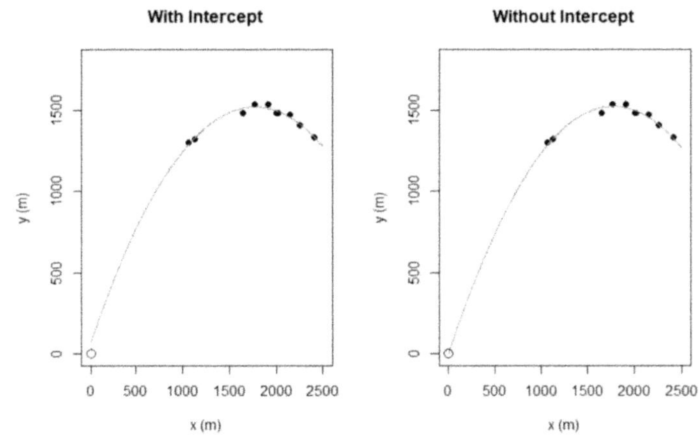

Figure 5.5. Estimated trajectory of the projectile of example 5.5 in the two model cases. In both images, the origin is indicated by the open circle.

The SSE is higher than before because now $a_3 = 0$ *forces the regression curve to be anchored at* $(0, 0)$ *so that full minimisation cannot be obtained. But the* a_1 *and* a_2 *coefficients are closer to their correct values. Thus, although is never advisable to carry out linear regression without the so-called intercept term* (a_3)*, when there is secure knowledge that this is true, then regression without the intercept term can eliminate part of the bias. On the other hand, to still try regression with the interception term in such circumstances can serve as validation of the model, the procedure and data quality when estimated values of* a_3 *are close to* 0.

The heightest point of the trajectory, (x_h, y_h)*, is found by taking the derivative of its equation and setting it to* 0:

$$\frac{\mathrm{d}y}{\mathrm{d}x} = 2a_1x + a_2 = 0 \Rightarrow x_h = -\frac{a_2}{2a_1}, \; y = a_1x_h^2 + a_2x_h + a_3$$

For the two cases of regression with and without intercept term we have, respectively:

$$x_h = 1775.91, \; y_h = 1519.70 \quad \text{and} \quad x_h = 1774.65, \; y_h = 1523.89$$

There is a difference of estimated heights of around four metres. This difference might be important for various reasons and therefore it is clear that deciding to retain or drop the intercept term can have meaningful consequences. A plot of the regression in both cases is shown in figure 5.5.

5.8.1 Transformation before regression

There are several classes of nonlinear functions for which it does not seem possible to apply the linear least squares regression. But an appropriate transformation of such functions yields a linear combination of linear or nonlinear expressions making it thus amenable to linear regression.

To see how the methodology just described works, let us consider a so-called *power-law model*, first. This is described by the following equation:

$$f(x) = ax^b, \quad a, b > 0$$

A linear model can be created once the logarithm of this expression is taken. It is customary to use the natural logarithm, ln, for the purpose. The new model is not linear in the parameters a and b, but in the transformed parameters α and β:

$$\ln(f) = \ln(a) + b\ln(x) \quad \Rightarrow \tilde{f} = \alpha + \beta\tilde{x},$$

where,

$$\tilde{f} \equiv \ln(f),\ \tilde{x} \equiv \ln x,\ \alpha \equiv \ln(a),\ \beta \equiv b$$

So, when data points, (x_i, y_i), suspected to follow a power law are provided, one should transform them applying the natural logarithm on both x_i and y_i, so to obtain $\tilde{x}_i$ and $\tilde{y}_i$. These are, next, subject to linear regression in order to estimate the parameters α and β. Finally, the original approximation is recovered with

$$a = \mathrm{e}^{\alpha}, \qquad b = \beta$$

Another example of a nonlinear function turned into a linear one is the *reciprocal model*,

$$f(x) = \frac{1}{ax + b}$$

The linearisation is achieved once the reciprocal of both sides is taken:

$$f = \frac{1}{ax + b} \quad \Rightarrow \frac{1}{f} = ax + b \quad \Rightarrow \tilde{f} = \alpha\tilde{x} + \beta,$$

where,

$$\tilde{f} \equiv \frac{1}{f},\ \tilde{x} \equiv x,\ \alpha \equiv a,\ \beta \equiv b$$

In the procedure to estimate a and b for the reciprocal model, the reciprocal of y_i is taken and then a linear regression is tried on x_i and $1/y_i$. The coefficients, α and β, obtained are this time equal to the original parameters:

$$a = \alpha, \qquad b = \beta$$

Many more *linearisations* of nonlinear functions are possible. A sample of the most frequently used ones is contained in the following table.

Name	**Model**	**Transformation**		**Parameters**	
Power Law	$y = ax^b$	$\tilde{y} = \ln(y)$	$\tilde{x} = \ln(x)$	$\alpha = \ln(a)$	$\beta = b$
Exponential 1	$y = ae^{bx}$	$\tilde{y} = \ln(y)$	$\tilde{x} = x$	$\alpha = \ln(a)$	$\beta = b$
Exponential 2	$y = ab^x$	$\tilde{y} = \ln(y)$	$\tilde{x} = x$	$\alpha = \ln(a)$	$\beta = \ln(b)$
Logarithmic	$y = \ln(ax^b)$	$\tilde{y} = y$	$\tilde{x} = \ln(x)$	$\alpha = \ln(a)$	$\beta = b$
Reciprocal 1	$y = 1/(ax + b)$	$\tilde{y} = 1/y$	$\tilde{x} = x$	$\alpha = a$	$\beta = b$
Reciprocal 2	$y = a + b/(1 + x)$	$\tilde{y} = y$	$\tilde{x} = 1/(1 + x)$	$\alpha = a$	$\beta = b$
Reciprocal 3	$y = 1/(a + bx)^2$	$\tilde{y} = 1/\sqrt{y}$	$\tilde{x} = x$	$\alpha = a$	$\beta = b$
Square root	$y = a + b\sqrt{x}$	$\tilde{y} = y$	$\tilde{x} = \sqrt{x}$	$\alpha = a$	$\beta = b$

5.9 Exercises on least squares

Exercise 01

Study the main function for linear regression, `solveLS`, in `comphy` and apply it to estimate the parameters, a_1, a_2, a_3, a_4, of the linear model

$$y = a_1x_1 + a_2x_2 + a_3x_3 + a_4$$

The data points are listed in the following table.

x_1	x_2	x_3	y
−0.959	0.829	0.419	4.327
−0.781	−0.218	0.752	6.513
−0.957	0.549	0.499	4.847
0.413	−0.184	0.919	9.838
−0.989	−0.683	0.819	6.931
0.460	−0.274	−0.362	4.606
−0.479	0.739	−0.156	2.738
−0.436	−0.229	−0.664	2.191
−0.329	0.289	0.663	6.657
−0.866	0.662	−0.252	1.630
0.988	−0.635	0.957	11.556
0.885	−0.422	−0.736	4.372
−0.760	−0.231	0.596	6.010
−0.338	0.705	0.502	5.397
−0.734	0.551	−0.991	−1.245

Exercise 02

Simulate a set of 15 data points coming from measuring x of the following physical law,

$$x(t) = 5t + \sin(2\pi t) + \cos(2\pi t), \quad 0 \leqslant t \leqslant 3$$

for the values of t equal to

$$t = 0.0, 0.2, 0.6, 1.8, 1.9, 2.5, 2.6$$

and where data are affected by random errors distributed according to the normal distribution with mean 0 and standard deviation 0.5. Use the seed `5522` for the simulation. Organise the data thus simulated into a data frame and use `solveLS` to estimate the physical law, assuming the following model,

$$x = a_1t + a_2 \sin(2\pi t) + a_3 \cos(2\pi t)$$

You should find that the estimate of a_1, a_2, a_3 is close to their theoretical value, $a_1 = 5$, $a_2 = 1$, $a_3 = 1$. What is the SSE if the model does not have the cosine components? On a same graph, plot the data points simulated and the two regression curves corresponding to the two models.

Exercise 03

Using the simulated data of exercise 02, find out which polynomial fits them best, using the variance σ_e^2 as criterion.

Exercise 04

Consider the dampened oscillations of a body of mass $m = 5$ kg, attached to a spring of unknown constant k, and oscillating horizontally, with angular frequency $\omega = 1.998$ (rad s^{-1}), on a smoothed plane which progressively decreases the amplitude of the oscillation. The motion of the body is described by the equation

$$x = x(t) = Ae^{-bt/(2m)}\cos(\omega t),$$

which describes the position of the body around its equilibrium position. In the above equation,

$$\omega = \sqrt{\frac{k}{m} - \left(\frac{b}{2m}\right)^2}$$

In order to calculate the spring constant, k, and the coefficient b that causes the dampening, a series of $n = 11$ regular measurements of the body's position is taken at regular time intervals between 0 s and 20 s. The values are summarised in the following table.

t (s)	x (m)
0	16.218
2	–11.947
4	–1.979
6	5.046
8	–4.620
10	2.821
12	2.026
14	–3.459
16	2.597
18	0.959
20	–0.982

Using linear regression, estimate the value of both k (N m^{-1}) and b (N s m^{-1}).

Exercise 05

Water is discharged from a container, according to the following dynamical formula:

$$h(t) = a + \frac{b}{1 + \sqrt{t}},$$

where h is the height of the discharging water surface above the ground, and a and b two positive constants related to the water's height before discharging starts (time $t = 0$) and after it has finished (time $t = \infty$).

The level has been measured at random, specified times and the values are reported in the following table.

t (s)	h (cm)
3.5	117.6
7.1	113.7
14.1	110.5
28.2	107.9
31.8	107.5
49.4	106.2
52.9	106.0
56.5	105.9

Using least squares regression, find the two constants a, b and the water's height before discharging starts and after it has finished.

5.10 Linear regression, the statistician's way. Function `lm`

In this section we will revisit data fitting using the concept of statistical modelling and statistical regression, rather than the much simpler idea of finding a curve passing as close as possible to data points. We will just scratch the surface of such a vast and deep topic, and the reader willing to spend more time on it is welcome to consult specialised texts on the subject (see for instance, [4–6]).

We are here interested in explaining how data fitting and the following data analysis can be carried out in R using the very important `lm` function (where `lm` stands for *linear model*). In the function, a specialised grammar to build statistical models is used. We have seen that in general such models are not known. Either their analytic form is guessed, or, if it is known, some of its parameters are unknown and have to be calculated with the regression. For statisticians, a model is simply an economic way to account for data variability. From this perspective, models containing the smallest number of parameters are to be preferred to models with more parameters. The exception is *mechanistic models*, where the analytic form, complicated as it might be, is justified by a law or a process based on principles or criteria outside statistics (for example a physics law governing a certain process).

5.10.1 The grammar of statistical modelling

The simplest model that can be thought to account for data variability, paradoxically as it might sound, is the *constant model*, $y = a_1$. The R grammar to identify and pass to the `lm` function the instruction to use the constant model is,

$$y \sim 1 \qquad \leftrightarrow \qquad y = a_1$$

The constant of the model must be the mean of y because this is the quantity that minimises the sum of squared differences between each data point, y_i, and the constant, a_1:

$$a_1 = \bar{y}$$

The following demonstration uses the data simulated with a cubic model, already used in section 5.7. The output of `lm` can be quickly checked using the `summary` function, but more details will be provided shortly.

```
# Load comphy package
library(comphy)

# Generate 30 data points for a cubic
# y = x^3/3 - x^2/2 -2x + 1
#
set.seed(4213)
x <- seq(-2,3,length=30)
eps <- rnorm(30,mean=0,sd=0.6)
y <- x^3/3-x^2/2-2*x+1+eps

# Linear regression with constant model


model1 <- lm(y ~ 1)

# Quick check of the results
# The first number close to "(Intercept)" is
# the estimated constant value of the model
summary(model1)
```

```
##
## Call:
## lm(formula = y ~ 1)
##
## Residuals:
##     Min      1Q  Median      3Q     Max
## -3.5968 -1.5941 -0.0814  1.5379  3.3585
##
## Coefficients:
##             Estimate Std. Error t value Pr(>|t|)
## (Intercept)  -0.1071     0.3415  -0.314    0.756
##
## Residual standard error: 1.871 on 29 degrees of freedom
```

```
# This value is equal to the mean of the data
print(mean(y))
```

```
## [1] -0.1071027
```

```
# The regression line found can be drawn overlapped
# to the data points, using function abline and the
# object returned by lm.
plot(x,y,pch=16)
abline(model1,lwd=2,col=2)
```

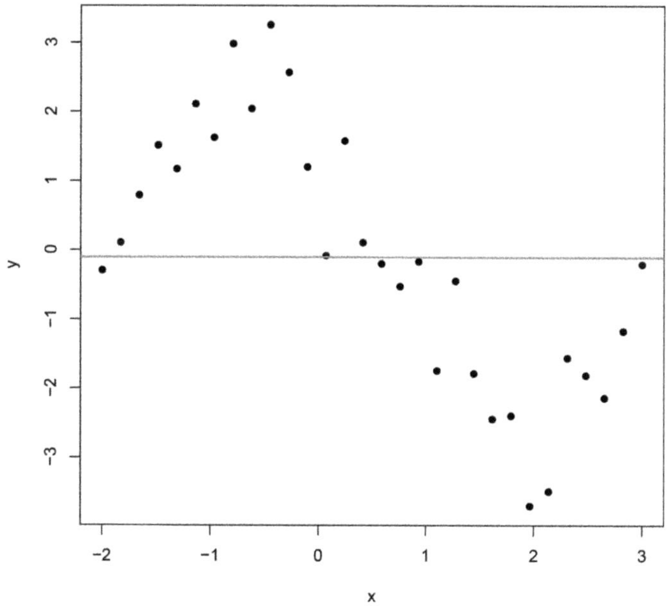

Next, in terms of complexity, is the linear model,

$$y = a_1 x + a_2$$

The grammar to describe this model is,

$$y \sim x \qquad \Leftrightarrow \qquad y = a_1 x + a_2$$

One could wonder why the one used to indicate the constant has disappeared in the above expression. This feature is part of the grammar we are describing. The one is, in fact, used to indicate absence of the intercept term. More specifically,

$$y \sim x - 1 \qquad \Leftrightarrow \qquad y = a_1 x$$

It is possible to omit the intercept term with the addition of `0`, rather than the subtraction of `1`. The use of `lm` and the quick reading of the output follows the same structure demonstrated earlier for the constant model. The difference is only that in addition to the estimate of the intercept, the coefficient a_1 is now also estimated (and it is close to the letter x). The intercept term is not estimated if forcibly omitted from the model.

```
# Linear model with intercept term
model2 <- lm(y ~ x)
summary(model2)
```

```
##
## Call:
## lm(formula = y ~ x)
##
## Residuals:
##      Min       1Q   Median       3Q      Max
## -2.40231 -0.96750  0.01261  0.77033  2.51970
##
## Coefficients:
##             Estimate Std. Error t value Pr(>|t|)
## (Intercept)   0.3352     0.2552   1.313      0.2
## x            -0.8846     0.1622  -5.455    8e-06 ***
## ---
## Signif. codes:  0 '***' 0.001 '**' 0.01 '*' 0.05 '.' 0.1 ' ' 1
##
## Residual standard error: 1.325 on 28 degrees of freedom
## Multiple R-squared:  0.5152, Adjusted R-squared:  0.4979
## F-statistic: 29.76 on 1 and 28 DF,  p-value: 8e-06
```

```
# Linear model without intercept term
model3 <- lm(y ~ x-1)
summary(model3)
```

```
##
## Call:
## lm(formula = y ~ x - 1)
##
## Residuals:
##     Min      1Q  Median      3Q     Max
## -2.0982 -0.5946  0.2858  1.1281  2.8852
##
## Coefficients:
##   Estimate Std. Error t value Pr(>|t|)
## x  -0.8169     0.1557  -5.248 1.27e-05 ***
## ---
## Signif. codes:  0 '***' 0.001 '**' 0.01 '*' 0.05 '.' 0.1 ' ' 1
##
## Residual standard error: 1.342 on 29 degrees of freedom
## Multiple R-squared:  0.4871, Adjusted R-squared:  0.4694
## F-statistic: 27.54 on 1 and 29 DF,  p-value: 1.273e-05
```

```
# Plot both regression lines
```

```
plot(x,y,pch=16)
abline(model2,lwd=2,col=2)
abline(model3,lwd=2,col=2,lty=2,add=TRUE)
```

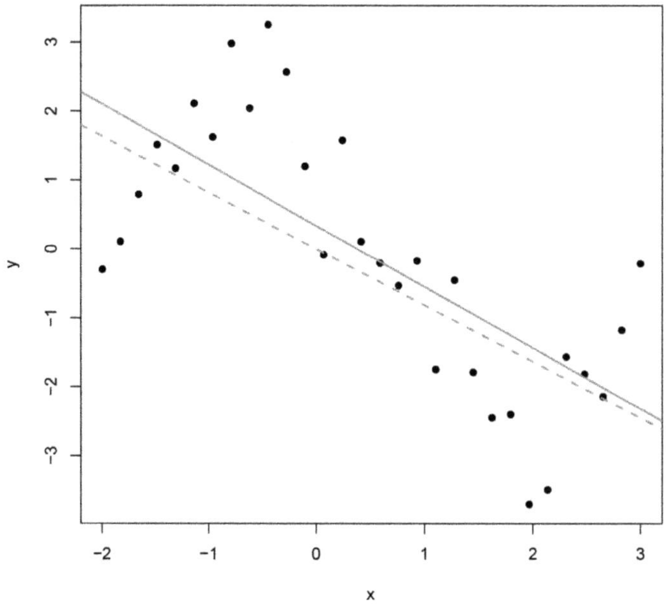

To add to the model terms of higher order (in x), we need to use the symbol `I()`, which protects its content from whatever hard-coded rule exists within R. So, for instance, `I(x^2)` and `I(x^3)` only indicates x^2 and x^3 such that,

$$y \sim I(x^2) \quad \Leftrightarrow \quad y = a_1x^2 \qquad \text{and} \qquad y \sim I(x^3) \quad \Leftrightarrow \quad y = a_1x^3$$

To model the full quadratic or cubic polynomial, the correct expressions are,

$$y \sim x + I(x^2) \quad \Leftrightarrow \quad y = a_1x^2 + a_2x + a_3$$

and

$$y \sim x + I(x^2) + I(x^3) \quad \Leftrightarrow \quad y = a_1x^3 + a_2x^2 + a_3x + a_4$$

In the summary output, the estimate of the coefficients will be printed close to the related symbol.

```
# In the next output, check where the estimate of the
# parameters are written out, close to the related symbol

# Full quadratic model
model4 <- lm(y ~ x+I(x^2))
summary(model4)
```

```
##
## Call:
## lm(formula = y ~ x + I(x^2))
##
## Residuals:
##      Min       1Q   Median       3Q      Max
## -2.31101 -0.93677  0.06633  0.75330  2.64703
##
## Coefficients:
##             Estimate Std. Error t value Pr(>|t|)
## (Intercept)   0.6003     0.3493   1.718 0.097155 .
## x            -0.7505     0.2019  -3.717 0.000932 ***
## I(x^2)       -0.1341     0.1212  -1.106 0.278306
## ---
## Signif. codes:  0 '***' 0.001 '**' 0.01 '*' 0.05 '.' 0.1 ' ' 1
##
## Residual standard error: 1.32 on 27 degrees of freedom
## Multiple R-squared:  0.5363, Adjusted R-squared:  0.5019
## F-statistic: 15.61 on 2 and 27 DF,  p-value: 3.123e-05
```

```
# Full cubic model
model5 <- lm(y ~ x+I(x^2)+I(x^3))
summary(model5)
```

```
##
## Call:
## lm(formula = y ~ x + I(x^2) + I(x^3))
##
## Residuals:
##      Min       1Q   Median       3Q      Max
## -1.32038 -0.36222 -0.04893  0.45007  1.22664
##
## Coefficients:
##             Estimate Std. Error t value Pr(>|t|)
## (Intercept)  1.38565    0.19811   6.994 1.99e-07 ***
## x           -2.11189    0.18441 -11.452 1.17e-11 ***
## I(x^2)      -0.76192    0.09371  -8.130 1.31e-08 ***
## I(x^3)       0.41854    0.04714   8.879 2.37e-09 ***
## ---
## Signif. codes:  0 '***' 0.001 '**' 0.01 '*' 0.05 '.' 0.1 ' ' 1
##
## Residual standard error: 0.6699 on 26 degrees of freedom
## Multiple R-squared:  0.885, Adjusted R-squared:  0.8717
```

```
## F-statistic:  66.7 on 3 and 26 DF,  p-value: 2.433e-12

# Plot both regression curves.
# This time we need something slightly more
# complicated than abline
plot(x,y,pch=16)
xp <- seq(-2,3,length=1000)                    ## prediction grid
yp4 <- predict.lm(model4,newdata=list(x=xp))  ## predicted values
yp5 <- predict.lm(model5,newdata=list(x=xp))  ## predicted values
lines(xp,yp4,lwd=2,col=2)
lines(xp,yp5,lwd=2,col=2,lty=2)
```

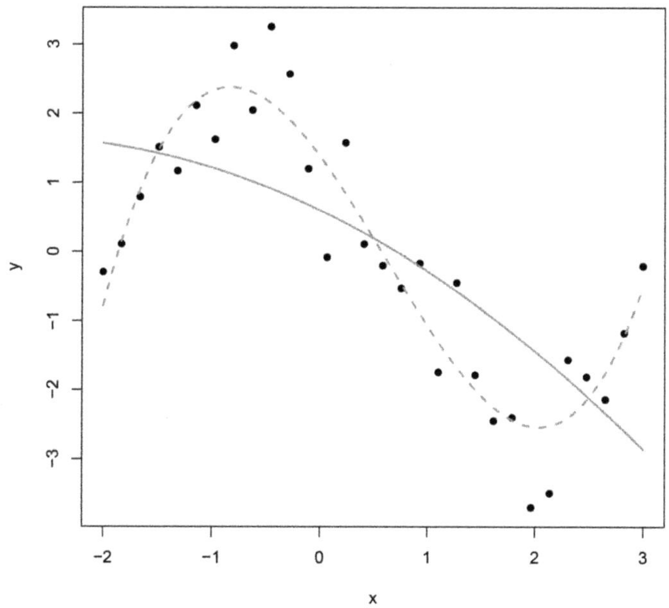

Models more complicated than those just described are possible. In fact, one should be able to use the grammatically correct R expression. A complete explanation can be found in the guide by the R Core Team [7]. We have, for example:

- $y \sim x + z \quad \Leftrightarrow \quad y = a_1x + a_2z + a_3$
- $y \sim x{*}z \quad \Leftrightarrow \quad y = a_1x + a_2z + a_3xz + a_4$
- $\log(y) \sim I(\log(x)) \Leftrightarrow \log(y) = \log(a) + b\log(x)$ transformed from $y = ax^b$

The first line indicates a linear model without interaction term. The *interaction term* is when the product xz of the two variables is present. The second line indicates a linear model with interaction term. Notice how just the expression `x*z` suffices because if there are both variables `x` and `z`, then the whole linear expression `x+z` must also be present. The last line shows how transformed data can be handled with the linear model. A demonstration follows, where new data have been simulated. The parameters of the model are nicely recovered by the regression.

```
# Simulation of y=3*x^2
set.seed(9912)
x <- seq(1,4,length=15)

# Errors are calculated with
# the propagation of errors
dx <- 0.1
dy <- rnorm(15,mean=0,sd=6*x*dx)

# Simulated data
y <- 3*x^2+dy

# Regression.
# a=exp(1.18711)=3.278, b=1.91474
# Close to 3 and 2, respectively
model6 <- lm(log(y) ~ I(log(x)))
summary(model6)
```

```
##
## Call:
## lm(formula = log(y) ~ I(log(x)))
##
## Residuals:
##       Min        1Q    Median        3Q       Max
## -0.22301 -0.06436  0.00951  0.05277  0.24719
##
## Coefficients:
##             Estimate Std. Error t value Pr(>|t|)
## (Intercept)  1.18711    0.06539   18.15 1.28e-10 ***
## I(log(x))    1.91474    0.06999   27.36 7.08e-13 ***
## ---
## Signif. codes:  0 '***' 0.001 '**' 0.01 '*' 0.05 '.' 0.1 ' ' 1
##
## Residual standard error: 0.1124 on 13 degrees of freedom
## Multiple R-squared:  0.9829, Adjusted R-squared:  0.9816
## F-statistic: 748.3 on 1 and 13 DF,  p-value: 7.084e-13
```

```
# Plot data points and regression curve
plot(x,y,pch=16)

# prediction grid
xp <- seq(1,4,length=1000)

# predicted values (linear)
yp <- predict.lm(model6,newdata=list(x=xp))

# predicted value
yp <- exp(yp)

# Plot regression curve
lines(xp,yp,lwd=2,col=2)
```

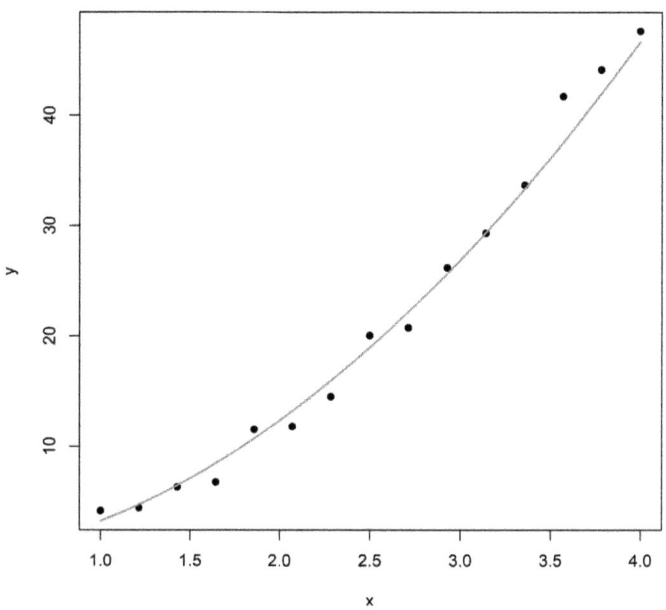

5.10.2 The output of `lm`

So far we have seen the summary output from the `lm` object, using the function `summary`, and we have learnt to identify the estimate of the coefficients of the regression. Their values and other quantities related to the fit are actually contained in the object created by the `summary` function. To be more specific, the object created by the `lm` belongs to the class `lm`. The object created applying `summary` to an object of class `lm` is another object, this time belonging to the class `summary.lm`. Objects of this class have a rich structure, as they include many other different objects. Let us look at this structure, using a practical demonstration, where data comes from the power law simulated in the last section.

```
# Simulation of y=3*x^2
set.seed(9912)
x <- seq(1,4,length=15)

# Errors are calculated with
# the propagation of errors
dx <- 0.1
dy <- rnorm(15,mean=0,sd=6*x*dx)

# Simulated data
y <- 3*x^2+dy

# Regression.
model <- lm(log(y) ~ I(log(x)))
print(class(model))
```

```
## [1] "lm"
```

```
# Object of class "summary.lm"
slm <- summary(model)
print(class(slm))
```

```
## [1] "summary.lm"
```

```
# What's in an object of class "summary.lm"?
names_slm <- names(slm)
print(names_slm)
```

```
##  [1] "call"          "terms"         "residuals"     "coefficients"
##  [5] "aliased"       "sigma"         "df"             "r.squared"
##  [9] "adj.r.squared" "fstatistic"    "cov.unscaled"
```

Most objects inside the object of class `summary.lm` relate to the important results and quantities of linear regression. The following list explains the most important ones.

- The object with name `coefficients` is a $p \times 4$ matrix with named rows and columns. p is the number of estimated model parameters, a_i, while the four columns contain the estimate of the parameters, the standard error of this estimate and two more columns on a hypothesis test about the parameters.

We will not delve here into the details of the statistical theory behind the mentioned quantities, but a sketchy summary is given in appendix C, while dedicated books on probability and statistics will have to be consulted for a detailed exposition (see for instance [8]). What can be perhaps highlighted is that the parameters of the linear model are treated, in statistics, as random variables with a given distribution. Under the assumptions normally valid for linear regression (see section C.2, i.e. that each value y_i comes from a normal distribution of the random variable Y_i, with mean $f(x_i)$ and a same constant variance σ^2, that each residual $\epsilon_i \equiv Y_i - f(x_i)$ comes also from a normal distribution, with mean 0 and same variance σ^2, and that all residuals are independent from each other), the expected value of each parameter is the one returned from the least squares regression, and its confidence interval is found using the following related random variable (see equation (C.14)),

$$(a_i - \hat{a}_i)/s_{\hat{a}_i},$$

in which $\hat{a}_i$ is the least squares' estimate and $s_{\hat{a}_i}$ is its standard error. This random variable is distributed according to the t-Student distribution with $n - (m + 1)$ degrees of freedom. The standard error is given by formulas like equation (C.11) or (C.12) and in which σ^2 is replaced by the sample version, s^2. It can be used to find out the confidence interval for $\hat{a}_i$ as follows. If t_α is the critical value corresponding to a given probability P_α, then the true value of $\hat{a}_i$ will be in the interval,

$$\hat{a}_i - t_\alpha s_{\hat{a}_i} \leqslant a_i \leqslant \hat{a}_i + t_\alpha s_{\hat{a}_i}$$

with probability P_α. An example will clarify the idea. Let us simulate $n = 15$ data points from the model,

$$y = 2x - 1$$

in which each residual ϵ_i has mean 0 and standard deviation $\sigma = 0.1$. After regression, these are the estimate of parameters a_1, a_2 and their standard errors, as calculated by `summary.lm`.

```
# Simulated data points
set.seed(7760)
x <- seq(2,5,length=15)
eps <- rnorm(15,mean=0,sd=0.1)
y <- 2*x-1+eps

# Regression
model <- lm(y ~ x)
slm <- summary(model)

# Estimates and standard errors of the parameters
cffs <- slm$coefficients
```

```
print(cffs[,1:2])
```

```
##              Estimate Std. Error
## (Intercept) -0.8777115  0.1007124
## x            1.9797104  0.0278182
```

The estimates and the standard errors can be used to calculate the confidence interval, say, at 95%. The same interval can be calculated using the function confint.

```
# Calculation of the 95% confidence interval by hand

# degrees of freedom
df <- slm$df[2]

# Critical value at 95% ( 0 -- 0.025 -- 0.50 -- 0.975 -- 1)
talpha <- qt(0.975,df)
print(talpha)
```

```
## [1] 2.160369
```

```
# Interval for first parameter
cinf1 <- c(cffs[1,1]-talpha*cffs[1,2],cffs[1,1]+talpha*cffs[1,2])
print(cinf1)
```

```
## [1] -1.0952874 -0.6601355
```

```
# Interval for second parameter
cinf2 <- c(cffs[2,1]-talpha*cffs[2,2],cffs[2,1]+talpha*cffs[2,2])
print(cinf2)
```

```
## [1] 1.919613 2.039808
```

```
# It's much easier with confint
confint(model)
```

```
##                  2.5 %     97.5 %
## (Intercept) -1.095287 -0.6601355
## x            1.919613  2.0398080
```

The last two columns display the results of an hypothesis test on each parameter. The null hypothesis is that the parameter is zero, while the alternative hypothesis is that the parameter is different from zero. The t-Student statistic is

$$t = \hat{a}_i / s_{\hat{a}_i},$$

and it is shown in column 3. The fourth column is the probability of a test statistic at least as unusual as $-|t|$ or $+|t|$ (because the t-Student distribution has two symmetric tails), if the null hypothesis were true. If this value is very small, then one can be very confident that the parameter is very different from zero (check the hypothesis testing framework in books on introductory statistics).

```
# The full "coefficients" matrix
print(cffs)
```

```
##               Estimate Std. Error   t value     Pr(>|t|)
## (Intercept) -0.8777115  0.1007124 -8.715026 8.654134e-07
## x            1.9797104  0.0278182 71.166012 3.097120e-18
```

The very small values for `Pr(>|t|)` suggest, in this case, that both parameters have mean values very different from zero.

- The object with name `residuals` is a vector filled with all the n residuals, ϵ_i. They correspond to the difference $y_i - \hat{a}_1 x_i - \hat{a}_2$, in the case of a straight line model.

```
# The residuals (print first four)
reds <- slm$residuals
print(class(reds))
```

```
## [1] "numeric"

print(length(reds))

## [1] 15

print(reds[1:4])

##          1          2           3          4
## 0.02598981 0.06687114 -0.17748529 0.15717467

# Let's check one of them, say the first
# It should be y_1-<a1>x_1-<a2>
print(y[1]-cffs[2,1]*x[1]-cffs[1,1])

## [1] 0.02598981

print(slm$residuals[1])

##          1
## 0.02598981
```

Residuals can also be extracted with the function `resid`, as in `resid (model)`, where `model` is an object of class `lm`.

- The object with name `sigma` is the square root of the estimate of the variance σ^2, it is, in other words, the s^2 explained in appendix C (see equation (C.9)). It is therefore related to the sum of squared residuals because,

$$s^2 = \sum_{i=1}^{n}(Y_i - \hat{g}(x_i))^2/(n - (m + 1)) = \sum_{i=1}^{n}\epsilon_i^2/(n - (m + 1))$$

And this is also the definition of σ_e^2 in equation (5.17)

```
# Estimate of variance sigma^2
print(slm$sigma^2)

## [1] 0.00994953

# Can be calculates using the sum of squared residuals
SSE <- sum(reds^2)
print(SSE/df)

## [1] 0.00994953
```

- The object with name `r.squared` represents r^2, which is a measure of how much data variability is described by the model. r^2 can be defined once the *famous five* quantities of the linear regression of $y = a_1 x + a_2$ have been introduced. They are:
  - SX=$\sum_{i=1}^{n} x_i$
  - SSX=$\sum_{i=1}^{n}(x_i - \bar{x})^2$
  - SY=$\sum_{i=1}^{n} y_i$
  - SSY=$\sum_{i=1}^{n}(y_i - \bar{y})^2$
  - SSXY=$\sum_{i=1}^{n}(x_i - \bar{x})(y_i - \bar{y})$

  These five quantities can be used to determine all the estimates of the regression, and more. In particular we have,

$$r^2 = \frac{SSY - SSE}{SSY}, \tag{5.18}$$

  where it is worth recalling that SSE is the sum of squared residuals. From the formula we infer that r^2 relates to that part of the variation that is explained by the model. When nothing is explained by the model (but it is only a product of chance) then SSY=SSE and $r^2 = 0$. When everything is explained by the model, then SSE=0 and $r^2 = 1$.

```
# Calculate r.squared as (SSY-SSE)/SSY
SSY <- sum((y-mean(y))^2)
SSE <- sum(reds^2)
my_r.squared <- (SSY-SSE)/SSY
```

```
# r.squared from lm
print(slm$r.squared)
```

```
## [1] 0.9974397
```

```
# Comparison
print(my_r.squared)
```

```
## [1] 0.9974397
```

r^2 is also in the printout from the `summary` command; it is the `"Multiple R-squared:"` value.

There are other objects in `summary.lm`, and their nature and usefullness can be investigated using the help pages (`help(ls)`), or consulting books on R.

5.10.3 Checking the model

The model chosen for regression can be not the appropriate model to justify the data. There are some tools to help checking whether the model chosen is the correct one to account for the available data. As one of the assumptions for the linear regression was that the residuals had to be distributed according to a normal with mean 0, then tools here chosen make use of the residuals. We will look at a couple of them, but a more complete list can be found in reference [4].

- The *residuals versus fitted-data* plot. The rationale behind this plot is that if the errors are distributed according to a normal distribution with mean zero and the same variance, and if they are independent, then a plot of these errors (the residuals ϵ_i) versus the fitted data $\hat{y}$ should resemble a uniform distribution of points in the plot window. The fitted values can be obtained with `fitted(model)`, where `model` is the `lm` object. They should be forming a regular grid if the x_i form a regular grid, because they are their projections on a straight line. The residuals should display no observable obvious pattern because of what we said earlier. The easiest way to obtain the residual versus fitted-data plot is via the use of `plot` on the object of class `lm`.

```
# Simulate data from straight line y=2x-1
set.seed(3310)
x <- seq(0,5,length=16)
```

```
y <- 2*x-1+rnorm(16,mean=0,sd=0.5)

# Regression
model <- lm(y ~ x)

# Functions "fitted" and "resid" can be used
# to quickly extract fitted data and residuals
yp <- fitted(model)
reds <- resid(model)

# Residual vs Fitted plot
plot(model,which=1,lwd=2)
```

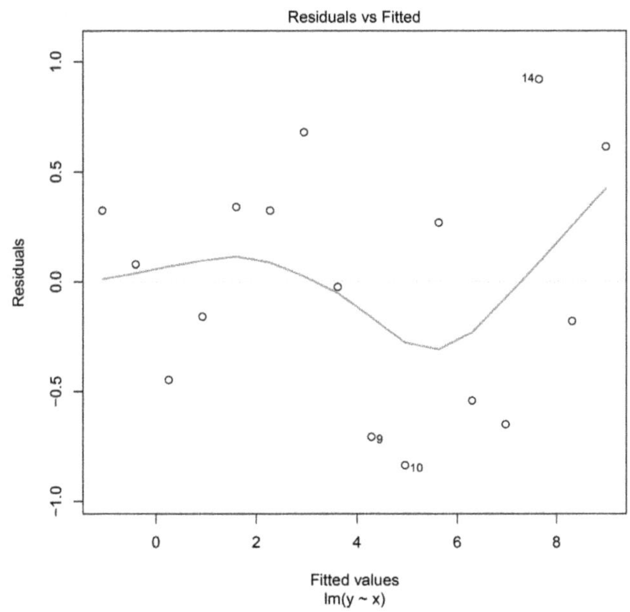

The red curve in the plot serves as an indicator of how much pattern the data have. The ideal case is when the line coincide with the horizontal zero line, means that the assumptions for regression hold in full. Small deviations from this behaviour are acceptable, but when these are larger or form visible trends, then at least a part of the assumptions does not hold. Data points with a number close to it might indicate outliers. The red curve is the result of an extreme smoothing of the data points using the function `lowess`, with *smoother span*[2] $f = 0.675$.

[2] In the help page for `lowess` we read: 'the smoother span'. This gives the proportion of points in the plot which influence the smooth at each value. Larger values give more smoothness.

```
# The red curve is a smoothing of the data, obtained using the
# smoother "lowess". The value f=0.675 controls how smooth the
# line is.
lo <- lowess(x=yp,y=reds,f=0.675)

# Overlap this to the previous curve to show they are the same
plot(model,which=1,lwd=2)
lines(lo,lty=2,col=4,lwd=3)
```

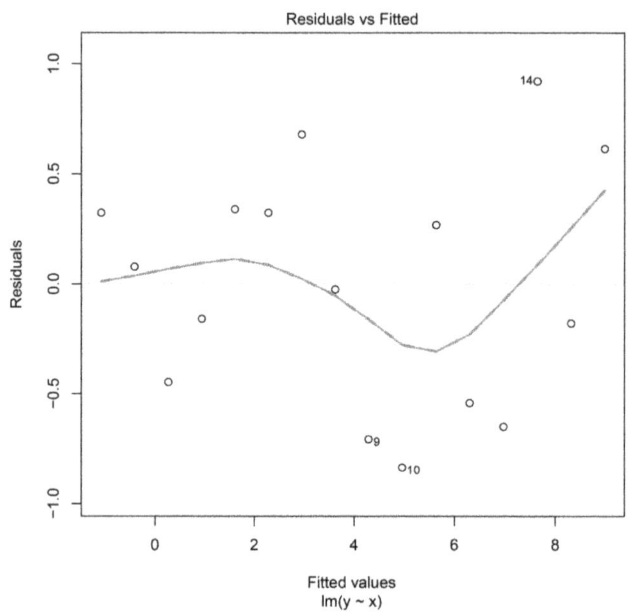

- The *normal QQ plot.* When data from the CDF (cumulative distribution function, see section C.1) of a normal distribution are compared with those coming from the CDF of another normal distribution, they tend to align along a straight line. The plot comparing the two CDFs is known as *normal QQ plot.* In R it can be created using the two functions qqnorm and qqline on the data whose distribution needs to be checked.

```
# Data from a generic normal distribution
x <- rnorm(100,mean=10,sd=0.1)
qqnorm(x)
qqline(x)
```

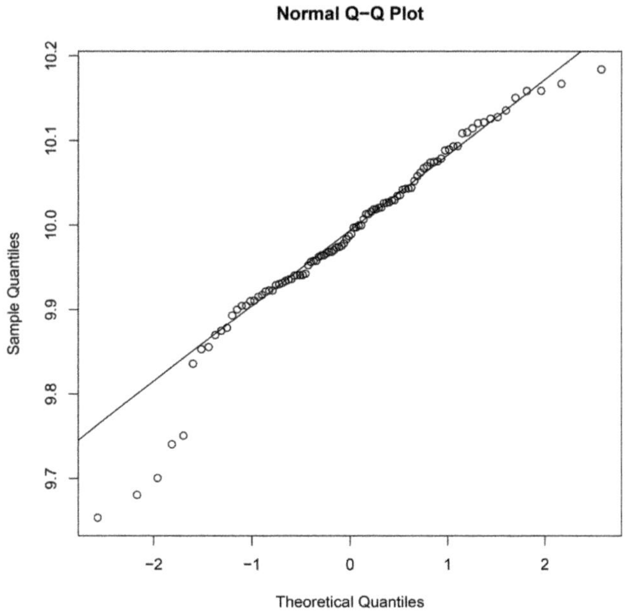

In the above plot, some of the data, in the lower distribution's tail, can be outliers, while most data belong to a normal distribution. The QQ plot is markedly different when the second distribution is not normal, as shown in the following demonstration in which the data come from a uniform distribution.

```
# Data from a generic normal distribution
x <- runif(100)
qqnorm(x)
qqline(x)
```

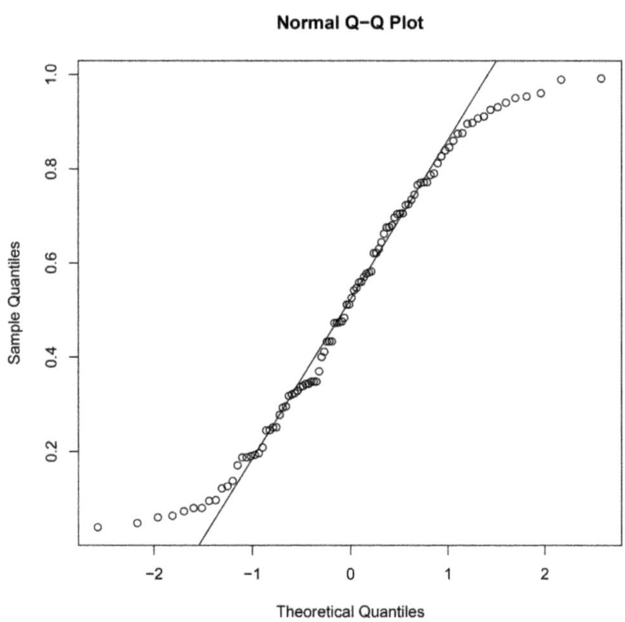

This S-shaped behaviour is typical of data who are not distributed according to a normal distribution. A U-shaped pattern also signals differences with a normal distribution. A normal QQ plot created for the residuals of a linear regression, gives an idea of whether they are distributed normally. In this case, S-shaped or U-shaped patterns indicate that the chosen model is possibly wrong. When an object of class `lm` is available, the best tool to create a normal QQ plot is `plot` with option `which=2`.

```
# Simulate data from straight line y=2x-1
set.seed(3310)
x <- seq(0,5,length=16)
y <- 2*x-1+rnorm(16,mean=0,sd=0.5)

# Linear regression
model <- lm(y ~ x)

# QQ plot (the residuals are distributed normally)
plot(model,which=2)
```

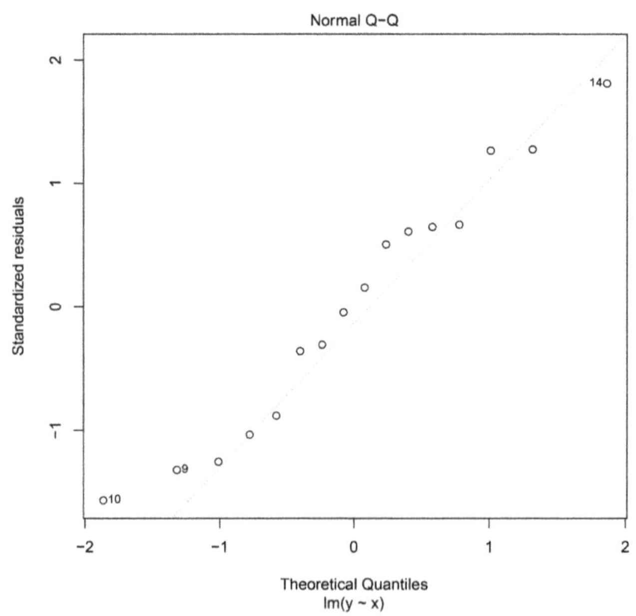

Here the QQ plot does not indicate any particularly evident deviation from a straight-line pattern, therefore the linear model chosen does not need to be changed.

5.11 Exercises on statistical linear regression

Exercise 06

Using the `lm` function, find the estimated parameters for the following models:

1. $y = 2x + 1$
2. $y = 6x_1 - 2x_2 + 3x_3 + 1$

The independent variables, x, x_1, x_2, x_3 have the following values:

$$
\begin{aligned}
x &= \{0.23, 0.31, 0.34, 0.64, 0.65, 1.42, 2.07, 3.03, 3.17, 3.27, 4.96\} \\
x_1 &= \{0.09, 0.34, 1.00, 1.04, 2.05, 2.05, 2.59, 3.44, 4.44, 4.59, 4.99\} \\
x_2 &= \{1.39, 4.51, 4.63, 5.12, 5.50, 7.89, 8.03, 9.48, 9.64, 9.64, 9.94\} \\
x_3 &= \{5.13, 5.41, 5.45, 5.49, 5.62, 5.67, 5.93, 6.41, 6.79, 6.97, 7.77\}
\end{aligned}
$$

The errors (residuals) for the first simulation are generated by a random distribution with mean 0 and variance $\sigma^2 = 0.01$ (use seed 1132), while for the second simulation the mean is still 0 and the variance $\sigma^2 = 0.04$ (use seed 2311).

Exercise 07

The estimation of parameters in multilinear regression can be problematic when one or more independent variables are correlated (*multicollinearity*). This exercise is about perfect collinearity between two independent variables in a linear model of the type,

$$y = a_1 x_1 + a_2 x_2 + a_3$$

The exercise (and its solution) provides insights on the problem of multicollinearity and on how this can be explained and managed.

The data points are listed in the following table:

x_1	x_2	y
0.0	–5.0	–7.16
0.5	–4.5	–3.06
1.0	–4.0	0.96
1.5	–3.5	5.19
2.0	–3.0	9.02
2.5	–2.5	13.02
3.0	–2.0	17.10
3.5	–1.5	21.17
4.0	–1.0	24.86
4.5	–0.5	29.27
5.0	0.0	32.82
5.5	0.5	37.00
6.0	1.0	40.90
6.5	1.5	44.95

(*Continued*)

7.0	2.0	48.91
7.5	2.5	53.08
8.0	3.0	56.95
8.5	3.5	60.81
9.0	4.0	65.18
9.5	4.5	69.22
10.0	5.0	73.04

(a) Try the least squares regression on these data, using the function `lm`. You will not be able to proceed. Can you explain why?
(b) Calculate, using the function `cor`, the correlation between x_1 and x_2.
(c) Using the result from part *b*, attempt a way out of the problem met in part *a*, and finally find an estimate for a_1, a_2, a_3.

References

[1] Golub G H and Van Loan C F 1996 *Matrix Computation* (John Hopkins University Press)
[2] Multicollinearityhttps://en.wikipedia.org/wiki/Multicollinearity (Accessed: 2022-01-15)
[3] Tipler P A and Mosca G 2008 *Physics for Scientists and Engineers* 6th edn (W. H. Freeman and Company)
[4] Crawley M J 2012 *The R book* (John Wiley & Sons)
[5] Mathai A M and Haubold H J 2017 *Probability and Statistics: a Course for Physicists and Engineers* (de Gruyter & Co)
[6] Barlow R J 2002 *Statistics. A Guide to the Use of Statistical Methods in the Physical Sciences* (Wiley)
[7] R Core Team An introduction to R https://cran.r-project.org/doc/manuals/R-intro.html (Accessed: 2022-01-12)
[8] Larson H J 1982 *Introduction to Probability Theory and Statistical Inference* (Wiley)

IOP Publishing

Computational Physics with R

James Foadi

Chapter 6

Numerical solution of nonlinear equations

The solutions of the generic equation,

$$f(x) = 0$$

are often called *roots* of the equation. But, as in the above case, any expression can be re-arranged so to have a zero at the right hand side, therefore the roots of an equation also correspond to the *zeros* of a specific function (f in the above generic example). Many methods are known to find analytic expressions of some nonlinear equations (see [1] for some examples), but very often these are not available, and one will have to resort to numerical approximations. In this chapter, the main emphasis will be on the bisection, secant and Newton methods. Some improvements to these methods will be also treated, but many more exist, quite often to solve specialised classes of nonlinear equations. A substantial albeit incomplete survey is contained in reference [2].

6.1 The bisection method

The zero of a single-variable function $f(x)$ corresponds to the curve representing the function on the Cartesian plane, crossing the x-axis. The bisection method stems from the simple observation that the values of the function at the two extremes of a small interval including the zero, have opposite signs (see figure 6.1). If the interval is divided exactly in two adjacent intervals with the introduction of a mid-point $c = (x_L + x_R)/2$, then the zero will be inside one of the two smaller intervals, or exactly at the mid-point c. When the last case is verified, then $f(c) = 0$ and the zero is c. If not, then only one of the smaller intervals will include the zero, the one for which the values of the function at its extremes have opposite signs. Once the new and smaller interval is identified, it becomes a new interval on which bisection can be performed once again. This procedure is repeated over and over again, until the width of the interval including the zero is smaller than a predefined width (*tolerance*).

doi:10.1088/978-0-7503-2632-2ch6

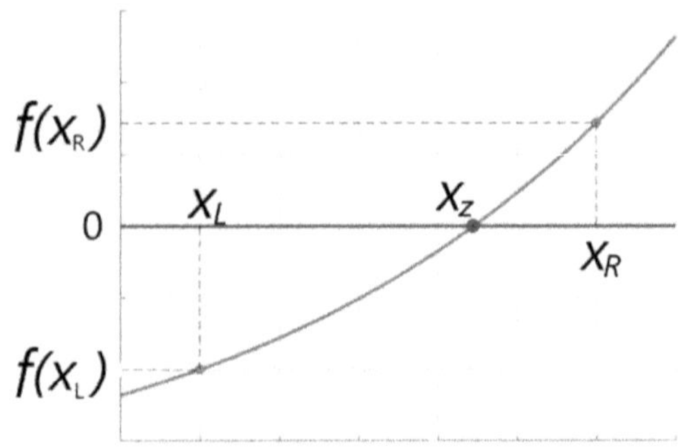

Figure 6.1. The zero of a function $f(x)$ corresponds to the point x_z at which the curve representing the function crosses the x-axis (blue point). For a small interval around the zero, here represented with x_z, the values of the function at its extremes x_L and x_R, have opposite signs.

In that case the zero will be approximated by the latest mid-point c, and the error obtained using such an approximation will be,

$$\epsilon = |x_z - c| < |x_L - x_R|/2^n, \tag{6.1}$$

where n is the number of times bisection has been applied.

Let us demonstrate the method using a simple quadratic function,

$$f(x) = x^2 - 5x + 6$$

which has two zeros, $x_z = 2$ and $x_z = 3$. If the starting interval for the search is $[-2.1, 2.1]$, the series of progressively smaller intervals selected by the method, together with their width, is displayed in the following table.

Iteration	Left	Right	Absolute difference
1	−2.100 000	2.100 000	4.200 000
2	0.000 000	2.100 000	2.100 000
3	1.050 000	2.100 000	1.050 000
4	1.575 000	2.100 000	0.525 000
5	1.837 500	2.100 000	0.262 500
6	1.968 750	2.100 000	0.131 250
7	1.968 750	2.034 375	0.065 625
8	1.968 750	2.001 563	0.032 813
9	1.985 156	2.001 563	0.016 406
10	1.993 359	2.001 563	0.008 203

We can observe how the interval selected includes always the root $x_z = 2$ and that such an interval has half the width of the interval in the previous iteration. At iteration 10, the width of the interval containing the root, [1.993 359, 2.001 563], is 0.008 203. If the tolerance had been set to, say, 0.01 the algorithm would have stopped at iteration 10, returning an approximated value for the root equal to $x_z = (1.993\ 359 + 2.001\ 563)/2 = 1.997\ 461$. The error committed by using this approximation is, in this case, $\epsilon = |2 - 1.997\ 461| = 0.002\ 539$. According to formula (6.1) this should be less than $4.2/2^{10} = 0.004\ 101$, and this is, indeed, the case.

In general, the bisection method proceeds as follows:

1. Choose two initial values which are suspected to be smaller and larger of the anticipated root and label these as x_L and x_R. In other words, these points are to be chosen such that the zero x_z is in the interval $[x_L, x_R]$.
2. Find the values of $f(x_L)$ and $f(x_R)$, then:
   - if $f(x_L)$ and $f(x_R)$ have the same sign, choose a different set of initial points and start again.
   - if $f(x_L)$ and $f(x_R)$ have different signs, proceed
3. Let $c = (x_L + x_R)/2$ and find $f(c)$, then:
   - if $f(c) = 0$ the zero is $x_z = c$.
   - if $f(c)$ and $f(x_L)$ have different signs let $x_{L1} = x_L$ and $x_{R1} = c$
   - if $f(c)$ and $f(x_R)$ have different signs let $x_{L1} = c$ and $x_{R1} = x_R$.
4. Go back to step 2 and repeat while replacing x_L and x_R with x_{L1} and x_{R1}, respectively.
5. Continue this process, taking care to update the indices of x_{Li} and x_{Ri}, until the width of the interval $[x_{Li}, x_{Li}]$, is less than some preassigned tolerance, or if the number of iterations has exceeded some given value.

Example 6.1 *Find the numerical approximation of the roots of the equation,*

$$2\sin(2\pi x) = \cos(2\pi x),$$

in the interval $[0, 1)$.

The equation can be re-arranged in the form,

$$2\sin(2\pi x) - \cos(2\pi x) = 0 \Leftrightarrow f(x) = 0$$

The function can cross the x-axis several times (multiple roots), therefore we must have a rough idea of where this happens. A possibility is given by observing the plot of the function $f(x) = 2\sin(2\pi x) - \cos(2\pi x)$ *in the given interval (see figure 6.2).*

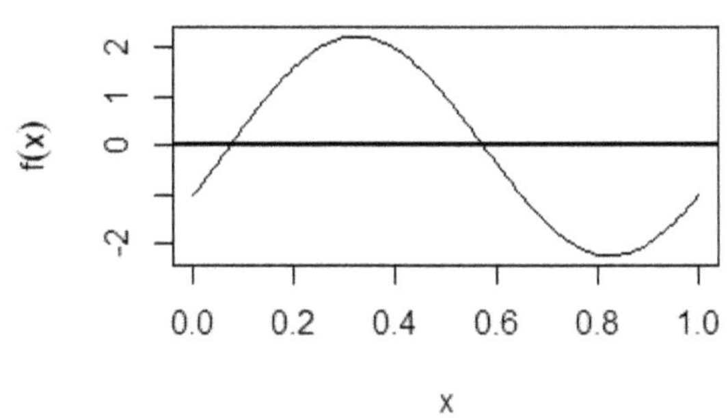

Figure 6.2. Plot of $f(x) = 2\sin(2\pi x) - \cos(2\pi x)$. The function crosses the x-axis two times.

From the plot it is clear that the function has two zeros and that one of them is somewhere between 0 *and* 0.5, *while the other is somewhere between* 0.5 *and* 1. *The bisection method to find the first root, using the search interval* [0, 0.5], *and using a tolerance (equivalent to the interval's width below which the algorithm stops) equal to* 0.001, *returns* $x_z = 0.073\,24$. *The error is, in this case, less than* 0.000 49; *this means that in the worst case the correct root could be* $0.073\,24 - 0.000\,49 = 0.072\,75$ *or* $0.073\,24 + 0.000\,49 = 0.073\,73$.

For the second root, the interval [0.5, 1] *is used, with the same tolerance as before,* 0.001. *The result is* $x_z = 0.573\,24$, *with the same upper limit for the error as before.*

6.2 Relevant R code. Function `roots_bisec`

This function is part of the `comphy` package. Its code, that can be viewed by typing `roots_bisec` in an R console, essentially carries through the bisection algorithm explained previously. The function can be demonstrated by finding the roots of the equation

$$e^x = e^{x^2 - 3}$$

The equation has been chosen because although the function resulting for the numerical search of the roots,

$$f(x) = e^x - e^{x^2 - 3},$$

is highly nonlinear, the solutions can be found analytically by taking the natural logarithm of both sides of the equation, so as to have

$$x^2 - x - 3 = 0 \quad \Rightarrow \quad x_1 = (1 + \sqrt{13})/2, \quad x_2 = (1 - \sqrt{13})/2$$

We have therefore their correct analytic expression and can test the precision of the numerical solutions found. `roots_bisec` takes as arguments:

- `f`, the name of the function whose roots have to be calculated.
- `x0`, a starting point, potentially close to a root. Not used if `lB` and `rB` have both valid numeric values.
- `lB,rB`, two extremes of an interval potentially including a root.
- `tol`, a small number charaterising the maximum error incurred when using the numerical approximation. If x is the numerical approximation to the root and x_t is the correct root, then

  $$|x - x_t| < \text{tol}$$

  The default value for `tol` is 10^{-9}.
- `imax`, an integer indicating the maximum number of bisections performed before stopping the algorithm. The default value is 10^6, although convergence (with `tol`) is normally reached with a number of bisections much smaller than `imax`.

- eps. When a single value, x0, is provided as input, the program searches for an interval around this value, such that the function has opposite signs at its extremes. The interval found is symmetric around x0, with a radius increasing in steps of eps, until the different signs are found. As the expansion of such an interval cannot continue indefinitely, the program stops with a warning message when the interval's radius reaches eps*imax. The default value for eps is 0.1.
- logg A logical variable to enable printing on detailed information on all iterations. Default is FALSE, i.e. no printing is enabled.

The first R chunk here displayed finds both roots of the nonlinear function above, starting from two different points, each one closer to one of the roots. The code also demonstrates how the numerical solutions are accurate to the default, which is the tenth decimal. Finally, search for one of the two roots is carried out using an interval, rather than a single point.

```
# Load comphy package
library(comphy)

# The equation is exp(x)=exp(x^2-3)
# Its solutions correspond to the two zeros
# of the function exp(x)-exp(x^2-3). These
# are (1 + sqrt(13))/2 and (1-sqrt(13))/2.
root1 <- (1+sqrt(13))/2
root2 <- (1-sqrt(13))/2

# Create function
f1 <- function(x) {return(exp(x)-exp(x^2-3))}

# Use roots_bisec with two different starting points
# (each one close to a different root)
xx <- roots_bisec(f=f1,x0=2)


## Searching interval: [1.600000,2.400000].
## The root is 2.302776. The error is less than 3.725290e-10.


# Increase accuracy to 10 digits to check
backup_options <- options()
options(digits=10)
```

```
print(root1)

## [1] 2.302775638

print(xx)

## [1] 2.302775637

# Back to default accuracy
options(backup_options)

# The other root
xx <- roots_bisec(f=f1,x0=-1)

## Searching interval: [-1.400000,-0.600000].
## The root is -1.302776. The error is less than 3.725290e-10.

print(xx)

## [1] -1.302776

# Now use an interval including one of the roots
xx <- roots_bisec(f=f1,lB=0,rB=5)

## Searching interval: [0.000000,5.000000].
## The root is 2.302776. The error is less than 2.910383e-10.

print(xx)

## [1] 2.302776
```

In the second demonstration, the precision is increased using a value of `tol` equal to 10^{-12}.

```
# Increase precision
xx <- roots_bisec(f=f1,x0=2,tol=1e-12,message=FALSE)
backup_options <- options()
options(digits=12)
print(root1)

## [1] 2.30277563773

print(xx)

## [1] 2.30277563773

# Back to default accuracy
options(backup_options)
```

The third R chunk explores special cases. In the first case, an interval including two roots rather than one is chosen. The algorithm usually returns only one of the two roots, the one closest to one of the interval's extremes. The second case displays the warning message when an interval cannot be found, in this case because there are no real roots.

```
# Choose an interval including both roots
# Only one root selected (as -2 closest to root2
# than 5 is to root1)
xx <- roots_bisec(f=f1,lB=-2,rB=5)

## Searching interval: [-2.000000,5.000000].
## The root is -1.302776. The error is less than 4.074536e-10.

# The quadratic equation x^2+1=0 has no real roots.
# The corresponding function f(x)=x^2+1 does not intersect
# the x axis. Starting from x0=0, a search interval cannot
# be found
f2 <- function(x) {return(x^2+1)}
xx <- roots_bisec(f=f2,x0=0)

Warning message:
In roots_bisec(f = f2, x0 = 0) :
## Failed to find starting interval within given imax.
```

6.3 The Newton–Raphson method

The only root of a straight line,

$$f(x) = a + bx,$$

is $x = -a/b$, assuming that the line is not parallel to the x-axis. Starting from any point x_0 different from $-a/b$, consider the new point x_1, calculated using the formula,

$$x_1 = x_0 - \frac{f(x_0)}{f'(x_0)}, \tag{6.2}$$

where $f'(x_0)$ is the first derivative of the function at x_0. As for the given straight line $f'(x_0) = b$, the formula just introduced yields

$$x_1 = x_0 - \frac{f(x_0)}{b} = x_0 - \frac{a + bx_0}{b} = -\frac{a}{b}.$$

Therefore, starting from any value of x, and using the iteration (6.2), one obtains the root of $f(x)$, if this is a linear function. As any smooth function can be approximated by a straight line in a small neighbourhood around any of its roots, we expect the same formula to provide, with successive iterations, approximations of the roots that are closer and closer to the root itself, provided the initial guess, x_0, is close enough to the root. At iteration $n + 1$, the formula generalising equation (6.2) is

$$x_{n+1} = x_n - \frac{f(x_n)}{f'(x_n)}. \tag{6.3}$$

This approach to finding the numerical value of a function's roots is known as *Newton–Raphson* method. In general, Newton–Raphson converges to a solution faster than the bisection method, as is discussed later in this chapter.

Example 6.2 *Let us find one of the roots of* $f(x) = x^2 - 5x + 6$, $x = 2$, *starting from the initial guess* $x_0 = 1$.

In order to us the method, one needs to work out the analytic expression of the first derivative. Here $f'(x) = 2x - 5$. *The succession of iterates, starting from* $x_0 = 1$, *is listed in the following table. The last column of the table reports the approximate root values, obtained using the bisection method, just for comparison.*

Iteration	$f(x_n)$	$f'(x_n)$	x_n	x_{n+1}	Bisection
0	2.000 000	−3.000 000	1.000 000	1.666 667	1.000 000
1	0.444 444	−1.666 667	1.666 667	1.933 333	1.550 000
2	0.071 111	−1.133 333	1.933 333	1.996 078	1.825 000
3	0.003 936	−1.007 843	1.996 078	1.999 985	1.962 500
4	0.000 015	−1.000 031	1.999 985	2.000 000	2.031 250

It is evident that the Newton–Raphson method converges to $x = 2$ faster than the bisection method. In fact, the value $x = 2.000\,000$, obtained with Newton–Raphson already at the fourth iteration, is reached with the bisection method only at the 21st iteration.

Although Newton–Raphson enjoys faster convergence, it suffers the drawback of needing the analytic expression of the first derivative which, when the starting function is complicated, can be tricky to calculate. Furthermore, the Newton–Raphson method does not converge in some special cases, for example when the first derivative at x_n is zero so that the expression (6.3) is undefined. There are also situations in which x_{n+1} is accidentally equal to one of the values obtained previously. In such a peculiar case, the algorithm will repeat all the values obtained up to and including x_{n+1}. This means that the algorithm remains stuck in a non-converging loop.

Example 6.3 *Consider the function* $f(x) = x^2 + 1$. *This function has no real roots (as* $x^2 + 1 = 0$ *yields* $x = \pm \mathrm{i}$*), but let us pretend we do not know it and apply Newton–Raphson, starting from* $x_0 = 1/\sqrt{3}$. *Using* $f'(x) = 2x$, *the first iteration yields,*

$$x_1 = \frac{1}{\sqrt{3}} - \frac{4/3}{2/\sqrt{3}} = -\frac{1}{\sqrt{3}}$$

The second iteration yields

$$x_2 = -\frac{1}{\sqrt{3}} - \frac{4/3}{-2/\sqrt{3}} = \frac{1}{\sqrt{3}},$$

i.e. $x_2 = x_0$. *The the cycle is repeated with exactly the same two values. This is indeed an example of the case in which the algorithm gets stuck in a non-converging loop.*

6.4 The secant method

One of the annoying features of the Newton–Raphson method is that the analytic expression of the first derivative must be provided, in addition to the function. The secant method obviates to this by replacing the first derivative with a finite difference approximation.

Consider then two points P_0, P_1 on the curve representing $f(x)$, see figure 6.3.

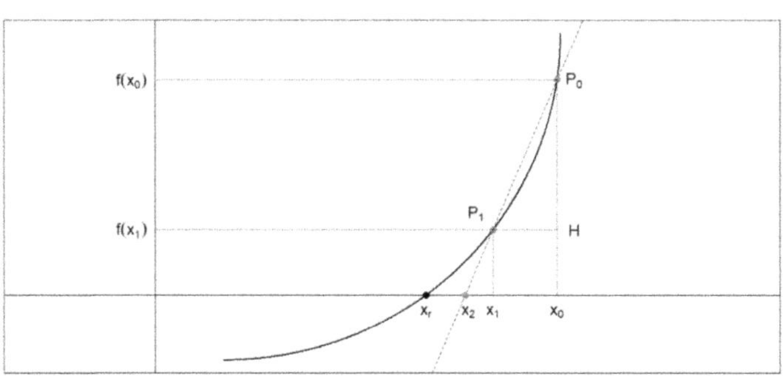

Figure 6.3. Construction used for the derivation of the algorithm in the secant method. The two triangles $x_2 P_1 x_1$ and $P_1 P_0 H$ are similar.

Due to the similitude of the triangles $x_2P_1x_1$ and P_1P_0H, the following identity holds,

$$\frac{x_2 - x_1}{f(x_1)} = \frac{x_0 - x_1}{f(x_0) - f(x_1)}$$

The value of x_2, which is an approximation to the root x_r better than x_0 and x_1, is then

$$x_2 = x_1 - f(x_1)\frac{x_0 - x_1}{f(x_0) - f(x_1)}$$

This means that two starting guesses, x_0 and x_1, are needed to calculate the next approximation to the root. It is important that $|f(x_1)| < |f(x_0)|$, for the approximation x_2 to be, in general, nearer to the root. If this is not the case then it will suffice, before starting the algorithm, to swap x_0 with x_1.

The form of the algorithm at iteration $n + 1$ will be

$$x_{n+1} = x_n - f(x_n)\frac{x_{n-1} - x_n}{f(x_{n-1}) - f(x_n)} \tag{6.4}$$

Example 6.4 *Consider again $f(x) = x^2 - 5x + 6$, which has roots $x = 2$ and $x = 3$. Let us find the numerical approximation to root $x = 2$ using the secant method, and compare its performance with both the bisection and Newton–Raphson methods. In order to make the comparison fair and meaningful, use two points around the root in the case of the secant and bisection method and one of the two points for Newton–Raphson; also limit the iteration to a tolerance equal to 10^{-6}.*

The values of the successive approximations to the root are given in the following table.

Iteration	Bisection	Secant	Newton–Raphson
1	1.800 000	2.500 000	1.100 000
2	2.150 000	2.321 429	1.710 714
3	1.975 000	1.100 000	1.946 986
4	2.062 500	2.183 258	1.997 459
5	2.018 750	2.096 073	1.999 994
6	1.996 875	1.975 570	2.000 000
7	2.007 812	2.002 528	2.000 000
8	2.002 344	2.000 060	2.000 000
9	1.999 609	2.000 000	2.000 000

Newton–Raphson converges first, followed by the secant method, which converges after 9 iterations. The bisection method reaches convergence only after 22 iterations (values not shown).

The example just described shows that the secant method is closer to the Newton–Raphson method than the bisection method, when speed of convergence is considered. It is, in fact, not difficult to demonstrate that the secant method is essentially equivalent to the Newton–Raphson method, where the first derivative is replaced by a first difference. It suffices to re-arrange formula (6.4) as follows

$$x_{n+1} = x_n - \frac{f(x_n)}{(f(x_n) - f(x_{n-1}))/(x_n - x_{n-1})}$$

It is thus immediate to recognise that the above formula is equal to Newton–Raphson, equation (6.3), with the first derivative $f'(x_n)$ replaced by the finite difference, $(f(x_n) - f(x_{n-1}))/(x_n - x_{n-1})$.

6.5 Relevant R code. Functions `roots_newton` and `roots_secant`

In `comphy`, these two functions implement the Newton–Raphson and secant methods, respectively. They have inputs that work in a way similar to the input explained for `roots_bisec`. A couple of major differences are that two functions (the original function and its first derivative) rather than one need to be defined as input to `roots_newton`, which can be at times annoying. The second difference is that for the secant method the convergence is measured by how small the function becomes, rather than how close to each other two successive approximations are. Practically, this means that in `roots_secant` only `ftol`, and not `tol`, is used.

In the following demonstration, both the secant and the Newton–Raphson methods will be used to find the only real root, $x = -3$, of the cubic function $f(x) = x^3 + 4x^2 + 4x + 3$.

```
# Load comphy package
library(comphy)

# Function whose only root, x=-3, has to be found:
# f(x)=x^3+4x^2+4x+3
f0 <- function(x) return(x^3+4*x^2+4*x+3)

# First derivative needed for Newton-Raphson
f1 <- function(x) return(3*x^2+8*x+4)

# Explore curve graphically, to "see" where the zero is
# (blue dot)
curve(f0(x),from=-5,to=5)
abline(h=0)
points(-3,0,pch=16,col=4)
```

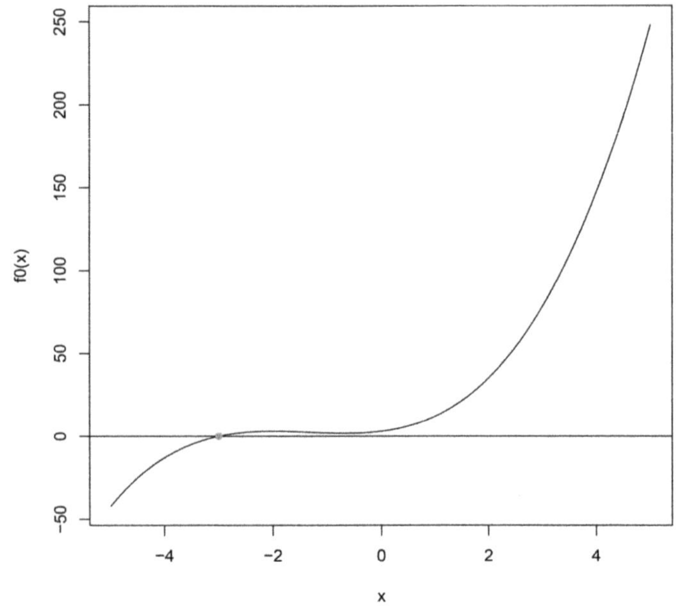


```
# Increase precision
backup_options <- options()
options(digits=12)

# Secant method, starting from -4 and 0
xx <- roots_secant(f0,x0=-4,x1=0,message=FALSE)
print(xx)

## [1] -2.99999999998

# Newton-Raphson method, starting from -4
xx <- roots_newton(f0,f1,x0=-4,message=FALSE)
print(xx)

## [1] -3.00000000002
```

Both functions return $x = -3$ to the required precision (10^{-9}), but it looks like Newton–Raphson is more accurate, and this is normally the case, as it converges faster than the other methods. The detailed break out in terms of successive approximations to the root can be obtain for all methods by enabling `logg`, with `logg=TRUE`, even if normal messages are inhibited.

```
# Secant and Newton-Raphson (again) with detailed output
xx <- roots_secant(f0,x0=-4,x1=0,message=FALSE,logg=TRUE)

##               x0             x1
## 1  -4.00000000000  0.00000000000
## 2   0.00000000000 -0.75000000000
## 3  -0.75000000000 -1.92000000000
## 4  -1.92000000000  1.09454141863
## 5   1.09454141863 -2.77827475937
## 6  -2.77827475937 -3.19762621936
## 7  -3.19762621936 -2.96850944135
## 8  -2.96850944135 -2.99591191924
## 9  -2.99591191924 -3.00009366505
## 10 -3.00009366505 -2.99999972593
## 11 -2.99999972593 -2.99999999998

XX <- roots_newton(f0,f1,x0=-4,message=FALSE,logg=TRUE)

##             Root             Shift
## 1 -4.00000000000                NA
## 2 -3.35000000000 6.50000000000e-01
## 3 -3.06425120773 2.85748792271e-01
## 4 -3.00276575572 6.14854520094e-02
## 5 -3.00000544836 2.76030735968e-03
## 6 -3.00000000002 5.44833921268e-06
```

6.6 Convergence

We have seen through some applications that Newton–Raphson converges to the correct solution faster than the other two methods. Such *rate of convergence* can be quantified exactly according to the following definition. Assume r is the zero of $f(x)$ and x_n, x_{n+1} are its approximations at cycle n, $n+1$, respectively. The errors occurring when x_n and x_{n+1} are used as approximations to the correct zero, can be defined as $\epsilon_n = x_n - r$ and $\epsilon_{n+1} = x_{n+1} - r$, respectively. A method is said to have degree of convergence p, with $p > 0$, if

$$|\epsilon_{n+1}| \quad \rightarrow K|\epsilon_n|^p \qquad \text{when} \qquad n \quad \rightarrow \infty, \tag{6.5}$$

where K is a positive constant. This means that the error at step $n+1$ has magnitude essentially equal to the magnitude of the error at the previous step, n, raised to the power p. As the error is supposed to be a small number (much smaller than 1), the higher the order p, the smaller the error at the subsequent step. This means that the convergence of a method is faster for higher values of p.

It is possible to prove rigorously that the order of convergence of the bisection method is 1, that of Newton–Raphson is 2, while $p = (1 + \sqrt{5})/2 \approx 1.618$ for the secant method. In one of the exercises of this chapter an empirical method to verify the order of convergence is also suggested. This might be useful when inquiring on a new method.

6.7 Exercises on the roots of nonlinear equations

Exercise 01

The only root of $f(x) = \cos(x)$ in the interval $[0, \pi]$ is $x = \pi/2$. Find the numerical value of this root with a precision of 12 digits, using `roots_bisec`, and compare it with the correct value $\pi/2$.

Exercise 02

When the search interval for `roots_bisec` is symmetric with respect to the two roots of a function, only one of the roots will be found, due to the way the algorithm works. This exercise is devoted to understanding such a mechanism, using the function $f(x) = x^2 - 1$, which has the two roots -1 and $+1$. Any interval symmetric with respect to 0, and including both -1 and $+1$, will trigger the output only of $x = -1$.

1. Try the claim using the search intervals $[-2, 2]$ and $[-3, 3]$.
2. Explore the code of `roots_bisec` and consider the following specific chunk, inside the main `while` loop:

```
if (f(a,...)*f(c,...) == 0 | f(b,...)*f(c,...) == 0) {
a <- c
b <- c
} else if (f(a,...)*f(c,...) < 0) {
b <- c
} else if (f(b,...)*f(c,...) < 0) {
a <- c
}
```

   a and b are the left extreme and right extreme of the search interval, respectively. c is their mid point. When the two extremes, $f(a), f(c)$, have different signs, a stays the same, but b becomes c; the line coming next in the `if` sequence is ignored as the current line satisfies it. Therefore, the second search interval will always be, in this case, on the left. This is, essentially, the reason why when we consider an interval symmetric around 0, for the function considered, the root found is always the one closest to the left of the interval.
3. Consider the function $f(x) = x^3 - 4x^2 + 2x$. It has three zeros at $x = 0, 1, 2$. Choose any two roots and try to find one of them with `roots_bisec`, using a search interval symmetric with respect to the mid point of the roots chosen. Do you observe the same problematic highlighted

earlier? What happens if you choose a search interval symmetric with respect to $x = 1$ and including both 0 and 2? Can you explain why?

Exercise 03

Finding the zeros of a function is also related to finding its optimal points. These can be found as zeros of the function's first derivative. Consider the fractional function

$$f(x) = \frac{x^3 + 6x^2 - x - 30}{x - 2}.$$

As this function is the ratio of a third degree and first degree polynomials, it has the same behaviour of a second degree polynomial. It has therefore only one optimal point. Find its optimal point using `roots_bisec`. Then plot the function between -20 and 20, and highlight its optimal point.

Exercise 04

Find the intersections between the curves C_1, given by the equation $y = 2\sin(6\pi x)$, and the curve C_2, given by the equation $y = \mathrm{e}^{-x}$, in the interval $x \in [0, 1]$. Use Newton–Raphson for the numerical solutions.

Exercise 05

Find the zeros of the function $f(x) = \sin(\cos(\mathrm{e}^x))$ between 0 and 2, using both Newton–Raphson and the secant method. Is there any difference between the sets of numerical solutions found? With what method was convergence reached first? How can it be demonstrated? What is the drawback in using Newton–Raphson, rather than the secant method?

Exercise 06

When the zero of a function is known, one can track the behaviour of the errors as n becomes bigger and bigger. It is then possible to plot the natural logarithm of $|\epsilon_{n+1}|$ versus the natural logarithm of $|\epsilon_n|$. The resulting graph should produce points with regression lines passing through them, having slopes equal to the specific convergence order.

Using a specific nonlinear function, say

$$f(x) = x^3 + (1 - \sqrt{3})x^2 + (1 - \sqrt{3})x - \sqrt{3},$$

create the plots suggested and compare the regression straight lines with lines passing through the origin,

$$y = px,$$

where p is 1, $(1 + \sqrt{5})/2$, 2 for the bisection, secant and Newton–Raphson methods, respectively. It can be of help to know that the only real zero of this function is $\sqrt{3}$.

6.8 Systems of nonlinear equations

Up to this point, we have focused on solving nonlinear equations of the form

$$f(x) = 0,$$

where f is a scalar function and x is a real variable. This setting covers many important problems, but in general one often encounters systems of nonlinear equations, typically written as

$$\mathbf{f}(\mathbf{x}) = \mathbf{0},$$

where $\mathbf{f} = (f_1, f_2, \ldots, f_n)$ is a vector-valued function of several variables $\mathbf{x} = (x_1, x_2, \ldots, x_n)$, and the goal is to find a point $\mathbf{x} \in \mathbb{R}^n$ such that all equations $f_i(\mathbf{x}) = 0$ are satisfied simultaneously.

While the methods discussed for scalar equations, such as the bisection method, the Newton–Raphson method, and the secant method, are essential for understanding how to tackle nonlinear problems, the generalisation to systems introduces a number of additional complications. These include the computation and handling of Jacobian matrices, convergence issues in multiple dimensions, and the increased sensitivity to initial values or ill-conditioned systems. Most methods for solving nonlinear systems are in fact based on an extension of Newton's method to higher dimensions. To see how this generalisation works, we begin with a first-order Taylor expansion of $\mathbf{f}(\mathbf{x})$ around the current estimate of the zero to be found, $\mathbf{x}^{(k)}$ (as before, this is an iterative method). For a small increment $\Delta\mathbf{x}$, we have:

$$\mathbf{f}(\mathbf{x}^{(k)} + \Delta\mathbf{x}) \approx \mathbf{f}(\mathbf{x}^{(k)}) + J(\mathbf{x}^{(k)}) \cdot \Delta\mathbf{x},$$

where $J(\mathbf{x}^{(k)})$ is the *Jacobian matrix*, containing all first-order partial derivatives of the components f_i,

$$J(\mathbf{x}^{(k)}) = \begin{pmatrix} \partial f_1/\partial x_1 & \partial f_1/\partial x_2 & \cdots & \cdots & \partial f_1/\partial x_n \\ \partial f_2/\partial x_1 & \partial f_2/\partial x_2 & \cdots & \cdots & \partial f_2/\partial x_n \\ \cdots & \cdots & \cdots & \cdots & \cdots \\ \partial f_n/\partial x_1 & \partial f_n/\partial x_2 & \cdots & \cdots & \partial f_n/\partial x_n \end{pmatrix},$$

and $\Delta\mathbf{x}$ is a vector of corrections to be applied to the current guess. To improve our approximation, we seek a new point $\mathbf{x}^{(k+1)} = \mathbf{x}^{(k)} + \Delta\mathbf{x}$ such that $\mathbf{f}(\mathbf{x}^{(k+1)}) \approx \mathbf{0}$. Substituting into the linearised form gives:

$$\mathbf{0} \approx \mathbf{f}(\mathbf{x}^{(k)}) + J(\mathbf{x}^{(k)}) \cdot \Delta\mathbf{x}.$$

Solving for $\Delta\mathbf{x}$, we obtain:

$$\Delta\mathbf{x} = -J(\mathbf{x}^{(k)})^{-1} \cdot \mathbf{f}(\mathbf{x}^{(k)}).$$

This leads to the iterative Newton scheme in multiple dimensions:

$$\mathbf{x}^{(k+1)} = \mathbf{x}^{(k)} - J(\mathbf{x}^{(k)})^{-1} \cdot \mathbf{f}(\mathbf{x}^{(k)}). \tag{6.6}$$

This method lies at the core of many numerical solvers for nonlinear systems. However, computing and inverting the Jacobian matrix at each step can be costly or unstable, especially for large systems. For this reason, practical solvers often replace the inverse with the solution of a linear system, or use approximate Jacobians that are updated iteratively (as in Broyden's method). The full development of such algorithms and their convergence theory is beyond the scope of this book. Instead, we will show how systems of nonlinear equations can be solved in practice using R. Two packages that provide user-friendly tools for this purpose are `nleqslv` and `rootSolve`.

6.8.1 Solving nonlinear systems with `nleqslv`

The function `nleqslv()` from the `nleqslv` package can be used to solve systems of nonlinear equations using a variety of methods. The user supplies **f**(**x**) and a starting guess for the solution. Some minimal knowledge of what is implied when using this function is actually needed.

- The *Newton* and *Broyden* methods are two popular techniques for solving systems of nonlinear equations. Both start from an initial guess and try to improve it step by step, using information about how the functions are changing. Newton's method uses the exact derivatives to guide each step, which can make it fast and accurate near a solution, but also sensitive to bad starting points. Broyden's method, on the other hand, avoids computing exact derivatives and instead builds up an approximate version as it goes, making it more flexible and often faster to compute. Despite their differences, both methods aim to move closer to a solution by combining the current guess with a correction based on how far off the equations still are.
- Although the Newton and Broyden methods are powerful tools for solving systems of nonlinear equations, they work best when the initial guess is already close to the true solution. When this is not the case, the step taken by these methods might be too long, point in the wrong direction, or even cause the iteration to diverge. To increase the chance of success in more difficult situations, an additional mechanism, called a global strategy, is introduced. This adjusts the step length before applying it, with the aim of ensuring that the method makes steady progress toward the solution. Rather than delving into the technical details of each global strategy, it is often more practical to experiment with different options (such as `dbldog`, `cline`, `gline`, or `none`) and observe how they affect the convergence in challenging cases. This trial-and-error approach can help improve robustness without requiring a deep understanding of the internal mechanisms.
- There are several parameters that tell the algorithm when to stop, meaning that it believes a solution has been found. Two of the most important are `xtol` and `ftol`. The parameter `xtol` controls how small the changes in the solution need to be from one step to the next, if the algorithm finds that the guesses for the solution are not changing much anymore, it assumes it is close enough and stops. The parameter `ftol`, on the other hand, checks

how close the function values are to zero. Since we are trying to solve $\mathbf{f}(\mathbf{x}) = 0$, this tells the algorithm whether it has actually reached a point where the system is nearly satisfied. In short: `xtol` watches how much the solution is moving, and `ftol` watches how small the errors are.

- After running a solver like `nleqslv()`, the user can check how good the solution is by looking at the function values at the final point. Specifically, the output list includes an element called `fvec`, which gives the values of $\mathbf{f}(\mathbf{x})$ at the computed solution. Since the user is trying to solve $\mathbf{f}(\mathbf{x}) = \mathbf{0}$, these numbers should ideally all be very close to zero. If they are small (say, all less than 10^{-6} or whatever tolerance is acceptable in the calculation), then the solution is likely good. The algorithm uses `ftol` to decide whether the solution is good enough to stop, but one can always double-check the actual `fvec` values to be sure. Another helpful output is `termcd` (*termination code*), which tells why the algorithm stopped, whether because the function values were small enough, the steps got tiny, or something went wrong (more details are present in the documentation).
- A default run of `nleqslv` uses Broyden method with the so-called *double dogleg* global strategy [3]: `nleqslv(x,f,method="Broyden", global="dbldog")`.

A couple of demonstrations will make it easier to start using this function. The first is represented by the following system

$$\begin{cases} x_1^2 + x_2^2 &= 1 \\ x_1 &= x_2, \end{cases}$$

that can be re-written in the form useful for roots search:

$$\begin{cases} f_1(x_1, x_2) &= x_1^2 + x_2^2 - 1 &= 0 \\ f_2(x_1, x_2) &= x_1 - x_2 &= 0. \end{cases}$$

The two components of $\mathbf{f}(\mathbf{x})$ are the output of a vector valued function. This must be defined at the beginning, prior to using `nleqslv`. A rough idea of where the roots are should also be given. In the current example there exist two sets of solutions, intersection of the unit circle and the bisector of the first and third quadrant. These are

$$(x_1, x_2) = \left(\frac{\sqrt{2}}{2}, \frac{\sqrt{2}}{2}\right) \quad \text{and} \quad (x_1, x_2) = \left(-\frac{\sqrt{2}}{2}, -\frac{\sqrt{2}}{2}\right).$$

Depending on where the initial point is placed, one or the other solution is found.

```
# Library needed
library(nleqslv)
```

```r
# Define the system as a function
f <- function(x) {
f1 <- x[1]^2+x[2]^2-1  # Circle equation
f2 <- x[1]-x[2]        # Line x = y

return(c(f1,f2))
}

# Solve starting near the known solution
res <- nleqslv(c(0.5,0.5),f)

# Print result
res$x      # Should be close to c(1/sqrt(2), 1/sqrt(2))

## [1] 0.7071068 0.7071068

res$fvec   # Should be close to c(0, 0)

## [1] -4.465724e-10  0.000000e+00

res$termcd # Check why it stopped

## [1] 1

# Second solution
res <- nleqslv(c(-0.5,-0.5),f)
res$x

## [1] -0.7071068 -0.7071068

res$fvec

## [1] -4.465728e-10  0.000000e+00

res$termcd

## [1] 1
```

When the system is not too hard to solve, even a distant starting point achieves convergence.

```
# Try a starting point far from the zero
x0 <- c(10,-10)
res <- nleqslv(x0,f)
res$x

## [1] 0.7071068 0.7071068

res$fvec

## [1] 1.278977e-13 0.000000e+00

res$termcd

## [1] 1
```

The value of `termcd` in all the runs above was always 1, which means that convergence was achieved.

The second demonstration seeks the solution of the following highly-nonlinear system of three equations and three unknowns:

$$\begin{cases} x^2 + y^2 & = 2 \\ e^{x-1} + y^3 & = 2 \\ \sin(z) + z^2 & = 1. \end{cases}$$

Changing the name to the variable leads to the following form of the system, ready for implementation in `nleqslv`:

$$\begin{cases} f_1(x_1, x_2, x_3) & = & x_1^2 + x_2^2 - 2 & = 0 \\ f_2(x_1, x_2, x_3) & = & e^{x_1 - 1} + x_2^3 - 2 & = 0 \\ f_3(x_1, x_2, x_3) & = & \sin(x_3) + x_3^2 - 1 & = 0. \end{cases}$$

If no global strategy is applied, the function fails to find the solution. But the default call includes the double dogleg strategy, which enables convergence.

```
# Define the system as a function
f <- function(x) {
f1 <- x[1]^2+x[2]^2-2
```

```
f2 <- exp(x[1]-1)+x[2]^3-2
f3 <- sin(x[3])+x[3]^2-1

return(c(f1,f2,f3))
}

# Arbitrary starting solution
xstart <- c(20,-30,-30)

# No global search (fails)
res <- nleqslv(x=xstart,fn=f,method="Broyden",global="none")
print(res$termcd)
```

```
## [1] 4
```

```
print(res$message)
```

```
## [1] "Iteration limit exceeded"
```

```
print(res$x)
```

```
## [1]   19.40203 -154.24617  -17.34353
```

```
print(res$fvec)
```

```
## [1] 2.416632e+04 9.448250e+07 3.007961e+02
```

```
# Double dogleg global search (default)
res <- nleqslv(x=xstart,fn=f)
print(res$termcd)
```

```
## [1] 1
```

```
print(res$message)
```

```
## [1] "Function criterion near zero"
```

```
print(res$x)
```

```
## [1]  1.000000  1.000000 -1.409624
```

```
print(res$fvec)
```

```
## [1] 2.500328e-09 2.675455e-09 3.042364e-10
```

The above code shows that for many difficult nonlinear systems, the default application of `nleqslv` will suffice to find a solution. A judicious and systematic use of different parameters can extend the default capabilities of the algorithm. Detailed information on various aspect of this and other search algorithms can be found in the excellent book by Dennis and Schnabel [3].

6.8.2 Solving nonlinear systems with `rootSolve`

Another practical way to solve systems of nonlinear equations in R is through the function `multiroot` from the `rootSolve` package. Unlike `nleqslv`, which offers a wide range of algorithmic options and tuning parameters, `multiroot` provides a simpler and more direct interface. It is particularly well-suited for small-to-moderate systems that are reasonably well-behaved. The underlying method is still based on Newton-like iterations, and the Jacobian is estimated numerically if the user does not supply it.

One of the strengths of `multiroot` is its ease of use: the user provides a function representing the system and a starting guess, and the solver takes care of the rest. However, even though the interface is simple, the function still allows some control over the convergence criteria through a set of tolerance parameters: `rtol`, `atol`, and `ctol`.

The parameters `rtol` (*relative tolerance*) and `atol` (*absolute tolerance*) determine how close the function values must be to zero before the algorithm declares convergence. Specifically, for each component of the system, the solver checks whether the estimated local error ϵ_i satisfies a condition of the form

$$\frac{\epsilon_i}{\mathrm{rtol} \times |f_i| + \mathrm{atol}} \leqslant 1.$$

The value ϵ_i is a rough estimate of the local error in the ith component of the function, typically based on how much that function component changes between iterations. That is, ϵ_i may be approximated by the difference between the current and previous function values, giving a sense of how stable or reliable the result is becoming. This test combines relative and absolute measures of accuracy: if the function value is large, relative error dominates; if the value is small or near zero, absolute tolerance becomes more important. This approach helps ensure robust behaviour across a wide range of problem scales.

The parameter `ctol` provides an additional stopping condition based on the size of the update in the solution. If, between two iterations, the largest change in any component of the solution vector is smaller than `ctol`, the algorithm assumes that it has stalled or reached sufficient accuracy and halts. This prevents the solver from continuing indefinitely when further progress is negligible.

The result returned by `multiroot` is a list containing several useful components. The most important are `root`, which gives the estimated solution vector, and `f.root`, which contains the function values at that solution (ideally close to zero). The component `iter` indicates how many iterations were performed, and `estim.precis` gives a rough estimate of the numerical precision achieved at the final step. This is, roughly, the absolute difference between the values of the function in the last two iterations.

Let us search the zeros for the two examples used to test `nleqlsv`. The first system is:

$$\begin{cases} f_1(x_1, x_2) &= x_1^2 + x_2^2 - 1 &= 0 \\ f_2(x_1, x_2) &= x_1 - x_2 &= 0, \end{cases}$$

and we choose two different starting points on the vertical axis.

```
# Library needed
library(rootSolve)

# Define the system as a function
f <- function(x) {
f1 <- x[1]^2+x[2]^2-1  # Circle equation
f2 <- x[1]-x[2]        # Line x = y

return(c(f1,f2))
}

# Solve starting from the top of vertical axis
res <- multiroot(f=f,start=c(0,10))

# Print result
print(res$root)
```

```
## [1] 0.7071068 0.7071068
```

```
print(res$f.root)

## [1] 5.846759e-08 0.000000e+00

print(res$iter)

## [1] 8

print(res$estim.precis)

## [1] 2.923379e-08

# Solve starting from the bottom of vertical axis
res <- multiroot(f=f,start=c(0,-10))

# Print result
print(res$root)

## [1] -0.7071068 -0.7071068

print(res$f.root)

## [1] 5.846e-08 0.000e+00

print(res$iter)

## [1] 8

print(res$estim.precis)

## [1] 2.923e-08
```

The other example is highly nonlinear:

$$\begin{cases} f_1(x_1, x_2, x_3) &= x_1^2 + x_2^2 - 2 &= 0 \\ f_2(x_1, x_2, x_3) &= e^{x_1 - 1} + x_2^3 - 2 &= 0 \\ f_3(x_1, x_2, x_3) &= \sin(x_3) + x_3^2 - 1 &= 0. \end{cases}$$

```
# Define the system as a function
f <- function(x) {
f1 <- x[1]^2+x[2]^2-2
f2 <- exp(x[1]-1)+x[2]^3-2
f3 <- sin(x[3])+x[3]^2-1

return(c(f1,f2,f3))
}

# Arbitrary starting solution
xstart <- c(20,-30,-30)

# Solve
res <- multiroot(f=f,start=xstart)

# Print result
print(res$root)

## [1] -0.7137474  1.2208868 -1.4096240

print(res$f.root)

## [1]  3.573941e-11  4.331202e-12 -1.110223e-16

print(res$iter)

## [1] 53

print(res$estim.precis)

## [1] 1.335691e-11
```

The algorithm converges but the root found is different from the one found by `nleqslv`. If this is true, `nleqslv` should also find the same zero, if the starting point is close to what `multiroot` has found. This is indeed the case, as demonstrated in the following demonstration.

```
# Load nleqslv
library(nleqslv)

# Use a starting point close to the solution from multiroot
xstart <- c(-0.71,1.22,-1.41)

# nleqlsv search
res2 <- nleqslv(fn=f,x=xstart)
print(res2$x) # Same as one from multiroot
```

```
## [1] -0.7137474  1.2208868 -1.4096240
```

```
print(res$root)
```

```
## [1] -0.7137474  1.2208868 -1.4096240
```

The example just seen highlights also a common feature of nonlinear systems: the potential occurrence of multiple solutions in a confined region. This is also one of the reasons why the search of zeros for nonlinear functions is very challenging.

6.9 Exercises on the roots of systems of nonlinear equations

Exercise 07

Consider the following system of nonlinear equations:

$$\begin{cases} x^2 + y^2 &= 4 \\ e^x + y &= 1. \end{cases}$$

Use a plot of the two curves represented by the above system to select starting points to find the solutions to the system. Use both `nleqslv` and `multiroot` for the task.

Exercise 08

Solve the system

$$\begin{cases} x^2 + y^2 + z^2 &= 9 \\ xyz &= 1 \\ x + y - z^2 &= 0 \end{cases}$$

using Newton's method to obtain the solution near (2.4, 0.2, 1.7).

Exercise 09

Given the system

$$\begin{cases} xyz - x^2 + y^2 &= 1.33 \\ xy - z^3 &= 0.1 \\ e^x - e^y + z &= 0.41, \end{cases}$$

find as many of its solutions as possible, in the range

$$-0.5 \leqslant x \leqslant 0.5, \quad -1 \leqslant y \leqslant 1, \quad -1 \leqslant z \leqslant 1.$$

Exercise 10

Solve the nonlinear system

$$\begin{cases} f_1(x, y) = x^2 + y^2 - 1 \\ f_2(x, y) = \exp(x) - y \end{cases}$$

using the `nleqslv` function in R. This system has a solution near $(x, y) \approx (0.0, 1.0)$, where the unit circle intersects the exponential curve. Use the starting point $(x_0, y_0) = (0.5, 0.5)$ and the default method and global strategy. Then:

1. Fix `ftol` but vary `xtol` in the following way:
   - Set `ftol = 1e-8`.
   - Try `xtol = 1e-1, 1e-4, 1e-8, 1e-12`.
   - For each run, record: the final result `res$x`, the number of iterations `res$iter`, the termination code `res$termcd`, the norm of the residual vector $\|\mathbf{f(x)}\|$.
2. Fix `xtol` but vary `ftol` in the following way:
   - Set `xtol = 1e-8`.
   - Try `ftol = 1e-1, 1e-4, 1e-8, 1e-12`.
   - Record the same outputs as in Step 1.

For large `xtol`, does the solver stop before the function values are small? For large `ftol`, does the solver stop even if the update step is still large? When both `xtol` and `ftol` are tight, does the result improve? Do more iterations occur? Based on the termination codes, which stopping condition triggered the exit in each case?

This exercise shows how `xtol` controls the size of the update step, while `ftol` governs how close the function values must be to zero. Either stopping condition can dominate, depending on how the tolerances are set. The exercise also illustrates the trade-off between accuracy and computational effort.

References

[1] Gullberg J 1997 *Mathematics - From the Birth of Numbers* (W. W. Norton & Company)
[2] Gerald C F and Wheatley P O 1999 *Applied Numerical Analysis* (Addison-Wesley)
[3] Dennis J E and Schnabel R B 1996 *Numerical Methods for Unconstrained Optimization and Nonlinear Equations (Classics in Applied Mathematics)* **vol 16** 1st edn (SIAM)

IOP Publishing

Computational Physics with R

James Foadi

Chapter 7

Differentiation and integration

Most physical laws are expressed as differential equations whose solutions describe the complete dynamics of a system. Unfortunately, only a few of these equations can be solved analytically, so numerical methods are typically required. In particular, techniques for approximating derivatives are essential, and the first part of this chapter focuses on such methods.

The second part of the chapter addresses numerical integration. As with differentiation, integrals are rarely solvable in closed form, making numerical approaches necessary.

Both differentiation and integration rely on discretizing the variables over a grid, which can be either regular or irregular. While part of the chapter is devoted to differentiation over irregular grids, in physics most differentiation is done over regular grids.

7.1 Differentiation over a regular grid

One of the best known formulas in mathematical analysis is the definition of first derivative of a function $f(x)$ via the limit of its *difference quotient*:

$$f'(x) = \lim_{h \to 0} \frac{f(x+h) - f(x)}{h}.$$

The difference quotient is a first-order divided difference, $f[x_1, x_2]$, where the two points are $x_1 = x$ and $x_2 = x + h$. We could use divided differences of increasing orders across all points of the grid to build up an interpolation of $f(x)$ and calculate the first derivative of the interpolation; this would be a numerical derivative based on the function's tabulated points. We will do this later but only for irregularly-spaced grids, as the calculation of divided differences is time consuming. When the grid is regular, it is convenient to use *forward* and *backward differences*, employing at the same time Taylor expansion to calculate numerical errors.

doi:10.1088/978-0-7503-2632-2ch7

Consider the Taylor expansion of a function $f(x)$ centred at x_i:

$$f(x) = f(x_i) + \frac{f'(x_i)}{1!}(x - x_i) + \frac{f''(x_i)}{2!}(x - x_i)^2 + \cdots + \frac{f^{(n)}(x_i)}{n!}(x - x_i)^n + R_n(x), \quad (7.1)$$

where the remainder $R_n(x)$ can be written as

$$R_n(x) = \frac{f^{(n+1)}(\xi)}{(n+1)!}(x - x_i)^{n+1}, \quad (7.2)$$

with ξ an unknown variable between x and x_i. A regular grid can be created starting from x_i and adding a same step $h > 0$. Depending on whether h is added or subtracted, we can create the so-called *forward difference*, $\Delta_F f_i$, and *backward difference*, $\Delta_B f_i$, defined as follows:

$$\Delta_F f_i \equiv f(x_i + h) - f(x_i), \qquad \Delta_B f_i \equiv f(x_i) - f(x_i - h). \quad (7.3)$$

Expansion (7.1) assumes specific forms depending on whether $x = x_i + h$ or $x = x_i - h$. More specifically

$$f(x_i + h) = f(x_i) + \frac{f'(x_i)}{1!}h + \frac{f''(x_i)}{2!}h^2 + \cdots$$

and

$$f(x_i - h) = f(x_i) - \frac{f'(x_i)}{1!}h + \frac{f''(x_i)}{2!}h^2 - \cdots$$

The first and second expressions yield, after elementary manipulations,

$$f'(x_i) = \frac{f(x_i + h) - f(x_i)}{h} - \frac{f''(x_i)}{2!}h - \cdots$$

and

$$f'(x_i) = \frac{f(x_i) - f(x_i - h)}{h} + \frac{f''(x_i)}{2!}h - \cdots$$

or, using forward and backward difference symbols,

$$f'(x_i) = \frac{\Delta_F f_i}{h} + O(h) \quad (7.4)$$

$$f'(x_i) = \frac{\Delta_B f_i}{h} + O(h). \quad (7.5)$$

These formulas approximate numerically the first derivative at a point x_i using knowledge of the function at two neighbouring points, x_i and $x_i + h$ or x_i and $x_i - h$, of a regular grid. The term $O(h)$ means that the error made using the numerical approximation is of the order of the increment (or step), h.

Example 7.1 *Consider the function* $f(x) = 2x + \cos(x) - e^x$. *We would like to calculate numerically the first derivative using an increment* $h = 0.01$. *We expect the error related to both forward and backward approximations to be of the order of* $h = 0.01$. *The correct derivative at* $x_i = 0$ *is* $f'(x_i) = f'(0) = 1$.

The forward approximation is

$$f'(0) \approx \frac{\Delta_f f_i}{h} = \frac{f(0 + 0.01) - f(0)}{0.01} = 0.989\,98.$$

The backward approximation is

$$f'(0) \approx \frac{\Delta_B f_i}{h} = \frac{f(0) - f(0 - 0.01)}{0.01} = 1.009\,98.$$

The error between the correct value and the approximations is $1 - 0.989\,98 = 0.010\,02$ *for the forward case and* $1 - 1.009\,98 = -0.009\,98$ *for the backward case; both values are comparable with* $h = 0.010$.

A better approximation for the numerical derivative can be obtained if the three points $x_i - h$, x_i, $x_i + h$, rather than the two points x_i, $x_i + h$ or x_i, $x_i - h$, are used for the calculation. Indeed, subtracting the Taylor expansions for $f(x_i - h)$ from that for $f(x_i + h)$ and rearranging, yields the following result:

$$\frac{f(x_i + h) - f(x_i - h)}{2h} = f'(x_i) + \frac{f'''(x_i)}{3!}h^2 + \cdots = f'(x_i) + O(h^2).$$

From this, using the so-called *centred difference*

$$\Delta_C f_i \equiv f(x_i + h) - f(x_i - h) = \Delta_F f_i + \Delta_B f_i, \tag{7.6}$$

we arrive at the following numerical approximation for the first derivative:

$$f'(x_i) = \frac{\Delta_C f_i}{2h} + O(h^2). \tag{7.7}$$

The result achieved with a centred difference derivative is more accurate than that achieved with forward or backward difference derivatives because the error is of the order of h^2.

Example 7.2 *Using the function and grid points of the previous example, the derivative calculated with the centred difference is*

$$f'(0) \approx \frac{f(0 + 0.01) - f(0 - 0.01)}{2(0.01)} = 0.999\,98.$$

The error associated with this approximation is $1 - 0.999\,98 = 0.000\,02$ *which is of the order of* $h^2 = 0.01^2 = 0.0001$, *as estimated by formula* (7.7).

7.2 Relevant R code. Function `deriv_reg`

This function relies on two separate functions to calculate forward and backward differences, `forwdif` and `backdif`, respectively. These replace the calculation of divided differences for regular grids. Execution time, especially when the grid includes thousands of points, is a lot shorter for forward and backward differences than for divided differences. The numerical results are clearly the same, once the relation between forward or backward difference and divided difference is considered. The following snippet demonstrates the time saved in execution; this is why divided differences are never used when the grid is regularly spaced.

```
# Load comphy
library(comphy)

# The function is 2sin(x) + cos(x)
# Tabulated values (regular grid)
x <- seq(0,pi,length.out=2001)
f <- 2*sin(x)+cos(x)

# Time measured to calculate divided differences
system.time(res1 <- divdif(x,f))
```

```
##    user  system elapsed
##    0.42    0.00    0.41
```

```
# Time taken to calculate forward differences
system.time(res2 <- forwdif(f))
```

```
##    user  system elapsed
##    0.04    0.03    0.09
```

```
# The calculation yields corresponding
# results: divdif = forwdif / h
print(res1[1:5,2])
```

```
## [1] 1.999214 1.997638 1.996057 1.994472 1.992881
```

```
print(res2[1:5,2]/(x[2]-x[1]))
```

```
## [1] 1.999214 1.997638 1.996057 1.994472 1.992881
```

The main function to calculate forward, backward, or centred difference derivatives is `deriv_reg`. Its arguments are `x0` which is a vector of grid points at which the derivative is calculated, `x` and `f` which are the tabulated x and $f(x)$. For this type of numerical derivative the grid must be regularly-spaced.

The coding structure of `deriv_reg` is relatively simple. First forward, backward, or centred differences are calculated with the two functions `forwdif` and `backdif`. Then division by the increment h or $2h$ yields the numerical derivatives. The centred differences are calculated as sums of the forward and backward differences, paying attention at the first and last grid points that for centred difference derivatives cannot be computed and are thus replaced by `NA`. For forward difference derivatives only the last grid point cannot be calculated, while for backward difference derivatives it is the first grid point not to be calculated. As explained before, centred difference derivatives are more accurate than forward or backward difference ones.

The following chunk of code calculates the first derivative of the function $f(x) = \sin(x) - 2\cos(2x)$ at points of a regular grid of step h close to 0.01, between 0 and π, and compares it with the correct values, $f'(x) = \cos(x) + 4\sin(2x)$. Centred differences are used throughout, as these provide more accurate results.

```
## Derivative of f(x)=sin(x)-2*cos(2*x) between 0 and pi
## The regular grid is made of 320 points, with h roughly
## equal to 0.01

   # Grid
   x <- seq(0,pi,length.out=320)

   # Tabulated values
   f <- sin(x)-2*cos(2*x)

   # Tabulated values for derivative
   Df <- cos(x)+4*sin(2*x)

# Plot function
plot(x,f,pch=16,cex=0.5,type="b",xlab="x",ylab="f(x)")
```

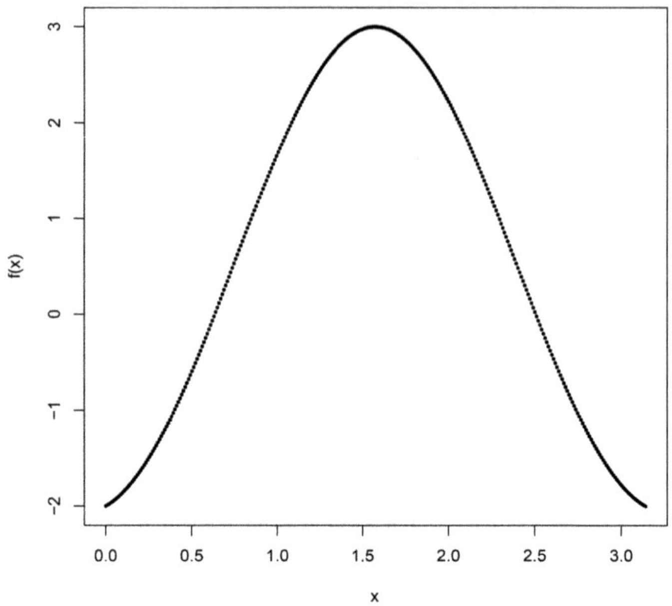


```
# Centred difference derivative (default)
Dcf <- deriv_reg(x,x,f)

# First and last value are not calculated
print(Dcf[c(1:2,319:320)])

## [1]        NA  1.078711 -1.078711        NA

# Plot true and numerical derivatives
# True and numerical derivative hardly
# different, given the accuracy (O(h^2))
Ry <- range(Dcf,Df,na.rm=TRUE)
plot(x,Dcf,pch=16,cex=0.5,col=1,ylim=Ry,xlab="x",ylab="Df")
points(x,Df,,col=2,cex=0.7)
```

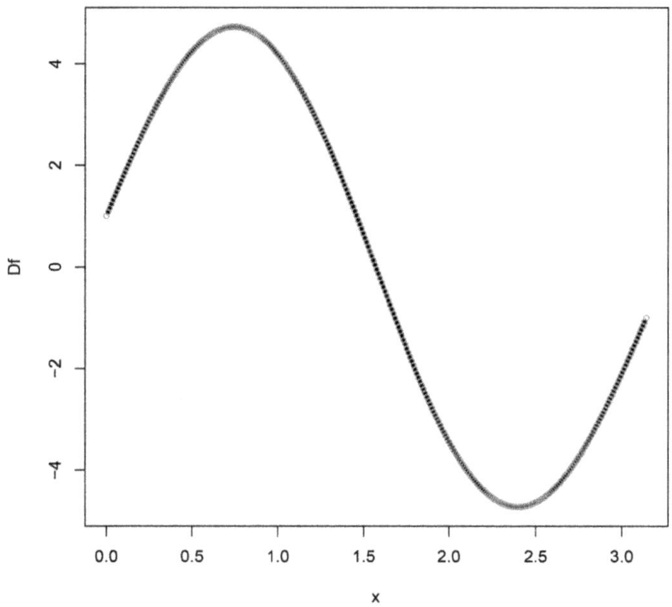


```
# Pick a few values to measure difference (less than 0.0001)
idx <- c(10,150,310)
Delta <- Df[idx]-Dcf[idx]
print(Delta)

## [1]  6.170835e-05  5.477617e-05 -6.669858e-05
```

7.3 Second-order differentiation

Although the first part of this chapter focused on approximating the first derivative of a function using finite differences, in many physical problems one also needs to compute the second derivative numerically. This is especially true in the context of second-order differential equations, where the second derivative appears explicitly. A standard and effective approximation for the second derivative is obtained using values of the function at three equally spaced points. Assuming the function is sampled at points $x - h$, x, and $x + h$, we can write:

$$f''(x) \approx \frac{f(x+h) - 2f(x) + f(x-h)}{h^2} \tag{7.8}$$

This is known as the *centred difference approximation for the second derivative*. Like the centred formula for the first derivative, it can be derived from Taylor expansions:

$$f(x+h) = f(x) + hf'(x) + \frac{h^2}{2}f''(x) + \frac{h^3}{6}f^{(3)}(x) + \frac{h^4}{24}f^{(4)}(x) + \cdots$$
$$f(x-h) = f(x) - hf'(x) + \frac{h^2}{2}f''(x) - \frac{h^3}{6}f^{(3)}(x) + \frac{h^4}{24}f^{(4)}(x) - \cdots$$

Adding these two expressions cancels the odd-order derivatives and gives:

$$f(x+h) + f(x-h) = 2f(x) + h^2 f''(x) + \frac{h^4}{12}f^{(4)}(x) + \cdots$$

Rearranging leads to the approximation above, with a truncation error proportional to h^2. This means the method is second-order accurate: halving the step size roughly quarters the error.

7.4 Differentiation over an irregular grid

When a function is known at a series of irregularly-spaced points in an interval of $\mathbb{R}$, it is not possible to use the methods described earlier for regular grids. In those cases, the first derivative can be approximated using the interpolation techniques introduced in chapter 3. Specifically, by constructing an interpolating polynomial of degree n through $n+1$ points, one can obtain an approximation of the derivative by simply differentiating this polynomial. If $P_n(x)$ is the polynomial, the function $f(x)$ will be different from the interpolating polynomial due to an error $E(x)$:

$$f(x) = P_n(x) + E(x), \qquad f'(x) = P_n'(x) + E'(x), \tag{7.9}$$

with the primed quantities symbolising derivatives.

With divided differences the polynomial (see section 3.8) is given by the following expression:

$$\begin{aligned} P_n(x) = {} & f[x_1] \\ & + f[x_1, x_2](x - x_1) \\ & + f[x_1, x_2, x_3](x - x_1)(x - x_2) \\ & + \cdots \\ & + f[x_1, \ldots, x_{n+1}](x - x_1) \cdots (x - x_n), \end{aligned} \tag{7.10}$$

with an expression for the error (3.10) reported here:

$$E(x) = \frac{f^{(n+1)}(\xi)}{(n+1)!}(x - x_1) \cdots (x - x_{n+1})$$

The approximated first derivative is therefore obtained as derivative of equation (7.10):

$$\begin{aligned} f'(x) \approx P_n'(x) = {} & f[x_1, x_2] \\ & + f[x_1, x_2, x_3]((x - x_2) + (x - x_1)) \\ & + \cdots \\ & + f[x_1, \ldots, x_{n+1}]((x - x_2) \cdots (x - x_{n+1}) + \cdots + (x - x_1) \cdots (x - x_n)) \end{aligned} \tag{7.11}$$

The error, $\Delta f'(x)$, associated with the approximation, is $\mathrm{d}E/\mathrm{d}x$. From (3.10), this means that

$$\begin{aligned}\Delta f'(x) = \frac{\mathrm{d}E}{\mathrm{d}x} = &\frac{\mathrm{d}(f^{(n+1)}(\xi)/(n+1)!)}{\mathrm{d}x}(x-x_1)\cdots(x-x_{n+1})\\ &+\frac{f^{(n+1)}(\xi)}{(n+1)!}\frac{\mathrm{d}((x-x_1)\cdots(x-x_{n+1}))}{\mathrm{d}x}.\end{aligned} \tag{7.12}$$

The derivative of $f^{(n+1)}(\xi)$ cannot be calculated exactly because ξ depends on x in an unknown way. Luckily, we need to know the error at $x = x_i$ (because these are the only available values for calculation), and the first part of $\Delta f'(x)$, which in equation (7.12) is the one associated with the derivative of $f^{(n+1)}(\xi)$, becomes zero at $x = x_i$. The expression for the error is then (pay attention, in the expression the error is calculated at $x = x_i$)

$$\Delta f'(x_i) = \frac{f^{(n+1)}(\xi)}{(n+1)!}(x_i - x_1)\cdots(x_i - x_{i-1})(x_i - x_{i+1})\cdots(x_i - x_{n+1}), \tag{7.13}$$

where the value of ξ is not known, although it has to be included in the interval containing all grid points (an example can be studied in one of the exercises on differentiation (7.6)). It might be worth noting that formula (7.13) yields an error different from zero at $x = x_i$, where the value of $f(x)$ is exactly known. The reason why this is reasonable is that the interpolating polynomial used to approximate the function has a shape different from the function, even if it coincides with the function at $x = x_i$.

Example 7.3 *Consider the function $f(x) = x^2 - x + 1$ and its tabulated points at $x_1 = -1$, $x_2 = 1$, $x_3 = 2$, $x_4 = 4$. The table of divided differences is*

i	x_i	$f(x_i)$	(1)	(2)	(3)
1	−1	3			
			−1		
2	1	1		1	
			2		0
3	2	3		1	
			5		
4	4	13			

and the headers in parenthesis indicate divided differences of order 1, 2 *and* 3. *In the table, all divided differences of order* 2 *are equal because the function is an exact polynomial of order* 2. *For the same reason, the divided difference of order* 3 *is null.*

Assume that all points are used for the interpolation. The corresponding polynomial should be of degree 3, *but as we have seen that the divided difference of order* 3 *is zero, the polynomial is of order* 2, *as expected. Using formulas (7.9) and (7.10), and the values in the table above, we obtain:*

$$P_3(x) = 3 - 1(x + 1) + 1(x + 1)(x - 1) + 0(x + 1)(x - 1)(x - 2) = x^2 - x + 1$$

$$\Downarrow$$

$$f(x) = x^2 - x + 1 + E(x)$$

In this case, $P_3(x)$ coincides with the function $f(x)$ and the error is expected to be zero. Indeed, using equation (3.10), *with $f^{(n+1)}(\xi)/(n + 1)!$ replaced by a divided difference (see equation* (3.20)), *we obtain that*

$$E(x) = f[-1, 1, 2, 4, x^*](x + 1)(x - 1)(x - 2)(x - 4),$$

where x^* is a value between -1 and 4. The divided difference of order 4 is zero since all divided differences with order 3 and above are in this case zero. Therefore $E(x) = 0$, as expected.

The approximation to the first derivative at points $\{-1, 1, 2, 4\}$ is

$$f'(x) \approx P_3'(x) = 2x - 1$$

which yields $\{-3, 1, 3, 7\}$. These are also the values of the exact derivative because, as seen before, the function in this case is a polynomial of order 2 (uniquely known when three tabulated points are available) and four tabulated points are available. The error must be zero at these tabulated points, as it is easy to verify.

A more interesting case is when the tabulated function is not a polynomial. Consider, for instance,

$$f(x) = 1 + (e^{2-x} - e^x)/(1 - e^2)$$

at the tabulated points $x_1 = 0$, $x_2 = 1/2$, and $x_3 = 1$. The table of divided differences is

i	x_i	$f(x_i)$	(1)	(2)
1	0	0.000 00		
			1.113 18	
2	0.5	0.555 59		−0.226 36
			0.886 82	
3	1	1.000 00		

This means that the approximating polynomial is the following quadratic (with only three points available cannot be any higher order polynomial):

$$P_2(x) = 0 + 1.113\ 18(x - 0) - 0.226\ 36(x - 0)(x - 1/2) = 1.226\ 36x - 0.226\ 36x^2.$$

The approximating first derivative is thus,

$$f'(x) \approx 1.226\ 36 - 0.452\ 72x.$$

The only points at which we can rightly use this approximation are x_1, x_2, x_3 because there we can estimate the derivative's error. In the current example we can also calculate the correct derivative because the true function is known. Its derivative is

$$f'(x) = (e^{2-x} + e^{x})/(e^{2} - 1).$$

The correct derivatives at points x_1, x_2, x_3 *are therefore*

$$f'(x_1) = 1.313\ 04, \quad f'(x_2) = 0.959\ 52, \quad f'(x_3) = 0.850\ 92,$$

while those calculated with the approximations are

$$P_2'(x_1) = 1.226\ 36, \quad P_2'(x_2) = 1, \quad P'(x_3) = 0.773\ 64.$$

The errors are

$$\begin{aligned}
\Delta f'(x_1) &= 1.313\ 04 - 1.226\ 36 &&= 0.086\ 68 \\
\Delta f'(x_2) &= 0.959\ 52 - 1 &&= -0.040\ 48 \\
\Delta f'(x_3) &= 0.850\ 92 - 0.773\ 64 &&= 0.077\ 28.
\end{aligned}$$

An estimate of the errors at x_1, x_2, x_3 *is formally given by* (7.13). *The word* formally *means that, in fact, we cannot calculate the* $n + 1$*-th derivative in the expression, unless a new tabulated point is added to the three existing, which is not always something feasible. Furthermore, thinking to the proof in appendix* A.4, *the introduction of a new tabulated point changes the value of* ξ; *therefore the next term rule gives an estimate rather than an exact equality, as we already know. Here, just to verify that reasonable values are produced, let us assume that a new tabulated point,* $x_4 = 0.25, f(x_4) = 0.300\ 28$, *is available. In this case, the derivative needed to calculate the errors is* $f'''(\xi)/3!$. *For example, to calculate the error at* $x_1 = 0$ *we have*:

$$\Delta f'(0) = \frac{f'''(\xi)}{3!}(0 - x_2)(0 - x_3) = \frac{f'''(\xi)}{3!}(0 - 0.5)(0 - 1)$$

and similar expressions can be written for $\Delta f'(0.5)$ *and* $\Delta f'(1)$. *An estimate of the unknown term, using the next term rule and the new tabulated point, yields*

$$\frac{f'''(\xi)}{3!} \approx f[x_1, x_2, x_3, x_4] = f[0, 0.5, 1, 0.25] = 0.16710.$$

We can now estimate the errors using the next term rule:

$$\begin{aligned}
\Delta_{\text{est}} f'(0) &\approx f[x_1, x_2, x_3, x_4](0 - 0.5)(0 - 1) &&= 0.167\ 10(-0.5)(-1) &&= 0.083\ 55 \\
\Delta_{\text{est}} f'(0.5) &\approx f[x_1, x_2, x_3, x_4](0.5 - 0)(0.5 - 1) &&= 0.167\ 10(0.5)(-0.5) &&= -0.041\ 78 \\
\Delta_{\text{est}} f'(1) &\approx f[x_1, x_2, x_3, x_4](1 - 0)(1 - 0.5) &&= 0.167\ 10(1)(0.5) &&= 0.083\ 55.
\end{aligned}$$

The comparison of $\Delta_{\text{est}} f'(x_i)$ *with the true errors,* $\Delta f'(x_i)$, *shows that formula* (7.13) *provides reasonable estimates.*

7.5 Relevant R code. Function `deriv_irr`

This function is part of `comphy`. The input is given by the available (tabulated) points of $f(x)$, provided as vectors of equal length `x` and `f`, and by a vector `x0` of values at which the first derivative is computed. The algorithm makes use of the function `divdif` to compute the table of divided differences. Then the crucial bit of the calculation is represented by the sums of quantities like

$$(x - x_i).$$

These are present as combinations of 2, 3, ..., n elements one at a time, two at a time, etc. `deriv_irr` uses the auxiliary (not available among the listed functions of `comphy`) function `aprod` to calculate the above combinations. The sum giving the derivative (equation (7.11)) has, in the code, the form

$$P[1, 2] + \sum_{i=3}^{n+1} \text{P[1,i]aprod(vdiff,i)},$$

where `vdiff` is the vector of differences between `x0` and each grid point `x`. This renders the code neater as in `aprod`, expressions like

$$(x - x_2) \cdots (x - x_n) + \cdots + (x - x_1) \cdots (x - x_{n-1})$$

are computed with a judicious use of combination of indices and products of selected vector elements (see the R code for details).

When $f(x)$ is a polynomial, the derivative returned by `deriv_irr` are exact, as shown in the following demonstration, where $f(x) = x^3 - 2x^2 + 4x + 5$ and this function is known at $x = -2, 0, 1, 3, 6$. The derivative, $f'(x) = 3x^2 - 4x + 4$, at, say, values $x_0 = 0.5, 3, 4$, yields 2.75, 19, 36; these exact values are what is expected using `deriv_irr`.

```
# Load comphy
library(comphy)

# The function is f(x)=x^3-2*x^2+4*x+5
# Tabulated values
x <- c(-2,0,1,3,6)
f <- x^3-2*x^2+4*x+5

# Values at which f' is computed
x0 <- c(0.5,3,4)

# First derivative
# (returns exact values because f(x) is a polynomial)
f1 <- deriv_irr(x0,x,f)
print(f1)

## [1]  2.75 19.00 36.00
```

If a polynomial of lower degree is used for the interpolation, for example because less tabulated points are available, then the first derivatives will not be exact, even with $f(x)$ a polynomial. This can be seen in what follows, where only three points are tabulated and therefore the interpolating polynomial has only degree 2.

```
# The function is f(x)=x^3-2*x^2+4*x+5
# Tabulated values (only 3 ot the 5 available)
x <- c(-2,1,6)
f <- x^3-2*x^2+4*x+5

# Values at which f' is computed
x0 <- c(0.5,3,4)

# First derivative
# (returns approximated values because interpolating
#  degree is less that degree of function)
f1 <- deriv_irr(x0,x,f)
print(f1)

## [1] 15 30 36
```

As clearly seen, the numerical error can be quite large when the interpolating polynomial does not match the function well.

With functions different from polynomials, the first derivative returned with this method is not in general exact. In addition to this, if the distance between contiguous tabulated points is relatively small, then rounding off issues can arise and yield values of the first derivative quite far from their correct value, especially when the calculation happens at the extremes of the interval. In the following demonstration, 100 values for the sine function are initially tabulated. The first derivative at $x_0 = 0$ turns out to be very large, and not equal to 1 (as it should be because if $f(x) = \sin(x)$ then $f'(x) = \cos(x)$). The number of sample points is next dropped to 10 and the derivative becomes a better approximation.

```
# The function is f(x)=sin(x)
# 100 tabulated values
x <- seq(0,pi/2,length=100)
f <- sin(x)

# First derivative at x0=0
# (value is swamped by rounding off errors)
f1 <- deriv_irr(0,x,f)
print(f1)

## [1] 4.18588e+12

# Things get better with only 10 tabulated points
```

```
x <- seq(0,pi/2,length=10)
f <- sin(x)
f1 <- deriv_irr(0,x,f)
print(f1)
```

```
## [1] 1
```

It might be therefore convenient to assess an optimal polynomial degree for the interpolation, before calculating the derivative, for instance with `decidepoly_n`. To conclude, it might be of help to remember that with divided differences, calculations done at grid points close to the extremes are more accurate when all points are re-arranged as to have the extremes in the middle.

7.6 Exercises on differentiation

Exercise 01

Consider the simple function $f(x) = x + e^x + \sin(x)$. Calculate the first derivative at $x = 0$ using forward, backward and centred differences. Use $h = 0.1$ and $h = 0.01$. Verify that the numerical errors are $O(h)$ and $O(h^2)$.

Exercise 02

Use the function $f(x) = e^x$ around $x = 1$, and values of h between 0.001 and 0.1 to show that the error goes like $O(h^2)$.

Exercise 03

Experimental or measurement errors on sampled points of a function can amplify the errors of a numerical derivative. In this exercise, create a function, $f(x) = \sin(x)$ at 33 regularly spaced points between 0 and π (h is then roughly 0.1). Then create a function $g(x) = f(x) + \epsilon$, where ϵ is a normal random variable extracted from a distribution with mean 0 and standard deviation 0.01. Plot the two functions together to appreciate their difference. Next, calculate the centred difference derivative at all points (clearly excluding 0 and π) for both $f(x)$ and $g(x)$. Plot both derivatives and appreciate how the differences for them are amplified, compared to the differences of the functions. Finally, find the maximum error for both derivatives with respect to the true derivative, $f'(x) = \cos(x)$.

Exercise 04

Find the numerical first derivative of $f(x) = x^2 - 3\cos(2x) + e^x$ at $x_0 = -\pi/6, 0, \pi/6$, when the function is tabulated at the 7 points $x = -\pi/2, -\pi/3, -\pi/4, 0, \pi/4, \pi/3, \pi/2$. Compare the values found with the values of the exact, analytic derivative.

Exercise 05

Considering the case in exercise 04, try and give an estimate of the errors associated with the numerical derivatives calculated at the 7 grid points.

Exercise 06

Using reasoning similar to the one used to derive the three-point formula for the centred difference derivative, derive a three-point formula for the second derivative. Apply the formula found to calculate numerically the second derivative of the function $f(x) = x^2$ on the regular grid from 0 to 1, using $h = 0.01$.

7.7 The Trapezoid and Simpson algorithms for numerical integration

The integral of a function $f(x)$ can be difficult to calculate analytically. Quite often, the integral is not available as a finite analytic expression. An example is the *incomplete elliptic integral of the first kind,*

$$F(\phi, k) = \int_0^\phi \frac{d\theta}{1 - k^2 \sin^2(\theta)}, \quad -1 < k < 1,$$

or the *error function,*

$$\text{erf}(x) = \frac{2}{\sqrt{\pi}} \int_0^x e^{-t^2} dt.$$

Furthermore, the analytic expression of the integrand $f(x)$ might not be available as it is only known at a finite number of points. In all these cases the numerical value of the definite integral can still be calculated with numerical methods.

When the interval of integration is regularly sampled, the expressions derived from the approximation of the function with interpolating polynomials and from the subsequent integration, are called *Newton–Cotes integration formulas.* The two best known Newton–Cotes formulas are the *trapezoid* and *Simpson* rules. In what follows, the integration interval $[a, b]$ will be divided into n regular intervals of length

$$h = \frac{b - a}{n}$$

with the $n + 1$ points defining these intervals written as

$$a \equiv x_1, \; x_2, \; x_3, \; \ldots, \; x_n, \; x_{n+1} \equiv b, \quad x_{i+1} - x_i = h.$$

7.7.1 The trapezoid algorithm

The rationale behind it is well represented in figure 7.1.

The area to compute, under the curve in the specific figure, is divided into partial areas of irregular shape. Each one can be approximated by a trapezium if the corresponding arc is approximated by a segment. This corresponds to a linear

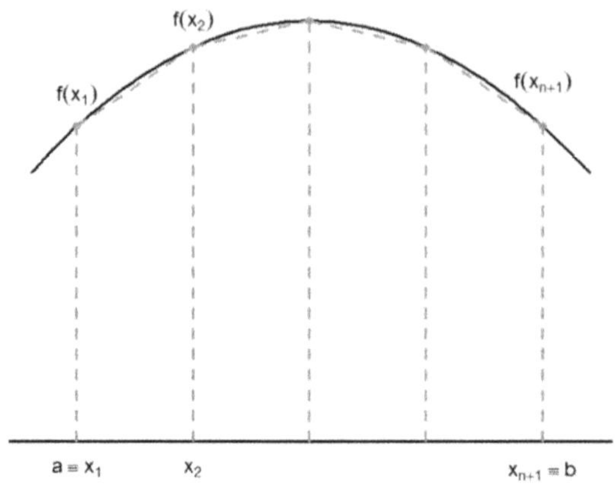

Figure 7.1. Construction related to the trapezoid algorithm for numerical integration. The interval $[a, b]$ is divided into n equal intervals of length h. Correspondingly, the area under the curve $f(x)$ is divided into n partial areas. Each area is unknown, but can be approximated by the area of a trapezium whose minor and major bases are the known heights of the function.

interpolation of $f(x)$ in each interval. The area ΔA_i of the trapezium between x_i and x_{i+1} is then[1]

$$\Delta A_i = h(f(x_i) + f(x_{i+1}))/2.$$

The approximation of the area between a and b is given by the sum of all the ΔA_i's:

$$\int_a^b f(x)\,\mathrm{d}x = \frac{h}{2}\sum_{i=1}^{n}(f_i + f_{i+1}) + \mathrm{O}(h^2), \tag{7.14}$$

where we have used the notation $f_i \equiv f(x_i)$ and where the approximation error will be explained in detail in section 7.8. Equation (7.14) describes the *trapezoid rule* of numerical integration.

7.7.2 The Simpson algorithm

The result in (7.14) can also be recovered by calculating the generic area ΔA_i as the integral between x_i and x_{i+1} of the linear interpolation,

$$f_i + \frac{f_{i+1} - f_i}{h}(x - x_i).$$

The integral is better calculated with the substitution

$$\frac{x - x_i}{h} = s, \quad \Rightarrow \quad x - x_i = sh,\ \mathrm{d}x = h\mathrm{d}s.$$

Thus

[1] It might be useful here to recall the area of a trapezium of height h, and bases b and B: $h(B + b)/2$.

$$\int_{x_i}^{x_{i+1}} f(x)\,\mathrm{d}x \approx \int_{x_i}^{x_{i+1}} \left(f_i + \frac{f_{i+1} - f_i}{h}(x - x_i)\right) \mathrm{d}x = \int_0^1 \left(f_i + \frac{f_{i+1} - f_i}{h} sh\right) h\,\mathrm{d}s$$

and

$$\int_{x_i}^{x_{i+1}} f(x)\,\mathrm{d}x \approx h \int_0^1 (f_i + (f_{i+1} - f_i)s)\,\mathrm{d}s = \frac{h}{2}(f_{i+1} + f_i),$$

exactly as written for the ΔA_i elements in equation (7.14). We can use this type of calculation to improve on the approximation of $f(x)$ and use a quadratic rather than a linear interpolation. We will need three rather than two points for the quadratic interpolation. The three points necessary for the interpolation are x_i, x_{i+1} and x_{i+2}. The approximation can be written using Lagrange interpolation (3.5) that, in this specific case, becomes

$$\begin{aligned} f(x) \approx\ & \frac{f_i}{(x_i - x_{i+1})(x_i - x_{i+2})}(x - x_{i+1})(x - x_{i+2}) \\ +\ & \frac{f_{i+1}}{(x_{i+1} - x_i)(x_{i+1} - x_{i+2})}(x - x_i)(x - x_{i+2}) \\ +\ & \frac{f_{i+2}}{(x_{i+2} - x_i)(x_{i+2} - x_{i+1})}(x - x_i)(x - x_{i+1}). \end{aligned}$$

To calculate the integral of this approximation between x_i and x_{i+2}, let us consider the first term. Similar calculations will apply to the other two terms For the first integral it is convenient to use the substitution $(x - x_{i+1})/h = s$; similar substitutions can be used for the second and third integral. We have then

$$\int_{x_i}^{x_{i+2}} f(x)\,\mathrm{d}x = \int_{x_i}^{x_{i+2}} \frac{f_i}{(x_i - x_{i+1})(x_i - x_{i+2})}(x - x_{i+1})(x - x_{i+2})\,\mathrm{d}x$$

or, using the new variable

$$\int_{x_i}^{x_{i+2}} f(x)\,\mathrm{d}x \approx \frac{hf_i}{2} \int_{-1}^{1} s(s - 1)\,\mathrm{d}s = \frac{hf_i}{2}\frac{2}{3} = \frac{h}{3}f_i.$$

For the other two integrals we obtain

$$\int_{x_i}^{x_{i+2}} \frac{f_{i+1}}{(x_{i+1} - x_i)(x_{i+1} - x_{i+2})}(x - x_i)(x - x_{i+2})\,\mathrm{d}x = \frac{4}{3}hf_{i+1}$$

and

$$\int_{x_i}^{x_{i+2}} \frac{f_{i+2}}{(x_{i+2} - x_i)(x_{i+2} - x_{i+1})}(x - x_i)(x - x_{i+1})\,\mathrm{d}x = \frac{1}{3}hf_{i+2}.$$

Adding the three terms together yields

$$\int_{x_i}^{x_{i+2}} f(x)\,\mathrm{d}x \approx \frac{h}{3}(f_i + 4f_{i+1} + f_{i+2}) \equiv \Delta S_i.$$

The integral between a and b, if we use a quadratic interpolation, can be approximated by a summation of areas like ΔS_i. As each area includes three interpolation points, the sum is possible only if n is even. The corresponding formula is known as *Simpson's 1/3 rule* because of the $h/3$ in front of the formula:

$$\int_a^b f(x)\,\mathrm{d}x \approx \frac{h}{3}\sum_{i=1}^{n/2}(f_{2i-1} + 4f_{2i} + f_{2i+1}) + \mathrm{O}(h^4), \tag{7.15}$$

where the error, again, will be discussed in section 7.8.

Example 7.4 *Let us write the Simpson's 1/3 rule for the integral of $f(x) = x^2$ between 0 and 2, using $n = 4$. As $n = 4$, the total number of points at which the function must be used is 5: $i = 1, 2, 3, 4, 5$. In this case we have:*

$$a \equiv x_1 = 0,\ x_2 = 0.5,\ x_3 = 1,\ x_4 = 1.5,\ x_5 \equiv b = 2$$

and

$$f_1 = 0^2 = 0, f_2 = 0.\ 5^2 = 0.25, f_3 = 1^2 = 1, f_4 = 1.\ 5^2 = 2.25, f_5 = 2^2 = 4.$$

The region between 0 and 2 is divided into two regions, the one supported by x_1, x_2, x_3 and the one supported by x_3, x_4, x_5. The summation of the Simpson's algorithm includes in this case only two terms Recalling that $h = 0.5$ here, the first term yields

$$\frac{0.5}{3}(0 + 4(0.25) + 1) = \frac{1}{3},$$

and the second term yields

$$\frac{0.5}{3}(1 + 4(2.25) + 4) = \frac{7}{3}.$$

The sum of these two contributions therefore gives 8/3. In this specific case the approximation is equal to the correct result because the function is quadratic and so $\mathrm{O}(h^3) = 0$.

A higher degree of accuracy is obtained if a cubic is used to interpolate $f(x)$. For a cubic, four points are needed, x_i, x_{i+1}, x_{i+2}, x_{i+3}, and n must be a multiple of 3. Let us indicate the partial area in this case with ΔR_i. Using an approach similar to the one used to derive Simpson's 1/3 rule, it is found that

$$\int_{x_i}^{x_{i+3}} f(x)\,\mathrm{d}x \approx \frac{3h}{8}(f_i + 3f_{i+1} + 3f_{i+2} + f_{i+3}) \equiv \Delta R_i.$$

Adding up all the ΔR_i's yields the *Simpson's 3/8 rule*, name given by the $3h/8$ term in front of the summation, with an $\mathrm{O}(h^4)$ error to be explained later, in section 7.8:

$$\int_a^b f(x)\,\mathrm{d}x \approx \frac{3h}{8}\sum_{i=1}^{n/3}(f_{3i-2} + 3f_{3i-1} + 3f_{3i} + f_{3i+1}) + \mathrm{O}(h^4). \tag{7.16}$$

Example 7.5 *Let us write the Simpson's 3/8 rule for the integral of* $f(x) = x^3$ *between* 0 *and* 3, *using* $n = 6$. *As* $n = 6$, *the total number of points at which the function must be used is* 7: $i = 1, 2, 3, 4, 5, 6, 7$. *In this case we have*:

$$a \equiv x_1 = 0,\ x_2 = 0.5,\ x_3 = 1,\ x_4 = 1.5,\ x_5 = 2,\ x_6 = 2.5,\ x_7 \equiv b = 3$$

and

$$f_1 = 0^3 = 0, f_2 = 0.\ 5^3 = 0.125, f_3 = 1^3 = 1, f_4 = 1.\ 5^3 = 3.375,$$

$$f_5 = 2^3 = 8, f_6 = 2.\ 5^3 = 15.625, f_7 = 3^3 = 27.$$

The region between 0 *and* 3 *is divided into two panels, the one supported by* x_1, x_2, x_3, x_4 *and the one supported by* x_4, x_5, x_6, x_7. *The summation of the Simpson's algorithm includes here two terms. Recalling that* $h = 0.5$, *the first term yields*

$$\frac{3(0.5)}{8}(0 + 3(0.125) + 3(1) + 3.375) = \frac{81}{64}$$

and the second term yields

$$\frac{3(0.5)}{8}(3.375 + 3(8) + 3(15.625) + 27) = \frac{1215}{64}.$$

The sum of these two contributions therefore gives 81/4. *In this specific case, as in the previous example, the approximation is equal to the correct result because the function is cubic and so* $O(h^4) = 0$.

7.7.3 Weights for trapezoid and Simpson formulas

The three formulas for numeric integration, equations (7.14), (7.15), (7.16), consist of the sum of groups of values of the function, two values for the trapezoid rule, three values for Simpson's 1/3 rule, and four values for Simpson's 3/8 rule. The coefficients of the function can be seen as weights multiplied to the function itself and they characterise the specific algorithm. The coefficients and the multiplicative h coefficient are summarised in the following table:

Name	h coefficient	w_i	w_{i+1}	w_{i+2}	w_{i+3}	Local Error	Global Error
Trapezoid	$h/2$	1	1	NA	NA	$O(h^3)$	$O(h^2)$
Simpson's 1/3	$h/3$	1	4	1	NA	$O(h^5)$	$O(h^4)$
Simpson's 3/8	$3h/8$	1	3	3	1	$O(h^5)$	$O(h^4)$

The meaning of 'local error' and 'global error' are explained in section 7.8.

7.8 Numerical errors for Newton–Cotes integration formulas

We have seen that the trapezoid and the Simpson's rules are two of the many possible Newton–Cotes integration formulas. In order to quantify the magnitude of the numerical errors arising from these formulas, we can use knowledge acquired when dealing with interpolation (see section 3.5). In all Newton–Cotes formulas, the numerical integration of $f(x)$ can be thought of as the integration of a polynomial approximation of $f(x)$, where we can indicate with $P_n(x)$ an interpolating polynomial of order n. The numerical integration is then just an approximation of the analytical integration because $P_n(x)$ is only an approximation of $f(x)$. The approximation error is related to the interpolation error that, in section 3.5.1, is known to be

$$\Delta P_n(x) = \frac{f^{(n+1)}(\xi)}{(n+1)!}(x - x_1)(x - x_2) \cdots (x - x_{n+1}),$$

where $f^{(n+1)}(x)$ is the $(n+1)$th derivative of $f(x)$ and where ξ, which is in the real interval (x_1, x_{n+1}), depends on the specific point x at which the error is calculated. The errors will then be obtained through integration of $\Delta P_n(x)$, that is

$$\text{error} = \int_a^b \Delta P_n(x)\, \mathrm{d}x = \frac{f^{(n+1)}(\xi)}{(n+1)!}\int_a^b (x - x_1) \dots (x - x_{n+1})\, \mathrm{d}x.$$

The integral is daunting and rather useless in this form. But we can exploit the regularity of the grid, where $x_{i+1} - x_i = h$, and introduce a variable $s = (x - x_i)/h$ instead of x, for any of the i between 1 and $n = 1$. For example, if we use $s = (x - x_1)/h$, the integral becomes

$$\text{error} = \frac{f^{(n+1)}(\xi)}{(n+1)!}\int_0^n h^{n+1} s(s-1) \cdots (s-n)\, h\mathrm{d}s,$$

or, using the expression of the binomial coefficient,

$$\text{error} = h^{n+2} f^{(n+1)}(\xi) \int_0^n \binom{s}{n+1} \mathrm{d}s. \tag{7.17}$$

Formula (7.12) is the starting point to work out the errors. The method to do that is to first calculate the so-called *local error*, that is the error caused by each integration element (x_i to x_{i+1} for the trapezoid, x_i to x_{i+2} for Simpson's 1/3 and x_i to x_{i+3} for Simpson's 3/8). Then the sum of the contribution due to all integration elements will provide the *global error*. We will now apply the method to estimate the errors for the trapezoid, Simpson's 1/3 and Simpson's 3/8 rules.

1. **Trapezoid**. The interpolation error, $\Delta P_1(x)$ is:

$$\Delta P_1(x) = \frac{f^{(2)}(\xi)}{2!}(x - x_i)(x - x_{i+1}).$$

Each integration element goes from x_i to x_{i+1}, so the local error is given by:

$$\text{local error} = \int_{x_i}^{x_{i+1}} \frac{f^{(2)}(\xi)}{2!}(x - x_i)(x - x_{i+1})\mathrm{d}x$$

We can then follow up with the same substitution used above, here $(x - x_i)/h = s$, and turn the integral into a tractable one:

$$\text{local error} = \frac{f^{(2)}(\xi)}{2!}h^3 \int_0^1 s(s-1)\mathrm{d}s.$$

The local error for the trapezoid rule is thus

$$\text{local error} = -\frac{1}{12}h^3 f^{(2)}(\xi). \tag{7.18}$$

The accumulation from the contributions of all the n local errors from a to b yields the global error:

$$\text{global error} = -\frac{1}{12}h^3 f^{(2)}(\xi)n = -\frac{1}{12}h^3 f^{(2)}(\xi)\frac{b-a}{h}$$

$$\Downarrow$$

$$\text{global error} = \mathrm{O}(h^2).$$

2. **Simpson's 1/3**. The interpolation error, $\Delta P_2(x)$ is:

$$\Delta P_2(x) = \frac{f^{(3)}(\xi)}{3!}(x - x_i)(x - x_{i+1})(x - x_{i+2}).$$

Accordingly, the local error, calculated integrating from x_i to x_{i+2}, can be estimated through the following expression:

$$\text{local error} = \frac{f^{(3)}(\xi)}{3!}h^4 \int_0^2 s(s-1)(s-2)\mathrm{d}s.$$

In this case, though, it turns out that the integral is equal to zero. Clearly, this does not mean that both the local and global errors are zero, because this numerical integration is still an approximation of the exact definite integral. It simply means that we must consider the next error term of the interpolation, which is

$$\frac{f^{(4)}(\xi)}{4!}(x - x_i)(x - x_{i+1})(x - x_{i+2})(x - x_{i+3}).$$

The reader should not be concerned having a $x - x_{i+3}$ term while the integration goes from x_i to x_{i+2}, the two facts are perfectly compatible in the calculation. Therefore, the non-zero element of the local error is(using integration variable $s = (x - x_i)/h$):

$$\text{local error} = \frac{f^{(4)}(\xi)}{4!}h^5 \int_0^2 s(s-1)(s-2)(s-3)\mathrm{d}s,$$

$$\text{local error} = -\frac{1}{90}h^5 f^{(4)}(\xi). \tag{7.19}$$

There are $n/2$ contributions of the local error to the global error in the numerical integration from a to b. Thus:

$$\text{global error} = -\frac{1}{90}h^5 f^{(4)}(\xi)\frac{n}{2} = -\frac{1}{90}h^5 f^{(4)}(\xi)\frac{b-a}{2h}$$

$$\Downarrow$$

$$\text{global error} = \mathrm{O}(h^4).$$

3. **Simpson's 3/8**. The local error for this rule turns out to be

$$\text{local error} = -\frac{3}{80}h^5 f^{(4)}(\xi). \qquad (7.20)$$

As for the simpson's 1/3 rule, the global error here is $\mathrm{O}(h^4)$. The calculation to derive the local error is requested as part of exercises 7.13.

7.9 Relevant R code. Function `numint_reg`

This function implements the three methods of numerical integration, the trapezoid and the two Simpson's. The default type is Simpson's 1/3 (parameter `scheme="sim13"`).

Any number of intervals, `n`, can be chosen to apply the trapezoid rule; Simpson's 1/3 needs an even number of intervals while for Simpson's 3/8 the number of intervals must be a multiple of 3. If `n` does not satisfy the specific requirements, the specific algorithm is applied for the appropriate number of intervals and in the remaining intervals integration is carried out using the trapezoid rule, with a warning being triggered to warn users about their inappropriate choice of intervals.

Input is formed of the regular grid, `x`, and the corresponding values of the function, `f`, in addition to the chosen algorithm with the parameter `scheme` that can take values `"trap"`, `"sim13"`, and `"sim38"`.

The most important part of the code used involves a convolution between $f(x)$ and the weights used in the scheme, convolution that is digitally equivalent to the product of two arrays, `w` and `f`. It is important to remember that values of the function at the end points are used twice. For example, thinking about Simpson 1/3, f_{i+2} is used both for the panel x_i, x_{i+1}, x_{i+2} and the panel x_{i+2}, x_{i+3}, x_{i+4}. So, although the weights for one panel are 1, 4, 1, across the two panels they become 1, 4, 2, 4, 1. The same, with different coefficients, happens for the other two algorithms.

A simple application of the function is presented in the following code, where integration of $\sin(x)$ between 0 and π is performed using all algorithms with a small and large number of intervals, to appreciate the error and the warning related to the inappropriate number of intervals (not a multiple of 3 or a multiple of 2, respectively). The correct value of this definite interval is 2, as the reader can quickly verify.

```
# Load comphy
library(comphy)

# The function to integrate is sin(x), between 0 and pi

# Tabulated values (regular grid)
x <- seq(0,pi,length.out=21) # Small n (even)


f <- sin(x)
y <- seq(0,pi,length.out=100) # Large n (multiple of 3)
g <- sin(y)

# Different algorithms (small n)
nvalue1 <- numint_reg(x,f,scheme="trap")
nvalue2 <- numint_reg(x,f,scheme="sim13")
nvalue3 <- numint_reg(x,f,scheme="sim38")

## Warning in numint_reg(x, f, scheme = "sim38"):  Last contribution to
integral is from trapezoid rule as the number of intervals is not a
multiple of 3.

print(c(nvalue1,nvalue2,nvalue3))

## [1] 1.995886 2.000007 1.999914

# Different algorithms (large n)
nvalue1 <- numint_reg(y,g,scheme="trap")
nvalue2 <- numint_reg(y,g,scheme="sim13")

## Warning in numint_reg(y, g, scheme = "sim13"):  Last contribution to
integral is from trapezoid rule as the number of intervals is not even.

nvalue3 <- numint_reg(y,g,scheme="sim38")
print(c(nvalue1,nvalue2,nvalue3))

## [1] 1.999832 2.000000 2.000000
```

It is perhaps worth remarking that, in the first case, the two values of the integral show some visible errors because of the relatively large h, while the last value has also the contribution from a trapezoid rule rather than a Simpson's 3/8 rule. In the second case the first value is close to 2 and the error is compatible with $O(h^2)$. The third value is the most accurate while the second still suffers from the last interval being calculated with the trapezoid rule because the number of intervals is not even, even though the number of digits displayed does not show that, as the error is much smaller.

7.10 Gaussian quadrature

When looking at the Newton–Cotes formulas for numerical integration (trapezoid and Simpson rules), one realises that the integral is approximated as a weighted combination of function values at certain points. These points are evenly spaced. The same approach can be used when the points are irregularly spaced. In this case the numerical integration is known as *Gaussian quadrature*.

7.10.1 Rationale for Gaussian quadrature

Suppose we want to approximate the integral of a function f over the specific interval $[-1, 1]$ (we will generalise to a generic interval $[a, b]$ later on). The central idea of Gaussian quadrature is to write

$$\int_{-1}^{+1} f(x)\,\mathrm{d}x \approx \sum_{i=1}^{n} w_i f(x_i),$$

where the values x_i are called *nodes* (or *evaluation points*) and the corresponding w_i's are called *weights*. This formula evaluates f at n points and returns a weighted sum. The crucial question is: how should we choose the points and weights? Our goal is to choose them so that this approximation is as accurate as possible. In particular, we want to make this formula exact for polynomials, that is, we want

$$\int_{-1}^{+1} p(x)\,\mathrm{d}x = \sum_{i=1}^{n} w_i p(x_i)$$

for all polynomials p up to the highest possible degree (it is key to observe that the approximation symbol was replaced by the equality symbol in the expression above). The reason why a request of exact calculation is done for polynomials is that continuous functions can be approximated well by polynomials, this is the content of *Weierstrass approximation theorem*, mentioned in section 3.11.

Each point x_i and each weight w_i is a free parameter. If we use n points, then we have $2n$ parameters: n values $x_1, \ldots, x_n$ and n corresponding weights $w_1, \ldots, w_n$. Therefore, we must impose up to $2n$ conditions, to find all parameters. A natural choice is to require the formula to integrate monomials x_k exactly, for $k = 0, 1, \ldots, 2n - 1$. These conditions guarantee that the rule is exact for any polynomial of degree up to $2n - 1$, which turns out to be the maximum possible degree for this kind of rule.

Let us construct such rules progressively, starting with the simplest cases.

1. **One point. Polynomials of degree less than or equal to 1**. Let us try one point x_1 with one weight w_1, and write:

$$\int_{-1}^{+1} f(x)\,\mathrm{d}x \approx w_1 f(x_1).$$

We want the formula to be exact for the monomials $x^0 = 1$ and $x^1 = x$. For the first monomial we have:

$$\int_{-1}^{+1} \mathrm{d}x = w_1 \quad \Rightarrow w_1 = 2.$$

For the second monomial we have:

$$\int_{-1}^{+1} x \, \mathrm{d}x = w_1 x_1 \quad \Rightarrow w_1 x_1 = 0 \quad \Rightarrow x_1 = 0,$$

as we had already found $w_1 = 2$. This means that integrals of polynomials of degree up to one can be integrated exactly with Gaussian quadrature according to the following formula:

$$\int_{-1}^{+1} f(x) \, \mathrm{d}x \approx 2f(0). \tag{7.21}$$

Example 7.6 *The integral of* $2x - 1$ *between* -1 *and 1 is* -2. *Let us calculate the integral numerically using Gaussian quadrature* (7.21).

$$\int_{-1}^{+1} (2x - 1) \, \mathrm{d}x = 2(2(0) - 1) = -2.$$

The result is exact as $f(x)$ *is in this case a polynomial of degree* 1. *For a generic function the result is only an approximation (a bad one because the quadrature uses only two parameters). For example, if we integrate* e^x *between* -1 *and* 1 *we obtain* $e - 1/e \approx 2.350$. *With Gaussian quadrature* (7.21) *we have, instead,*

$$\int_{-1}^{+1} \mathrm{e}^x \, \mathrm{d}x \approx 2\mathrm{e}^0 = 2.$$

2. **Two points. Polynomials of degree less than or equal to 3**. Let us try, next, two points and two weights. Four parameters can be accommodated when using a polynomial of degree 3, as this contains four parameters. The quadrature formula is:

$$\int_{-1}^{+1} f(x) \, \mathrm{d}x \approx w_1 f(x_1) + w_2 f(x_2).$$

The monomials to use to find the parameters are now $x^0 = 1$, $x^1 = x$, x^2, x^3:

$$\begin{aligned}
\int_{-1}^{+1} \mathrm{d}x &= w_1 + w_2 &\Rightarrow w_1 + w_2 &= 2 \\
\int_{-1}^{+1} x \, \mathrm{d}x &= w_1 x_1 + w_2 x_2 &\Rightarrow w_1 x_1 + w_2 x_2 &= 0 \\
\int_{-1}^{+1} x^2 \, \mathrm{d}x &= w_1 x_1^2 + w_2 x_2^2 &\Rightarrow w_1 x_1^2 + w_2 x_2^2 &= 2/3 \\
\int_{-1}^{+1} x^3 \, \mathrm{d}x &= w_1 x_1^3 + w_2 x_2^3 &\Rightarrow w_1 x_1^3 + w_2 x_2^3 &= 0.
\end{aligned}$$

The system can be solved with some level of difficulty, yielding

$$w_1 = w_2 = 1 \qquad x_1 = -\frac{1}{\sqrt{3}}, \; x_2 = -x_1 = \frac{1}{\sqrt{3}}.$$

The formula for the two points Gaussian quadrature is, therefore:

$$\int_{-1}^{+1} f(x)\,\mathrm{d}x \approx f\left(-\frac{1}{\sqrt{3}}\right) + f\left(\frac{1}{\sqrt{3}}\right). \tag{7.22}$$

Example 7.7 *The two-points Gaussian quadrature returns exact integrals for polynomials up to degree* 3. *Furthermore, it is a better approximation for integral of other functions, compared to the one-point quadrature. For example, we have seen previously that the integral of* $f(x) = \mathrm{e}^x$ *between* -1 *and* 1 *yields roughly* 2.350. *With the one-point formula we have obtained an approximate value of* 2. *With the two-points formula we have*:

$$\int_{-1}^{+1} \mathrm{e}^x\,\mathrm{d}x \approx \mathrm{e}^{-1/\sqrt{3}} + \mathrm{e}^{1/\sqrt{3}} \approx 0.561 + 1.781 = 2.342,$$

certainly a better approximation to 2.350.

3. **More than two points. Polynomials of degree higher than 3**. Proceeding in a fashion similar to the one demonstrated earlier, it is possible to build Gaussian quadrature formulas with a higher accuracy. We will see soon, though, that a better and general result can be described using *Legendre polynomials*, shortly described in appendix F.

7.10.2 Gaussian quadrature and Legendre polynomials

The weights, w_i, and points, x_i, of any Gaussian quadrature can be systematically found using *Legendre polynomials* (see appendix F). More specifically, in appendix A (section A.11) it is proved that the nodes x_i of an n-point Gaussian quadrature are the zeros of the Legendre polynomial of order n, while in section A.11 the formula

$$w_k = \frac{2}{(1 - x_k^2)(P'(x_k))^2} \tag{7.23}$$

is given, relating the Gaussian quadrature's weights to the zeros and first derivative of the Legendre polynomial $P_n(x)$.

Proving the above results can be rather daunting and requires knowledge of the definition and properties of orthogonal polynomials. Although self-contained proofs are included in appendix A, they are not required for applications. A couple of examples will be sufficient to demonstrate how nodes and weights can be derived from Legendre polynomials.

Example 7.8 *Let us derive weights and nodes for the one-point Gaussian quadrature. We have seen (equation (7.21)). In this case* $n = 1$ *and we will need to consider Legendre polynomial* $P_1(x) = x$, *whose zero in* $[-1, 1]$ *is* $x_1 = 0$. *The derivative of the polynomial is* $P_1'(x) = 1$ *and therefore formula (7.23) in this case yields*:

$$w_1 = \frac{2}{(1 - 0)^2(1)} = 2.$$

This corresponds to the result previously found.

Example 7.9 *Let us derive weights and nodes for the two-points Gaussian quadrature. Here* $n = 2$ *and* $P_2(x) = (1/2)(3x^2 - 1)$. *The two zeros can be found easily*:

$$P_2(x) = 0 \quad \Leftrightarrow \frac{1}{2}(3x^2 - 1) = 0 \quad \Rightarrow x_1 = -\frac{1}{\sqrt{3}},\ x_2 = \frac{1}{\sqrt{3}}.$$

They correspond to what was found earlier. To calculate the corresponding weights, we need the first derivative. It is $P_2'(x) = 3x$. *We have, accordingly*:

$$w_1 = \frac{2}{(1 - (-1/\sqrt{3})^2)(3(-1/\sqrt{3}))^2} = 1 = w_2$$

which, again, is what we found earlier.

Example 7.10 *Let us derive weights and nodes for the three-points Gaussian quadrature. This allows an exact estimate of definite integrals of polynomials up to the fifth degree, between* -1 *and 1. Here* $n = 3$ *and* $P_3(x) = (1/2)(5x^3 - 3x)$. *The three zeros are found algebrically*:

$$\frac{1}{2}(5x^3 - 3x) = 0 \quad \Rightarrow x_1 = -\sqrt{\frac{3}{5}},\ x_2 = 0,\ x_3 = \sqrt{\frac{3}{5}}.$$

The derivative of the polynomial is $P_3'(x) = (1/2)(15x^2 - 3)$. *At* $x_3 = -x_1 = \sqrt{3/5}$, *its value is* $P_3'(x_1) = P_3'(x_3) = 3$, *while at* $x_2 = 0$ *its value is* $P_3'(x_2) = -3/2$. *Therefore, the weights are*:

$$w_1 = w_3 = \frac{2}{(1 - 3/5)9} = \frac{5}{9}, \qquad w_2 = \frac{2}{(1)(9/4)} = \frac{8}{9}.$$

7.10.3 Gaussian quadrature for arbitrary intervals

The derivations for the nodes and weights of Gaussian quadrature, as presented in the preceding sections, are based on the standard integration interval of $[-1, 1]$. However, in practical applications, integrals often need to be evaluated over arbitrary finite intervals $[a, b]$. This section explains how to adapt the Gaussian quadrature formulas to such intervals through a simple change of variables.

Consider an integral of a function $f(t)$ over an arbitrary interval $[a, b]$,

$$\int_a^b f(t)\ \mathrm{d}t.$$

To apply Gaussian quadrature, we introduce a linear transformation that maps the variable x from the standard interval $[-1, 1]$ to the variable t in the desired interval $[a, b]$. The transformation is given by

$$t = \frac{b - a}{2}x + \frac{a + b}{2}.$$

From this transformation, we can also find the differential relationship by differentiating both sides with respect to x:

$$\mathrm{d}t = \left(\frac{b-a}{2}\right)\mathrm{d}x.$$

Next, we substitute these expressions for t and dt into the integral over $[a, b]$:

$$\int_a^b f(t)\,\mathrm{d}t = \int_{-1}^{1} f\left(\frac{b-a}{2}x + \frac{a+b}{2}\right)\left(\frac{b-a}{2}\right)\,\mathrm{d}x.$$

Let $g(x) = f\left(\frac{b-a}{2}x + \frac{a+b}{2}\right)$. The integral then becomes:

$$\int_a^b f(t)\,\mathrm{d}t = \frac{b-a}{2}\int_{-1}^{1} g(x)\,\mathrm{d}x.$$

Recall that the standard n-point Gaussian quadrature rule for the interval $[-1, 1]$ is given by:

$$\int_{-1}^{1} g(x)\,\mathrm{d}x \approx \sum\nolimits_{i=1}^{n} w_i g(x_i),$$

where x_i are the standard nodes (roots of $P_n(x)$) and w_i are the standard weights (calculated using $P_n(x)$ and $P'(x)$). To apply this to our transformed integral, we define new nodes and weights that are appropriate for the interval $[a, b]$:

1. **Transformed nodes** (t_i). Each standard node x_i is mapped to a corresponding node t_i in the interval $[a, b]$ using the transformation formula:

$$t_i = \frac{b-a}{2}x_i + \frac{a+b}{2} \tag{7.24}$$

2. **Transformed weights** (W_i). Each standard weight w_i is scaled by the factor $\left(\frac{b-a}{2}\right)$, which arises from the change in the differential $\mathrm{d}t$:

$$W_i = \left(\frac{b-a}{2}\right)w_i. \tag{7.25}$$

Using these transformed nodes t_i and transformed weights W_i, the Gaussian quadrature formula for an arbitrary interval $[a, b]$ becomes:

$$\int_a^b f(t)\,\mathrm{d}t \approx \sum\nolimits_{i=1}^{n} W_i f(t_i). \tag{7.26}$$

In essence, one first determines the standard nodes x_i and weights w_i for the interval $[-1, 1]$, and then applies the linear transformation to obtain the corresponding nodes t_i and weights W_i for the specific interval $[a, b]$.

Example 7.11 *Consider the two-point Gaussian quadrature to calculate*

$$\int_1^4 (x^2 - 1)\,\mathrm{d}x.$$

The definite integration yields 18, *let us check that the two-point Gaussian quadrature returns the same value, as the integrand is a polynomial of degree* 2 *and the two-point quadrature returns exact results for polynomials of up to degree* $2(2) - 1 = 3$.

The nodes, as we know, are $\pm 1/\sqrt{3}$ *and the weights are both equal to* 1. *Using transformation (7.24), and noting that* $a = 1$, $b = 4$, *the new nodes are*

$$t_1 = \frac{3}{2}\left(-\frac{1}{\sqrt{3}}\right) + \frac{5}{2}, \quad t_2 = \frac{3}{2}\left(\frac{1}{\sqrt{3}}\right) + \frac{5}{2}.$$

Then, using transformation (7.25), the new weights are

$$w_1 = w_2 = \frac{3}{2}(1) = \frac{3}{2}.$$

The quadrature will then be:

$$\frac{3}{2}\left(\left(\frac{5}{2} - \frac{3}{2\sqrt{3}}\right)^2 - 1\right) + \frac{3}{2}\left(\left(\frac{5}{2} + \frac{3}{2\sqrt{3}}\right)^2 - 1\right).$$

The result, as it can be easily checked through a thorough calculation, is 18, *as expected.*

7.11 Relevant R code. Function `Gquad`

This function calculates the nodes and weights of the quadrature using Legendre polynomials, and returns the approximate value of the given definite integral. To avoid using external packages, these numbers are obtained through use of the well-known and stable *Golub–Welsch algorithm* [1].

In chapter 4 we introduced tridiagonal matrices and discussed their relevance to solving linear systems and computing eigenvalues. In the context of Gaussian quadrature, these matrices make a surprising reappearance. Using the Golub–Welsch algorithm to compute the nodes x_i and weights w_i for the n-point Gaussian quadrature, makes it possible to avoid evaluating Legendre polynomials directly. Instead, the algorithm constructs a symmetric tridiagonal matrix known as the *Jacobi matrix*, whose entries are derived from the recurrence relation of Legendre polynomials (see [1] for details). This $n \times n$ matrix, T, has zeros on the diagonal, while the two lines adjacent to the diagonal are given by

$$\beta_k = \frac{1}{\sqrt{4k^2 - 1}}, \quad \text{for} k = 1, 2, \ldots, n - 1.$$

That is, T has the form

$$T = \begin{pmatrix} 0 & \beta_1 & 0 & \cdots & 0 \\ \beta_1 & 0 & \beta_2 & \ddots & \vdots \\ 0 & \beta_2 & 0 & \ddots & 0 \\ \vdots & \ddots & \ddots & \ddots & \beta_{n-1} \\ 0 & \cdots & 0 & \beta_{n-1} & 0 \end{pmatrix}.$$

The eigenvalues of T are the quadrature nodes x_i, all lying in the interval $[-1, 1]$. The corresponding *weights* are obtained from the squares of the first components of the normalised eigenvectors $v^{(i)}$:

$$w_i = 2\left(v_1^{(i)}\right)^2.$$

This approach turns out to be both stable and efficient [1]. As we know (see chapter 4), the R function `eigen()` computes eigenvalues and eigenvectors for symmetric matrices like T very reliably. Once the nodes and weights on $[-1, 1]$ are known, they can be rescaled to any finite interval $[a, b]$ using a simple affine transformation.

`Gquad()` takes as input a function `f` that must be previously defined, the extremes of integration `a` and `b`, and the order of the quadrature (default is `n=5`). The output is a list with three elements. The first is `xt`, a numeric vector containing the nodes of the quadrature. The second is `wt`, a numeric vector containing the corresponding weights, and the third, `itg`, is the approximate value of the integral.

Let us illustrate use of `Gquad()` using some of the examples previously introduced. The first is example 7.6, where $f(x) = 2x - 1$ and the extremes of integration are -1 and 1. Only one point is used for the quadrature in this example. Zeros and weights turn out to be what expected, and the value of the integral is exact.

```
# Load comphy
library(comphy)

# Function 2x - 1
f <- function(x) {ff <- 2*x-1; return(ff)}

# 1-point Gaussian quadrature
ltmp <- Gquad(f,-1,1,n=1)

# Zeros (just one) and weights (just one)
print(ltmp$xt)

## [1] 0

print(ltmp$wt)

## [1] 2

# Approximate integral (exact in this case)
print(ltmp$itg)
## [1] -2
```

The second integral is taken from the two-point quadrature in example 7.7. The function is $f(x) = e^x$ and the integration interval is still $[-1, 1]$. With two points the approximation is good, but with the default $n = 5$ it is much better.

```
# Function exp(x) does not need to be re-defined

# 2-point Gaussian quadrature
ltmp <- Gquad(exp,-1,1,n=2)

# Zeros and weights, two values each
print(ltmp$xt)

## [1]  0.5773503 -0.5773503

print(ltmp$wt)

## [1] 1 1

# Approximate integral (not bad)
print(ltmp$itg)

## [1] 2.342696

# 5-point quadrature(default)
ltmp <- Gquad(exp,-1,1)

# Approximate integral (better)
print(ltmp$itg)

## [1] 2.350402
```

Finally, let us treat an example where the integration interval is not $[-1, 1]$. The calculation is taken from example 7.11, in which $f(x) = x^2 - 1$, $a = 1$, $b = 4$, and a two-point formula is used. The result in this case is exact.

```
# Function
f <- function(x) {ff <- x^2-1; return(ff)}

# 2-point Gaussian quadrature
ltmp <- Gquad(f,1,4,n=2)

# Approximate integral (exact)
print(ltmp$itg)

## [1] 18
```

7.12 Multiple integrals

In physics and engineering, many problems require the evaluation of integrals over two, three, or even higher dimensions. These are known as *multiple integrals*. While analytical solutions exist for some simple cases, the vast majority of multiple integrals arising from real-world phenomena, especially those involving complex integrands or irregular domains, necessitate numerical approximation. This field is a rich and active area of research, with continuous development of faster and more efficient algorithms to tackle increasingly challenging problems. In this section we will only show how 2D integration can be carried out using the numerical techniques adopted in 1D, more specifically Gaussian quadrature. The reader interested to deepen this topic can consider browsing through some of the suggested, specialised literature [2–4].

Extending one-dimensional numerical integration techniques (such as Newton–Cotes formulas or Gaussian quadrature) to higher dimensions introduces a significant computational challenge often referred to as the *curse of dimensionality*. As the number of dimensions increases, the number of function evaluations required by traditional grid-based methods grows exponentially. For instance, if a one-dimensional rule uses N points, a direct extension to d dimensions using a simple tensor product would require N^d function evaluations. This rapid growth quickly renders such methods computationally prohibitive even for moderately high dimensions (e.g. $N = 10$ points in $d = 5$ dimensions already means 10^5 evaluations).

For integrals over hyper-rectangular domains (e.g. a square in 2D, a cube in 3D, or a hypercube in higher dimensions), the most straightforward and widely used approach is to construct a *tensor product rule*. This method involves applying a one-dimensional quadrature rule sequentially for each dimension. Consider a two-dimensional integral over a rectangular region $R = [a_x, b_x] \times [a_y, b_y]$:

$$\iint_R f(x, y)\, \mathrm{d}x\mathrm{d}y = \int_{a_y}^{b_y} \left(\int_{a_x}^{b_x} f(x, y)\, \mathrm{d}x \right) \mathrm{d}y$$

To implement this using Gaussian quadrature, we first apply the 1D Gaussian quadrature rule to the inner integral with respect to x. This requires transforming the

x-interval $[a_x, b_x]$ to the standard $[-1, 1]$ interval. Let (x_i, w_i) be the standard nodes and weights for the 1D Gaussian quadrature rule. The transformed nodes for the x-dimension will be $X_i = ((b_x - a_x)/2)x_i + (a_x + b_x)/2$, and the transformed weights $W_{X_i} = ((b_x - a_x)/2)w_i$. The inner integral is then approximated as:

$$\int_{a_x}^{b_x} f(x, y)\, \mathrm{d}x \approx \sum_{i=1}^{n_x} W_{X_i} f(X_i, y)$$

This approximation effectively turns the 2D problem into a sum of 1D integrals with respect to y, each evaluated at a specific X_i node. We then substitute this approximation into the outer integral:

$$\iint_R f(x, y)\, \mathrm{d}x\mathrm{d}y \approx \int_{a_y}^{b_y} \left(\sum_{i=1}^{n_x} W_{X_i} f(X_i, y)\right) \mathrm{d}y$$

Since the summation and weights W_{X_i} are constant with respect to y, we can move them outside the integral:

$$\approx \sum_{i=1}^{n_x} W_{X_i}\left(\int_{a_y}^{b_y} f(X_i, y)\, \mathrm{d}y\right)$$

Finally, we apply the 1D Gaussian quadrature rule to each of these remaining integrals with respect to y. Let (y_j, w_j) be the standard nodes and weights for the y-dimension. The transformed nodes for the y-dimension will be $Y_j = \frac{by - ay}{2} y_j + \frac{ay + by}{2}$, and the transformed weights $W_{Y_j} = \frac{by - ay}{2} w_j$. The full 2D approximation becomes:

$$\iint_R f(x, y)\, \mathrm{d}x\mathrm{d}y \approx \sum_{i=1}^{n_x} W_{X_i}\left(\sum_{j=1}^{n_y} W_{Y_j} f(X_i, Y_j)\right)$$

This can be rewritten as a double summation:

$$\iint_R f(x, y)\, \mathrm{d}x\mathrm{d}y \approx \sum_{i=1}^{n_x} \sum_{j=1}^{n_y} W_{X_i} W_{Y_j} f(X_i, Y_j)$$

It is relatively straightforward to appreciate that this concept extends directly to three or more dimensions simply by adding more summation indices.

Example 7.12 *Calculate the 2D integral*

$$\iint_R (x^2 + y^2)\, \mathrm{d}x\mathrm{d}y,$$

when $R = [-1, 1] \times [-1, 1]$, *using a two-point Gaussian quadrature.*

It is straightforward to calculate the double integral analytically and find the value 8/3. As both integration intervals are the canonical ones, zeros and weights are given by

$$x_1 = y_1 = -\frac{1}{\sqrt{3}}, \quad x_2 = y_2 = \frac{1}{\sqrt{3}}, \quad w_{x_1} = w_{x_2} = w_{y_1} = w_{y_2} = 1.$$

The quadrature for this integral is thus

$$w_{x_1}w_{y_1}f(x_1, y_1) + w_{x_1}w_{y_2}f(x_1, y_2) + w_{x_2}w_{y_1}f(x_2, y_1) + w_{x_2}w_{y_2}f(x_2, y_2) = 4\left(\frac{1}{3} + \frac{1}{3}\right) = \frac{8}{3}.$$

The result is in this case exact because the two-point quadrature yields exact integration for polynomials of up to degree 3 *in both* x *and* y.

Not all integration domains are hyper-rectangular. For integrals over non-rectangular regions (e.g. a circle, a triangle, or an arbitrarily shaped volume), applying tensor product rules directly is not immediately possible. In such cases, a common strategy is to perform a *change of variables* (coordinate transformation) to map the complex domain onto a simpler, standard domain (such as a unit square or unit cube). Once the transformation is applied, the integral can then be evaluated using tensor product rules on the transformed, simpler domain. The Jacobian of the transformation must be included in the integrand after the change of variables. The complexity of these transformations can vary significantly depending on the geometry of the original domain.

Example 7.13 *Evaluate the same integral shown in the previous example, this time in the region* R *represented by the unit circle centred at the origin.*

Given the circular symmetry of the region, it is possible to transform the 2D integral into a 1D integral. The variables transformation will be

$$x = \rho\cos(\theta), \qquad y = \rho\sin(\theta)$$

and the Jacobian of the transformation is ρ. *Therefore, the integral becomes*

$$\iint_R (x^2 + y^2)\,\mathrm{d}x\mathrm{d}y = \int_0^1 \int_0^{2\pi} \rho^2\,\rho\,\mathrm{d}\rho\mathrm{d}\theta = 2\pi\int_0^1 \rho^3\,\mathrm{d}\rho.$$

The last 1D integral can then be solved analytically or numerically without any difficulty. The final value of the double integral is $\pi/2$.

It is, in general, more complicated to calculate numerically 2D integral with Gaussian quadrature when the integration R is not compatible with any useful symmetry. But systematic methods exist in this case too, although they will not be explored in this textbook. Algorithmic implementation of the 2D integration suggested above over a rectangular integral, will not be illustrated here and no function related to 2D (or any dimension larger than 1) integration is included in the `comphy` package. For a multi-dimensional integral it is nearly always more advantageous to resort to Monte Carlo calculations (see a later chapter), or to an ad hoc algorithm that depends on the type of integration required. An exercise to implement the rectangular 2D integration described above is suggested in section 7.13.

7.13 Exercises on integration

Exercise 07

Calculate the following integral,

$$\frac{1}{\sqrt{2\pi}} \int_{-1}^{+1} e^{-x^2/2} \, dx,$$

numerically using the trapezoid, and Simpson's 1/3 and 3/8 rules. Compare the results obtained with those displayed with the R function `pnorm`.

Exercise 08

Consider the following *complete elliptic integral of the first kind*,

$$K(k) \equiv \int_0^{\pi/2} \frac{d\theta}{1 - k^2 \sin^2(\theta)},$$

where $k \in (-1, 1)$. Calculate $K(0.5)$ numerically using `numint_reg` and compare your result with that calculated using the R package `elliptic`.

Exercise 09

Calculate the local error for Simpson's 3/8 rule and verify that the result is $-(3/80)h^5 f^{(4)}(\xi)$.

Exercise 10

The global error when applying the trapezoid rule is

$$\frac{a - b}{12} f^{(2)}(\xi) h^2.$$

While it is not possible to know the value of $f^{(2)}(\xi)$, the quantity $(a - b)/12$ is constant and, even though $f^{(2)}(\xi)$ varies across the interval (x_1, x_{n+1}), it will be bounded by a finite number, in general comparable with the values that the function takes in the integration interval. Accordingly, the global error should show a square dependency on h.

Use the trapezoid rule to integrate $f(x) = x^2 - 1$ in the interval $[-1, 1]$, for many values of h when $[-1, 1]$ is divided into $n = 20, 21, \ldots, 39, 40$ equal intervals. Plot the global error versus the corresponding values of h. You should verify visually that the set of points in the plot follows a curved, rather than straight, pattern. Can we ascertain that the curve is a quadratic?

Exercise 11

Use two-point and three-point Gaussian quadrature to estimate numerically the integral

$$\int_{-2}^{3} x^4 \, dx.$$

What difference do you observe in going from the two-point to the three-point quadrature?

Exercise 12

Calculate numerically the integral

$$\frac{1}{\sqrt{2\pi}} \int_{-1}^{+1} \mathrm{e}^{-x^2/2}\,\mathrm{d}x,$$

using Gaussian quadrature, what order n is necessary for the result to be comparable with the one given by the function `pnorm`?

Exercise 13

Adapt the code of function `Gquad` to write an algorithm to calculate numerically 2D integrals over rectangular domains. Use the algorithm developed to calculate

$$\iint_R y\mathrm{e}^x\,\mathrm{d}x\mathrm{d}y, \qquad R = [0,\,2] \times [0,\,3].$$

References

[1] Golub G H and Welsch J H 1969 Calculation of gauss quadrature rules *Math. Comput.* **23** 221–30

[2] Davis P J and Rabinowitz P 2014 *Methods of Numerical Integration* 2nd edn (Academic)

[3] Engels H 1980 *Numerical Quadrature and Cubature* 1st edn (Academic)

[4] Krommer A R and Ueberhuber C W 1998 *Computational Integration* 1st edn (SIAM)

IOP Publishing

Computational Physics with R

James Foadi

Chapter 8

Ordinary differential equations

8.1 Introduction

Ordinary differential equations (ODEs) play a central role in modelling physical systems. Whether describing the motion of particles, the flow of heat, or the evolution of wave functions, ODEs capture the way in which physical quantities change over time and space. While some of these equations can be solved analytically, most real-world problems require numerical methods. Most first-year undergraduate textbooks include good introductions to the notation and theory of ODEs. Students reading the current chapter are assumed to be familiar with this theory.

There are three main types of problems involving ODEs:

- **Initial-value problems (IVPs)**. Where the solution is determined by the value of the unknown function and its derivatives at a starting point. These are typical in time-dependent problems, such as classical mechanics or radioactive decay.
- **Boundary-value problems (BVPs)**. Where the solution is subject to conditions specified at two or more points, often at the boundaries of a spatial domain. Examples include steady-state temperature profiles or deflection of beams.
- **Eigenvalue problems** Where the ODE and its boundary conditions admit non-trivial solutions only for specific values of a parameter (the eigenvalue). These appear frequently in quantum mechanics, vibration analysis, and wave propagation.

This chapter begins with initial-value problems and introduces several key numerical methods to solve them. We start with the Euler method, then move to improved versions such as the advanced Euler and the Runge–Kutta methods. As the chapter progresses, we will also explore techniques for solving boundary-value and eigenvalue problems. Along the way, particular attention will be paid to the stability, accuracy, and computational cost of the different approaches, with all methods illustrated through practical implementation in R.

doi:10.1088/978-0-7503-2632-2ch8

In terms of initial-value problems, the methods described below solve the first-order ODE:

$$\frac{dy}{dt} = f(t, y), \quad y(t_0) = y_0. \tag{8.1}$$

Higher-order ODEs are reconducible to the form (8.1) (see later), so this ODE will be of primary importance for the development of all methods described in this chapter.

8.2 Initial value problems (IVPs)

In many physical problems, one is interested in determining the future behaviour of a system given its current state. Mathematically, this leads to an initial value problem (IVP) for an ODE, typically of the form

$$\frac{dy}{dt} = f(t, y), \qquad y(t_0) = y_0,$$

where the function f defines the rate of change of y at time t, and y_0 is the known initial value at time t_0. The goal is to compute an approximation of the solution $y(t)$ over a time interval starting from t_0, using the information contained in the differential equation and the initial condition. The methods described in the following five sections are all designed to achieve this. These techniques are local in nature: they proceed step-by-step from the known value y_0, updating the solution incrementally using approximations to the derivative. Accuracy and stability are key considerations in the choice and implementation of these methods.

8.3 The Euler method

The Euler method is the simplest numerical technique for solving ODEs, and forms the foundation for more advanced methods. Although it is not very accurate and may suffer from instability, its conceptual clarity makes it an essential starting point for the study of ODEs algorithms. In a sense, it can be seen as the *mother* of all modern step-based integration methods.

To understand where the Euler method comes from, consider the first-order differential equation of the form (8.1). The goal is to approximate the solution $y(t)$ over an interval $[t_0, t_f]$, starting from the initial condition y_0. The basic idea is to discretise time using a uniform step size h, and to compute successive approximations $y_1, y_2, \ldots$ to the exact solution at points $t_1 = t_0 + h$, $t_2 = t_0 + 2h, \ldots$.

The theoretical basis of the method is provided by the Taylor expansion of $y(t)$ around t_i:

$$y(t_{i+1}) = y(t_i) + h\frac{dy}{dt}(t_i) + \frac{h^2}{2}\frac{d^2y}{dt^2}(\xi),$$

for some $\xi \in [t_i, t_{i+1}]$. Since the differential equation tells us that $dy(t_i)/dt = f(t_i, y_i)$, the right-hand side becomes entirely computable. By discarding the second-order term, we obtain the Euler approximation

$$y_{i+1} \approx y_i + hf(t_i, y_i). \tag{8.2}$$

This is the essence of the Euler method: the next value is predicted using the current value and the slope, which is known from the differential equation. In algorithmic terms, the Euler method is simple:

- Set y_0 to the initial condition.
- For each $i = 0, 1, 2, \ldots$, compute

$$y_{i+1} = y_i + hf(t_i, y_i). \tag{8.3}$$

The simplicity of Euler's method is reflected in its limited accuracy. It is only first-order accurate, meaning that the global error behaves like O(h). Its success depends on the step size h being reasonably small. If h is too large, not only does the accuracy deteriorate, but the method may also become unstable, especially in the case of *stiff equations*[1]. These issues will be discussed further, later in the chapter. Nevertheless, the Euler method is a crucial pedagogical tool. It introduces the key idea of numerical integration as a stepwise construction of approximate solutions, and it forms the basis for improved methods such as Heun's method and Runge–Kutta schemes.

Example 8.1 *Consider the differential equation:*

$$\frac{dy}{dt} + 2y = 6, \quad y(0) = 0, \quad t \in [0, 2].$$

This can be rewritten in standard form as:

$$\frac{dy}{dt} = f(t, y) = 6 - 2y.$$

The analytic solution of this IVP is

$$y(t) = 3(1 - e^{-2t}).$$

Knowing the exact solution is in general not an option as the numerical method exists because most ODEs are too difficult or impossible to solve analytically. But the example chosen is a simple one where the analytic solution is known, so that it can be used to assess the validity of the numerical method. In the current example, we apply Euler's method with initial condition $y_0 = 0$*, step size* $h = 0.4$*, and compute the approximate solution at*

[1] Informally, an IVP is *stiff* when it contains widely separated time scales, e.g. fast, strongly damped transients together with slow dynamics, so that explicit methods are forced to use step sizes dictated by stability rather than accuracy. Equivalently, a local linearisation has eigenvalues with large negative real parts, making stable explicit steps prohibitively small even though the true solution is smooth. Suitable remedies (often implicit schemes) are discussed later in the chapter.

$$t_1 = 0.4,\ t_2 = 0.8,\ t_3 = 1.2,\ t_4 = 1.6,\ \ldots$$

For example, y_1 *is calculated as follows:*

$$y_1 = y_0 + hf(t_0, y_0) = 0 + 0.4(6 - 2(0)) = 2.4$$

Starting from this value, we can then calculate y_2*:*

$$y_2 = y_1 + hf(t_1, y_1) = 2.4 + 0.4(6 - 2(2.4)) = 2.4 + 0.4(1.2) = 2.88,$$

and so on for all subsequent points. Each new value, y_{i+1}*, will be different from the correct value,* $y(t_{i+1})$*, of the solution because Euler's is a numerical method offering approximate solutions. The first few Euler values, compared with the correct ones, are listed in the following table.*

i	t_i	y_i	t_{i+1}	y_{i+1}	$y(t_{i+1})$
0	0.0	0.0000	0.4	2.4000	1.6520
1	0.4	2.4000	0.8	2.8800	2.3943
2	0.8	2.8800	1.2	2.9760	2.7278
3	1.2	2.9760	1.6	2.9952	2.8777

In general, the first few approximations are very different from their correct counterpart. But later on even the approximate solutions converge to the correct and limiting value, 3 *(see analytic solution). The correct solution,* $y(t)$*, and the approximated points for this example are visually compared in figure* 8.1.

8.4 Local and global error for ODEs

The approximate value of the solution found at $t = t_i$ was indicated as y_i in the previous section. This is in general different from the corresponding value of the correct solution, $y(t_i)$. The difference between the two is called *local error*, if we assume that up to step $i - 1$ the result is exact. That is, if $y_{i-1} = y(t_{i-1})$, the quantity

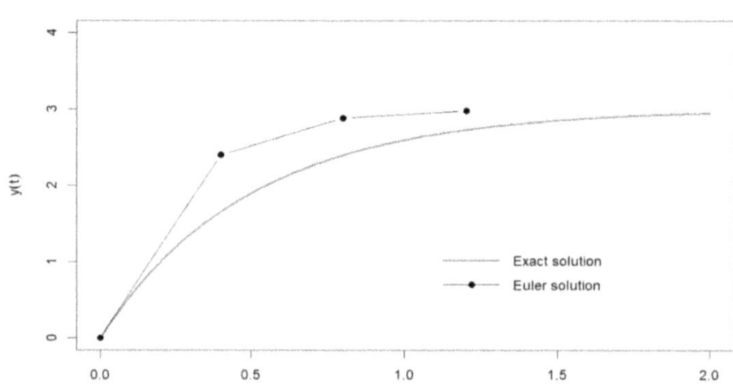

Figure 8.1. Correct solution of the ODE in example 8.1 and approximate solutions obtained with the Euler method, where the step is $h = 0.4$.

$$\epsilon \equiv y(t_i) - y_i \tag{8.4}$$

is defined as the *local error* at step i. All local errors then accumulate to produce a *global error*, E.

8.5 Local and global errors for the Euler method

Consider the Taylor expansion around t_{i-1}:

$$y(t_i) = y(t_{i-1}) + h\frac{\mathrm{d}y(t_{i-1})}{\mathrm{d}t} + \mathrm{O}(h^2).$$

In the Euler method, the first derivative $\mathrm{d}y(t_{i-1})/\mathrm{d}t$ is replaced by the gradient calculated at $t = t_{i-1}$ but this is exactly equal to $\mathrm{d}y(t_{i-1})/\mathrm{d}t$ because it is assumed that $y(t_{i-1}) = y_{i-1}$. Therefore the local error is:

$$\begin{aligned}
\epsilon &= y(t_i) - y_i \\
&= y(t_{i-1}) + h\mathrm{d}y(t_{i-1})/\mathrm{d}t + \mathrm{O}(h^2) - (y_{i-1} + hf(t_{i-1}, y_{i-1})) \\
&= y(t_{i-1}) + h\mathrm{d}y(t_{i-1})/\mathrm{d}t + \mathrm{O}(h^2) - y(t_{i-1}) - h\mathrm{d}y(t_{i-1})/\mathrm{d}t \\
&= \mathrm{O}(h^2).
\end{aligned}$$

This means that the local error is $\mathrm{O}(h^2)$.

After n steps, the Euler method will approximate $y(t_n)$ with y_n. As each step will contribute $\epsilon = \mathrm{O}(h^2)$ to the overall error, this will be given by the product

$$E = n\mathrm{O}(h^2).$$

But $n = (t_n - t_0)/h$. Therefore,

$$E = \frac{t_n - t_0}{h}\mathrm{O}(h^2) = \mathrm{O}(h).$$

To summarise, for the Euler method we have:

$$e = \mathrm{O}(h^2), \qquad E = \mathrm{O}(h). \tag{8.5}$$

When calculating the local or global error for a specific value of h, it is common to find that the result does not appear to match the expected $\mathrm{O}(h^2)$ or $\mathrm{O}(h)$ behaviour. This is not a contradiction. The notation $\mathrm{O}(h^p)$ describes how the error behaves *as* $h \to 0$, it does not mean that the error is equal to h^p, or even necessarily close to it, for any fixed value of h. In practice, the actual error behaves like Ch^p, where the constant C depends on the differential equation, the solution's derivatives, and the interval of integration. If C is large, or if h is not small enough, the observed error may seem inconsistent with the theory.

Example 8.2 *A concrete example illustrates this. Consider Euler's method applied to the equation*

$$\frac{\mathrm{d}y}{dt} = 6 - 2y, \quad y(0) = 0,$$

with step size $h = 0.4$ *(example 8.1). The exact solution at* $t_1 = 0.4$ *is* $y(t_1) = 3(1 - e^{-0.8}) \approx 1.6520$*, while Euler's method gives* $y_1 = 2.4$*. The error is therefore*

$$y(t_1) - y_1 \approx -0.748,$$

which may seem too large to be consistent with an $O(h^2)$ *estimate since* $h^2 = 0.16$*. However, the local error is more accurately given by*

$$\epsilon \approx \frac{h^2}{2} y''(\xi),$$

for some $\xi \in [0, 0.4]$*. Now, using the fact that*

$$\frac{dy}{dt} = 6 - 2y \quad \Rightarrow \quad \frac{d^2y}{dt^2} = -2\frac{dy}{dt} = -2(6 - 2y),$$

we find $y''(0) = -12$*, hence*

$$\epsilon \approx \frac{0.4^2}{2} \cdot (-12) = -0.96.$$

The observed error of -0.748 *lies well within this theoretical bound, confirming that the local error is indeed* $O(h^2)$*, though the constant is large and the step size is not particularly small. This reinforces the idea that to verify convergence rates, one must compare errors for multiple values of h, not just inspect their size for a single choice of step.*

8.6 The Heun method (improved Euler)

Euler's method offers a straightforward and intuitive way to approximate solutions of ODEs, but its accuracy and stability are limited. One natural way to improve on Euler's idea is to take into account not just the slope at the beginning of the interval, but also an estimate of the slope at the end. This leads to what is known as the *Heun method,* also referred to as the *improved Euler method.*

The basic idea is simple: instead of relying solely on the slope at t_i, we use Euler's method to make a preliminary prediction $\tilde{y}_{i+1}$, then evaluate the slope again at t_{i+1} using this predicted value. The final value y_{i+1} is then computed as the average of the two slopes. Methods working in two separate steps like this, where some kind of prediction is done in the first step and a correction to the prediction is done in the second step, are commonly called *predictor–corrector methods*. The Heun method is a predictor–corrector method, consisting of the two following steps:

- **Predictor step**. Essentially the application of Euler's method:

$$\tilde{y}_{i+1} = y_i + hf(t_i, y_i)$$

- **Corrector step** This step 'amends' the previous calculation by averaging the two slopes:

$$y_{i+1} = y_i + \frac{h}{2}(f(t_i, y_i) + f(t_{i+1}, \tilde{y}_{i+1}))$$

The correction step leads to significantly better accuracy, as we will soon see. Heun's method preserves the step-by-step nature of Euler's scheme but improves its reliability by accounting for the change in slope over each interval. It is also easy to implement in code, and serves as a conceptual bridge between Euler's method and more advanced multi-slope techniques like the Runge–Kutta methods, described later.

Example 8.3 *Let us consider the same ODE of example* 8.1 *and the same step size,* $h = 0.4$, *and let us calculate the first few steps of the approximate solution, using the Heun method. Starting with* $i = 1$, *the predictor step yields*

$$\tilde{y}_1 = y_0 + hf(t_0, y_0) = 0 + 0.4(6 - 2(0)) = 2.4.$$

For the corrector step, the gradient at $t = t_1$ *and* $y = \tilde{y}_1$ *is calculated, and this is averaged with the gradient of the predictor step*:

$$y_1 = y_0 + \frac{h}{2}(f(t_0, y_0) + f(t_1, \tilde{y}_1)) = 0 + 0.2((6 - 2(0)) + (6 - 2(2.4))) = 1.44.$$

The next predictor value, at $t = t_2$, *is calculated similarly, this time remembering that* $y_1 = 1.44$ *and not* 2.4:

$$\tilde{y}_2 = y_1 + hf(t_1, y_1) = 1.44 + 0.4(6 - 2(1.44)) = 2.688.$$

And the corresponding corrector value, y_2, is:

$$y_2 = y_1 + \frac{h}{2}(f(t_1, y_1) + f(t_2, \tilde{y}_2)) = 1.44 + 0.2((6 - 2(1.44)) + (6 - 2(2.688))) = 2.1888.$$

Proceeding in a similar manner, we obtain the values included in the following table:

i	t_i	y_i	t_{i+1}	$\tilde{y}_{i+1}$	y_{i+1}	$y(t_{i+1})$
0	0.0	0.0000	0.4	2.4000	1.4400	1.6520
1	0.4	1.4400	0.8	2.6880	2.1888	2.3943
2	0.8	2.1888	1.2	2.8378	2.5782	2.7278
3	1.2	2.5782	1.6	2.9156	2.7801	2.8777

It is immediately apparent that the approximated values are closer to the true value, if compared with the values calculated with the Euler method. A visual comparison is also depicted at figure 8.2.

8.7 Local and global errors for the Heun method

The level of difficulty involved in calculating local errors varies from method to method. For some algorithms, the derivation is straightforward; for others, it requires more algebraic work and care with Taylor expansions. In the case of the

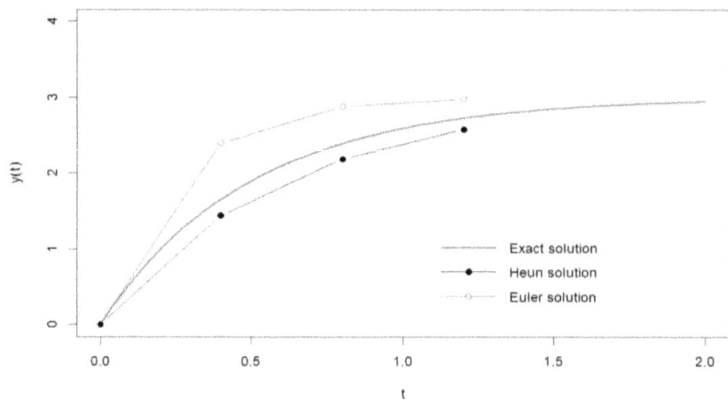

Figure 8.2. Correct solution of the ODE in example 8.3 and approximate solutions obtained with the Heun method, where the step is $h = 0.4$. The open circles correspond to the numerical solution calculated using the simple Euler method.

Heun method, the calculation of the local error is rather lengthy and has been moved to appendix A, section A.12. In the remainder of this chapter, we will not derive the local and global errors for every method. The examples given here for Euler's method and, in the appendix, for Heun's method, should be sufficient to provide a sense of how these errors are estimated and what order of accuracy one can typically expect.

As seen in the appendix, the local error for Heun is $O(h^3)$. This means that the global error is $O(h^2)$. The calculation for the global error starting from the local error is similar to that performed for the Euler method. To summarise, for the Heun method:

$$e = O(h^3), \qquad E = O(h^2). \tag{8.6}$$

This means that both in intermediate steps and for the full calculation, the Heun method is more accurate than the Euler method. As the global error is $O(h^2)$, the Heun method is said to be a *second order* method, while Euler's is a *first order method*, being its global error equal to $O(h)$.

8.8 The Runge–Kutta methods

While the Heun's method was introduced earlier as a predictor–corrector method, using an Euler step to predict the solution and then refining it using a corrected slope, it is also a member of a broader family known as *Runge–Kutta methods*. These methods aim to improve the accuracy of the basic Euler step by sampling the gradient $f(t, y)$ at multiple points within the interval $[t_i, t_{i+1}]$, and combining these samples into a weighted average. The key feature is that they achieve higher-order accuracy using only information from within the current step, making these methods explicit.

Formally, a general *explicit Runge–Kutta method* with s stages has the following structure:

$$\begin{aligned}
k_1 &= f(t_i, y_i) \\
k_2 &= f(t_i + c_2 h, y_i + h a_{21} k_1) \\
k_3 &= f(t_i + c_3 h, y_i + h(a_{31} k_1 + a_{32} k_2)) \\
&\vdots \\
k_s &= f\left(t_i + c_s h, y_i + h \sum_{j=1}^{s-1} a_{sj} k_j\right) \\
y_{i+1} &= y_i + h \sum_{j=1}^{s} b_j k_j
\end{aligned} \tag{8.7}$$

Here, the constants a_{ij}, b_j, and c_j define the specific Runge–Kutta method.

We have mentioned at the beginning that Heun's method is in fact a Runge–Kutta method. More specifically, it is a *second-order* Runge–Kutta method (RK2); let us see why. Recall that Heun's performs:

1. A predictor: $k_1 = f(t_i, y_i)$
2. A corrector: $k_2 = f(t_i + h, y_i + hk_1)$
3. Then: $y_{i+1} = y_i + \frac{h}{2}(k_1 + k_2)$.

This matches the Runge–Kutta structure with:

$$a_{21} = 1, \quad c_2 = 1, \quad b_1 = b_2 = \frac{1}{2},$$

and all other $a_{ij} = 0$. The fact that Heun is a particular case of RK2 method (sometimes referred to as the *explicit trapezoidal method*) demonstrates how the Runge–Kutta framework generalises simple methods by designing combinations of slope evaluations that increase accuracy without requiring derivative expressions beyond $f(t, y)$.

8.8.1 The fourth-order Runge–Kutta method (RK4)

Among all Runge–Kutta methods, the most widely used in practice is the classical fourth-order method, commonly referred to as *RK4*. It strikes an excellent balance between accuracy and computational efficiency, and is often the default method in many scientific computing environments. The RK4 method evaluates the gradient four times per step, using different points within the interval. These evaluations are then combined into a weighted average that approximates the solution at the next time point with fourth-order accuracy. The method is given by:

$$
\begin{aligned}
k_1 &= f(t_i, y_i),\\
k_2 &= f\left(t_i + \frac{h}{2}, y_i + \frac{h}{2}k_1\right),\\
k_3 &= f\left(t_i + \frac{h}{2}, y_i + \frac{h}{2}k_2\right),\\
k_4 &= f(t_i + h, y_i + hk_3),
\end{aligned}
\tag{8.8}
$$

and the update formula is:

$$
y_{i+1} = y_i + \frac{h}{6}(k_1 + 2k_2 + 2k_3 + k_4). \tag{8.9}
$$

This formula may appear more complex than previous methods, but it is entirely explicit, that is, each stage depends only on previously computed values, and it avoids the need to compute higher derivatives directly.

The RK4 method is accurate to order four: the local truncation error is $O(h^5)$, and the global error is $O(h^4)$. This means that reducing the step size by half typically reduces the global error by a factor of sixteen, making RK4 significantly more accurate than Euler or Heun for a given step size.

Example 8.4 *Using the same ODE and step size of example* 8.1, *Let us calculate the first step of RK4 to produce* y_1. *We need to calculate first the four coefficients* k_1, k_2, k_3, k_4. *The calculation needs to be carried out in order as* k_2 *depends on* k_1, k_3 *on* $k2$, *and* k_4 *on* k_3. *We have*:

$$
k_1 = f(t_0, y_0) = 6 - 2(0) = 6,
$$

$$
k_2 = f\left(t_0 + \frac{h}{2}, y_0 + \frac{h}{2}k_1\right) = 6 - 2\left(0 + \frac{0.4}{2}6\right) = 3.6,
$$

$$
k_3 = f\left(t_0 + \frac{h}{2}, y_0 + \frac{h}{2}k_2\right) = 6 - 2\left(0 + \frac{0.4}{2}3.6\right) = 4.56,
$$

$$
k_4 = f(t_0 + h, y_0 + hk_3) = 6 - 2(0 + 0.4(4.56)) = 2.352.
$$

With the four parameters just obtained, the value y_1 *can then be calculated using formula (??)*:

$$
y_1 = y_0 + \frac{h}{6}(k_1 + 2k_2 + 2k_3 + k_4) = 0 + \frac{0.4}{6}(6 + 2(3.6) + 2(4.56) + 2.352) = 1.6448.
$$

The correct value of the solution, $y(t_1)$, *is* 1.6520. *This clearly show how accurate RK4 is.*

8.8.2 A rationale for Runge–Kutta methods

Runge–Kutta methods will not be justified in detail in this book, but a rationale to illustrate their validity goes as follows. The key idea is to ensure that the numerical method approximates the Taylor expansion of the exact solution as closely as possible. Consider the Taylor expansion of the exact solution around t_i:

$$y(t_i + h) = y(t_i) + hy'(t_i) + \frac{h^2}{2}y''(t_i) + \frac{h^3}{6}y'''(t_i) + \cdots$$

Since $y'(t) = f(t, y(t))$, and successive derivatives $y''(t)$, $y'''(t)$, ... can be expressed in terms of f and its partial derivatives, one can in principle match the series term by term.

A Runge–Kutta method with s stages defines intermediate slopes:

$$k_1 = f(t_i, y_i), \quad k_2 = f(t_i + c_2h, y_i + ha_{21}k_1), \quad \ldots, \quad k_s = f\left(t_i + c_sh, y_i + h\sum_{j=1}^{s-1} a_{sj}k_j\right),$$

and then updates the solution using:

$$y_{i+1} = y_i + h\sum_{j=1}^{s} b_jk_j.$$

The constants a_{ij}, b_j, c_j are chosen so that this formula matches the Taylor expansion of $y(t_i + h)$ up to a certain order of h. This process leads to a set of algebraic equations known as the *order conditions*, which guarantee that the method achieves the desired accuracy. For example, in the classical fourth-order method (RK4), the coefficients are chosen so that the approximation agrees with the Taylor expansion up to terms of order h^4. This results in a local error of $O(h^5)$, and therefore a global error of $O(h^4)$. This approach provides a systematic way to construct numerical methods with higher-order accuracy, while relying only on evaluations of the gradient and not on its derivatives with respect to t or y.

8.9 Stability of IVPs

When studying numerical algorithms for the solution of initial value problems, it is not enough to consider only accuracy. A method can be locally and globally accurate, but if it is applied to the wrong type of problem, or with a step size that is too large, it may produce solutions that diverge or oscillate in an unphysical way. This behaviour is what we mean by *instability*.

To introduce the concept of stability, it is customary to test algorithms on a very simple differential equation whose solution is well known: a linear decay or growth equation. This 'problem function' is chosen because, in a neighbourhood, many differential equations behave like an exponential. By applying a numerical method to this test problem, one can study whether the method reproduces the correct long-term behaviour. In practice, this leads to the notion of a *region of stability* for a given method: the set of values of the step size (combined with the characteristics of the problem) for which the algorithm behaves in a stable way.

Different methods have different regions of stability, and this has direct consequences on how useful they are for stiff or rapidly decaying problems. In the following subsections we will study stability for the Euler method, Heun method, and the classical fourth-order Runge–Kutta method.

8.9.1 Stability for the Euler method

We begin with the simplest algorithm, Euler. To test its stability we apply it to the problem function, that is the differential equation

$$y' = \lambda y, \quad y(0) = 1,$$

whose exact solution is an exponential. Applying the Euler method gives the recursive formula

$$y_{n+1} = y_n + h\lambda y_n = (1 + h\lambda)\, y_n.$$

After n steps the numerical solution is therefore

$$y_n = (1 + h\lambda)^n.$$

The behaviour of the approximation is controlled by the factor $1 + h\lambda$. For the method to be stable, the size of this factor must be less than one in absolute value, otherwise the numerical solution will grow without bound. The stability condition is therefore

$$|1 + h\lambda| < 1.$$

This inequality describes a disk in the complex plane, centred at -1 and with radius 1. This disk is called the *region of absolute stability* of the Euler method. If the step size h and the parameter λ are such that $h\lambda$ lies inside this region, the numerical solution decays correctly. If they are outside, the approximation either oscillates or diverges, even when the true solution tends to zero.

For real negative values of λ and for real h, the stability condition simplifies to

$$h < -\frac{2}{\lambda}. \tag{8.10}$$

This shows that the step size must be chosen carefully: too large a step will produce an unstable approximation, while a sufficiently small step reproduces the expected exponential decay.

Example 8.5 *Consider the initial value problem*

$$y' = -5y, \qquad y(0) = 1.$$

The exact solution is $y(t) = \mathrm{e}^{-5t}$, *which decays smoothly to zero as* t *increases. Applying Euler with step size* $h > 0$ *gives the recurrence*

$$y_{n+1} = y_n + h(-5)y_n = (1 - 5h)\, y_n,$$

so after n steps,

$$y_n = (1 - 5h)^n \, y_0.$$

Euler's region of absolute stability for the test equation $y' = \lambda y$ *requires*

$$|1 + h\lambda| < 1.$$

Here $\lambda = -5$*, hence*

$$|1 - 5h| < 1 \quad \Leftrightarrow -1 < 1 - 5h < 1 \quad \Leftrightarrow \quad 0 < h < \frac{2}{5} = 0.4.$$

For example, with $h = 0.1$, $|1 - 5h| = |1 - 0.5| = 0.5 < 1$, *Euler is stable and the numerical solution decays correctly. But with* $h = 0.6$, $|1 - 5h| = |1 - 3| = 2 > 1$, *Euler is unstable; the numerical solution diverges even though the true solution decays.*

In the previous example, the parameter λ could be read directly from the differential equation, which was already in the form $y' = \lambda y$. In more general situations, however, it is not always immediate to determine the value of λ that controls stability. Let us briefly describe two important cases.

- **Linear systems.** We will illustrate the IVP methods for linear systems later on in this chapter. Here we want just to consider the stability issue. For a system of the form

$$\mathbf{y}' = A\mathbf{y},$$

  the relevant values of λ are the eigenvalues of the matrix A. The stability condition for the Euler method then requires that every eigenvalue λ_i satisfies

$$|1 + h\lambda_i| < 1.$$

- **Nonlinear equations.** For a nonlinear problem

$$y' = f(t, y),$$

  there is no single λ. One usually examines the Jacobian $\partial f / \partial y$ near the solution. This local slope provides an effective λ that can be used to test stability. Large negative slopes correspond to rapidly decaying behaviour, which can make the Euler method unstable unless h is very small.

Example 8.6 *Consider the nonlinear problem*

$$y' = -20\,(y - \sin t).$$

Here the exact solution combines exponential decay with oscillations. For the purpose of stability, we notice that the coefficient -20 *multiplies* y*, so the system behaves locally like* $y' = -20y$*. Thus the relevant effective parameter is* $\lambda = -20$*. Here the condition*

$$|1 + h\lambda| < 1$$

becomes

$$|1 - 20h| < 1 \quad \Leftrightarrow \quad 0 < h < 0.1.$$

Therefore, for example, a step size $h = 0.05$ *ensures stability, while* $h = 0.2$ *produces instability. This example shows how the concept of stability extends to cases where* λ *is not immediately evident in the formulation of the IVP.*

8.9.2 Stability for the Heun method

We now turn to Heun. Applying this method to the test equation

$$y' = \lambda y, \qquad y(0) = 1,$$

with step size h gives

$$y_{n+1} = y_n + \frac{h}{2}(\lambda y_n + \lambda(y_n + h\lambda y_n)).$$

After simplification this becomes

$$y_{n+1} = \left(1 + h\lambda + \frac{1}{2}(h\lambda)^2\right) y_n.$$

The factor multiplying y_n is called the *amplification factor*, here equal to

$$R(z) = 1 + z + \frac{1}{2}z^2, \qquad z = h\lambda.$$

Stability requires that

$$|R(z)| < 1.$$

Thus the region of absolute stability of the Heun method is the set of complex values $z = h\lambda$ for which

$$\left|1 + z + \frac{1}{2}z^2\right| < 1.$$

Compared with the Euler's circular stability region, Heun's method has a larger region extending further into the left half-plane, which means that it can handle larger step sizes before instability sets in. But for real, negative eigenvalues and real h, it is not difficult to verify that the stability region is

$$h < \frac{2}{\lambda}. \tag{8.11}$$

This is the same stability region as the one for Euler, equation (8.10), so no advantage in terms of stability is gained by using Heun instead of Euler for real-valued IVPs.

8.9.3 Stability for the fourth-order Runge–Kutta method

For the classical fourth-order Runge–Kutta method (RK4) we analyse stability exactly as for Euler and Heun: apply the method to the test equation

$$y' = \lambda y, \qquad z \equiv h\lambda,$$

which yields a stability function (amplification factor)

$$R(z) = 1 + z + \frac{1}{2}z^2 + \frac{1}{6}z^3 + \frac{1}{24}z^4.$$

The region of stability is $\{\ z \in \mathbb{C}: \quad |R(z)| < 1\ \}$. Considering real negative eigenvalues (so that odd powers of λh would gain a negative sign, leaving $\lambda > 0$) and real h yields the following double inequality describing the stability region:

$$-1 < 1 - \lambda h + \frac{1}{2}\lambda^2 h^2 - \frac{1}{6}\lambda^3 h^3 + \frac{1}{24}\lambda^4 h^4 < 1. \tag{8.12}$$

The solution to this double inequality is found in section A.14 of appendix A and it is reported here:

$$h < \frac{\gamma}{\lambda} \approx \frac{2.785\,293\,563}{\lambda}, \tag{8.13}$$

where,

$$\gamma = \frac{4}{3} + \sqrt[3]{\frac{172 + 36\sqrt{29}}{27}} + \sqrt[3]{\frac{172 - 36\sqrt{29}}{27}}.$$

For comparison, Euler and Heun yield $0 < h < 2/\lambda$ on the real axis; RK4's interval is wider but still bounded as h must be smaller than a given quantity.

8.9.4 Concluding remarks on stability

The three examples above (Euler, Heun, RK4) illustrate a general recipe that applies to any newly proposed one-step method. In principle, the stability region can always be determined by applying the method to the linear test problem

$$y' = \lambda y, \qquad z = h\lambda,$$

deriving the corresponding amplification factor $R(z)$, and defining the region of stability as $\{z \in \mathbb{C}: \quad |R(z)| < 1\}$. In practice, a computational physicist seldom needs to derive stability regions: for all major time-stepping schemes they are well known and tabulated. Day-to-day work typically proceeds with a sensible (often small) step size and simple checks. These typically are: (i) monitor whether the numerical solution behaves qualitatively as expected (e.g. decays when the physics dictates decay, preserves smoothness when appropriate); (ii) if unusual oscillations or spurious growth appear, first reduce the step size; (iii) confirm by a quick accuracy check (e.g. halve h once and compare); (iv) if instability or impractically small steps persist, the problem is likely stiff and a different scheme may be more appropriate.

Thus, while the stability function viewpoint offers a systematic way to analyse any method, routine computation relies on known regions, conservative step sizes, and simple empirical diagnostics to detect and address instability.

8.10 Implicit methods

So far we have only considered *explicit* methods, in which the new value y_{n+1} is given directly in terms of known information at previous time steps. A common drawback of such methods is their limited region of stability: when the step size h is too large, or when the underlying differential equation is stiff, explicit schemes may become unstable regardless of their formal accuracy. This motivates the study of *implicit* methods.

The simplest implicit scheme is the *backward Euler method*[2], given by

$$y_{n+1} = y_n + hf(t_{n+1}, y_{n+1}). \tag{8.14}$$

Unlike the forward Euler method, here y_{n+1} appears on both sides of the equation. To compute the new value one must therefore solve an algebraic equation at each step. If f is nonlinear in y, this requires the use of a root-finding procedure, typically Newton's method. This makes implicit methods more costly per step, but they provide significant advantages in terms of stability.

To illustrate the point, consider once again the linear test equation

$$y'(t) = \lambda y(t), \qquad \lambda < 0,$$

where $f(t, y) = \lambda y$. Backward Euler yields the recurrence

$$y_{n+1} = y_n + h\lambda y_{n=1} \quad \Rightarrow (1 - h\lambda)y_{n+1} = y_n,$$

leading to the update step:

$$y_{n+1} = \frac{1}{1 - h\lambda} y_n.$$

The associated stability function is therefore

$$R(z) = \frac{1}{1 - z}, \qquad z = h\lambda.$$

It is straightforward to check that $|R(z)| < 1$ for all real negative z. In fact, $|R(z)| < 1$ for all complex z with $\Re(z) < 0$. This means that backward Euler is stable for any step size h when applied to decaying exponentials, in sharp contrast with forward Euler which requires $h < -2/\lambda$. The backward Euler method is only the simplest example of this class. Implicit generalisations exist for higher order, such as implicit Runge–Kutta methods and backward differentiation formulas, which are widely used in scientific computing. In this book we will not implement

[2] Backward Euler is obtained in the same way as the forward Euler method, except that the slope is taken at the end of the step rather than at the beginning. More generally, both can be seen as special cases of the *θ-method*:

$$y_{n+1} = y_n + h[(1 - \theta)f(t_n, y_n) + \theta f(t_{n+1}, y_{n+1})],$$

with $\theta = 0$ giving forward Euler and $\theta = 1$ giving backward Euler.

implicit methods, but it is important for the student to be aware of their role: they complement explicit schemes by providing robustness in precisely those situations where explicit methods fail.

Example 8.7 *Consider the nonlinear equation*

$$y'(t) = ry(t)(1 - y(t)), \qquad r > 0.$$

Applying (8.14) *gives*

$$y_{n+1} = y_n + h\ r\ y_{n+1}(1 - y_{n+1}).$$

Rearranging terms we obtain a quadratic equation for y_{n+1}:

$$hr\ y_{n+1}^2 - (1 + hr)\ y_{n+1} + y_n = 0.$$

This can be solved explicitly:

$$y_{n+1} = \frac{(1 + hr)\ \pm\ \sqrt{(1 + hr)^2 - 4hry_n}}{2hr}.$$

Among the two possible roots, only the one that remains between 0 *and* 1 *(the natural range of the logistic solution) is acceptable. This example shows how implicit methods typically lead to an equation in the unknown* y_{n+1}, *which may be nonlinear but can sometimes be solved in closed form. In more complicated cases, iterative solvers such as Newton's method must be employed.*

8.11 Stiff ODEs

We have assumed that the step size h can be chosen on the basis of accuracy. However, in some problems the permitted step size is dictated instead by stability. These problems are called stiff An initial value problem is said to be *stiff* if it contains widely separated time scales: the solution exhibits both fast, strongly damped transients and slow, smooth dynamics. Explicit methods (Euler, Heun, RK4, etc) then require prohibitively small step sizes to remain stable, even though the true solution varies slowly on the time interval of interest.

Example 8.8 *Consider the IVP*

$$y'(t) = -1000(y(t) - \cos(t)) - \sin(t), \qquad y(0) = 1.$$

The exact solution is

$$y(t) = \cos(t),$$

which is perfectly smooth. Yet the term $-1000(y - \cos(t))$ *introduces a very fast transient. If we apply RK4 with a moderate step size such as* $h = 0.1$, *the numerical solution quickly diverges instead of settling to* $\cos(t)$. *Only when h is reduced to below* 10^{-3} *does RK4 behave correctly, an unreasonably small step size given that the solution itself varies on the scale of* 1.

An R demonstration follows where the solver RK4ODE *is used with decreasing step sizes. Only when* $h = 0.001$ *is the correct solution found.*

```
# Define the gradient
f <- function(t,y) {
ff <- -1000*(y-cos(t))-sin(t)

return(ff)}

# Solution interval
t0 <- 0
tf <- 10

# Initial conditions
y0 <- 1

# Stepsize 1: too coarse
h <- 0.1

# Solution with RK4
ltmp <- RK4ODE(f,t0,tf,y0,h)

# Plot
plot(ltmp$t,ltmp$y,type="b",pch=16,cex=0.5,
xlab=expression(t),ylab=expression(y(t)))
```

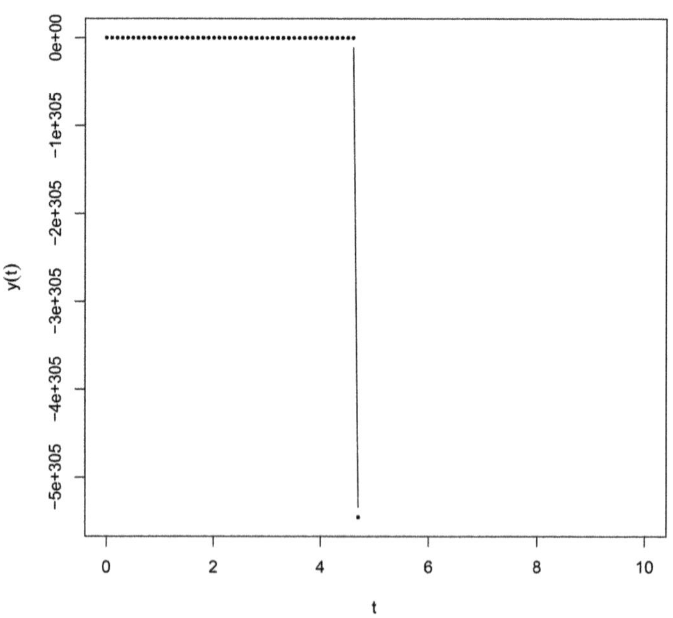

```
# Stepsize 2: still coarse
h <- 0.01

# Solution with RK4
ltmp <- RK4ODE(f,t0,tf,y0,h)

# Plot
plot(ltmp$t,ltmp$y,type="b",pch=16,cex=0.5,
xlab=expression(t),ylab=expression(y(t)))
```

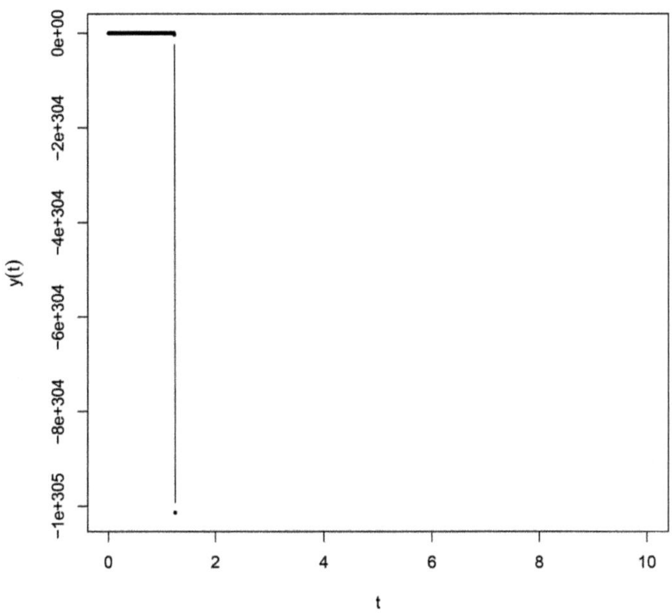


```
# Stepsize 3: just about right
h <- 0.001

# Solution with RK4
ltmp <- RK4ODE(f,t0,tf,y0,h)

# Plot
plot(ltmp$t,ltmp$y,type="b",pch=16,cex=0.5,
xlab=expression(t),ylab=expression(y(t)))
```

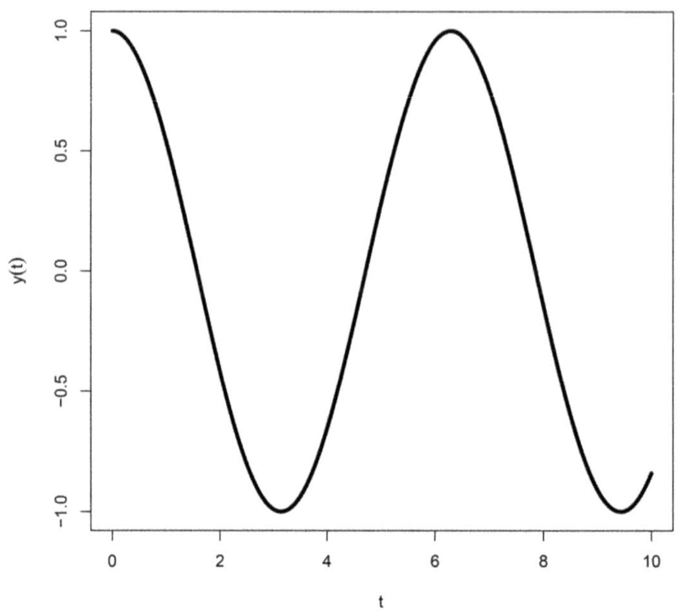

Example 8.9 *The* nonlinear Van der Pol oscillator

$$y'' - \mu(1 - y^2)y' + y = 0, \qquad y(0) = 2, \, y'(0) = 0,$$

is not stiff for small μ, *but becomes stiff when* $\mu \gg 1$. *For* $\mu = 100$, *explicit schemes require extremely small h to avoid instability.*

An R demonstration with the solver `RK4ODE` *follows here too. Only when* $h = 0.001$ *is the correct solution found, but a better solution is found wit h even smaller. The drawback here is that the algorithm takes longer to complete the task, given the high number of grid points required. For reference, times of execution have been measured with the base R function* `system.time`.

```
# Define the gradient
f <- function(t,y) {
dy1 <- y[2]
dy2 <- 100*(1-y[1]^2)*y[2]-y[1]

return(c(dy1,dy2))
}

# Solution interval
t0 <- 0
tf <- 100

# Initial conditions
```

```
y0 <- c(2,0)

# Stepsize 1: too coarse
h <- 0.1

# Solution with RK4 (with timings)
t_solve <- system.time ({
ltmp <- RK4ODE(f,t0,tf,y0,h)
})
cat("Timings with h=0.1\n")

## Timings with h=0.1

print(t_solve)

##    user  system elapsed
##    0.02    0.00    0.01

# Plot
plot(ltmp$t,ltmp$y[,1],type="b",pch=16,cex=0.3,
xlab=expression(t),ylab=expression(y(t)))
```

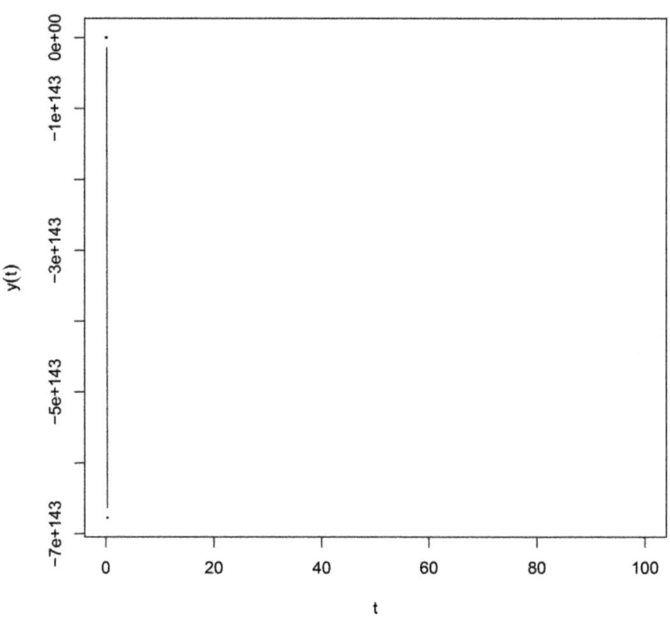

```
# Stepsize 2: still coarse
h <- 0.01

# Solution with RK4 (with timings)
t_solve <- system.time ({
ltmp <- RK4ODE(f,t0,tf,y0,h)
})
cat("Timings with h=0.01\n")
```

```
## Timings with h=0.01
```

```
print(t_solve)
```

```
##    user  system elapsed
##    0.08    0.00    0.08
```

```
# Plot
plot(ltmp$t,ltmp$y[,1],type="b",pch=16,cex=0.3,
xlab=expression(t),ylab=expression(y(t)))
```

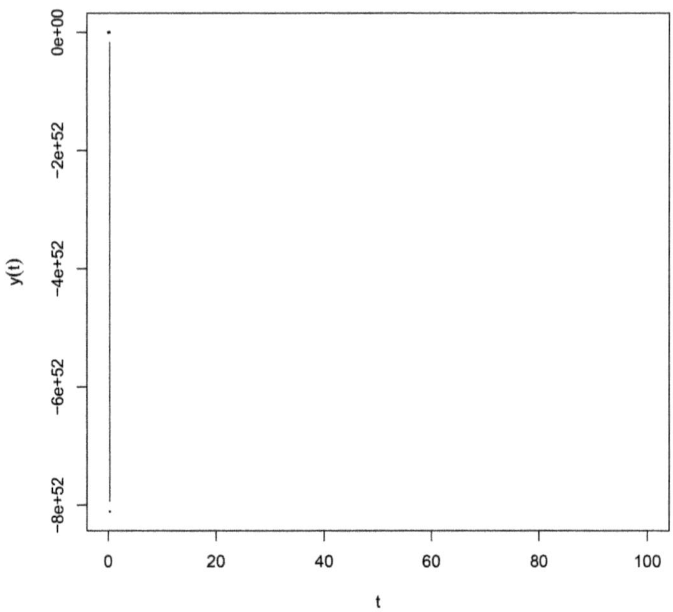

```
# Stepsize 3: OK
h <- 0.001

# Solution with RK4 (with timings)
t_solve <- system.time ({
ltmp <- RK4ODE(f,t0,tf,y0,h)
})
cat("Timings with h=0.001\n")
```

```
## Timings with h=0.001
```

```
print(t_solve)
```

```
##    user  system elapsed
##    0.72    0.01    0.74
```

```
# Plot
plot(ltmp$t,ltmp$y[,1],type="b",pch=16,cex=0.3,
xlab=expression(t),ylab=expression(y(t)))
```

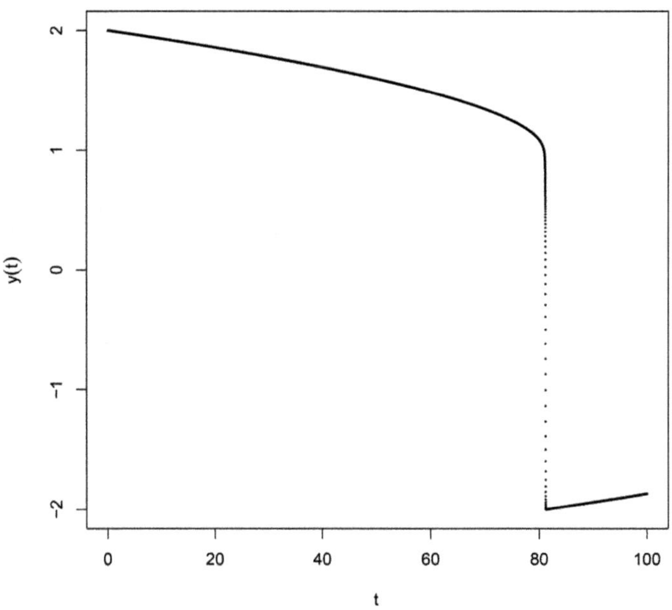

```
# Stepsize 4: Even better
h <- 0.0001

# Solution with RK4 (with timings)
t_solve <- system.time ({
ltmp <- RK4ODE(f,t0,tf,y0,h)
})
cat("Timings with h=0.0001\n")
```

```
## Timings with h=0.0001
```

```
print(t_solve)
```

```
##    user  system elapsed
##    9.72    0.03    9.83
```

```
# Plot
plot(ltmp$t,ltmp$y[,1],type="b",pch=16,cex=0.3,
xlab=expression(t),ylab=expression(y(t)))
```

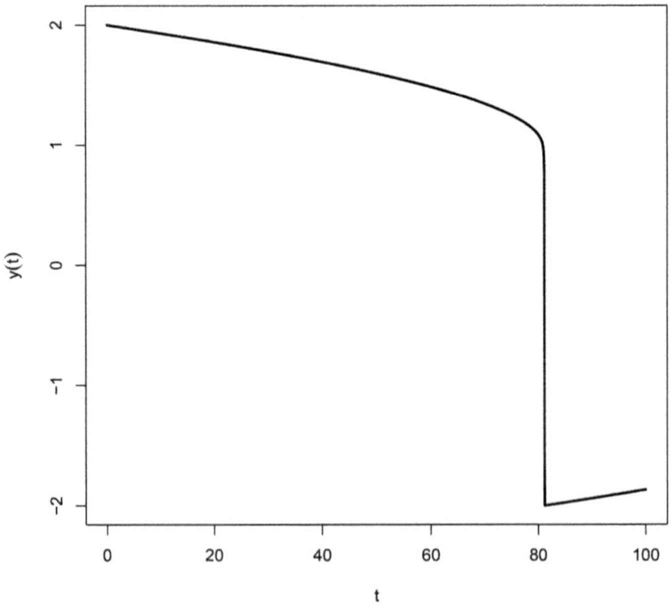

These examples highlight the typical signature of stiffness: explicit methods (even high-order RK4) may completely fail unless the step is reduced to impractical levels. To deal with such problems, one must use specialised solvers designed for stiff equations. Appropriate solvers will be introduced when treating the R `deSolve` package in chapter 10.

8.12 Relevant R code. Functions `EulerODE`, `HeunODE`, `RK4ODE`

This family of functions present all the same coding structure to implement three different algorithms, the Euler, Heun and fourth-order Runge–Kutta methods. The gradient

$$\frac{d\mathbf{y}}{dt} = f(t, \mathbf{y})$$

must be defined in advance. This is not surprising as every ODE is characterised by a specific gradient and initial conditions. The gradient is a function of the independent variable, t, and of as many dynamic variables,

$$\mathbf{y}(t) \equiv (y_1(t), y_2(t), \ldots, y_m(t)),$$

as implied by the system of ODEs. We have not yet treated systems of ODEs in this chapter, but the code was written to handle both single ODEs or a system of ODEs. We will see how the code solves these systems later. For the purpose of this section, you can think of `y` as containing just one dynamic function.

The algorithms/functions used to solve ODEs like `EulerODE` numerically, are called in jargon *solvers*. The three solvers presented here all accept as input:

- A function, `f`, describing the gradient of the ODE. `f` is in general a function of `t` and `y`.
- The starting, `t0`, and final, `tf`, value of the independent variable. `t0` coincides with the value of the independent variable for the initial conditions,

$$\mathbf{y}(t_0) = \mathbf{y}_0.$$

- The value/s of the initial condition/s, `y0`.
- The step size, `h`.

The variables `y`, in `f`, and `y0`, will be single numbers or vectors, depending on whether a single ODE or a system of ODEs are going to be solved. Further parameters, different from `t` and `y`, can be passed to the gradient, as this is taken care of in the `comphy` code through ellipses (`...`). This is a very useful mechanism in R as it avoids the unnecessary complication of setting up hard-coded gradients, when parameters are needed.

The output is a list with two elements called `t` and consisting of all grid points spaced by h (t_0, $t_1 = t_0 + h$, $t_2 = t_0 + 2h$, ...), and `y`, a matrix whose column j consists of the numerical solution for the dynamic variable $y_j(t)$.

The m initial conditions ($m = 1$ for just one ODE) are allocated, in the code, to the first column of an $m \times n$ matrix. n is the number of steps carried out by the algorithm. It is calculated as `(tf - t0)/h`. The remaining $n - 1$ columns of the same matrix will contain the numeric solutions at steps 1, 2, 3, etc. The heart of the algorithm consists of a variable number of lines of code, depending on the accuracy and, accordingly, complexity of the solution. These lines have been extracted here:

```
# EulerODE
for (i in 1:nsteps) {
t[i+1] <- t[i]+h
y[i+1,] <- y[i,] + h*f(t[i],y[i,],...)
}

# HeunODE
for (i in 1:nsteps) {
t[i+1] <- t[i]+h
yp <- y[i,]+h*f(t[i],y[i,],...)                    # Predictor
y[i+1,] <- y[i,]+h/2*(f(t[i],y[i,],...)+
f(t[i+1],yp,...))  # Corrector
}

# RK4ODE
for (i in 1:nsteps) {
t[i+1] <- t[i]+h

k1 <- f(t[i], y[i,],...)
k2 <- f(t[i] + h/2, y[i,] + h/2 * k1,...)
k3 <- f(t[i] + h/2, y[i,] + h/2 * k2,...)
k4 <- f(t[i] + h,   y[i,] + h   * k3,...)

y[i+1,] <- y[i,] + (h/6)*(k1 + 2*k2 + 2*k3 + k4)
}
```

All code can be understood once it is considered that `t[1]` contains `t0` and `y[1]` contains `y0`. Each step will thus depend on the knowledge from the previous step, as it is typical of explicit methods.

Let us demonstrate these solvers with the ODE

$$\frac{\mathrm{d}y}{\mathrm{d}t} = (1 - 2t)y, \quad y(0) = 1, \, t \in [0, 3].$$

The analytic solution is $e^{t(1-t)}$ and it will be useful to gauge the numerical result. The code for solving this ODE is displayed here. The second comparison plot is simply a zoomed-in version of the first one.

```
# Define the gradient
f <- function(t,y) {ff <- (1-2*t)*y; return(ff)}

# Solution interval
t0 <- 0
tf <- 3

# Initial conditions
y0 <- 1

# Stepsize
h <- 0.2

# Solution with Euler
ltmpE <- EulerODE(f,t0,tf,y0,h)

# Solution with Heun
ltmpH <- HeunODE(f,t0,tf,y0,h)

# Solution with RK4
ltmpR <- RK4ODE(f,t0,tf,y0,h)

# Exact solution
tt <- seq(0,3,length.out=100)
yy <- exp(tt*(1-tt))

# Visual comparison of methods
plot(tt,yy,type="l",
xlab=expression(t),ylab=expression(y(t)),
ylim=c(-0.1,1.6))
points(ltmpE$t,ltmpE$y,type="b",pch=16,col=2,lty=2)
points(ltmpH$t,ltmpH$y,type="b",pch=16,col=3,lty=3)
points(ltmpR$t,ltmpR$y,type="b",pch=2,col=4,lty=4)
legend(0,0.5,legend=c("Exact","Euler","Heun","RK4"),
pch=c(-1,16,16,2),lty=c(1,2,3,4),col=c(1,2,3,4))
```

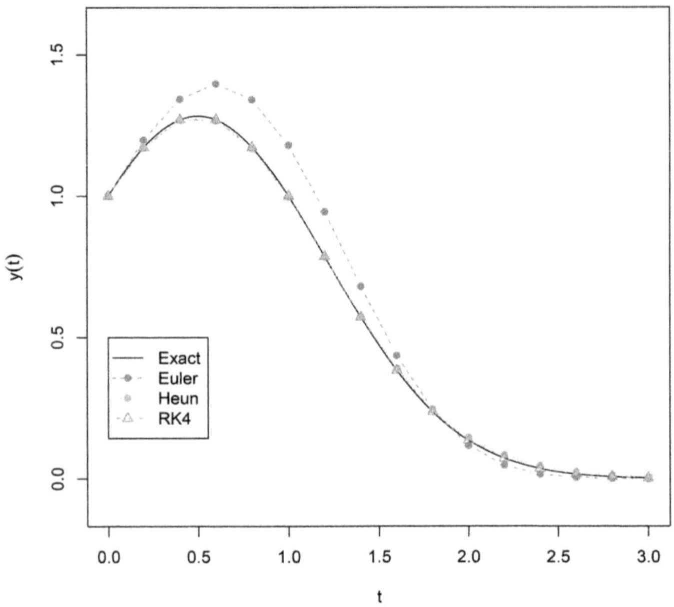


```
# Closer look around peak
plot(tt,yy,type="l",
xlab=expression(t),ylab=expression(y(t)),
xlim=c(0,1),ylim=c(1.1,1.4))
points(ltmpE$t,ltmpE$y,type="b",pch=16,col=2,lty=2)
points(ltmpH$t,ltmpH$y,type="b",pch=16,col=3,lty=3)
points(ltmpR$t,ltmpR$y,type="b",pch=2,col=4,lty=4)
legend(0,0.5,legend=c("Exact","Euler","Heun","RK4"),
pch=c(-1,16,16,2),lty=c(1,2,3,4),col=c(1,2,3,4))
```

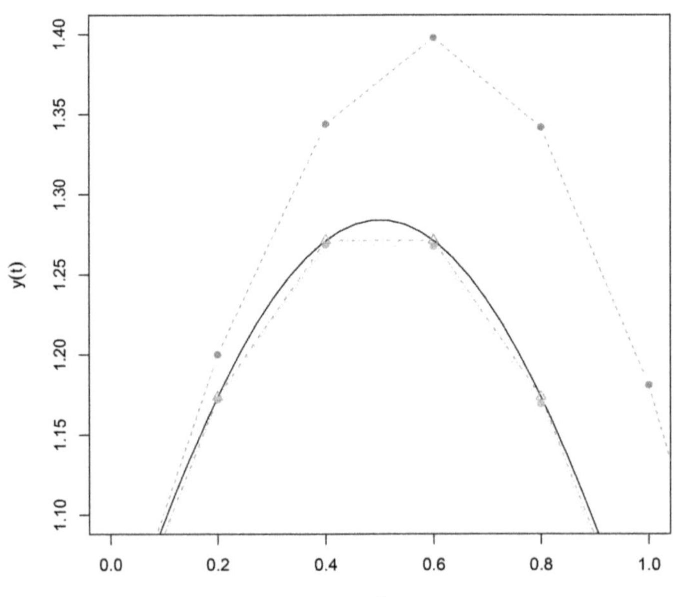

Quite often, as in this case, the accuracy of the Heun method yields solutions with an accuracy close to those of fourth-order Runge–Kutta.

8.13 Systems of ODEs

A system of first-order ODEs consists of m ODEs for the m dynamic variables $y_1(t), y_2(t), \ldots, y_m(t)$:

$$\begin{cases} \mathrm{d}y_1/\mathrm{d}t &= f_1(t, y_1, \ldots, y_m) \\ \mathrm{d}y_2/\mathrm{d}t &= f_2(t, y_1, \ldots, y_m) \\ & \ldots \\ \mathrm{d}y_m/\mathrm{d}t &= f_m(t, y_1, \ldots, y_m) \end{cases} \tag{8.15}$$

or, more synthetically,

$$\frac{\mathrm{d}\mathbf{y}}{\mathrm{d}t} = \mathbf{f}(t, \mathbf{y}). \tag{8.16}$$

For an IVP, this set of equations must be accompanied by initial conditions:

$$y_1(t_0) = y_{10}, y_2(t_0) = y_{20}, \ldots, y_m(t_0) = y_{m0} \tag{8.17}$$

$$\Updownarrow$$

$$\mathbf{y}(t_0) = \mathbf{y}_0. \tag{8.18}$$

The numerical solution of the system of ODEs (8.15) uses the same solvers used by single ODEs. In general, all values $y_j(t)$ at time t_i will be calculated by the values available at time t_{i-1}, through the gradient $\mathbf{f}(t, y_1(t), y_2(t), \ldots, y_m(t))$. Thus, algorithmically, one will simply have to make sure that all components $y_j(t)$ will be updated using the specific solver. In computer implementations, this is normally achieved using vectors and matrices instead of scalars and vectors.

Example 8.10 *A simple example with two dynamic functions will illustrate the concept. Let us try and proceed to the first step of the solution of*

$$\begin{cases} \mathrm{d}y_1/\mathrm{d}t &= y_1 + 2y_2 \\ \mathrm{d}y_2/\mathrm{d}t &= (3/2)y_1 - y_2, \qquad y_1(0) = 1,\ y_2(0) = 0, \end{cases}$$

using the Euler solver. The initial conditions correspond to vector

$$\mathbf{y}(t_0) \equiv \begin{pmatrix} y_{10} \\ y_{20} \end{pmatrix} = \begin{pmatrix} 1 \\ 0 \end{pmatrix}.$$

This will make a transition to vector $\mathbf{y}(t_1)$ *when both its components are derived through the Euler algorithm. More specifically, we have*:

$$\begin{aligned} y_1(t_1) &= y_1(t_0) + hf_1(t_0, y_{10}, y_{20}), \Leftrightarrow y_1(t_1) &= 1 + h(1 + 2(0)) &= 1 + h \\ y_2(t_1) &= y_2(t_0) + hf_2(t_0, y_{10}, y_{20}), \Leftrightarrow y_2(t_1) &= 0 + h((3/2)1 - 0) &= (3/2)h. \end{aligned}$$

Once the numerical value for the step size h is chosen, the value of $\mathbf{y}(t_1)$ *can be determined. The routine is then ready to proceed and determine* $\mathbf{y}(t_2)$*, etc.*

Example 8.11 *Solve numerically the system*

$$\begin{cases} \mathrm{d}y_1/\mathrm{d}t &= y_2 \\ \mathrm{d}y_2/\mathrm{d}t &= -y_1 \end{cases}, \quad \begin{cases} y_1(0) &= 1 \\ y_2(0) &= 0 \end{cases}, \quad t \in [t_0 = 0,\ t_f = \pi],$$

using fourth-order Runge–Kutta, with h=0.2.

We will use the solver RK4ODE *after having defined the gradient, initial conditions, and the step size. These actions are illustrated in the following code. Furthermore, we can compare the numerical solution with the analytic one because the system is well known to be satisfied by* $y_1(t) = \cos(t)$ *and* $y_2(t) = -\sin(t)$.

```
# Define the gradient
f <- function(t,y) {
dy1 <- y[2]
dy2 <- -y[1]

return(c(dy1,dy2))
}

# Solution interval
t0 <- 0
tf <- pi


# Initial conditions
y0 <- c(1,0)

# Stepsize
h <- 0.2

# Solution with RK4
ltmp <- RK4ODE(f,t0,tf,y0,h)

# Exact solution
tt <- seq(0,pi,length.out=100)
yy1 <- cos(tt)
yy2 <- -sin(tt)

# Visual comparison
plot(tt,yy1,type="l",
xlab=expression(t),ylab=expression(y[1](t),y[2](t)),
ylim=c(-1,1))
points(tt,yy2,type="l",lty=2)
points(ltmp$t,ltmp$y[,1],type="b",pch=16,cex=0.5,col=2)
points(ltmp$t,ltmp$y[,2],type="b",pch=16,cex=0.5,col=2,lty=2)
legend(0,0.5,legend=c("Exact y1(t)","Exact y2(t)",
"RK4 y1(t)","RK4 y2(t)"),
pch=c(-1,-1,16,16),lty=c(1,2,1,2),col=c(1,1,2,2))
```

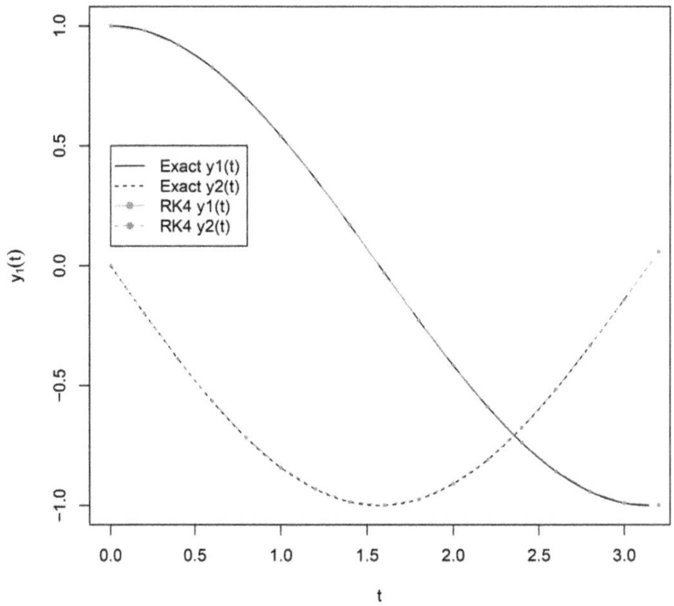

The lines corresponding to the numerical solution are hardly distinguishable from those corresponding to the exact solutions, given the $O(h^4)$ *accuracy of the fourth-order Runge–Kutta method.*

8.14 Higher-order ODEs

Many physical problems are naturally described by differential equations of second or higher order. However, numerical methods such as Euler, Heun, or Runge–Kutta are designed for systems of first-order ODEs. To apply these methods, we must first rewrite higher-order equations as equivalent systems of first-order equations. The transformation is straightforward: we introduce new variables to represent each derivative of the unknown function, up to one order less than the original equation.

Example 8.12 *Consider the second-order ODE describing a simple harmonic oscillator*:

$$\frac{\mathrm{d}^2x}{\mathrm{d}t^2} = -x.$$

We define a new variable:

$$v(t) = \frac{\mathrm{d}x}{\mathrm{d}t}.$$

This gives the system:

$$\begin{cases} \mathrm{d}x/\mathrm{d}t = v, \\ \mathrm{d}v/\mathrm{d}t = -x. \end{cases}$$

This system is now in a form suitable for numerical solution: a pair of coupled first-order ODEs.

In general, an nth order differential equation can always be rewritten as a system of n first-order equations. This approach is widely used in mechanics, electronics, and many other fields of applied science.

8.15 Exercises on IVPs

Exercise 01

The solution to the following IVPs,

$$ty' + y = 2t, \quad y(1) = 0,$$

is

$$y(t) = t - \frac{1}{t}, \quad t \neq 0.$$

Solve this ODE numerically using `EulerODE` and compare the result visually with the exact solution for $t \in [1, 3]$, when using step size $h = 0.4, 0.2, 0.1$. What error is expected for the solution at $t = 3$? Is this reasonable?

Exercise 02

The solution to the following IVPs,

$$y' + \frac{1}{x}y = xy^2, \quad y(4) = -\frac{1}{4},$$

is

$$y(x) = \frac{1}{3x - x^2}.$$

Solve this ODE numerically using `RK4ODE` and compare the result visually with the exact solution fot $t \in [4, 10]$, when using step size $h = 0.5, 0.25, 0.1$. How can you extract the number of steps used by the method?

Exercise 03

In a closed environment, a biological population may grow quickly at first, but then slow down as resources become limited. This process can be modelled by the *logistic growth equation*:

$$\frac{dn}{dt} = rn\left(1 - \frac{n}{K}\right),$$

where:

- $n(t)$ is the population size at time t,
- $r > 0$ is the intrinsic growth rate,
- $K > 0$ is the *carrying capacity* (the maximum population that the environment can sustain).

The equation assumes that the population grows approximately exponentially when small ($n \ll K$), but that growth slows and eventually stops as n approaches K.

In this exercise, we consider the case:

- Growth rate: $r = 0.5$,
- Carrying capacity: $K = 1000$,
- Initial population: $n(0) = 50$.

The analytical solution to the logistic equation is:

$$n(t) = \frac{Kn_0 e^{rt}}{K + n_0(e^{rt} - 1)}.$$

1. Use the analytical solution to compute the exact population values at $t = 5$, $t = 10$, and $t = 20$.
2. Implement Euler's method to solve the equation numerically on the interval $t \in [0, 20]$ using step size $h = 1$. Compare your numerical results with the exact values from part 1.
3. Repeat the numerical solution using the Heun method and the classical Runge–Kutta method (RK4). Report and compare the results at $t = 5$, $t = 10$, and $t = 20$.
4. What do your numerical methods predict for large t? Does the population approach the expected limiting value K?

Exercise 04

Consider the classical *Lotka–Volterra system*, modelling the interaction between a prey population $x(t)$ and a predator population $y(t)$. The governing equations are:

$$\mathrm{d}x/\mathrm{d}t = \alpha x - \beta xy$$
$$\mathrm{d}y/\mathrm{d}t = \delta xy - \gamma y$$

The meaning of the variables and parameters is as follows:

- $x(t)$: number of *prey* (e.g. rabbits) at time t
- $y(t)$: number of *predators* (e.g. foxes) at time t
- α: natural *growth rate of prey* in the absence of predators
- β: *predation rate coefficient* (how often predators encounter and eat prey)
- δ: *growth rate of predators* per prey eaten
- γ: *natural death rate of predators* in the absence of prey.

The task to be carried out in this exercise are:

1. Implement a gradient function `f(t,u)` where `u=c(x,y)` and `f` returns the derivatives $\mathrm{d}x/\mathrm{d}t$ and $\mathrm{d}y/\mathrm{d}t$.
2. Use the `RK4ODE` solver to solve the system numerically over the interval $t \in [0, 30]$, where the parameters are

$$\alpha = 1.0, \quad \beta = 0.1, \quad \delta = 0.075, \quad \gamma = 1.5,$$

   and the initial conditions are

$$x(0) = 40, \quad y(0) = 9.$$

   Use a step size $h = 0.1$.
3. Plot both populations as functions of time.
4. Plot the *phase portrait*, i.e. a plot of $y(t)$ versus $x(t)$.
5. Try changing the parameters and observe whether the solution remains periodic or tends to a steady state.

Exercise 05

In mechanics and engineering, the third derivative of position with respect to time is known as *jerk*, and it plays an important role in the modelling of motion where smoothness or mechanical stress is a concern. For instance, jerk arises in motion planning for vehicles or robotic arms, where sudden changes in acceleration must be avoided.

Consider the following third-order differential equation:

$$\frac{\mathrm{d}^3x}{\mathrm{d}t^3} + 4\frac{\mathrm{d}x}{\mathrm{d}t} + x = 0,$$

with initial conditions:

$$x(0) = 1, \quad \frac{\mathrm{d}x(0)}{\mathrm{d}t} = 0, \quad \frac{\mathrm{d}^2x(0)}{\mathrm{d}t^2} = 0.$$

In this exercise, you must:

1. Introduce new variables

$$y_1 = x, \quad y_2 = \frac{\mathrm{d}x}{\mathrm{d}t}, \quad y_3 = \frac{\mathrm{d}^2x}{\mathrm{d}t^2},$$

   and rewrite the equation as a system of three coupled first-order differential equations.
2. Write down the corresponding initial conditions for the variables y_1, y_2, y_3.
3. Use a numerical method (e.g. Euler or Runge–Kutta of order 4) to solve the system on the interval $t \in [0, 10]$ with a step size of your choice.
4. Plot the numerical solution for position $x(t)$, velocity $\mathrm{d}x/\mathrm{d}t$, and acceleration $\mathrm{d}^2x/\mathrm{d}t^2$. Discuss the overall behaviour of the system.
5. What role does the jerk term (third derivative) appear to play in the motion over time?

8.16 Boundary-value problems (BVPs)

In IVPs, the solution of a differential equation is determined by specifying all necessary conditions at a single point, typically the initial time. In contrast, *boundary-value problems (BVPs)* involve conditions that are imposed at two or more distinct points in the domain (see appendix G). This changes both the nature of the problem and the numerical techniques required to solve it. A standard example is the second-order differential equation:

$$\frac{d^2y}{dt^2} = -y,$$

together with boundary conditions such as:

$$y(0) = 0, \quad y\left(\frac{\pi}{2}\right) = 1.$$

This type of problem arises frequently in physics and engineering, for example, in the vibration of strings, steady-state heat flow, and beam deflection. The goal is to find a function $y(t)$ that satisfies the differential equation throughout the interval, while also satisfying the prescribed values at both endpoints.

Unlike IVPs, BVPs cannot be solved simply by stepping forward from a single point. Instead, they require either an iterative approach (as in shooting methods) or a global approximation (as in the Galerkin method). In the sections that follow, we introduce a simple but often used strategy for solving BVPs numerically: the shooting method. Other strategies exist but will not be treated here.

8.17 The shooting method

The *shooting method* converts a BVP into an IVP by replacing the unknown initial slope $y'(t_0)$ with a parameter s. For a fixed value of s, the corresponding IVP becomes

$$\frac{d^2y}{dt^2} = f(t, y, y'), \qquad y(t_0) = y_0, \quad y'(t_0) = s.$$

This problem can be solved numerically to produce a solution $y(t, s)$, which depends on the chosen initial slope s. In general, the solution will not satisfy the second boundary condition $y(t_f) = y_f$. However, it enables us to define, in principle, a function

$$F(s) = y(t_f, s) - y_f,$$

which measures the discrepancy between the computed solution and the desired boundary value at t_f. The aim of the shooting method is to find a value of s, s_0, such that $F(s_0) = 0$. Importantly, $F(s)$ is not an explicit analytical function: it is defined *implicitly*, via numerical integration of the differential equation with initial slope s. Each evaluation of $F(s)$ requires solving the IVP from t_0 to t_f using a numerical method (typically, fourth-order Runge–Kutta).

To find the value of s such that $F(s) = 0$, one typically applies a root-finding algorithm such as the bisection method (see section 6.1). This is done either by selecting two values of s, say s_1 and s_2, such that $F(s_1)$ and $F(s_2)$ have opposite signs, or by selecting a single value that triggers the algorithm to search for the two values just mentioned. The root is then approximated by iteratively narrowing the interval $[s_1, s_2]$ until the value of s satisfying $F(s) = 0$ is found to within a desired tolerance. The corresponding solution $y(t; s)$ then satisfies both boundary conditions and solves the original problem.

Example 8.13 *Consider the boundary-value problem*

$$y'' = -y, \qquad y(0) = 1, \qquad y\left(\frac{\pi}{2}\right) = 0.$$

This equation can be solved analytically, and the exact solution is $y(t) = \cos(t)$, *which satisfies the boundary conditions and shows that the correct initial slope is*

$$y'(0) = -\sin(0) = 0.$$

To apply the shooting method, we convert the second-order equation into a system of two first-order equations:

$$\begin{cases} \mathrm{d}y_1/\mathrm{d}t = y_2 \\ \mathrm{d}y_2/\mathrm{d}t = -y_1, \end{cases} \quad \text{with} \quad y_1(0) = 1, \quad y_2(0) = s,$$

where $y_1 = y$, $y_2 = y'$, *and s is an unknown parameter representing the initial slope. For each guess of s, we solve the system numerically using a method such as Runge–Kutta and define the function*

$$F(s) = y_1\left(\frac{\pi}{2}, s\right),$$

which measures the deviation from the desired boundary condition at $t = \pi/2$.

Let us now illustrate the first few iterations of the bisection method:

- *For* $s = 0$, *we find* $F(0) = -0.009\,204$.
- For $s = 0.5$, *we obtain* $F(0.5) = 0.490\,775$.

Since $F(0) < 0$ *and* $F(0.5) > 0$, *there is a sign change, so a root* s_0 *must lie within the interval* $[0, 0.5]$. *We continue with the bisection method:*

- *Try* $s = 0.25$: $F(0.25) = 0.240\,786$.
- *Now try* $s = 0.125$: *the value is closer to zero but still positive.*

We continue halving the interval until $|F(s)|$ *falls below a prescribed tolerance. In this way, we converge numerically to the correct initial slope* $s_0 = 0$ *without relying on knowledge of the exact solution.*

Note that the function $F(s)$ *is not known in closed form: each evaluation of* $F(s)$ *requires solving an IVP numerically for the corresponding value of s. This is a typical feature of shooting methods, which combine integration and root-finding in a nested procedure.*

8.18 Relevant R code. Function BVPshoot2

As explained in the previous section, the shooting method makes use of an algorithm to find the zeros of the function $F(s)$. There are many such algorithms in comphy and we have chosen a relatively simple one, the bisection algorithm, which in comphy is implemented in the function roots_bisec. The implementation of the function thus consists of a gradient (of the first-order system, equivalent to the second order ODE), of the function $F(s)$ (which includes the RK4 ODE solver), and of a loop inside the bisection function that stops when the tolerance is reached.

BVPshoot2 takes in the gradient $f(t, y, y')$ of the second-order ODE[3], the start and end of the solution interval, t_0, t_f, the corresponding boundary values, $y_0 = y(t_0)$, $y_f = y(t_f)$, a value for the step size, h, and one or two initial values for the search of the root. These correspond to the same option present in roots_bisec. The function returns a list with two elements. The first, t, is the grid of values of the integration variable, t. The second is an $n \times 2$ matrix, called y, containing the solution $y(t)$ (first column) and its derivative $y'(t)$ (second column).

To show how to use BVPshoot2, let us attempt a solution of the following (non-homogeneous) BVP:

$$\frac{\mathrm{d}^2 y}{\mathrm{d}t^2} + y = -9\sin(2t), \qquad y(0) = -1, \, y\left(\frac{3\pi}{2}\right) = 0.$$

The unique analytical solution of this BVP is

$$y(t) = 3\sin(2t) - \cos(t).$$

The gradient associated with this ODE is

$$\frac{\mathrm{d}^2 y}{\mathrm{d}t^2} = f(t, y, y') = -9\sin(2t) - y.$$

The solution via BVPshoot2 is shown in the following code. The default value s_guess=1 has been left for the solution; roots_bisec takes care of starting from that value, $s = 1$, and find an interval at the extremes of which $F(s)$ has different signs.

```
# Define the gradient
f <- function(t,y,dy) {
ff <- -9*sin(2*t)-y

return(ff)
}

# Interval
```

[3] By *gradient* here is meant, of course, the right hand side in the expression $\mathrm{d}^2y/\mathrm{d}t^2 = f(t, y, y')$, even if this would not normally be classified as gradient, being the derivative of the second order.

```
t0 <- 0
tf <- 3*pi/2

# BVPs
y0 <- -1
yf <- 0

# Step size
h <- 0.01

# Solution. Shooting method
ltmp <- BVPshoot2(f,t0,tf,y0,yf,h)

# Exact (analytical) solution
tt <- ltmp$t
yy <- 3*sin(2*tt)-cos(tt)

# Compare numerical solution with exact one
plot(tt,yy,type="l",
xlab=expression(t),ylab=expression(y(t)))
points(ltmp$t,ltmp$y[,1],type="b",pch=16,cex=0.5,col=2)
```

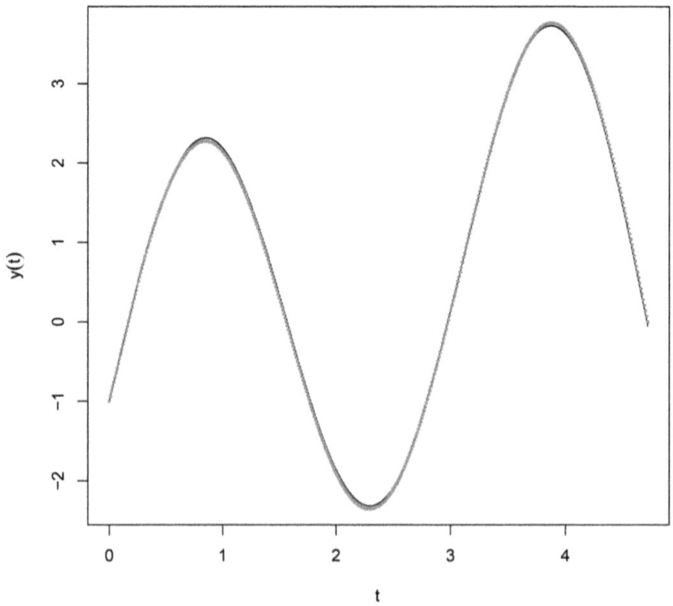

In this plot it is not easy to see how close the numerical solution is to the analytical one. We can *zoom in* to observe some details. In the following code, only some of the solution points found are selected for plot and this gives us the level of detail wanted.

```
# Select values around t=0.7
idx <- which(tt >= 0.7 & tt <= 1.1)
plot(tt[idx],yy[idx],type="l",
xlab=expression(t),ylab=expression(y(t)))
points(ltmp$t[idx],ltmp$y[idx,1],type="b",pch=16,cex=0.5,col=2)
```

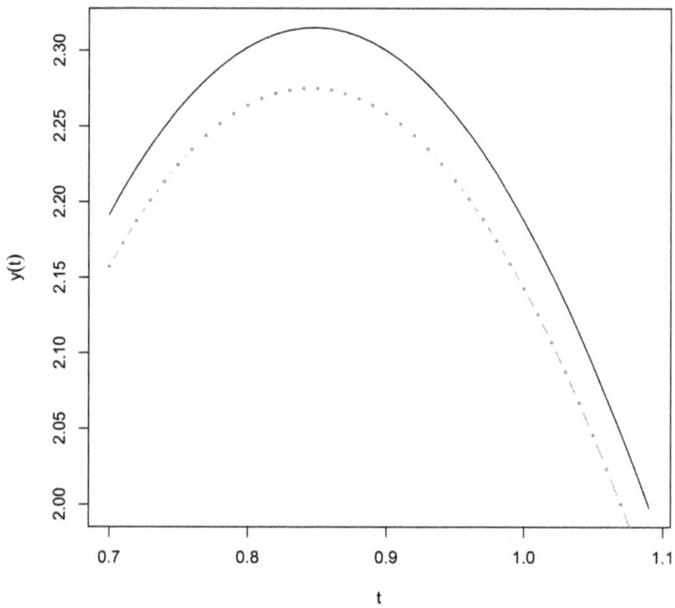

The difference is substantial, the error is certainly larger than the O(h^4) global error promised by the fourth-order Runge–Kutta. In fact, we could calculate a *mean local error* that is the average of all local errors.

```
# Mean local error
MLerror <- mean(abs(ltmp$y[,1]-yy))
print(MLerror)
```

```
## [1] 0.03393272
```

The value 0.0339 is O(h) rather than the O(h^5) promised by RK4. Part of the problem can be the accuracy with which the zero is found with the bisection method but this is, in fact, not the most important reason. The BVP shown above has as solution a trigonometric function and the values chosen for t are expressed as fractions of π. At the same time, though, the step size is a finite decimal, 0.01, not a fraction of π. When 0.01 is chosen as step size, the last interval falls over and beyond the last boundary value, as shown here.

```
# Extremes of last interval
print(tail(ltmp$t,2))
```

```
## [1] 4.71 4.72
```

```
# Second extreme of the original interval
print(3*pi/2)
```

```
## [1] 4.712389
```

As seen, the extremes of the last interval, [4.71, 4.72], include the second boundary point. This means that the solver considers $t = 4.72$ and not $3\pi/2 \approx 4.712389$ as second boundary point; this clearly introduces an extra inaccuracy that turns out to be significant.

We can try and remove such an inaccuracy by choosing the second boundary point to be a multiple of h. So, let us re-work the BVP modifying the interval and the boundary conditions slightly.

```
# Original BVP with tf=3*pi/2
t01 <- 0
tf1 <- 3*pi/2
y01 <- -1
yf1 <- 0
ltmp1 <- BVPshoot2(f,t01,tf1,y01,yf1,h=0.01)

# New (slightly modified) BVP with tf=4.72
t02 <- 0
tf2 <- 4.72
y02 <- 3*sin(2*t02)-cos(t02)
print(y02) # Still -1
```

```
## [1] -1
```

```
yf2 <- 3*sin(2*tf2)-cos(tf2)
print(yf2) # Not 0, but close to it
```

```
## [1] -0.0532753

ltmp2 <- BVPshoot2(f,t02,tf2,y02,yf2,h=0.01)

# Correct solution
tt <- ltmp1$t
yy <- 3*sin(2*tt)-cos(tt)

# Plot details for both BVPs
idx <- which(tt >= 0.7 & tt <= 1.1)
plot(tt[idx],yy[idx],type="l",
xlab=expression(t),ylab=expression(y(t)))
points(ltmp1$t[idx],ltmp1$y[idx,1],pch=16,cex=0.5,col=2)
points(ltmp2$t[idx],ltmp2$y[idx,1],pch=16,cex=0.5,col=3)
```

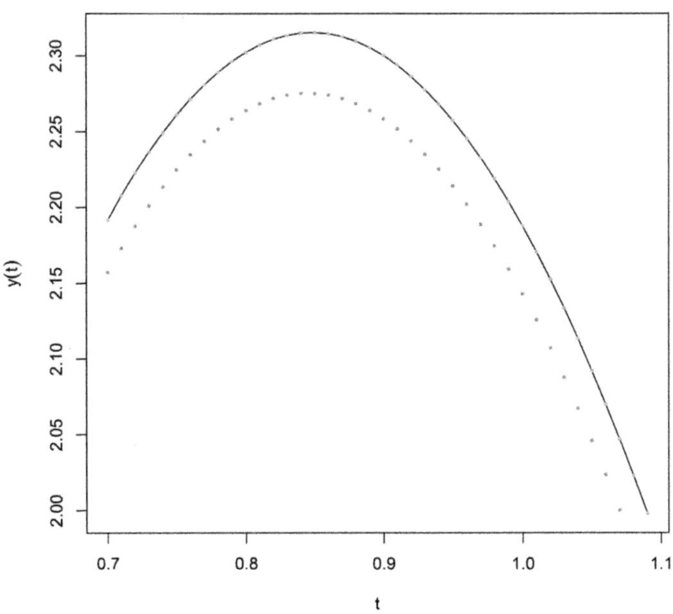

```
# Mean local errors
MLerror1 <- mean(abs(ltmp1$y[,1]-yy))
print(MLerror1)

## [1] 0.03393272

MLerror2 <- mean(abs(ltmp2$y[,1]-yy))

print(MLerror2)

## [1] 1.180797e-09
```

Amazing! The local error is now what it is supposed to be ($h^5 = 0.01^5 = 10^{-10}$). We can infer that essentially the shooting method does not contribute to the numerical error of RK4, as long as the tolerance used in the bisection method is kept small.

The example treated consisted of a linear BVP, for which a slight variant of the shooting method (see next section) is more appropriate. The method treated by `BVPshoot2` is valid for general, even nonlinear, second-order BVPs. Let us try and find the numerical solution of

$$\frac{d^2y}{dt^2} - y + 2y^3 = 0, \quad y(0) = 1, \, y(1) = \frac{1}{\cosh(1)}.$$

The BVP is nonlinear and second order; thus the shooting method that uses RK4 and the bisection algorithm, is appropriate. The gradient is

$$f(t, y, y') = y - 2y^3.$$

The analytic solution is

$$y(t) = \frac{1}{\cosh(t)}.$$

Let us apply the method and check that the numerical solution obtained is accurate.

```
# Define the gradient
f <- function(t,y,dy) {
ff <- y-2*y^3

return(ff)
}

# Interval
t0 <- 0
tf <- 1

# BVPs
y0 <- 1
yf <- 1/cosh(1) # Approx 0.6480543

# Step size
h <- 0.01
```

```
# Solution. Shooting method
ltmp <- BVPshoot2(f,t0,tf,y0,yf,h)

# Exact (analytical) solution
tt <- ltmp$t
yy <- 1/cosh(tt)

# Compare numerical solution with exact one
plot(tt,yy,type="l",
xlab=expression(t),ylab=expression(y(t)))
points(ltmp$t,ltmp$y[,1],pch=16,cex=0.5,col=2)
```

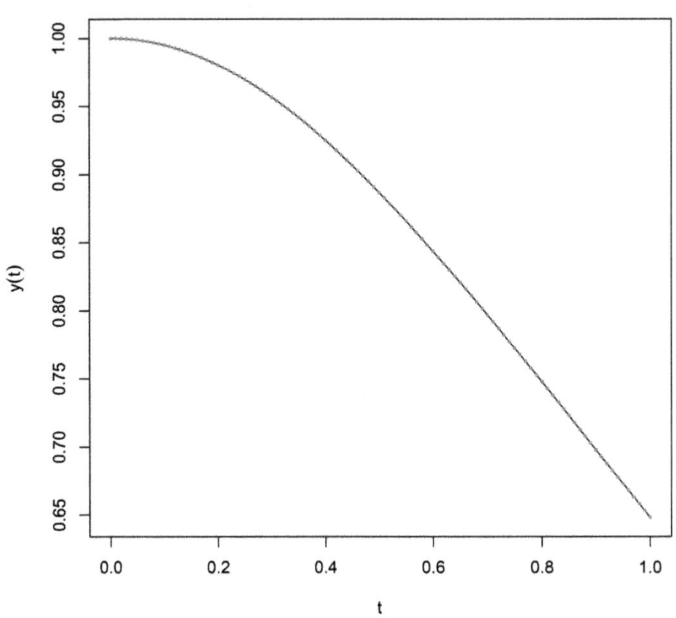

```
# Mean local errors
MLerror <- mean(abs(ltmp$y[,1]-yy))
print(MLerror)
```

```
## [1] 2.530239e-10
```

The solution is accurate to the expected $O(h^5)$.

8.19 The shooting method for second-order linear BVPs

In the two previous sections, we described a general shooting method applicable to all second-order BVPs, including nonlinear ones. That method involved guessing the initial slope, solving the corresponding IVP, and iteratively adjusting the guess until the final boundary condition was satisfied. While flexible and widely applicable, this method requires a root-finding procedure and repeated integrations. When the BVP is *linear*, however, the situation simplifies considerably. We can avoid iteration entirely by exploiting the principle of superposition for linear differential equations.

Consider the linear second-order BVP with Dirichlet boundary conditions:

$$y'' + P(x)y' + Q(x)y = R(x), \quad y(a) = \alpha, \quad y(b) = \beta$$

It is a standard result in the theory of differential equations that the general solution of a linear inhomogeneous ODE is given by

$$y(x) = u(x) + sv(x)$$

where:

- $u(x)$ is any particular solution of the inhomogeneous equation;
- $v(x)$ is a solution of the associated homogeneous equation:

$$v'' + P(x)v' + Q(x)v = 0;$$

- s is a constant to be determined from the boundary conditions. This plays exactly the same role played by the s in section 8.17, where $y'(a) = s$.

To construct such a solution that satisfies both boundary conditions, we proceed as follows.

1. We numerically integrate two IVPs from $x = a$ to $x = b$:
    - **Particular solution** $u(x)$ (to impose the left boundary condition $y(a) = \alpha$):

$$u'' + P(x)u' + Q(x)u = R(x), \quad u(a) = \alpha, \quad u'(a) = 0$$

- **Homogeneous solution** $v(x)$ (to span the solution space of the homogeneous problem):

$$v'' + P(x)v' + Q(x)v = 0, \quad v(a) = 0, \quad v'(a) = 1$$

After solving these numerically (e.g. using the Runge–Kutta method), we obtain the values $u(b)$ and $v(b)$.

The initial derivative $u'(a) = 0$ is chosen arbitrarily, since any particular solution suffices; the value $v'(a) = 1$ is selected to make $y'(a) = s$, as expected by the general shooting method.

2. We now impose the second boundary condition $y(b) = \beta$ by solving for s:

$$y(b) = u(b) + sv(b) = \beta \quad \Rightarrow \quad s = \frac{\beta - u(b)}{v(b)}$$

Provided $v(b) \neq 0$, this equation uniquely determines s, and hence

$$y(x) = u(x) + sv(x)$$

is the full solution to the BVP.

This approach is valid because the differential operator is linear. The particular solution $u(x)$ satisfies the inhomogeneous equation and the left boundary condition, while the homogeneous solution $v(x)$ allows us to adjust the slope to satisfy the right boundary condition. The constant s is calculated directly from the numerically computed endpoint values $u(b)$ and $v(b)$, with no need for iteration or root-finding.

This version of the shooting method is elegant, efficient, and numerically stable. It should be the method of choice when the underlying ODE is linear and has Dirichlet (or other simple) boundary conditions. If $v(b) = 0$, however, the BVP may not admit a unique solution, and further analysis is required.

Example 8.14 *To demonstrate how this variant of the shooting method works, consider the following linear BVP:*

$$\frac{d^2y}{dx^2} - \frac{3}{x}\frac{dy}{dx} + \frac{4}{x^2}y = x, \quad y(1) = 0, y(2) = 4(\ln(2) + 1).$$

This is a linear BVP whose unique solution is

$$y(x) = x^2(\ln(x) - 1) + x^3.$$

To apply the linear shooting method, we rewrite the second-order equation as a system of two first-order equations by setting $y_1 = y$, $y_2 = y'$. The resulting gradient system is:

$$\begin{cases} y' = y_2 \\ y' = \dfrac{3}{x}y_2 - \dfrac{4}{x^2}y_1 + x \end{cases}$$

We solve two initial-value problems on the interval $x \in [1, 2]$, one for the inhomogeneous equation and one for the associated homogeneous equation.

The inhomogeneous IVP is:

$$\begin{cases} u' = u_2 \\ u' = \dfrac{3}{x}u_2 - \dfrac{4}{x^2}u_1 + x \end{cases}, \quad u_1(1) = 0, \quad u_2(1) = 0$$

The homogeneous IVP is:

$$\begin{cases} v' = v_2 \\ v' = \dfrac{3}{x}v_2 - \dfrac{4}{x^2}v_1 \end{cases}, \quad v_1(1) = 0, \quad v_2(1) = 1.$$

After numerically solving these systems (e.g. using the Runge–Kutta method), we evaluate the solutions at $x = 2$ and compute:

$$s = \frac{4(\ln(2) + 1) - u_1(2)}{v_1(2)}$$

The final solution is then reconstructed as:

$$y(x) = u_1(x) + sv_1(x)$$

This satisfies the original boundary conditions by construction and approximates the exact solution to within the accuracy of the numerical integrator.

8.20 Relevant R code. Function `BVPlinshoot2`

This function takes in:

- the gradient $f(t, y, y')$ of the second-order, and linear, ODE;
- the initial and final value of the solution interval, `t0` and `tf`;
- the two boundary values, `y0,yf`, corresponding to `t0,tf`;
- the step size, h.

The output is a list with two elements: a vector `t` containing the t points and a matrix, `y`, with two columns containing the solution $y(t)$ and its derivative, $y'(t)$. Ellipses are included in the function, catering for possible parameters included in the gradient.

The code follows the theory explained above quite closely. The most peculiar and important part of the code is the internal (automatic) extraction of the gradient for the homogeneous ODE, which is defined as follows:

```
# Define system of ODEs for homogeneous part (v)
fhom <- function(t,y,dy,...) {
f(t,y,dy,...)-f(t,0,0,...)
}
```

To demonstrate how this function works, let us attempt solving the linear BVP introduced in example 8.14.

```
# Define the gradient
f <- function(x,y,dy) {
ff <- (3/x)*dy-(4/x^2)*y+x

return(ff)
}

# Interval
t0 <- 1
tf <- 2

# BVPs
y0 <- 0
yf <- 4*(log(2)+1)

# Step size
h <- 0.01

# Solution. Linear shooting method
ltmp <- BVPlinshoot2(f,t0,tf,y0,yf,h)

# Analytic solution
xx <- ltmp$t
yy <- xx^2*(log(xx)-1)+xx^3

# Comparison
plot(xx,yy,type="l",
xlab=expression(x),ylab=expression(y(x)))
points(ltmp$t,ltmp$y[,1],pch=16,cex=0.5,col=2)
```

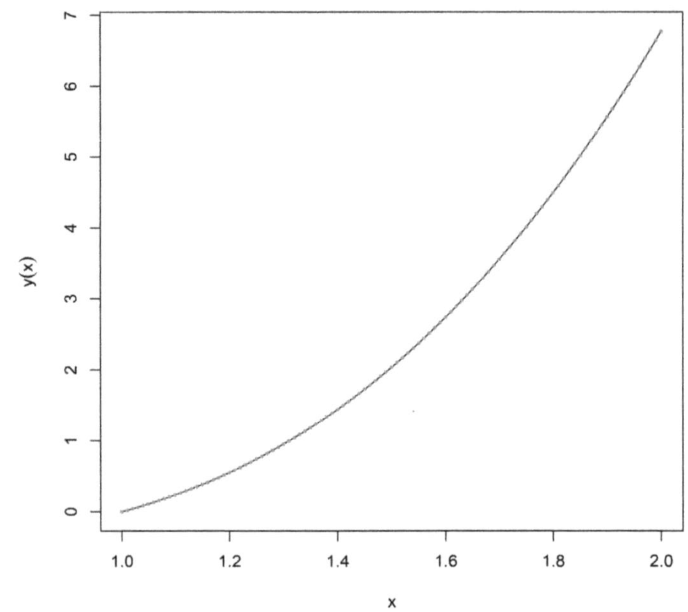


```
# Mean local error
MLerror <- mean(abs(ltmp$y[,1]-yy))
print(MLerror)
```

```
## [1] 7.439809e-11
```

The average local error is in line with the $O(h^5)$ expectation. The linear shooting algorithm is generally faster than the standard one using bisection, because it involves no iteration.

8.21 Exercises on BVPs

Exercise 06

Consider the following linear BVP:

$$\frac{d^2y}{dx^2} - \frac{2}{x}\frac{dy}{dx} + y(x) = x, \qquad y(1) = 0, y(2) = 1.$$

1. Use BVPlinshoot2 to compute the solution on [1, 2]. Try with step sizes $h = 0.01$, $h = 0.005$, $h = 0.0025$ and record the values of $y(1.5)$.
2. Plot the solution for the three different step sizes, using different colours/line traits.

Exercise 07

Solve the following BVP analytically,

$$\frac{d^2y}{dx^2} = -e^x, \qquad y(0) = 0, y(1) = 0,$$

and compare your solution with the numerical solution obtained with the linear shooting method. Do errors behave according to expectation?

Exercise 08

A rod of length $L = 10$ m has its ends held at temperatures

$$T(0) = T_A = 300 \text{ K}, \qquad T(L) = T_B = 350 \text{ K}.$$

Assume steady one–dimensional conduction with

$$\kappa(T) = 1 + \alpha T, \qquad \alpha = 10^{-3} \quad \text{K}^{-1},$$

where κ is the variable *thermal conductivity* of the rod (measured in W m^{-1} K^{-1}) and where no internal heat sources are present. The steady state equation is

$$\frac{d}{dx}\left(\kappa(T)\frac{dT}{dx}\right) = 0,$$

which, written in the form $d^2T/dx^2 = f(x, T, T')$, becomes

$$\frac{d^2T}{dx^2} = -\frac{\alpha}{1 + \alpha T}\left(\frac{dT}{dx}\right)^2.$$

1. Define $f(x, T, T') = -\alpha/(1 + \alpha T)\,(T')^2$ and solve on [0, 10] with `BVPshoot2`, using $T_A = 300$, $T_B = 350$.
2. Compute and report the numerical value of $T(5)$.
3. Produce a line plot of $T(x)$ on [0, 10].
4. Repeat with $\alpha = 0$ (constant conductivity). Plot both solutions together and comment briefly on how the temperature-dependent conductivity bends the profile away from the straight line.

8.22 Eigenvalue problems (EPs)

A short introduction to this type of ODEs and to the kind of problems and solutions they relate to is presented in appendix G. In this section we will mainly deal with the numerical solution of those EPs collectively known as *Sturm–Liouville problems* (see appendix G). The aim is to approximate the eigenvalues and eigenfunctions of a differential operator by converting the continuous problem into a discrete one. This is typically done by replacing derivatives with finite differences, resulting in a matrix eigenvalue problem.

We begin with Sturm–Liouville problems subject to homogeneous Dirichlet boundary conditions, which lead to standard symmetric matrix eigenvalue problems and provide a clear starting point for illustrating the discretisation process. We will then consider cases where the weight function implicit in Sturm–Liouville problems (see appendix G) differs from one, requiring a generalised formulation, and finally turn to nonhomogeneous Dirichlet boundary conditions. Other types of boundary conditions will not be treated in this book.

8.23 A simple Sturm–Liouville problem

We start illustrating the method using the very simple Sturm–Liouville problem

$$\frac{d^2y}{dx^2} + \lambda y = 0, \qquad y(0) = 0,\ y(\pi) = 0.$$

As discussed in appendix G, this equation has an infinite sequence of exact eigenvalues $\lambda_n = n^2$ and corresponding eigenfunctions $y_n(x) = \sin(nx)$. Our goal is to recover approximations to these eigenpairs numerically.

To begin, we discretise the interval $[0, \pi]$ using a uniform grid of $n + 1$ points, x_i, for $i = 1, \ldots, n + 1$. The spacing between points is $h = \pi/n$. The boundary conditions require that the solution is zero at the ends of the interval, so we can ignore x_1 and x_{n+1} in our system of equations, and only consider the interior points $x_2, \ldots, x_n$.

At each internal grid point x_i, we approximate the second derivative with the centred difference formula (see equation (7.8)):

$$\left.\frac{\mathrm{d}^2y}{\mathrm{d}x^2}\right|_{x=x_i} \approx \frac{y_{i-1} - 2y_i + y_{i+1}}{h^2}$$

Substituting this into the differential equation gives us a system of linear equations for the unknown values of the function, y_i, at each grid point:

$$\frac{y_{i-1} - 2y_i + y_{i+1}}{h^2} + \lambda y_i = 0$$

Rearranging this, we obtain

$$\frac{1}{h^2}(y_{i-1} - 2y_i + y_{i+1}) = -\lambda y_i$$

The Dirichlet boundary conditions $y(0) = 0$ and $y(\pi) = 0$ mean that $y_1 = 0$ and $y_{n+1} = 0$. This gives us a system of $n - 1$ linear equations for the unknowns $y_2, \ldots, y_n$. We can express this system in the form of a standard matrix eigenvalue problem, $A\mathbf{y} = \lambda\mathbf{y}$, where $\mathbf{y} = [y_2, y_3, \ldots, y_n]^T$ is the vector of numerical solution values. The matrix A is an $(n - 1) \times (n - 1)$ tridiagonal matrix given by

$$A = -\frac{1}{h^2}\begin{pmatrix} -2 & 1 & 0 & \cdots & 0 \\ 1 & -2 & 1 & \cdots & 0 \\ 0 & 1 & -2 & \cdots & 0 \\ \vdots & \vdots & \vdots & \ddots & \vdots \\ 0 & 0 & 0 & \cdots & -2 \end{pmatrix} \tag{8.19}$$

The eigenvalues of this matrix provide numerical approximations to the true eigenvalues λ_n, and the corresponding eigenvectors give the numerical values of the eigenfunctions at the grid points. These can then be plotted to visualise the shape of the eigenfunctions.

Example 8.15 *For the Sturm–Liouville problem with $n + 1 = 5$ points, the interval $[0, \pi]$ is divided into $n = 4$ subintervals. The spacing is $h = \pi/4$. The internal grid points are x_2, x_3, x_4. The unknown values are y_2, y_3, y_4. The system of equations from the finite difference approximation, with $y_1 = 0$ and $y_5 = 0$, gives:*

$$\begin{aligned} -2y_2 + y_3 &= -\lambda h^2 y_2 \\ y_2 - 2y_3 + y_4 &= -\lambda h^2 y_3 \\ y_3 - 2y_4 &= -\lambda h^2 y_4 \end{aligned}$$

This forms a 3×3 matrix eigenvalue problem, $A\mathbf{y} = \lambda\mathbf{y}$, where $\mathbf{y} = [y_2, y_3, y_4]^T$ and $h = \pi/4$.

$$A = -\frac{1}{h^2}\begin{pmatrix} -2 & 1 & 0 \\ 1 & -2 & 1 \\ 0 & 1 & -2 \end{pmatrix}$$

The eigenvalues of the matrix in parentheses, let's call it T, are $-2 + \sqrt{2}$, -2, and $-2 - \sqrt{2}$. To find the numerical approximations for λ, we multiply these by the factor $-\frac{1}{h^2} = -\frac{16}{\pi^2}$. The numerical eigenvalues are thus:

$$\begin{aligned} \lambda_1 &= -\frac{16}{\pi^2}(-2 + \sqrt{2}) \approx 0.950 \quad (\text{Exact}: \lambda_1 = 1) \\ \lambda_2 &= -\frac{16}{\pi^2}(-2) \approx 3.242 \quad (\text{Exact}: \lambda_2 = 4) \\ \lambda_3 &= -\frac{16}{\pi^2}(-2 - \sqrt{2}) \approx 5.535 \quad (\text{Exact}: \lambda_3 = 9) \end{aligned}$$

As we can see, the approximations are not particularly close to the first three exact eigenvalues, although a correlation among them can be observed. This is an expected outcome of the finite difference method, which replaces the continuous problem with a discrete model, akin to approximating a perfectly smooth vibrating string with a chain of connected beads. The discrete model can accurately represent the simple, low-frequency oscillations (the first few modes), but it struggles to capture the complex, higher-frequency oscillations with many nodes. The number of internal grid points, in this case three, fundamentally limits the number of modes that can be accurately resolved.

8.24 Sturm–Liouville with nonconstant weight function

To explore a more general situation, we now consider Sturm–Liouville problems where the weight function $w(x)$ is not equal to 1. These problems still involve homogeneous Dirichlet boundary conditions, but the differential equation takes the more general form

$$-\frac{\mathrm{d}}{\mathrm{d}x}\left(p(x)\frac{\mathrm{d}y}{\mathrm{d}x}\right) + q(x)y = \lambda w(x)y, \qquad y(a) = y(b) = 0.$$

Let the interval $[a, b]$ be divided into n subintervals of equal length $h = (b - a)/n$, giving $n - 1$ interior grid points $x_2, x_3, \ldots, x_n$. The unknowns will be the values of the function $y_i \approx y(x_i)$ at those internal points. We denote by A the matrix obtained by discretising the left-hand side of the equation using centred differences, just as in the simpler case where $w(x) = 1$. The right-hand side, however, is no longer simply $\lambda\mathbf{y}$, but must be modified to include a diagonal matrix W, whose entries are $w(x_i)$ evaluated at the internal points. This leads to the generalised eigenvalue problem:

$$A\mathbf{y} = \lambda W\mathbf{y}. \tag{8.20}$$

The differential operator

$$-\frac{\mathrm{d}}{\mathrm{d}x}\left(p(x)\frac{\mathrm{d}y}{\mathrm{d}x}\right) + q(x)y$$

is approximated at each internal point x_i by:

$$-\frac{1}{h^2}[p_{i+1/2}(y_{i+1} - y_i) - p_{i-1/2}(y_i - y_{i-1})] + q_i y_i, \tag{8.21}$$

where

$$p_{i+1/2} = p\left(x_i + \frac{h}{2}\right), \quad p_{i-1/2} = p\left(x_i - \frac{h}{2}\right), \quad q_i = q(x_i), \quad w_i = w(x_i)$$

(see section A.13 in appendix A for details). As we do not have grid points at $x_{i\pm1/2}$, the quantities $p_{i+1/2}$ and $p_{i-1/2}$ are approximated by

$$p_{i+1/2} \approx \frac{p(x_i) + p(x_{i+1})}{2}, \qquad p_{i-1/2} \approx \frac{p(x_{i-1}) + p(x_i)}{2}. \tag{8.22}$$

This is still a second-order accurate approximation, provided that $p(x)$ is sufficiently smooth.

The discretisation carried out leads to a tridiagonal matrix A, and a diagonal matrix W. The generalised eigenvalue problem takes the form

$$A\mathbf{y} = \lambda W\mathbf{y}, \tag{8.23}$$

with

$$A = \begin{bmatrix} \alpha_2 & \beta_2 & & & \\ \gamma_3 & \alpha_3 & \beta_3 & & \\ & \ddots & \ddots & \ddots & \\ & & \gamma_{n-1} & \alpha_{n-1} & \beta_{n-1} \\ & & & \gamma_n & \alpha_n \end{bmatrix}, \qquad W = \begin{bmatrix} w_2 & & & & \\ & w_3 & & & \\ & & \ddots & & \\ & & & w_{n-1} & \\ & & & & w_n \end{bmatrix},$$

and

$$\beta_i = -\frac{p_{i+1/2}}{h^2}, \quad \gamma_i = -\frac{p_{i-1/2}}{h^2}, \quad \alpha_i = \frac{p_{i+1/2} + p_{i-1/2}}{h^2} + q_i. \tag{8.24}$$

The matrix A is symmetric if $p(x)$ and $q(x)$ are regular[4], and the weight matrix W is diagonal and positive under standard assumptions. To convert the generalised problem to a standard one solvable with `eigen` in R, we multiply both sides by W^{-1}, obtaining

$$W^{-1}A\mathbf{y} = \lambda\mathbf{y}. \tag{8.25}$$

The matrix $W^{-1}A$ is not symmetric, but for moderate problem sizes this formulation is sufficient and numerically stable.

Example 8.16 *Let us illustrate the discretisation process described above using the* modified Legendre equation:

$$-\frac{\mathrm{d}}{\mathrm{d}x}\left((1 - x^2)\frac{\mathrm{d}y}{\mathrm{d}x}\right) = \lambda(1 - x^2)y, \qquad y(-1) = y(1) = 0.$$

Here, the functions are

$$p(x) = 1 - x^2, \qquad q(x) = 0, \qquad w(x) = 1 - x^2.$$

We discretise the interval $[-1, 1]$ *using* $n + 1 = 5$ *equally spaced grid points*:

$$x_1 = -1, \quad x_2 = -0.5, \quad x_3 = 0, \quad x_4 = 0.5, \quad x_5 = 1,$$

with step size $h = 0.5$. *The interior points are* x_2, x_3, x_4, *and we seek approximate values* y_2, y_3, y_4 *of the solution.*

The weight matrix W *is diagonal, with entries*

$$W = \begin{pmatrix} 0.75 & 0 & 0 \\ 0 & 1 & 0 \\ 0 & 0 & 0.75 \end{pmatrix},$$

since

$$w(x_2) = 1 - (-0.5)^2 = 0.75, \quad w(x_3) = 1, \quad w(x_4) = 0.75.$$

Next, we compute matrix A, *whose entries are defined as*

$$\beta_i = -\frac{p_{i+1/2}}{h^2}, \qquad \gamma_i = -\frac{p_{i-1/2}}{h^2}, \qquad \alpha_i = -(\beta_i + \gamma_i) + q_i.$$

[4] In this context, 'regular' means that $p(x)$ is sufficiently smooth (typically at least continuously differentiable) and strictly positive on the interval, and that $q(x)$ is continuous. These conditions ensure that the differential operator is well-defined and that the resulting matrix A is symmetric and suitable for numerical discretisation.

The midpoint values of p(x) are approximated by averaging:

$$
\begin{aligned}
p(x_2) &= 0.75, \quad p(x_3) = 1, \quad p(x_4) = 0.75,\\
p_{2+1/2} &\approx \frac{p(x_2)+p(x_3)}{2} = 0.875,\\
p_{2-1/2} &\approx \frac{p(x_1)+p(x_2)}{2} = \frac{0+0.75}{2} = 0.375,\\
p_{3+1/2} &\approx \frac{p(x_3)+p(x_4)}{2} = 0.875,\\
p_{3-1/2} &\approx \frac{p(x_2)+p(x_3)}{2} = 0.875,\\
p_{4+1/2} &\approx \frac{p(x_4)+p(x_5)}{2} = \frac{0.75+0}{2} = 0.375,\\
p_{4-1/2} &\approx \frac{p(x_3)+p(x_4)}{2} = 0.875.
\end{aligned}
$$

With $h = 0.5$, *so that* $h^2 = 0.25$, *we compute*:

$$
\begin{aligned}
\beta_2 &= -\frac{0.875}{0.25} = -3.5, \quad \gamma_2 = -\frac{0.375}{0.25} = -1.5, \quad \alpha_2 = 5.0,\\
\beta_3 &= -\frac{0.875}{0.25} = -3.5, \quad \gamma_3 = -\frac{0.875}{0.25} = -3.5, \quad \alpha_3 = 7.0,\\
\beta_4 &= -\frac{0.375}{0.25} = -1.5, \quad \gamma_4 = -\frac{0.875}{0.25} = -3.5, \quad \alpha_4 = 5.0.
\end{aligned}
$$

This yields:

$$
A = \begin{pmatrix} 5.0 & -3.5 & 0 \\ -3.5 & 7.0 & -3.5 \\ 0 & -3.5 & 5.0 \end{pmatrix}, \qquad W = \begin{pmatrix} 0.75 & 0 & 0 \\ 0 & 1 & 0 \\ 0 & 0 & 0.75 \end{pmatrix}.
$$

We then form the matrix $W^{-1}A$ *and solve the eigenvalue problem*

$$
W^{-1}A\mathbf{y} = \lambda \mathbf{y}
$$

using the `eigen` *function in R*.

```
# Grid and function definitions
x <- seq(-1,1,length.out=5)
h <- x[2]-x[1]

# Internal grid points
x_int <- x[2:4]
```

```
# Evaluate p(x) and w(x)
p <- function(x) 1 - x^2
w <- function(x) 1 - x^2
p_vals <- p(x)
w_vals <- w(x_int)

# Approximate p at midpoints by averaging
p_half <- function(i_left,i_right) {
(p_vals[i_left]+p_vals[i_right])/2
}

# Compute entries of matrix A
beta <- c(-p_half(2,3),-p_half(3,4),-p_half(4,5))/h^2
gamma <- c(-p_half(1,2),-p_half(2,3),-p_half(3,4))/h^2
alpha <- -(beta+gamma)

# Construct matrix A (3x3)
A <- matrix(0,3,3)
diag(A) <- alpha
A[row(A) == col(A)-1] <- beta[1:2]
A[row(A) == col(A)+1] <- gamma[2:3]
print(A)
```

```
##      [,1] [,2] [,3]
## [1,]  5.0 -3.5  0.0
## [2,] -3.5  7.0 -3.5
## [3,]  0.0 -3.5  5.0
```

```
# Construct diagonal weight matrix W
W <- diag(w_vals)
print(W)
```

```
##      [,1] [,2] [,3]
## [1,] 0.75    0 0.00
## [2,] 0.00    1 0.00
## [3,] 0.00    0 0.75
```

```
# Form W^{-1} A
W_inv_A <- solve(W,A)  # Equivalent to W^{-1} %*% A

# Solve the eigenvalue problem
```

```
eig <- eigen(W_inv_A)

# Print eigenvalues and eigenvectors
print(eig$values)
```

```
## [1] 12.551239  6.666667  1.115428
```

```
print(eig$vectors)
```

```
##            [,1]          [,2]      [,3]
## [1,] -0.5277747 -7.071068e-01 0.5411312
## [2,]  0.6655132  3.362042e-16 0.6437033
## [3,] -0.5277747  7.071068e-01 0.5411312
```

The differential equation considered in this example is closely related to the classical Legendre equation, whose eigenfunctions are the Legendre polynomials $P_n(x)$ *and whose eigenvalues are given by* $\lambda_n = n(n+1)$. *However, those polynomials satisfy a different eigenvalue problem: the weight function on the right-hand side is constant, and the boundary conditions are typically non-Dirichlet (they require finiteness rather than vanishing at the endpoints). In contrast, the present problem involves a nonconstant weight function* $w(x) = 1 - x^2$ *and homogeneous Dirichlet conditions. As a result, the eigenfunctions are not the Legendre polynomials, and the eigenvalues are not exactly* $n(n+1)$. *Nonetheless, the computed eigenvalues (e.g.* $1.12, 6.67, 12.55$) *follow a similar hierarchical pattern and reflect the same qualitative structure: higher eigenvalues correspond to more oscillatory modes, and the solutions remain orthogonal with respect to the inner product weighted by* $w(x)$.

8.25 Sturm–Liouville problems with nonhomogeneous Dirichlet conditions

We now consider the case where the Dirichlet boundary conditions are nonhomogeneous:

$$y(a) = \alpha, \qquad y(b) = \beta,$$

with at least one of α or β nonzero. The differential equation retains its standard form:

$$-\frac{\mathrm{d}}{\mathrm{d}x}\left(p(x)\frac{\mathrm{d}y}{\mathrm{d}x}\right) + q(x)y = \lambda w(x)y,$$

but the discretisation must now account for the specified boundary values.

This can be handled by introducing a known function $\phi(x)$ satisfying the boundary conditions exactly:

$$\phi(a) = \alpha, \qquad \phi(b) = \beta.$$

A natural and simple choice is, for example, the linear interpolant:

$$\phi(x) = \alpha\frac{b-x}{b-a} + \beta\frac{x-a}{b-a}. \tag{8.26}$$

We then write the unknown solution as

$$y(x) = u(x) + \phi(x), \tag{8.27}$$

where $u(x)$ satisfies homogeneous Dirichlet conditions:

$$u(a) = 0, \qquad u(b) = 0.$$

Substituting this decomposition into the differential equation yields[5]

$$-\frac{\mathrm{d}}{\mathrm{d}x}\left(p(x)\frac{\mathrm{d}u}{\mathrm{d}x}\right) + q(x)u = \lambda w(x)u + f(x),$$

where the function $f(x)$ is completely determined by the known shift $\phi(x)$ and its derivatives:

$$f(x) = \frac{\mathrm{d}}{\mathrm{d}x}\left(p(x)\frac{\mathrm{d}\phi}{\mathrm{d}x}\right) - q(x)\phi(x) + \lambda w(x)\phi(x).$$

Once the function $\phi(x)$ is fixed, its values and derivatives can be computed explicitly. However, the right-hand side of the equation for $u(x)$ typically still contains a term proportional to the eigenvalue parameter λ, such as $\lambda w(x)\phi(x)$. This occurs, for example, when the weight function $w(x)$ is constant and nonzero, in which case the

[5] To reduce the boundary conditions to a homogeneous form, we write the solution as $y(x) = u(x) + \phi(x)$, where $\phi(x)$ is a known function satisfying the boundary conditions, and $u(x)$ is the unknown correction satisfying $u(a) = u(b) = 0$. Substituting into the Sturm–Liouville equation

$$-\frac{\mathrm{d}}{\mathrm{d}x}\left(p(x)\frac{\mathrm{d}y}{\mathrm{d}x}\right) + q(x)y = \lambda w(x)y,$$

we obtain

$$-\frac{\mathrm{d}}{\mathrm{d}x}\left(p(x)\frac{\mathrm{d}u}{\mathrm{d}x}\right) + q(x)u + \left[-\frac{\mathrm{d}}{\mathrm{d}x}\left(p(x)\frac{\mathrm{d}\phi}{\mathrm{d}x}\right) + q(x)\phi\right] = \lambda w(x)u + \lambda w(x)\phi.$$

Grouping all terms involving u on the left-hand side, and the known terms on the right, we obtain:

$$-\frac{\mathrm{d}}{\mathrm{d}x}\left(p(x)\frac{\mathrm{d}u}{\mathrm{d}x}\right) + q(x)u = \lambda w(x)u + f(x),$$

where the source term is

$$f(x) = \lambda w(x)\phi(x) - \frac{\mathrm{d}}{\mathrm{d}x}\left(p(x)\frac{\mathrm{d}\phi}{\mathrm{d}x}\right) + q(x)\phi(x).$$

This is still a second-order linear ODE in the unknown function $u(x)$. The inhomogeneous term $f(x)$ is entirely determined by $\phi(x)$ and the coefficients of the differential equation, and can be evaluated explicitly once $\phi(x)$ is chosen.

inhomogeneous term remains λ-dependent and cannot be absorbed into a fixed source.

As a result, the transformed problem for $u(x)$ is no longer a standard eigenvalue problem. Although the boundary conditions are now homogeneous and the equation remains linear in $u(x)$, its discretisation leads to an algebraic system where λ appears on both sides. The resulting system does not have the standard form $A\mathbf{y} = \lambda\mathbf{y}$ or $A\mathbf{y} = \lambda B\mathbf{y}$, and cannot be solved using typical eigenvalue routines. Instead, such problems lead to nonlinear algebraic systems that must be handled with care. Numerical approaches often involve root-finding in λ, residual minimisation, or continuation methods. Each candidate value of λ typically defines a linear system in the unknowns, which must be solved (for instance, via LU decomposition or iterative solvers) within the search procedure. These methods are more computationally involved and depend on the specific structure of the problem. For this reason, we do not provide an implementation of this case in the `comphy` package.

Example 8.17 *We consider the differential equation*

$$-\frac{\mathrm{d}^2y}{\mathrm{d}x^2} = \lambda y(x), \qquad x \in [0, \pi],$$

subject to nonhomogeneous Dirichlet boundary conditions:

$$y(0) = \alpha, \qquad y(\pi) = \beta.$$

To reduce the problem to one with homogeneous boundary conditions, we write $y(x) = u(x) + \phi(x)$, *where* $\phi(x)$ *is the linear function* (8.26). *Then the unknown function* $u(x)$ *satisfies homogeneous boundary conditions:*

$$u(0) = 0, \qquad u(\pi) = 0.$$

Substituting into the differential equation gives

$$-\frac{\mathrm{d}^2u}{\mathrm{d}x^2} = \lambda u(x) + \lambda\phi(x),$$

since $\phi(x)$ *is linear and* $\frac{\mathrm{d}^2\phi}{\mathrm{d}x^2} = 0$. *We discretise the interval* $[0, \pi]$ *using* $n + 1 = 5$ *equally spaced grid points:*

$$x_1 = 0, \quad x_2 = \frac{\pi}{4}, \quad x_3 = \frac{\pi}{2}, \quad x_4 = \frac{3\pi}{4}, \quad x_5 = \pi,$$

with step size $h = \pi/4$. *The unknowns are* u_2, u_3, u_4, *and the second derivative is approximated by the centred difference formula:*

$$\frac{-u_{i-1} + 2u_i - u_{i+1}}{h^2} = \lambda u_i + \lambda\phi(x_i), \qquad i = 2, 3, 4.$$

This leads to the matrix equation

$$A\mathbf{u} = \lambda\mathbf{u} + \lambda\boldsymbol{\phi},$$

where $\mathbf{u} = [u_2, u_3, u_4]^T$, *and* $\phi = [\phi(x_2), \phi(x_3), \phi(x_4)]^T$. *The matrix* A *is the standard finite difference discretisation of the negative second derivative*:

$$A = \frac{1}{h^2}\begin{pmatrix} 2 & -1 & 0 \\ -1 & 2 & -1 \\ 0 & -1 & 2 \end{pmatrix}.$$

Rewriting the system:

$$(A - \lambda I)\mathbf{u} = \lambda\phi.$$

This is not a standard eigenvalue problem, because the parameter λ *appears on both sides of the equation. One possible strategy for solving it is to*:

- *Choose trial values of* λ.
- *For each value, solve the linear system*
$$(A - \lambda I)\mathbf{u}_\lambda = \lambda\phi.$$
- *Reconstruct the full approximation to the solution via* $y_i = u_i + \phi(x_i)$.
- *Assess the accuracy of the result by computing the residual or evaluating an additional constraint.*

This example illustrates how nonhomogeneous Dirichlet conditions lead to a nonlinear algebraic system. While solvable in small-scale demonstrations, these problems do not fit the standard eigenvalue framework and are not handled by standard routines.

8.25.1 Sturm–Liouville as BVP

It is worth noting that not all Sturm–Liouville-type problems with nonhomogeneous boundary conditions must be treated as eigenvalue problems. In particular, if the spectral parameter λ is fixed, or entirely absent from the equation, the problem reduces to a standard second-order BVP. Such problems can be linear or nonlinear and are typically solved using finite difference schemes, shooting methods, or collocation techniques.

Even when λ is present, introducing a change of variables $y(x) = u(x) + \phi(x)$, where ϕ satisfies the nonhomogeneous boundary conditions, can convert the problem into one with homogeneous boundary conditions for u. However, this transformation often introduces λ-dependent terms into the source, so that the resulting equation does not have the form $Au = \lambda Bu$. In such cases, the problem no longer fits the standard eigenvalue framework, and must instead be solved as a BVP with a parameter, potentially requiring iterative root-finding or continuation methods to determine suitable values of λ.

8.26 Relevant R code. Function `EPSturmLiouville2`

This solves the eigenvalue problem associated with the Sturm–Liouville equation, on a closed interval $[a, b]$, and subject to homogeneous Dirichlet boundary conditions, as described in the two previous sections. The equation is discretised on a uniform grid using finite differences, and the resulting matrix eigenvalue problem is solved numerically. The function is general in that the coefficient functions $p(x)$, $q(x)$, and $w(x)$ may be supplied either as actual functions or as precomputed numeric vectors. Its structure closely reflects the theory developed earlier in the preceding chapters. As this function includes several complex steps, a detailed explanation of the code follows.

After verifying that the grid `x` is numeric and contains at least three values (two boundary points and at least one interior point), the function checks for uniform spacing, which is required by the finite difference scheme. The grid step is denoted by h, and the function will halt if the grid deviates from uniformity by more than a specified tolerance. The variable `n_plus_1` records the number of grid points, and $n = n + 1 - 1$ is the number of intervals. Because the solution is required to vanish at the endpoints, the number of unknowns is $m = n - 1$, corresponding to the interior points. These are selected using `x[2L:n]`, and the midpoints (for use with $p(x)$) are computed as the average of adjacent nodes.

The next stage is to evaluate the coefficient functions at the appropriate locations. The potential and weight functions $q(x)$ and $w(x)$ are evaluated at the grid nodes x_i, while $p(x)$ is evaluated at midpoints $x_{i\pm 1/2}$. This corresponds to the discretisation of the term

$$-\frac{\mathrm{d}}{\mathrm{d}x}\left(p(x)\frac{\mathrm{d}y(x)}{\mathrm{d}x}\right),$$

which, when approximated using centred differences, naturally involves values of p between nodes. If the user supplies a function (e.g. `function(s) 1`), the function is evaluated at the relevant grid locations; if it returns a scalar, it is extended by repetition to a full vector of the expected length. This extension is sometimes referred to as 'broadcasting.' If instead the user provides a numeric vector, it must have the correct length: p must be of length n, corresponding to midpoints between nodes; q and w must be of length $n + 1$, corresponding to the full grid of nodes. The helper function `eval_on()` enforces this logic.

The function then extracts the interior portions of the coefficient arrays for use in the matrix construction: `qi` and `wi` are the values of $q(x)$ and $w(x)$ at the interior nodes, while `p_minus` and `p_plus` store the values $p_{i-1/2}$ and $p_{i+1/2}$ associated with each interior node. These midpoint values are required in the finite difference formula for the discretised operator. When p is supplied as a function, these values are computed on the fly by evaluating the function at midpoints. However, if p is precomputed and passed as a numeric array with values already aligned to the midpoints, this can result in slightly higher accuracy and avoids repeated function calls. If `check_inputs` is set to `TRUE`, the function verifies that all extracted values are finite and that $p(x) > 0$, $w(x) > 0$ throughout the interior of the interval.

Next, the function assembles the so-called *stiffness matrix* K[6], which represents the finite difference discretisation of the differential operator

$$-\frac{\mathrm{d}}{\mathrm{d}x}\left(p(x)\frac{\mathrm{d}}{\mathrm{d}x}\right) + q(x).$$

[6] The name K is used in the code for what we have called A in all previous formulas. This, in fact, follows a widespread symbol convention for the stiffness matrix. So, in the code presented, `K` is matrix A of the formulas, while `A` is not related to matrix A of the formulas. The reader should pay attention to this detail to avoid confusion.

The main diagonal contains the terms

$$(p_{i-1/2} + p_{i+1/2})/h^2 + q_i,$$

and the sub- and super-diagonals contain the off-centred differences

$$-p_{i\pm 1/2}/h^2.$$

The weight matrix W is a diagonal matrix containing the interior values of $w(x)$. The generalised eigenvalue problem is then

$$K\mathbf{u} = \lambda W\mathbf{u},$$

where $\mathbf{u}$ is the vector of interior unknowns. To convert this into a standard symmetric eigenvalue problem, the function performs the transformation

$$A = W^{-1/2}KW^{-1/2},$$

so that the problem becomes

$$A\mathbf{z} = \lambda\mathbf{z},$$

with

$$\mathbf{u} = W^{-1/2}\mathbf{z}.$$

This transformation preserves the eigenvalues but produces a symmetric matrix A, which is better suited to numerical solvers. It differs from the alternative form $W^{-1}K$, discussed earlier in the text, which is generally non-symmetric unless W is the identity. Symmetric matrices have real eigenvalues and orthogonal eigenvectors, and their computation is numerically more stable.

Once A is assembled, its eigenvalues and eigenvectors are computed using R's standard `eigen` function. The results are sorted in ascending order. If only a specified number of eigenpairs is requested (via `nev`), the output is truncated accordingly. The matrix indexing uses `drop=FALSE` to ensure that single-column matrices remain matrices, rather than being simplified to vectors.

The eigenvectors of A are then back-transformed to yield the eigenvectors of the original problem, using

$$\mathbf{u} = W^{-1/2}\mathbf{z}.$$

If the `normalize` flag is `TRUE`, the code scales each column of the eigenvector matrix so that it has unit norm in the discrete weighted $L^2(w)$ sense:

$$\sum_i h\, w_i\, u_i^2 = 1.$$

This is done using the function `sweep`, which divides each column by its corresponding norm.

The final step pads the eigenvectors with zeros at the two endpoints, thereby enforcing the homogeneous Dirichlet boundary conditions explicitly. An internal check ensures that these boundary values are indeed zero. The function returns a list containing the eigenvalues, the interior and full eigenvectors, the grid and step size, and (optionally) the matrices K and W.

The following short demo illustrates how to solve a basic Sturm–Liouville eigenvalue problem using `EPSturmLiouville2`. We consider the simple equation

$$-\frac{d^2y}{dx^2} = \lambda y$$

on the interval $[0, \pi]$, with homogeneous Dirichlet boundary conditions $y(0) = y(\pi) = 0$. This is the same problem treated earlier using manual discretisation in example 8.15. The coefficients $p(x) = 1$, $q(x) = 0$, and $w(x) = 1$ are constant, and the grid is made of $n + 1 = 5$ uniformly spaced nodes, yielding three interior unknowns. We instruct the function to compute the first three eigenvalues and eigenfunctions using `nev=3`.

```
# comphy must be loaded in memory
require(comphy)

# Define the interval and number of grid points
a <- 0
b <- pi
n <- 4   # number of subintervals => n+1 = 5 nodes
x <- seq(a,b,length.out=n+1)

# Define constant coefficient functions
p <- function(s) 1   # p(x) = 1 at midpoints
q <- function(s) 0   # q(x) = 0
w <- function(s) 1   # w(x) = 1

# Solve the Sturm-Liouville eigenproblem
ep <- EPSturmLiouville2(p,q,w,x,nev=3,

normalize=TRUE,return_matrices=TRUE)

# Print the eigenvalues
print(round(ep$values,3))


## [1] 0.950 3.242 5.535


# Display full eigenfunctions
print(round(ep$vectors_full,3))


##        [,1]   [,2]   [,3]
## [1,] 0.000  0.000  0.000
## [2,] 0.564 -0.798  0.564
## [3,] 0.798  0.000 -0.798
## [4,] 0.564  0.798  0.564
## [5,] 0.000  0.000  0.000
```

The printed eigenvalues are 0.950, 3.242, 5.535. These match those obtained earlier when the system was discretised manually and solved using base R routines.

The matrix ep$vectors_full contains the computed eigenfunctions, evaluated at all five grid nodes. As expected, the values at the endpoints (i.e. the first and last rows) are zero, in accordance with the homogeneous Dirichlet boundary conditions. The interior values correspond to discretised sine-like profiles associated with the first few eigenmodes. For example, the second row of the matrix contains the values of the first three eigenfunctions at $x = \pi/4$, the first interior node. These values closely match the normalised values of $\sin(kx)$ for $k = 1, 2, 3$, evaluated at $x = \pi/4$, up to an overall sign. This confirms that the numerical eigenfunctions approximate the exact ones, which are $\sin(kx)$, both in shape and scale, consistent with the theory of this classical eigenproblem.

```
# Extract interior eigenvectors
U <- ep$vectors_interior

# First interior point corresponds to x[2] = pi/4
xval <- pi/4
k <- 1:3

# Compute sin(k * x) at x = pi/4, and normalise
svals <- sin(k*xval)
svals <- svals/sqrt(sum(svals^2))

# Compare to first row of U (values at x = pi/4)
cat("Discrete eigenvector values at x=pi/4:\n")
```

```
## Discrete eigenvector values at x=pi/4:
```

```
print(round(U[1,],3))
```

```
## [1]  0.564 -0.798  0.564
```

```
cat("Normalised sin(k*pi/4) values:\n")
```

```
## Normalised sin(k*pi/4) values:
```

```
print(round(svals,3))
```

```
## [1] 0.500 0.707 0.500
```

A second demonstration can be based on what was calculated manually in example 8.16. But there the values of $p(x)$ were available at the five grid points and approximated as averages at the four middle points. If we use the analytic $p(x) = 1 - x^2$, `EPSturmLiouville2` will calculate $p(x)$ exactly at the four midpoints and the eigenvalues returned will be different. So we need to provide the same averages as in the manual example. This also provides the chance to show how `EPSturmLiouville2` works with sampled p, q, w rather than functions.

```
# Define the interval and number of grid points
a <- -1
b <- 1
n <- 4
x <- seq(a,b,length.out=n+1)    # nodes

# p(x) evaluated at nodes
p_node <- 1-x^2                   # same used in the manual example
# 5 rather than 4

# Midpoint values computed by averaging neighboring node values
p_vals <- (p_node[-1]+p_node[-(n+1)])/2  # length = 4

# q and w also at nodes
q_vals <- rep(0,length(x))
w_vals <- 1-x^2

# Call EPSturmLiouville2 with numeric vectors (not functions)
ep <- EPSturmLiouville2(
p=p_vals,q=q_vals,w=w_vals,
x=x,
nev=3,
normalize=TRUE,
return_matrices=TRUE
)

# Print eigenvalues to compare
print(round(ep$values,3))
```

```
## [1]  1.115  6.667 12.551
```

```
# Also print stiffness matrix
print(ep$K)
```

```
##      [,1] [,2] [,3]
## [1,]  5.0 -3.5  0.0
## [2,] -3.5  7.0 -3.5
## [3,]  0.0 -3.5  5.0
```

The first three eigenvalues are exactly equal to those found in example 8.16. The stiffness matrix, which corresponds to matrix A, is also the same. A more accurate result makes use of the exact function $p(x) = 1 - x^2$.

```
# Define the interval and number of grid points
a <- -1
b <- 1
n <- 4  # number of subintervals => n+1 = 5 nodes
x <- seq(a,b,length.out=n+1)

# Define functions
p <- function(s) 1-s^2  # p(x)
q <- function(s) 0  # q(x)
w <- function(s) 1-s^2  # w(x)


# Solve the Sturm-Liouville eigenproblem
ep <- EPSturmLiouville2(p,q,w,x,nev=3,normalize=TRUE)

# Print the eigenvalues
print(round(ep$values,3))


## [1]  1.292  7.333 13.541
```

We can verify that now the first three eigenvalues are slightly different than those calculated manually; in fact, they are more accurate.

8.27 Exercises on EPs

Exercise 09

The stationary Schrödinger equation for a particle of mass m in one dimension is

$$-\frac{\hbar^2}{2m}\frac{\mathrm{d}^2\psi}{\mathrm{d}x^2} + V(x)\psi = E\,\psi.$$

Consider the *infinite square well* potential

$$V(x) = \begin{cases} 0, & 0 < x < L, \\ +\infty, & \text{otherwise}, \end{cases} \qquad \psi(0) = 0,\ \psi(L) = 0.$$

Inside the well ($0 < x < L$), this reduces to the Sturm–Liouville problem

$$\frac{\mathrm{d}^2\psi}{\mathrm{d}x^2} + \lambda\,\psi = 0, \qquad \psi(0) = \psi(L) = 0,$$

with parameter $\lambda = 2mE/\hbar^2$ (λ has dimensions of inverse length squared, L^{-2}; physical energies are recovered via $E = (\hbar^2/2m)\lambda$).

It is known that the eigenpairs are

$$\lambda_n = \left(\frac{n\pi}{L}\right)^2, \quad \psi_n(x) = \sqrt{\frac{2}{L}}\sin\left(\frac{n\pi x}{L}\right), \quad E_n = \frac{\hbar^2\pi^2}{2mL^2}n^2, \qquad n = 1, 2, 3, \ldots$$

1. Use `EPSturmLiouville2` to compute the first four numerical eigenvalues and eigenfunctions on $[0, L]$ with $L = 1$.
2. Compare the numerical eigenvalues with the exact values $\lambda_n = (n\pi)^2$. Report the relative errors for $n = 1, \ldots, 4$.
3. Plot the first four eigenfunctions obtained numerically and overlay the exact sine functions for visual comparison.

Exercise 10

The BVP related to the *modified Legendre equation*,

$$-\frac{\mathrm{d}}{\mathrm{d}x}\left((1 - x^2)\frac{\mathrm{d}y}{\mathrm{d}x}\right) = \lambda(1 - x^2)y, \qquad y(-1) = y(1) = 0.$$

is a Sturm–Liouville problem with homogeneous Dirichlet conditions but non-constant weight function:

$$p(x) = 1 - x^2, \qquad q(x) = 0, \qquad w(x) = 1 - x^2.$$

1. Use `EPSturmLiouville2` to find the first four eigenvalues and eigenvectors of the problem, using a step size $h = 0.01$.
2. Plot the eigenvectors in the interval $[-1, 1]$.
3. Repeat using a coarser grid with $h = 0.1$. Compare eigenvalues and eigenvectors with those of the first numerical solution.

Part III

Computational physics with R

This final part of the book shifts focus from algorithm design and low-level implementation to the *practical use of R and its packages* for solving problems in computational physics. While parts I and II were designed for students aiming to understand and code numerical methods from scratch, this part adopts a different perspective: *how a physicist can use R as a modern computational toolbox.*

The chapters in this part present examples of how computational physics can be approached in R through existing tools, with particular emphasis on the tidyverse. Every effort has been made to base the code and workflows on tidyverse principles, so that data handling and visualisation follow a consistent, modern grammar. The three main themes are Monte Carlo methods, ordinary differential equations, and machine learning. The emphasis is on doing physics with code that already exists, not developing new algorithms or extending the `comphy` package introduced earlier in the book.

There are no exercises in this part, and the code is provided only as demonstration, to support understanding of the underlying physical models and to illustrate typical workflows used in research and applied settings. Readers are expected to have a good grasp of R, or to read chapter 2 to acquire the necessary expertise. They must also have an understanding of the numerical techniques introduced in earlier parts of the book. However, those already familiar with computational physics methods, perhaps through other languages or courses, can start directly from this part to learn *how R, and especially the tidyverse, can be used for physics problems.*

By the end of part III, the reader will have a hands-on overview of how the R ecosystem can be leveraged for computational physics, covering both analytical and simulation-based approaches, and laying the groundwork for further independent exploration.

IOP Publishing

Computational Physics with R

James Foadi

Chapter 9

Monte Carlo methods

9.1 Historical introduction

During the Second World War, while working on the Manhattan Project at Los Alamos, physicists were confronted with the formidable problem of *neutron transport* inside fissionable material. For a nuclear device to work, neutrons produced by fission events had to propagate through a dense medium, colliding with nuclei, sometimes being absorbed, sometimes scattering in new directions, sometimes inducing further fissions.

The difficulty lay in predicting whether a sufficient chain reaction would develop before neutrons escaped from the system. Mathematically, the problem translated into solving integro–differential equations with extremely complicated boundary conditions, accounting for the random motion of neutrons through matter, their scattering angles, and absorption probabilities. Analytic solutions were intractable, and even approximate deterministic numerical schemes were daunting given the limited computational resources of the 1940s[1].

At this juncture, *Stanisław Ulam* had a key insight. Recuperating from illness, he played endless games of solitaire and began wondering about estimating outcomes by sampling many possible card configurations at random rather than trying to enumerate all possibilities exactly. He realized that a similar philosophy could be applied to neutron transport: instead of solving the equations directly, one could simulate the random life history of a neutron. With *John von Neumann*, Ulam developed a systematic way of using the early ENIAC computer to generate sequences of pseudo-random numbers and use them to mimic the stochastic scattering and absorption processes that neutrons undergo. By following many such 'virtual neutrons' through matter, they could obtain statistical estimates of fluxes, escape probabilities, and criticality conditions [3, 4].

[1] For an interesting and anecdotal account of the events of the Manhattan project you can read the interesting books by Rhodes [1] and Metropolis [2].

doi:10.1088/978-0-7503-2632-2ch9

This was the birth of the *Monte Carlo method*, a revolutionary approach in which problems that are deterministic but analytically insoluble are reformulated in probabilistic terms and solved by large-scale random sampling. The method not only addressed the neutron transport question at Los Alamos but also opened an entire field of computational science, still central today in physics, finance, biology, and beyond.

In essence, a *Monte Carlo simulation* is a computational experiment in which a complicated process is represented by a probabilistic model, and outcomes are simulated by a *pseudo-random* generation of data (see later). Rather than attempting to solve deterministic equations directly, one constructs an ensemble of possible 'histories' of the system, each driven by random numbers mimicking the underlying stochastic behaviour. By averaging over many such trials, one obtains approximate answers to questions that would otherwise be analytically inaccessible. The accuracy of Monte Carlo methods increases systematically with the number of samples, and their power lies in their generality: any problem that can be described in probabilistic terms is, in principle, amenable to Monte Carlo treatment. This combination of simplicity, flexibility, and robustness explains why the method has become a cornerstone of modern computational physics.

9.2 A first simple example: the calculation of π

A classic 'hello world' for Monte Carlo is to estimate π by random sampling in the unit square. Consider the quarter of the unit circle

$$\mathcal{C} = \{(x, y) \in \mathbb{R}^2\colon\ 0 \leqslant x \leqslant 1,\, 0 \leqslant y \leqslant 1,\, x^2 + y^2 \leqslant 1\}.$$

Its area is $\pi/4$. Consider also the square

$$\mathcal{S} = \{(x, y) \in \mathbb{R}^2\colon\ 0 \leqslant x \leqslant 1,\, 0 \leqslant y \leqslant 1\}.$$

Its area is 1. Now suppose we generate N random points $(X_1, Y_1), \ldots, (X_N, Y_N)$, each uniformly distributed in $\mathcal{S}$. For each point we can check whether it lies inside the quarter circle, i.e. whether $X_k^2 + Y_k^2 \leqslant 1$. If M of the N points fall inside, then the fraction

$$\hat{p}_N = \frac{M}{N}$$

is an estimator of p. Multiplying by 4 we obtain the *Monte Carlo estimate* of π,

$$\hat{\pi}_N = 4\hat{p}_N\,.$$

By the law of large numbers, the estimate $\hat{\pi}_N$ converges to the true value π as the number of samples N goes to infinity. More precisely, if M denotes the number of points that fall inside the quarter circle, then

$$M \sim \text{Binomial}(N, p), \qquad p = \pi/4.$$

Hence, considering that the random variable $\hat{\pi}_n$ is equal to $4M/N^2$,

$$\mathbb{E}(\hat{\pi}_N) = \frac{4}{N}\,\mathbb{E}(M) = \frac{4}{N}\,N\frac{\pi}{4} = \pi,$$

and

$$\mathrm{Var}(\hat{\pi}_N) = \frac{16}{N^2}\mathrm{Var}(M) = \frac{16}{N^2}Np(1-p) = \frac{16}{N}\frac{\pi}{4}\left(1-\frac{\pi}{4}\right).$$

This shows that the method is unbiased and that the variability of the estimate decreases like $N^{-1/2}$. Furthermore, for large N, the central limit theorem justifies an approximate $(1-\alpha)$ confidence interval of the form

$$\hat{\pi}_N \quad \pm z_{1-\alpha/2}\sqrt{\frac{16}{N}\,\hat{p}_N(1-\hat{p}_N)},$$

where $\hat{p}_N = M/N$ and $z_{1-\alpha/2}$ is the corresponding quantile of the standard normal distribution.

The following **R** code performs the simulation, reports $\hat{\pi}_N$, its estimated standard error, and a 95% confidence interval. For the sake of reproducibility, we fix the seed of the pseudo-random number generator with `set.seed`. This guarantees that the same sequence of random numbers (and hence the same result) will be produced every time the code is run (essential for reproduction in this book). In general applications, however, fixing the seed is not necessary, and one may allow the generator to vary freely to obtain independent replications.

```
# set seed for reproducibility
set.seed(9195)

# Monte Carlo estimator of pi with N samples
mc_pi <- function(N) {
x <- runif(N)
y <- runif(N)
inside <- (x*x+y*y) <= 1
phat <- mean(inside)  # estimate of p = pi/4
pi_hat <- 4*phat # Mean
se <- 4*sqrt(phat*(1-phat)/N) # Standard deviation
ci <- pi_hat+c(-1,1)*1.96*se # 95% confidence interval
list(est=pi_hat,se=se,ci=ci)
}

# Try out estimators with a few N's
N <- 1e4
res <- mc_pi(N)
```

[2] If $X \sim$ Bernoulli(p), then $\mathbb{E}(X) = p$ and $\mathrm{Var}(X) = p(1-p)$. More generally, if $M \sim$ Binomial(N, p), which can be seen as the sum of N independent Bernoulli(p) trials, then $\mathbb{E}(M) = Np$ and $\mathrm{Var}(M) = Np(1-p)$.

```
print(round(c(res$est,res$se,res$ci[1],res$ci[2]),5))
```

```
## [1] 3.14560 0.01639 3.11347 3.17773
```

```
N <- 1e5
res <- mc_pi(N)
print(round(c(res$est,res$se,res$ci[1],res$ci[2]),5))
```

```
## [1] 3.14388 0.00519 3.13371 3.15405
```

```
N <- 1e6
res <- mc_pi(N)
print(round(c(res$est,res$se,res$ci[1],res$ci[2]),5))
```

```
## [1] 3.14304 0.00164 3.13982 3.14626
```

It is clear that with increasing N the variance and, accordingly, the 95% confidence interval, decreases. This means that the estimate is more accurate with increasing N. The drawback is increasing execution time and limits of the random arrays created in the working memory. For this reason, and this must be a priority when implementing Monte Carlo techniques, each specific algorithm will have to be optimised in terms of speed of execution and memory efficiency.

For example, the limit of the previous code is that we cannot keep increasing N because sooner or later we will run into memory limits (you can try this yourself). Alternatives must be found, for example partial executions of the same code can be run on smaller-size arrays (chunks) and the partial results blended cumulatively. An implementation of this idea is provided in the next code snippet.

```
set.seed(9195)

# MC estimator of pi for large N
mc_pi_stream <- function(N,chunk=1e6) {
M_inside <- 0L        # running count of points inside
drawn    <- 0L        # running count of points drawn
while (drawn < N) {
m <- min(chunk,N-drawn)   # size of this chunk
x <- runif(m)
y <- runif(m)
```

```
M_inside <- M_inside+sum(x*x+y*y <= 1)
drawn    <- drawn+m
}
phat   <- M_inside/N
pi_hat <- 4*phat
se     <- 4*sqrt(phat*(1-phat)/N)
ci     <- pi_hat+c(-1, 1)*1.96*se
list(est=pi_hat,se=se,ci=ci,N=N)
}

# Examples

# 1,000,000 draws in chunks of 1e6
t_solve <- system.time({
res <- mc_pi_stream(1e6)
})
print(round(c(res$est,res$se,res$ci[1],res$ci[2]),5))
```

```
## [1] 3.14328 0.00164 3.14007 3.14650
```

```
print(t_solve)
```

```
##    user  system elapsed
##    0.05    0.01    0.06
```

```
# 10,000,000 draws in chunks of 1e6
t_solve <- system.time({
res <- mc_pi_stream(1e7)
})
print(round(c(res$est,res$se,res$ci[1],res$ci[2]),5))
```

```
## [1] 3.14179 0.00052 3.14078 3.14281
```

```
print(t_solve)
```

```
##    user  system elapsed
##    0.64    0.02    0.65
```

```
# 100,000,000 draws in chunks of 1e6
t_solve <- system.time({
res <- mc_pi_stream(1e8)
})
print(round(c(res$est,res$se,res$ci[1],res$ci[2]),5))
```

```
## [1] 3.14180 0.00016 3.14148 3.14212
```

```
print(t_solve)
```

```
##    user  system elapsed
##    8.22    0.58    8.79
```

The execution time, as expected, increases with increasing N, but at least execution is possible even with large arrays, and the accuracy of the π estimate with such large numbers is self-evident: the correct value of π to the first five decimals is $\pi = 3.141\ 59$.

9.3 A second example: the Gaussian integral

We now look at the well-known *Gaussian integral*,

$$I = \int_0^1 e^{-x^2}\, dx.$$

This integral has no closed form in elementary functions. It is instead expressed in terms of the *error function*, defined as

$$\text{erf}(x) = \frac{2}{\sqrt{\pi}} \int_0^x e^{-t^2}\, dt.$$

With this definition we can write

$$I = \int_0^1 e^{-x^2}\, dx = \frac{\sqrt{\pi}}{2}\, \text{erf}(1) \approx 0.746\ 824.$$

Monte Carlo methods can be used to calculate numerically integrals like this. An initial approach uses the same argument applied for the calculation of π. In fact, we can think of this integral as the area under the curve $y = e^{-x^2}$ between 0 and 1[3]. If we draw points (X_i, Y_i) uniformly from the unit square $\mathcal{S}$, then the fraction that land below the curve (i.e. those with $Y_i \leqslant e^{-X_i^2}$) provides an estimate of I. This approach is simple and intuitive, but it only records yes/no information (point below or above

[3] The function e^{-x^2} takes its highest value, 1, at $x = 0$ and it is decreasing from there to its lowest value $e^{-1} \approx 0.368$, so the area under it is well within the unit square $\mathcal{S}$.

the curve), so the variability of the estimates is similar to the one found for the estimate of π, which is relatively high for Monte Carlo techniques.

There is a more efficient way to use uniform samples. If X is uniform on [0, 1], then the integral we want to calculate is exactly equal to the expected value of e^{-X^2}:

$$E(e^{-X^2}) = \int_0^1 e^{-x^2}\,dx \equiv I.$$

So instead of generating pairs (X, Y), we only need to generate an i.i.d. sample, $\{X_i\}$ uniformly distributed between 0 and 1, and average the values e^{-X^2}:

$$\hat{I}_N = \frac{1}{N}\sum_{i=1}^{N} e^{-X_i^2},$$

This estimator is unbiased and makes fuller use of the information in the function values, leading to much lower variance than the hit-or-miss method.

The function e^{-x^2} is largest near $x = 0$ and decreases quickly as x increases to 1. Uniform sampling spreads points evenly across the interval, which means many samples fall where the function is small and contribute little. A better idea is to sample more points near 0, where the function contributes most to the integral. A convenient choice is found to be an exponential distribution truncated to [0, 1]. Its density is

$$g(y) = \frac{\lambda e^{-\lambda y}}{1 - e^{-\lambda}}, \qquad 0 \leqslant y \leqslant 1,$$

where $\lambda > 0$ controls how concentrated the samples are near 0. To keep the estimate unbiased, we correct for the new sampling law by reweighting:

$$I = \int_0^1 \frac{e^{-y^2}}{g(y)}\, g(y)\,dy = \mathbb{E}_g\left(\frac{e^{-Y^2}}{g(Y)}\right),$$

so with $Y_1, \ldots, Y_N \sim g$, the estimator becomes

$$\hat{I}_N = \frac{1}{N}\sum_{i=1}^{N} \frac{e^{-Y_i^2}}{g(Y_i)}.$$

This estimate is still unbiased, but it typically has much smaller variance because the sampling distribution g directs more points where the integrand is largest.

To summarise, we have three unbiased Monte Carlo estimators of the same integral (variances have been calculated in section A.15 of appendix A).

1. Hit-or-miss with uniform points in the square. The variance (highest of the three) is:

$$\mathrm{Var}(\hat{I}_N) \approx \frac{0.189\,08}{N}.$$

2. Averaging e^{-X^2} with uniform X. The variance (smaller than for case 1) is:

$$\mathrm{Var}(\hat{I}_N) \approx \frac{0.040\,40}{N}.$$

3. Reweighted averaging with truncated exponential samples. The variance (the lowest of the three), when the parameter λ in $g(y)$ is 1, yields:

$$\mathrm{Var}(\hat{I}_N) \approx \frac{0.003\,03}{N}.$$

From the formula presented, it is clear that in Monte Carlo methods, variability (and thus accuracy) can be governed with a judicious choice of sampling distribution functions.

The following R code provides a practical demonstration of what is described above. It carries out several Monte Carlo simulations (300 in this specific case) to estimate the variance empirically.

```
# set seed for reproducibility
## Monte Carlo for I = int_0^1 exp(-x^2) dx
## Focus: comparing variances across three estimators
##    (1) Hit-or-miss (uniform on the unit square)
##    (2) Uniform averaging of exp(-X^2), X ~ Unif[0,1]
##    (3) Reweighted averaging with truncated exponential
##        g_lambda on [0,1] (lambda = 1)

set.seed(20250826) # reproducibility for the demo
# (not required in general)

# Error function via the Normal CDF. Handy to have one ready.
erf <- function(x) 2*pnorm(x*sqrt(2))-1

# True value of the integral
I_true <- sqrt(pi)/2*erf(1)
print(I_true)


## [1] 0.7468241


## Theoretical variance "constants" c so that Var(estimator) = c/N
# (1) Hit-or-miss: Bernoulli with p = I_true
c_hit <- I_true*(1-I_true)

# (2) Uniform averaging: Var(exp(-X^2)), X ~ Unif[0,1]
```

```
E_e2   <- sqrt(pi)/(2*sqrt(2))*erf(sqrt(2))
c_unif <- E_e2-I_true^2

# (3) Truncated exponential g_lambda on [0,1], here lambda = 1
c3_int <- {
a <- 3*sqrt(2)/4
b <- sqrt(2)/4
pref <- (1-1/exp(1))*(sqrt(pi)/(2*sqrt(2)))*exp(1/8)
pref*(erf(a)+erf(b))
}
c_trun <- c3_int-I_true^2

# Show the theoretical constants for reference
print(round(c(hit_or_miss=c_hit,
uniform_avg=c_unif,
trunc_exp=c_trun),5))


## hit_or_miss uniform_avg   trunc_exp
##     0.18908     0.04040     0.00303


# Now the actual Monte Carlo simulation: here done several
# times (i.e. we produce several estimates of the integral
# hat{I}_N because we also need to estimate the variance
# and check it's close to the theoretical values).
#-------------------------------
# Simulation parameters
#-------------------------------
N <- 5e4      # samples per replication
B <- 300      # number of independent replications
# (for variance estimation)

# Containers for the three different methods
est_hit  <- numeric(B)
est_unif <- numeric(B)
est_trun <- numeric(B)

# Inverse CDF sampler for truncated exponential on [0,1]
# (to be explained later on in the text book)
rinv_truncexp01 <- function(n,lambda) {
u <- runif(n)
-log(1-u*(1-exp(-lambda)))/lambda
}
```

```
# Density g_lambda(y) on [0,1]
dtruncexp01 <- function(y,lambda)
lambda*exp(-lambda*y)/(1-exp(-lambda))

#--------------------------------
# Replications
#--------------------------------
for (b in seq_len(B)) {
# (1) HIT-OR-MISS using uniform points in the unit square
x1 <- runif(N)
y1 <- runif(N)
est_hit[b] <- mean(y1 <= exp(-x1^2))

# (2) UNIFORM AVERAGING: mean of exp(-X^2), X ~ Unif[0,1]
x2 <- runif(N)
est_unif[b] <- mean(exp(-x2^2))

# (3) TRUNCATED EXPONENTIAL (lambda = 1): reweighted average
lambda <- 1
y3 <- rinv_truncexp01(N,lambda)
g3 <- dtruncexp01(y3,lambda)
w3 <- exp(-y3^2)/g3
est_trun[b] <- mean(w3)
}

# Empirical variances across replications
var_hit_emp  <- var(est_hit)
var_unif_emp <- var(est_unif)
var_trun_emp <- var(est_trun)
print(cbind(var_hit_emp=var_hit_emp,
var_unif_emp=var_unif_emp,
var_trun_emp=var_trun_emp))
```

```
##       var_hit_emp var_unif_emp var_trun_emp
## [1,] 3.826277e-06 7.866993e-07 6.010458e-08
```

```
# Theoretical estimates
print(cbind(theor_hit=c_hit/N,
theor_unif=c_unif/N,
theor_trun=c_trun/N))
```

```
##         theor_hit   theor_unif   theor_trun
```

```
## [1,] 3.781557e-06 8.079544e-07 6.053353e-08
```

The empirical values measured for the variances are quite close to their theoretical estimate. When investigating a process with a Monte Carlo method, it is advisable to run several simulations (like with the above code) and produce empirical estimates of the variances. These will then be normally used to calculate confidence intervals.

We have not printed the integral's value and its confidence interval. We can do that using the tidyverse (see section 2.21), to practice with this modern and widespread R tool. As explained, the primary component to start the output process is a *tibble*, i.e. a table containing the data. Here a tibble including lower extreme, mean, and upper extreme of the confidence interval, is constructed. As tibbles have standardised formats for print out, here only three decimal digits are displayed, which is clearly not enough for what we are doing. We can then use `mutate` to change the formatting. When we do that, double precision numbers will be changed into character strings, as it is clear by looking at the heading of the columns: `<dbl>` is changed into `<chr>`.

```
# Compute means
m_hit  <- mean(est_hit)
m_unif <- mean(est_unif)
m_trun <- mean(est_trun)

# Means and 95% confidence intervals
ci_hit <- m_hit+c(-1,1)*1.96*sqrt(var_hit_emp/B)
ci_unif <- m_unif+c(-1,1)*1.96*sqrt(var_unif_emp/B)
ci_trun <- m_trun+c(-1,1)*1.96*sqrt(var_trun_emp/B)

# Assemble into a tibble and print
ci_tbl <- tibble(
method = c("Hit",
"Uniform",
"Truncated"),
lower  = c(ci_hit[1],  ci_unif[1],  ci_trun[1]),
mean   = c(m_hit,      m_unif,      m_trun),
upper  = c(ci_hit[2],  ci_unif[2],  ci_trun[2])
) |>
print()
```

```
## # A tibble: 3 x 4
##   method    lower  mean upper
##   <chr>     <dbl> <dbl> <dbl>
## 1 Hit       0.747 0.747 0.747
## 2 Uniform   0.747 0.747 0.747
```

```
## 3 Truncated 0.747 0.747 0.747
```

```
# Increase digits to display
ci_tbl |>
mutate(across(where(is.numeric),
~ formatC(.x,format="f",digits=7))) |>
print(n=Inf)
```

```
## # A tibble: 3 x 4
##   method    lower     mean      upper
##   <chr>     <chr>     <chr>     <chr>
## 1 Hit       0.7466007 0.7468221 0.7470434
## 2 Uniform   0.7467223 0.7468226 0.7469230
## 3 Truncated 0.7467935 0.7468212 0.7468489
```

The results clearly show that all Monte Carlo methods calculate the integral with some accuracy, and that the accuracy depends on the method chosen.

9.4 The elements of Monte Carlo methods

The two examples we have just discussed already capture the essence of what a Monte Carlo method is about. Although they differ in form, both share the same broad structure: they rely on the use of random numbers to produce samples from a model, and on statistical ideas to turn those samples into meaningful estimates. In fact, all Monte Carlo methods can be understood as a combination of a few fundamental elements. These are:

1. **Random number generation**: the production of uniform pseudo–random numbers and their transformation into values distributed according to a desired probability law.
2. **Sampling, estimation, and uncertainty quantification**: the use of random numbers to reproduce the characteristics of the process under study. Starting from these numbers, various estimates of interest will be calculated, together with an assessment of their accuracy.
3. **Efficiency and variance reduction**: methods that make simulations converge faster or require fewer samples.
4. **Reporting results**: presenting estimates together with a measure of their uncertainty, and making clear what conclusions can and cannot be drawn from them.

In the following sections, we will look at each of these elements in turn.

9.5 Random number generation

Monte Carlo simulations rely on the production of random numbers. Since computers are deterministic machines, they cannot produce truly random values on their own. Instead, they generate sequences of numbers that imitate randomness, called *pseudo–random numbers*. These sequences are produced by deterministic algorithms that start from an initial value, the *seed*, and evolve according to a recurrence rule. The seed determines the entire sequence: using the same seed leads to the same numbers being generated again.

For a Monte Carlo simulation, the starting point is usually the uniform distribution on the interval [0, 1]. Once uniform numbers are available, they can be transformed into numbers distributed according to many other probability laws.

An important practical issue is reproducibility. Setting the seed ensures that the same pseudo–random sequence can be reproduced on the same system. However, this does not guarantee that identical sequences will be obtained on all computers, even if the same seed is used. The reason is that different versions of R may use different algorithms as their default random number generator. To illustrate this, consider the following simple experiment:

```
set.seed(123)
runif(5)

## [1] 0.2875775 0.7883051 0.4089769 0.8830174 0.9404673

R.version.string

## [1] "R version 4.5.1 (2025-06-13 ucrt)"

RNGkind()

## [1] "Mersenne-Twister" "Inversion"        "Rejection"
```

On one computer this may produce a certain sequence of five numbers, while on another computer with a different R version installed the sequence may be different, despite the same seed. The command `R.version.string` (or `getRversion`) reveals which version of R is being used, while `RNGkind` shows which generator is currently active. If two systems have different defaults, the same seed leads to different sequences.

To enforce reproducibility across platforms, one can explicitly specify the generator, for example:

```
RNGkind(kind = "Mersenne-Twister", normal.kind = "Inversion")
set.seed(123)
runif(5)
```

```
## [1] 0.2875775 0.7883051 0.4089769 0.8830174 0.9404673
```

By doing this, the same sequence will be obtained across machines that implement the same generator.

In the next subsections we will discuss in more detail the uniform random number generator, and the acceptance–rejection technique to generate pseudo–random numbers from specific probability distributions.

9.5.1 The uniform random number generator

The cornerstone of all Monte Carlo simulations is the ability to generate numbers that are uniformly distributed in the interval [0, 1]. In practice this is achieved using a deterministic recurrence. One of the simplest and most widely used algorithms for this is the *linear congruential generator* (LCG). It produces a sequence of integers $\{X_n\}$ according to the recurrence

$$X_{n+1} = (aX_n + c) \mod m, \tag{9.1}$$

where a, c, and m are integer parameters. The output is then scaled to the unit interval by setting

$$U_{n+1} = \frac{X_{n+1}}{m}.$$

For example, if we use $a = 3$, $c = 1$, $m = 10$ and start from seed 4, we obtain the sequence

$$0.3, 0, 0.1, 0.4, 0.3, 0, 0.1, 0.4, \ldots$$

which only generates four pseudo–random numbers. Here obviously a, c, and m are ridicolously small and that explains why only four different numbers are generated. In fact, the modulus m fixes the maximum size of the sequence before it repeats, while the multiplier a and increment c control how evenly the numbers are spread across the interval. If $a = 131$, $c = 1$, $m = 20$ were chosen in the previous example and we started the sequence still with seed 4, the sequence would have become:

$$0.25, 0.80, 0.85, 0.40, 0.45, 0.00, 0.05, 0.60, 0.65, 0.20, 0.25, 0.80, 0.85, \ldots$$

where 10, rather than 4, pseudo–random numbers are generated. With good parameter choices the sequence cycles only after a very long period and has statistical properties close to those of independent uniforms; with poor choices, the values may fall into visible patterns or short cycles, making the generator unsuitable for Monte Carlo use.

Modern generators improve on the LCG in various ways to increase the period (the length of the sequence before it repeats) and to enhance statistical properties. In R, the default generator is based on the *Mersenne Twister* algorithm [5], which has a very long period ($2^{19937} - 1$) and good uniformity properties.

A clear way to assess quality is to inspect successive pairs (U_n, U_{n+1}). To make artifacts visibly obvious here, we implement a deliberately poor LCG with a *tiny modulus* and small multiplier so that strong regularity/correlation appears already in 2D. (These parameters are intentionally unrealistic and chosen only to accentuate the effect.)

```
# Modern R; use ggplot2
require(ggplot2)

# Bad LCG (RANDU-like)
bad_lcg <- function(n,seed=1L,a=131L,c=7L,m=2^14) {
x <- numeric(n+1L)
x[1] <- seed %% m
for (i in 1:n) {
x[i+1] <- (a*x[i]+c) %% m
}
u <- x[-1L]/m

return(u)
}

set.seed(123)
N <- 5000

# Good generator
u_good <- runif(N)

# Bad generator
u_bad <- bad_lcg(N,seed=123)

# Build successive pairs
pairs_from <- function(u) {
data.frame(U_n=u[-length(u)],U_next=u[-1L])
}
df_good <- pairs_from(u_good)
df_bad  <- pairs_from(u_bad)

# Add source of generation and combine data for later plotting
```

```
df_good$Source <- "R default (good)"
df_bad$Source  <- "Bad LCG (unrealistic m,a,c)"
df <- rbind(df_good,df_bad)

# ggplot2 side-by-side plots with facet_wrap
ggplot(df,aes(x=U_n,y=U_next)) +
geom_point(size=0.3,alpha=0.6) +
coord_fixed(ratio=1) +
facet_wrap(~ Source,ncol=2,scales="fixed") +
labs(x=expression(U[n]),y=expression(U[n+1])) +
theme_minimal() +
theme(strip.text=element_text(hjust=0.5)) # centers facet titles
```

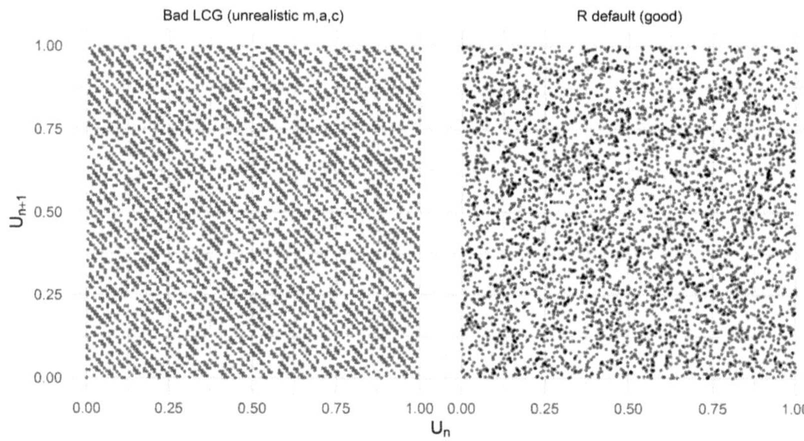

The unit square is regularly covered by the pseudo-random generated points, differently than in the case of the Mersenne Twister algorithm, thus exposing the random generation to a dangerously systematic pattern.

The function `runif` also allows generation of uniforms on an arbitrary interval $[a, b]$ by specifying the arguments `min=a` and `max=b`.

9.5.2 Non-uniform generation

Once uniform pseudo-random numbers are available, the next step is to transform them into values that follow a different distribution. Many specialised algorithms exist for this task: for example, the normal distribution can be sampled very efficiently by the Box–Muller method [6] or by more sophisticated transformations. The random number generators included in R are up to date with the most widely used techniques, so that users can simply call functions such as `rnorm`, `rexp`, or `rgamma` to obtain samples from the corresponding laws. Here, however, we describe one general-purpose method that can be applied when direct transformations are not obvious: the *acceptance–rejection method.*

This is a versatile procedure that uses uniform random numbers together with a proposal distribution that is easy to sample from. Suppose we wish to generate samples from a distribution with density $f(x)$, but it is not straightforward to obtain $X \sim f$ directly. We proceed as follows:

1. Choose a proposal density $g(x)$ from which it is easy to generate samples, and a constant M such that

$$f(x) \leqslant Mg(x) \quad \text{for all } x.$$

2. Generate $Y \sim g$ and $U \sim \text{Unif}(0, 1)$ independently.
3. Accept Y as a sample from f if $U \leqslant \frac{f(Y)}{Mg(Y)}$; otherwise, reject Y and repeat the procedure.

It is not difficult to see why the accepted values follow the target density $f(x)$. When a candidate value Y is generated from the proposal distribution g, the probability that it falls near some point x is proportional to $g(x)$. The rule of the algorithm is to accept this candidate with probability $f(x)/(Mg(x))$. Hence the overall chance that a value near x is both proposed and accepted is proportional to $g(x)f(x)/(Mg(x)) = f(x)/M$. This quantity no longer depends on g and, apart from the constant factor $1/M$, has the same shape as $f(x)$. After renormalisation, to ensure total probability one, the distribution of the accepted values coincides exactly with $f(x)$.

The efficiency of the method depends on the choice of g and M: the closer $Mg(x)$ is to $f(x)$, the higher the acceptance rate. In practice, acceptance–rejection is used both as a stand–alone technique and as a building block within more elaborate Monte Carlo algorithms.

In the following R demonstration we are going to use the acceptance-rejection method to generate pseudo–random numbers from the distribution

$$f(x) = \begin{cases} x + 1 & \text{for } \; -1 \leqslant x < 0 \\ -x + 1 & \text{for } \quad 0 \leqslant x < 1 \\ 0 & \quad\;\; \text{otherwise.} \end{cases}$$

Any probability density that covers the $[-1, 1]$ interval can be chosen. Given its simplicity, we will use the uniform distribution,

$$g(x) \sim \text{Unif}(-1, 1).$$

It does not make the generation mechanism very effective because its shape does not follow the triangular shape of $f(x)$ closely, but for this demonstration it will do. With this choice of $g(x)$, it is clear that

$$f(x) \leqslant Mg(x) \quad \text{for all } x$$

when $M \geqslant 2$. $M = 2$ is the quantity that minimises rejection and so it characterises the optimal choice for this proposal, $g(x)$. In the code presented, though, M must be found empirically (inside the algorithm) and extra care must be taken if the target and proposal distributions are very close. In such a circumstance, it might accidentally (due to statistical fluctuations) happen that a significant representation of uniformly-generated values is not close enough to the extremes of the target distribution. By allowing an extra margin (see R code) one makes sure that such a bias never presents itself.

After the independent generation of Y from the distribution $g(y)$ and U from the uniform distribution Unif(0, 1), Y is accepted if $U \leqslant f(Y)/(Mg(Y))$. Accepted numbers are progressively stored in an array, until the array is completely filled. It logically takes many more random generations than the number of values wanted but this number is limited by the shape of the proposal: the closer $g(y)$ is to the target function, the smaller this number will be.

For demonstration purposes, the code presented displays the total number of points created, divided into accepted and rejected. To be consistent with the goal of part III of the book of using the tidyverse (see section 2.21), the plot is done by `ggplot` and uses the standard accessories connected with it.

```
suppressPackageStartupMessages({
library(ggplot2)
library(viridis)    # for scale_color_viridis_c
})

## Target: triangular density on [-1,1]. f(x)
f_tri <- function(x) {
y <- numeric(length(x))
y[x >= -1 & x < 0] <- x[x >= -1 & x < 0]+1
y[x >= 0 & x <= 1] <- -x[x >= 0 & x <= 1]+1

return(y)
}
```

```
## Proposal: g(x) ~ Unif(-1,1)
g_dens <- function(x) ifelse(abs(x) <= 1,0.5,0)

## Compute minimal M >= sup_x f(x)/g(x) over support [-1,1]
ratio <- function(x) {
fx <- f_tri(x); gx <- g_dens(x)
out <- ifelse(gx > 0,fx/gx,0)

return(out)
}
xs <- seq(-1,1,length.out=20001)
M_est <- max(ratio(xs))
M <- 1.05*M_est   # small safety margin
print(M)
```

```
## [1] 2.1
```

```
## Acceptance{rejection draw of N candidates (for visualisation)
set.seed(9123)
N <- 8000
Y <- runif(N,min=-1,max=1)
U <- runif(N)
alpha <- f_tri(Y)/(M*g_dens(Y)) # acceptance probability
accept <- (U <= alpha)

## For the standard \under the curve" picture,
## plot points at height U * M * g(Y)
df <- data.frame(
y=Y,
uMg=U*M*g_dens(Y),
accept = accept
)

## Curves for f and M g
xx  <- seq(-1.5,1.5,length.out=1000)
cur <- data.frame(
x=xx,
f=f_tri(xx),
Mgc=M*g_dens(xx)
)

## Plot: accepted in magenta; rejected coloured with viridis
p <- ggplot() +
```

```
geom_line(data=cur,aes(x=x,y=Mgc),linetype=2) +
geom_line(data=cur,aes(x=x,y=f),linewidth=1) +
geom_point(data=subset(df,!accept),
aes(x=y,y=uMg,colour=y),
size=0.5,alpha=0.6,na.rm=TRUE) +
scale_color_viridis_c(option="viridis",end=0.9,
name="Rejected\nY") +
geom_point(data=subset(df,accept),
aes(x=y,y=uMg),
colour="magenta",size=0.5,alpha=0.7,na.rm=TRUE) +
coord_cartesian(xlim=c(-1.5,1.5),
ylim=c(0,max(cur$Mgc,1))) +
labs(x="y (candidate from g)",y="u * M * g(y)",
title =
"Acceptance{rejection for triangular target f(x) on [-1,1]",
subtitle =
"Accepted (magenta) under f(x); rejected (viridis) under M g(x)") +
theme_minimal()

p
```

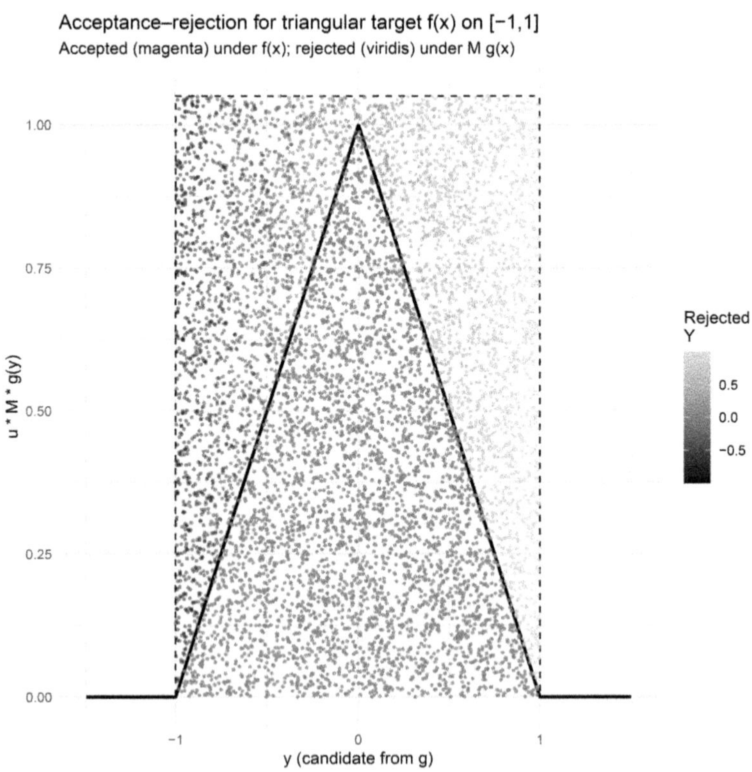

In real applications, the above demonstration, or similar code, aims at understanding and testing the correct code for the random generation. Once this is validated (via numerical results and/or plots), the actual code is encapsulated into a function that becomes the actual pseudo-random generator for the target distribution function. This function is described, together with a small demonstration of its use, in the code below.

```
# After the previous demonstration, here is the actual pseudo-
# random number generator based on acceptance-rejection
rtri_ar <- function(n) {
out <- numeric(n);
k <- 0L

# Keep going until vector out is full
while (k < n) {
m <- min(n-k,1e6L)
y <- runif(m,-1,1)
u <- runif(m)

# Accepted points
acc <- (u <= f_tri(y))  # acceptance prob = f(y)

# Transfer accepted points to vector out
if (any(acc)) {
keep <- y[acc]
nk <- min(length(keep),n-k) # Cannot overspill "out"
out[(k+1L):(k+nk)] <- keep[seq_len(nk)]
k <- k+nk  # Update reference to fill "out"
}
}
return(out)
}

# Example of generation
set.seed(42)
samples <- rtri_ar(5)
round(samples,5)
```

```
## [1] -0.42772  0.28349 -0.08452  0.43822 -0.05001
```

9.6 Sampling, estimation, and uncertainty quantification

Once pseudo-random numbers are generated from the model of interest (specific target distribution), the next ingredient of a Monte Carlo method is sampling: drawing a finite collection of realisations and using them to approximate quantities we care about, like expectations, probabilities, quantiles, integrals. The process of drawing a sample is performed by a Monte carlo simulation. This shows what the advantage of Monte Carlo methods is, that they replace potentially expensive real sampling with simulated sampling.

In sampling, we face one or more random variables with probability distributions that are known as the model (or target distribution). The process studied with Monte Carlo methods typically includes a function of the random variables, with finite variance under the target distribution. Sampling consists of selecting N independent random variables with that same target probability distribution.

Example 9.1 *In section* 9.3, *one way to calculate the Gauss integral consisted in sampling from a uniform distribution between 0 and 1. We selected N independent random variables,* $X_1, \ldots, X_N$, *all with target distribution* Unif(0, 1). *Another way involved selecting N independent and identically distributed (i.i.d.) random variables,* $Y_1, \ldots, Y_N$, *from a different target distribution,* $g(Y) = \lambda e^{-\lambda Y}/(1 - e^{-\lambda})$.

The sample is then used to estimate a key quantity of the specific problem under study. This quantity is a function of the random variables of the problem, although it does not necessarily possess an explicit analytic expression.

Example 9.2 *Still referring to section* 9.3, *the key quantity of interest corresponding to the sample choice* $X_1, \ldots, X_N$ *with* $X_i \sim$ Unif(0, 1), *was*

$$h(X) = e^{-X^2}.$$

For the sample choice $Y_1, \ldots, Y_N$ *with* $Y \sim g(Y)$, *the key quantity of interest was*

$$h(Y) = \frac{e^{-Y^2}}{g(Y)}.$$

The key quantity must be estimated using the sample selected. The most important idea of Monte Carlo methods is that this estimate reveals the true value of the key quantity, with some uncertainty.

Example 9.3 *Continuing with reference to section* 9.3, *in the first case the estimate is*

$$\mathbb{E}(h(X)) = \frac{1}{N}\sum_{i=1}^{N} e^{-X_i^2}.$$

In the second case it is

$$\mathbb{E}(h(Y)) = \frac{1}{N}\sum_{i=1}^{N}\frac{e^{-Y_i^2}}{g(Y_i)}.$$

These estimators, generally called *Monte Carlo estimators*, are themselves random variables subject to variability. There will be, therefore, a certain uncertainty connected with the evaluation of the quantity of interest. In fact, the central limit theorem states that these estimators are normally distributed, with a mean and variance depending on the details of the process.

Example 9.4 *In section* 9.3, *the evaluation of* $h(X)$ *(which coincided with the value of the Gauss integral) using various simulations (each one obtained with samples of size* $N = 50\ 000$), *returned an estimate equal to* 0.746 82, *with an uncertainty represented by a 95% confidence interval equal to* [0.746 72, 0.746 92]. *Similarly, the evaluation of* $h(Y)$ *returned an estimate equal to* 0.746 82 *and an uncertainty represented by a 95% confidence interval equal to* [0.746 79, 0.746 85].

When dealing with simulations in which the random variables can take just two values, FALSE/TRUE or 0/1, a final sample is still selected, but the estimate and uncertainty are here governed by the Bernoulli and binomial distributions.

Example 9.5 *When considering the calculation of* π *in section* 9.2, *although we generated two independent samples from uniform distributions, the random variable of interest is the 'hit or miss' the quarter of the circle. In section* 9.2 *it was not specifically declared, but we can do it here and define a sample of N Bernoulli variables,* $L_1, \ldots, L_N$, *where each* L_i *can have outcome 0 ('miss') or 1 ('hit'). In the section, we defined a variable* M, *which is the number of points falling inside the circle. We have therefore*

$$M = L_1 + \cdots + L_N \sim \text{Binomial}(N, p), \ , p = \frac{\pi}{4}.$$

The key quantity of interest was

$$h(M) = 4\frac{M}{N} \equiv \hat{\pi}.$$

The estimate of this quantity yields[4]:

$$\mathbb{E}(h(M)) = \frac{4}{N}\mathbb{E}(M) = \frac{4}{N}Np = \frac{4}{N}N\frac{\pi}{4} = \pi.$$

The uncertainty can be quantified recalling the expression for the variance of a binomial variable:

[4] Recall that M is a binomial variable.

$$\mathrm{Var}(h(M)) = \mathrm{Var}\left(4\frac{M}{N}\right) = \frac{16}{N^2}\mathrm{Var}(M) = \frac{16}{N^2}Np(1-p) = \frac{16}{N}\frac{\pi}{4}\left(1-\frac{\pi}{4}\right).$$

Due to the central limit theorem, $h(M)$ is normally distributed and, accordingly, confidence intervals can be used to quantify its uncertainty.

The hardest part of Monte Carlo work is often not the average itself, but working out how uncertain that average is. With a large number of samples, things are easier: the central limit theorem tells us that the result will usually look normal, and we can estimate the variance directly from the data. The real trouble comes when the samples are not independent. Correlations can make the spread of results look smaller or larger than it really is. In those cases, one has to be more careful, for example by grouping data into blocks or using other tricks to get a trustworthy measure of the uncertainty

9.7 Efficiency and variance reduction

A Monte Carlo method is *more efficient* if it delivers the same accuracy with less computing, or more accuracy with the same computing. A good practical way of framing efficiency is to think of accuracy as the tightness of error bars, and of computing as CPU time, memory, or simply the number of samples N. More specifically:

- **Accuracy**. For many Monte Carlo estimators, the *standard error* (SE) characterises accuracy. It is approximately defined as:

$$\mathrm{SE} \approx \frac{\sigma}{\sqrt{N}}$$

  where σ is the per-sample standard deviation. If each draw can be made less noisy (by reducing σ), the standard error shrinks faster. As a rule of thumb, halving σ typically means one needs about 4 times fewer samples to reach the same SE.
- **Cost**. Faster code, vectorisation, good random-number usage, and parallel runs reduce the time to generate and process each draw. This improves efficiency even if the variance per sample stays the same.

Most of the time, making Monte Carlo methods more efficient means reducing variance, i.e. getting tighter error bars without simply throwing more samples at the problem. A variety of techniques are available to achieve that [7] but here we will only discussed *importance sampling*. The remaining part of this section contains a quantitative treatment of efficiency and contains many equations. The reader who is not inclined to an equations-ridden paragraph can skip it and just remember that an appropriate choice of the proposal distribution, $g(y)$, can reduce variance and make the method more efficient.

To simplify the treatment, let us consider a Monte Carlo method with one variable. The generic Monte Carlo estimator is

$$\theta = \mathbb{E}_f(h(X)) = \int_{\mathcal{D}} h(x)f(x)\,\mathrm{d}x,$$

where the index f refers to a target density on the domain $\mathcal{D}$ and h is the key quantity of interest. A plain Monte Carlo simulation draws a sample directly from f, that is $X_i \sim f$. With importance sampling, we draw samples from an easier proposal density g (instead of f), and correct the definition of estimate by a weight:

$$w(x) = \frac{f(x)}{g(x)}. \tag{9.2}$$

Accordingly, the *importance-sampling* (IS) *estimator* is defined as

$$\hat{\theta}_{\text{IS}} = \frac{1}{N}\sum\nolimits_{i=1}^{N} h(Y_i)\, w(Y_i), \qquad Y_i \overset{\text{i.i.d.}}{\sim} g. \tag{9.3}$$

The IS estimator is unbiased because:

$$\mathbb{E}_g(h(Y)w(Y)) = \int h(x)f(x)\,\mathrm{d}x = \theta.$$

The variance is given by:

$$\text{Var}(\hat{\theta}_{\text{IS}}) = \frac{1}{N}(\mathbb{E}_g[h(Y)^2 w(Y)^2] - \theta^2).$$

Thus it follows that the efficiency improves when g places more mass where $|h(x)|f(x)$ is large[5].

Example 9.6 *We now revisit the example of section* 9.3 *in light of the concepts just introduced. The initial target density is*

$$f(x) = \text{Unif}(0, 1),$$

while the key important quantity $h(x)$ *is*

$$h(x) = \mathrm{e}^{-x^2}, \qquad \text{with domain } \mathcal{D} = [0, 1].$$

Thus the plain Monte Carlo estimator is

$$\hat{\theta}_{\text{unif}} = \frac{1}{N}\sum\nolimits_{i=1}^{N} \mathrm{e}^{-X_i^2}, \qquad X_i \overset{\text{i.i.d.}}{\sim} \text{Unif}[0, 1].$$

To enact importance sampling, we choose the proposal density, still on domain $\mathcal{D}$,

$$g(y) = \frac{\lambda \mathrm{e}^{-\lambda y}}{1 - \mathrm{e}^{-\lambda}}, \qquad \lambda > 0,$$

[5] Indeed,

$$\mathbb{E}_g(h(Y)^2 w(Y)^2) = \int_{\mathcal{D}} h(x)^2 \frac{f(x)^2}{g(x)}\,\mathrm{d}x.$$

Thus increasing $g(x)$ exactly where $|h(x)|f(x)$ is large reduces the integrand $h(x)^2 f(x)^2/g(x)$ in those regions (the large numerator is divided by a larger g), while making g small there would inflate the term and blow up the variance.

which concentrates draws near 0*, where* $|h(x)|f(x) = \mathrm{e}^{-x^2}$ *is largest. Adopting*

$$w(y) = \frac{f(y)}{g(y)},$$

the IS estimator becomes

$$\hat{\theta}_{\mathrm{IS}} = \frac{1}{N}\sum_{i=1}^{N} h(Y_i)\, w(Y_i) = \frac{1}{N}\sum_{i=1}^{N} \frac{\mathrm{e}^{-Y_i^2}}{g(Y_i)}, \qquad Y_i \overset{\text{i.i.d.}}{\sim} g.$$

This estimator is unbiased because

$$\mathbb{E}_g[h(Y)w(Y)] = \int_0^1 h(y)\frac{f(y)}{g(y)}g(y)\,\mathrm{d}y = \int_0^1 h(y)f(y)\,\mathrm{d}y = \frac{\sqrt{\pi}}{2}\mathrm{erf}(1) \equiv \theta.$$

Let us now turn to the variance. We have:

$$\mathrm{Var}(\hat{\theta}_{\mathrm{IS}}) = \frac{1}{N}(\mathbb{E}_g(h^2w^2) - \theta^2).$$

This expression can be re-written as:

$$\mathrm{Var}(\hat{\theta}_{\mathrm{IS}}) = \frac{1}{N}(C(\lambda) - \theta^2), \qquad C(\lambda) = \int_0^1 \frac{\mathrm{e}^{-2y^2}}{g(y)}\,\mathrm{d}y,$$

that is, we focus on the new quantity $C(\lambda)$*. Completing the square yields an explicit closed form which has been already calculated in section* A.15 *of appendix* A*. Thus*:

$$\begin{aligned} C(\lambda) &= \frac{1 - \mathrm{e}^{-\lambda}}{\lambda}\mathrm{e}^{\lambda^2/8}\int_0^1 \mathrm{e}^{-2(y-\lambda/4)^2}\,\mathrm{d}y \\ &= \frac{1 - \mathrm{e}^{-\lambda}}{\lambda}\frac{\sqrt{\pi}}{2\sqrt{2}}\,\mathrm{e}^{\lambda^2/8}\left(\mathrm{erf}\left(\sqrt{2}\left(1 - \frac{\lambda}{4}\right)\right) - \mathrm{erf}\left(-\sqrt{2}\,\frac{\lambda}{4}\right)\right). \end{aligned}$$

Therefore,

$$\mathrm{Var}(\hat{\theta}_{\mathrm{IS}}) = \frac{1}{N}\left(\frac{1 - \mathrm{e}^{-\lambda}}{\lambda}\frac{\sqrt{\pi}}{2\sqrt{2}}\,\mathrm{e}^{\lambda^2/8}\left[\mathrm{erf}\left(\sqrt{2}\,(1 - \frac{\lambda}{4})\right) + \mathrm{erf}\left(\frac{\sqrt{2}\,\lambda}{4}\right)\right] - \theta^2\right),$$

with $\theta = (\sqrt{\pi}/2)\mathrm{erf}(1)$*. The expression for the new variance is rather complicated (and this tends to be the case for importance sampling quite often) but we have seen in* section 9.3 *that thanks to the new choice* $g(y)$*, the 95% confidence interval shrinks from* [0.746 72, 0.746 92] *to* [0.746 79, 0.746 85]*, thus making the method more efficient.*

9.8 Reporting results

Communication style is personal: there is no single 'right' voice for reporting Monte Carlo results. Still, some light-touch conventions can help you stay on track. The general approach used by R users is the one adopted in the *tidyverse* (see section 2.21). This is what is suggested:

- Use one short sentence naming the target (e.g. $\theta = \mathbb{E}[h(X)]$) and the method (plain MC or importance sampling, including the proposal g and any parameters tuning).
- Report the point estimate $\hat{\theta}$ with a standard error or a 95% confidence interval; match the shown digits to the SE (avoid false precision).
- Give the sample size N and/or wall-clock time so readers can judge efficiency.
- Present a compact, well-labelled table (or a small plot) that can be scanned in seconds; keep outputs in a clean, tabular format consistent with the tools in section 2.21.
- Show minimal code to generate the table/figure, record the random seed, and (if relevant) note key package versions.
- When contrasting methods (e.g. plain versus IS), hold the budget fixed and place results side by side; let the numbers speak, then add a one-line interpretation.
- Note independence/correlation assumptions and modelling choices that affect the estimator (e.g. proposal family in IS).
- Use clear labels, units, and consistent formatting; prefer simple wording over jargon.
- End with a sentence that interprets the result in context (e.g. 'this precision is sufficient for …', 'further gains would require …').

Here it is not suggested to standardise scientific communication thus flattening it. The computational scientist should keep their own tone and style but should try and anchor it with clarity about the target and method, honest uncertainty, visible cost, tidy presentation, and reproducibility, very much like what is written in the suggested guidelines.

9.9 Monte carlo simulation of a nuclear reactor

We conclude this chapter by returning to the story that opened it: the origins of the Monte Carlo method during the Manhattan Project. At that time, scientists were faced with the problem of understanding the behavior of neutrons inside nuclear reactors, where their motion and interactions were too complex to describe analytically. Random sampling techniques provided a new way forward, laying the foundation of what is now called Monte Carlo simulation.

In this section we attempt a simplified but instructive version of such an analysis. We will simulate the dynamics of a group of neutrons generated inside a spherical reactor, composed of a mixture of uranium and graphite (carbon). The aim is not to reproduce the full physical reality, but to capture the essential processes in a way that illustrates both the power and the limitations of the Monte Carlo approach. The treatment follows closely that described in the book by Woolfson and Pert [8].

9.9.1 The journey of neutrons towards fission

Consider a spherical reactor of radius R, composed of a homogeneous mixture of uranium and graphite (carbon). Neutrons are spontaneously released by a few uranium nuclei, with initial energies distributed according to a prescribed probability law. From that moment each neutron begins a journey, subject to the random processes that govern nuclear interactions.

The first stage of this journey is *free flight*. A neutron travels in a straight line until it either collides with a nucleus or escapes from the spherical boundary. The probability that the next collision occurs within a given distance is governed by the macroscopic total cross section of the medium[6]. For many neutrons this may result in a quick loss through leakage, especially if the initial point of emission is close to the surface. For those neutrons that remain inside, several outcomes are possible at each interaction, including a loss of energy and the triggering of nuclear fission, with the ensuing production of new neutrons and extra energy. Over many such events, the fate of each neutron is decided by probabilities tied to the physical cross sections of the materials. The balance between these processes determines whether the system is *sub-critical, critical,* or *super-critical*[7].

In the following subsection, the mathematics and related code of the various parts involved in the simulation will be explained in detail, with related chunks of code interspersed with the text.

9.9.2 R code describing the reactor simulation

In our simplified Monte Carlo simulation, these microscopic rules are implemented in modern R code divided into different modules (R files). The complete set of files is included in appendix I. The separate files, functions, and blocks of code tend to follow closely the above narrative. Furthermore, we have attempted to use a tidyverse structure as much as possible, in order to demonstrate that a modern use of R is also possible in computational physics.

At any step, the neutrons generated in the reactor are contained in a *tibble* structure (a table, see section 2.21). Each row of the tibble represents a neutron. The row includes the neutron's id, the x, y, z coordinates of its current position, the three cosine directors, dx, dy, dz, of the direction it is heading to, the neutron's current energy in MeV, a character string called `status` that holds the label characterising whether the neutron is just `"born"`, whether it has scattered elastically (`elasticM"`, `"elasticU"`), or it has been absorbed (`"absorbedM"`, `"absorbedU238"`, `"absorbedU235"`), and the three distances s, s_{int}, and s_{bnd}. s is the actual free flight distance traveled by the neutron before the next event occurs; s_{int} is the sampled distance to the next nuclear interaction, obtained from the exponential law with parameter Σ_t; s_{bnd} is the distance to the spherical boundary

[6] The *macroscopic total cross section* Σ_t is the probability per unit path length that a neutron will undergo any kind of interaction (scattering or absorption) in the medium. It is defined as $\Sigma_t = N\sigma_t$, where N is the number density of nuclei and σ_t is the microscopic total cross section of a single nucleus. Since N is measured in nuclei per cm^3 and σ_t in cm^2, the unit of Σ_t is cm^{-1}.

[7] These terms refer to the neutron multiplication factor k_{eff}, defined as the average number of neutrons in one generation divided by the average number in the preceding generation. If $k_{\text{eff}} < 1$, the chain reaction dies out and the reactor is said to be *sub-critical*; if $k_{\text{eff}} = 1$, the reaction is self-sustaining (*critical*); if $k_{\text{eff}} > 1$, the reaction grows exponentially and the reactor is *super-critical*.

along the neutron's current direction of motion. At each step the neutron advances by

$$s = \min(s_{\text{int}}, s_{\text{bnd}}).$$

If $s = s_{\text{int}}$, a physical interaction occurs inside the reactor; if $s = s_{\text{bnd}}$, the neutron escapes through the boundary. Although the original number of rows in the tibble is decided by the user at the start of the simulation, this number can potentially change after each free flight, due to absorption, fission, and leakage outside the reactor.

The advantage of using a tibble to hold the current snapshot of neutrons is that queries and manipulations are easily carried out through verbs and piping (|>). For example, to eliminate from the tibble rows corresponding to neutrons leaking out of the reactor, the following code is used,

```
neutrons <- neutrons |> dplyr::filter(!id %in% ids)
```

where filtering is applied so to retain only the rows which have not been selected for deletion. A different but related advantage is the possibility of querying the status of thousands of neutrons quickly and to produce graphical plots easily with `ggplot2`.

9.9.3 Neutrons' initial production

R code for this part is in `neutronProduction.R`. The simulation begins with the generation of an initial set of neutrons. Three ingredients must be specified: their directions of motion, their initial positions inside the spherical reactor, and their kinetic energies. The direction of each neutron is chosen uniformly on the unit sphere. To achieve this, we draw the polar cosine $\cos\theta$ uniformly from $[-1, 1]$, and the azimuthal angle ϕ uniformly from $[0, 2\pi)$. The direction cosines are then given by

$$dx = \sin\theta\cos\phi, \qquad dy = \sin\theta\sin\phi, \qquad dz = \cos\theta.$$

This guarantees an isotropic distribution, with probability density

$$f(\theta, \phi) = \frac{1}{4\pi}, \qquad 0 \leqslant \theta \leqslant \pi, \ 0 \leqslant \phi < 2\pi.$$

```
# sample isotropic directions
u  <- runif(N,-1,1)
phi <- runif(N,0,2*pi)
s  <- sqrt(pmax(0,1-u^2)) # sin(theta), safe against
# tiny negatives
dx <- s * cos(phi)
dy <- s * sin(phi)
dz <- u
```

It is important to recall that in **R** uniform random numbers are generated with the function `runif`. To sample positions uniformly in the reactor volume, we use spherical coordinates (r, θ, ϕ). The key is that the radial coordinate r must follow the distribution

$$f(r) = \frac{3r^2}{R^3}, \qquad 0 \leqslant r \leqslant R,$$

so that the probability of being in a spherical shell of thickness dr is proportional to its volume. In practice, this is obtained by sampling a uniform random variable $U \sim \text{Uniform}(0, 1)$ and setting

$$r = R\ U^{1/3}.$$

The angular coordinates are again chosen isotropically.

```
# sample uniform points in sphere of radius R
r    <- R*runif(N)^(1/3)
# reuse independent isotropic directions for placement
u2   <- runif(N,-1,1)
phi2 <- runif(N,0,2*pi)
s2   <- sqrt(pmax(0,1-u2^2))
x <- r*s2*cos(phi2)
y <- r*s2*sin(phi2)
z <- r*u2
```

The neutron kinetic energies are sampled from an exponential law with prescribed mean $\overline{E}$:

$$f(E) = \frac{1}{\overline{E}}\, e^{-E/\overline{E}}, \qquad E \geqslant 0.$$

This distribution ensures that most neutrons start with energies around $\overline{E}$, while higher energies are possible with decreasing probability.

```
# energies (exponential with mean = meanE)
E <- rexp(N,rate=1/meanE)
```

In R, random numbers from the exponential family are generated with the function `rexp`. The two constants `R` (radius of the reactor) and `meanE` (average energy of the initial neutrons produced) are available in memory as a consequence of the file `FixedData.R` being source at the beginning of the simulation. Indeed, all constants needed in the code are available in memory.

9.9.4 Free flight

R code for this part is in `neutronTravel.R`. At any given step, a neutron at position $\mathbf{r} = (x, y, z)$ with velocity (direction) $\hat{\mathbf{v}} = \left(\hat{d}_x, \hat{d}_y, \hat{d}_z\right)$ travels in a straight line until either (i) it undergoes a nuclear interaction in the medium, or (ii) it reaches the spherical boundary of radius R and escapes. The step length s is the minimum of two competing distances:

$$s = \min\{\ s_{\text{int}},\quad s_{\text{bnd}}\ \}.$$

Numerically, we first ensure the direction is a unit vector:

$$\hat{\mathbf{v}} = \frac{\left(d_x, d_y, d_z\right)}{\left\|\left(d_x, d_y, d_z\right)\right\|} = \left(\hat{d}_x, \hat{d}_y, \hat{d}_z\right), \qquad \left\|\left(d_x, d_y, d_z\right)\right\| = \sqrt{d_x^2 + d_y^2 + d_z^2}.$$

This guards against accumulated floating–point drift and zero/invalid directions.

Assuming a homogeneous medium with macroscopic total cross section Σ_t (units cm^{-1}), the free–flight distance to the next interaction is exponentially distributed:

$$f_{s_{\text{int}}}(s) = \Sigma_t\, e^{-\Sigma_t s}, \qquad s \geqslant 0.$$

Sampling uses the inversion method[8]: if $U \sim \text{Uniform}(0, 1)$,

$$s_{\text{int}} = -\frac{\ln U}{\Sigma_t}.$$

(When $\Sigma_t \leqslant 0$ or invalid, the code treats $s_{\text{int}} = \infty$, i.e., no interaction in the medium.)

The neutron travels along the ray $\mathbf{r}(s) = \mathbf{r} + s\hat{\mathbf{v}}$. The escape condition is $\|\mathbf{r}(s)\| = R$, i.e.

$$\|\mathbf{r} + s\hat{\mathbf{v}}\|^2 = (\mathbf{r}\cdot\mathbf{r}) + 2s\,(\mathbf{r}\cdot\hat{\mathbf{v}}) + s^2\,(\hat{\mathbf{v}}\cdot\hat{\mathbf{v}}) = R^2.$$

Since $\|\hat{\mathbf{v}}\| = 1$, this gives the quadratic

$$s^2 + 2\,(\mathbf{r}\cdot\hat{\mathbf{v}})\,s + \left(\|\mathbf{r}\|^2 - R^2\right) = 0.$$

Define

$$r^2 = \|\mathbf{r}\|^2, \qquad r_v = \mathbf{r}\cdot\hat{\mathbf{v}}, \qquad \Delta = r_v^2 - \left(r^2 - R^2\right).$$

For a neutron inside the sphere, $\Delta \geqslant 0$, and the forward intersection (nonnegative s) is

[8] The *inversion method* is a general technique for generating random samples from a given distribution. If $F(x)$ is the cumulative distribution function (CDF) of the target distribution, and $U \sim \text{Uniform}(0, 1)$, then the random variable $X = F^{-1}(U)$ has distribution F. For example, the exponential law with density $f(x) = \lambda e^{-\lambda x}$ $(x \geqslant 0)$ has CDF $F(x) = 1 - e^{-\lambda x}$, so its inverse is $F^{-1}(u) = -(1/\lambda)\ln(1 - u)$ or, as $1 - u$ is as uniformly random as u, $F^{-1}(u) = -(1/\lambda)\ln(u)$. Thus one can generate exponential samples simply by computing $-\ln(u)/\lambda$, as done in the code.

$$s_{\text{bnd}} = - \ r_v + \sqrt{\Delta}.$$

(In code, Δ is clamped to $\max\{\Delta, 0\}$ to protect against tiny negative round-off.)

With $s = \min\{s_{\text{int}}, s_{\text{bnd}}\}$, the new position is

$$(x', y', z') = (x, y, z) + s\left(\hat{d}_x, \hat{d}_y, \hat{d}_z\right).$$

If $s = s_{\text{bnd}}$ the neutron escapes; if $s = s_{\text{int}}$ the neutron undergoes an interaction, after which the event type (scattering/absorption/fission) and any kinematic updates are sampled according to the models described in the following subsections.

```
dnorm <- sqrt(neutrons$dx^2 + neutrons$dy^2 + neutrons$dz^2)
bad_dir <- which(!is.finite(dnorm) | dnorm <= 0)
if (length(bad_dir))
stop("neutron_travel():
 found non-finite/zero direction vectors.")

dxu <- neutrons$dx / dnorm
dyu <- neutrons$dy / dnorm
dzu <- neutrons$dz / dnorm

rv   <- neutrons$x*dxu + neutrons$y*dyu + neutrons$z*dzu
r2   <- neutrons$x^2 + neutrons$y^2 + neutrons$z^2
disc <- pmax(rv^2 - (r2 - R^2), 0)
s_bnd_new <- -rv + sqrt(disc)

# Repeat Sigma_t into a vector
sig_t <- rep_len(as.numeric(Sigma_t), n)

if (any(!is.finite(sig_t) | sig_t <= 0)) {
warning("neutron_travel():


 non-positive/invalid Sigma_t; using Inf for s_int there.")
}
u <- runif(n)
s_int_new <- ifelse(is.finite(sig_t) & sig_t > 0,
-log(u)/sig_t, Inf)

s_new <- pmin(s_int_new, s_bnd_new)

x_new <- neutrons$x + dxu * s_new
y_new <- neutrons$y + dyu * s_new
z_new <- neutrons$z + dzu * s_new
```

9.9.5 Collisions

When a neutron completes a free flight and does not escape, it undergoes a collision with the material inside the reactor. This is the central step of the simulation, because the outcome of the collision determines whether the neutron continues its journey, is lost, or generates new neutrons through fission.

The decision is handled by the function `select_event()`. Its role is limited but crucial: for each neutron it updates the `status` column of the tibble, recording what kind of collision has taken place. No other physical quantities are changed at this stage. Those changes are applied later by the function `update()`, which interprets the `status` and acts accordingly.

The possible collision outcomes are:

- **Elastic scattering with the moderator (carbon)**. The neutron bounces off a carbon nucleus, losing some energy and changing direction. This is recorded as `status = "elasticM"`.
- **Elastic scattering with uranium nuclei (U-235 or U-238)**. The neutron collides elastically with a uranium nucleus, again changing direction but losing relatively little energy because of the heavier target. This is recorded as `status = "elasticU"`.
- **Absorption by U-235**. The neutron is absorbed by a fissile nucleus. Here `status = "absorbed235"`. In later steps, `update` decides whether this leads to radiative capture (no new neutrons) or fission (2 or 3 new neutrons).
- **Absorption by U-238**. The neutron is absorbed but no fission occurs. This is recorded as `status = "absorbed238"`.
- **Absorption by the moderator**. A rarer process in which the neutron is absorbed by carbon. This is recorded as `status = "absorbedM"`.

The choice among these outcomes is made probabilistically, using the macroscopic cross sections of the materials. For a given neutron energy E, the total macroscopic cross section is

$$\Sigma_t(E) = \Sigma_s(E) + \Sigma_a(E),$$

and the probability of each event is proportional to its contribution to Σ_t. For example, the probability of U-235 absorption is

$$P\left(\text{absorbed by U}_{235}\right) = \frac{\Sigma_{a,235}(E)}{\Sigma_t(E)}.$$

In practice, `select_event()` computes all these probabilities, draws a uniform random number, and assigns the corresponding `status`. The tibble of neutrons is thus updated with a clear record of what has happened to each particle. More details can be found by inspecting the actual code in appendix I. Later, when `update` is called, the physical consequences of the assigned `status` are carried out: elastic scatterings modify directions and energies, absorptions remove neutrons, and fission events create new ones.

9.9.6 Elastic collisions

When a neutron undergoes an elastic collision, it scatters off a nucleus without being absorbed. In this process two things happen:

1. the neutron changes direction;
2. the neutron loses part of its kinetic energy, with the fraction lost depending on the target nucleus and on the scattering angle.

The function `elastic_scattering` in file `elasticScattering.R` implements these changes. In the simulation the collision is assumed to be isotropic in the centre-of-mass frame. This means that

$$\mu = \cos\theta \sim \text{Uniform}(-1, 1), \qquad \phi \sim \text{Uniform}(0, 2\pi),$$

where θ is the polar angle and ϕ the azimuthal angle. Let E be the neutron energy before collision and A the mass ratio of the target nucleus to the neutron ($A \approx 12$ for carbon, $A \approx 235, 238$ for uranium). The neutron energy after collision is given by the kinematic relation

$$E' = E\,\frac{A^2 + 1 + 2A\cos(\theta_{\text{CM}})}{(A + 1)^2},$$

where θ_{CM} is the scattering angle in the centre-of-mass frame.

- For large A (uranium) the fraction $(A^2 + 1 + 2A\cos(\theta_{\text{CM}}))((A + 1)^2)$ is close to 1, so the neutron loses little energy.
- For small A (carbon) the energy loss can be substantial, especially for backward scatterings ($\cos(\theta_{\text{CM}}) < 0$).

```
# Energy fraction bounds from isotropic CM scattering
alpha <- ((A_vec - 1)^2) / (A_vec + 1)^2  # per-row alpha in [0,1]

# Sample g = E'/E ~ Uniform[alpha, 1]
g <- alpha + (1 - alpha) * runif(length(idx))

# Update energies
E_new <- neutrons$E
E_new[idx] <- g * E_new[idx]
```

The neutron's outgoing direction is determined by the sampled angles (θ, ϕ). In the code, this direction is first computed in the centre-of-mass frame and then rotated back to the laboratory frame to produce new Cartesian components (dx, dy, dz) for the neutron's velocity vector. After the collision the neutron's `status` is set to `"elastic"`, and the step-length fields `s`, `s_int`, and `s_bnd` are reset for the next free flight.

```
# New isotropic LAB directions
udir <- runif(length(idx), -1, 1)       # cos(theta)
phi  <- runif(length(idx), 0, 2*pi)
sdir <- sqrt(pmax(0, 1 - udir^2))
dx_n <- sdir * cos(phi)
dy_n <- sdir * sin(phi)
dz_n <- udir

dx_new <- neutrons$dx
dy_new <- neutrons$dy
dz_new <- neutrons$dz
dx_new[idx] <- dx_n
dy_new[idx] <- dy_n
dz_new[idx] <- dz_n

# Clear step fields for scattered rows
s_int_new <- neutrons$s_int; s_int_new[idx] <- NA_real_
s_bnd_new <- neutrons$s_bnd; s_bnd_new[idx] <- NA_real_
s_new     <- neutrons$s;     s_new[idx]     <- NA_real_

# Keep status as is ("elasticM"/"elasticU"/...)
status_new <- neutrons$status
```

In summary, `elastic_scattering` takes the current energy and direction of each neutron that has scattered, samples a random angle, applies the kinematic formula to compute the new energy, and updates the direction vector accordingly.

9.9.7 Nuclear fission

When a neutron is absorbed by a U_{235} nucleus, the outcome may be nuclear fission. In this process the parent neutron disappears, the uranium nucleus splits into two lighter nuclei, and a few new neutrons are emitted. These new neutrons are the engine of the chain reaction. The function `fission` in file `fission.R` handles this step in the simulation. Its role is to remove the absorbed parent neutron and to create new 'child' neutrons at the same spatial location. The number of neutrons produced in a fission event is a random variable. In our simplified model the outcome is either 2 or 3 new neutrons with equal probability, reproducing the fact that the average number of neutrons per fission, $\overline{\nu}$, is a little above 2.

Each newborn neutron is assigned a kinetic energy sampled from an exponential distribution,

$$f(E) = \frac{1}{\overline{E}_{\text{fission}}}\, e^{-E/\overline{E}_{\text{fission}}}, \qquad E \geqslant 0,$$

with mean $\overline{E}_{\text{fission}} \approx 2$ MeV. This reflects the typical energy spectrum of fission neutrons: most are born in the MeV range.

```
# Multiplicity: 2 or 3 with equal probability
k_vec <- sample(c(2L, 3L), size = length(parents_rows),
replace = TRUE)
Ktot  <- sum(k_vec)

# Defaults for fast fission spectrum (simple placeholder)
meanE_fast <- if (exists("meanE_fission"))
meanE_fission else 2.0  # MeV
............
............
............
# Fast energies (placeholder exponential spectrum)
E  <- rexp(Ktot, rate = 1/meanE_fast)
```

The emission directions of the newborn neutrons are chosen isotropically, that is, the cosine of the polar angle is drawn uniformly from $[-1, 1]$ and the azimuthal angle from $[0, 2\pi)$. Finally, the parent neutron is marked as removed from the system, while the newborn neutrons are appended to the tibble, each with a fresh identifier, position equal to the parent's location, isotropic direction, fast energy sampled as above, and status set to "born".

```
# Build newborns at parents' positions
# Build newborns at parents' positions
rep_idx <- rep(seq_along(parents_rows), times = k_vec)
x <- parents$x[rep_idx]


y <- parents$y[rep_idx]
z <- parents$z[rep_idx]

# Isotropic directions
u  <- runif(Ktot, -1, 1)         # cos(theta)
ph <- runif(Ktot, 0, 2*pi)
s  <- sqrt(pmax(0, 1 - u^2))
dx <- s * cos(ph)
dy <- s * sin(ph)
dz <- u
```

9.10 Demonstrating the code

We now have all the components needed to run a full Monte Carlo simulation of neutron dynamics inside the spherical reactor. The work proceeds by calling the functions in a natural sequence. The starting point is the function `neutron_production`, which generates an initial tibble of neutrons with positions, directions, energies, and identifiers. Once the initial neutrons are created, the simulation advances by alternating between `neutron_travel` and `update`. Each call to `neutron_travel` moves all active neutrons forward by one free flight, computing whether they escape the reactor or reach a collision. The subsequent call to `update` interprets what has happened: neutrons may scatter elastically, be absorbed, or undergo fission with the creation of new neutrons.

This cycle can be repeated as many times as desired. At every stage the state of the system is represented by the neutrons tibble, which is progressively modified by the simulation. Through inspection of the final tibble one can see how the population has evolved: how many neutrons have escaped, how many were absorbed, how many fission events took place, and how many newborn neutrons were created.

9.10.1 Monitoring energies and number of neutrons

As a first example, let us produce in the spherical reactor with radius 50 cm, 1000 neutrons with average energy equal to 2 MeV. The presence of the moderator and the fuel in the homogeneous mixture is represented by

$$\phi_C = 1.8 \quad \text{and} \quad \phi_U = 0.9.$$

Only the relative strength of ϕ_C and ϕ_U matters, whatever their absolute value. It can be useful to use absolute values from a practical point of view, though. For example, it is rather natural to say ‘two parts of carbon and one part of uranium’, so that $\phi_C = 2$ and $\phi_U = 1$. These quantity will be scaled anyway to obey

$$\phi_C + \phi_U = 1.$$

The parameters to be used in the experiment must be changed in the `FixedData.R` file. For the current simulation:

```
# Energy of ignition neutrons
meanE <- 2

# Radius of spherical reactor (cm)
R <- 50

# Importance of presence of moderator and uranium
# in mixture (will be normalised)
phi_C <- 1.8
phi_U <- 0.9
```

It is also important to remember to source `all.R` to load everything in memory. In the following code, you can see all functions and parameters present in memory for the simulation.

```
# Load all parameters and functions
source("all.R")

# Functions and parameters loaded (print to avoid margin overspill)
print(matrix(ls(),ncol=2,byrow=TRUE)) # ls() only in 2 columns
```

```
##       [,1]                   [,2]
##  [1,] "A_235"                "A_238"
##  [2,] "A_C"                  "c_max"
##  [3,] "E_therm"              "E0_res"
##  [4,] "elastic_scattering"   "f235"
##  [5,] "f238"                 "fission"
##  [6,] "H"                    "K"
##  [7,] "meanE"                "meanE_fission"
##  [8,] "missing_nts"          "neutron_production"
##  [9,] "neutron_travel"       "nts"
## [10,] "nts0"                 "nu_bar"
## [11,] "partition_vector"     "phi_C"
## [12,] "phi_U"                "R"
## [13,] "remove_neutrons"      "rho_C"
## [14,] "rho_U"                "select_event"
## [15,] "sig_a_235"            "sig_a_238"
## [16,] "sig_a_C"              "sig_f_235"
## [17,] "sig_s_235"            "sig_s_238"
## [18,] "sig_s_C"              "sigma_res"
## [19,] "Sigma_t"              "update"
```

The simulation starts with the generation of the initial batch of neutrons, 1000 in this example, which are stored in a tibble called `nts`.

```
# Use the tidyverse (modern R)
library(dplyr)
library(ggplot2)

# Produce 1000 neutrons with average energy=2 MeV
# The reactor has radius 50 cm
nts0 <- neutron_production(N=1000,seed=8123)

# Average energy of neutrons (should be around 2)
nts0 |> summarise(mean_E=mean(E,na.rm=TRUE))
```

```
## # A tibble: 1 x 1
##   mean_E
##    <dbl>
## 1   1.97
```

```
# Initial number of neutrons (must be 1000)
nts0 |> filter(!is.na(E)) |> nrow()
```

```
## [1] 1000
```

The so called *grammar of data* (see section 2.22) has been used to query about average energy and current number of free neutrons in the reactor. As expected, the average energy is close to 2 MeV, while there currently are 1000 free neutron because they have just been created.

Now we proceed forward with the simulation by allowing the 1000 neutrons to free flight until collision. This is implemented as succession of the functions `neutron_travel` and `update`. As a consequence, the status of all neutrons changes from `"born"` before interaction, to `"elasticM"` or `"elasticU"` after interaction.

```
# At the beginning, status of all neutron is "born"
nts0 |> select(id,status) |> head(5)
```

```
## # A tibble: 5 x 2
```

```
##          id status
##       <int> <chr>
## 1         1 born
## 2         2 born
## 3         3 born
## 4         4 born
## 5         5 born
```

```
# One cycle of free flight + collisions
nts <- neutron_travel(nts0,seed=623) # free flight
nts <- update(nts,seed=128) # Collisions

# All neutrons have collided
nts |> select(id,status) |> head(5)
```

```
## # A tibble: 5 x 2
##          id status
##       <int> <chr>
## 1         1 elasticM
## 2         2 elasticU
## 3         3 elasticU
## 4         4 elasticM
## 5         6 elasticM
```

In this specific instance, the first four neutrons have collided elastically, two with the moderator and two with the fuel. The fifth neutron has been absorbed or leaked out the reactor, as we can deduce by noticing that `id=5` is missing from the new tibble. Information on escaped or absorbed neutrons is lost, going from tibble to tibble, and can be recovered only by storing every tibble that comes out after each cycle of free flight and update.

```
# Which neutrons have been absorbed?
missing_nts <- setdiff(nts0$id,nts$id)

# Print avoiding margin overspill
print(matrix(missing_nts,ncol=7,byrow=TRUE)) # 7 columns
```

```
##         [,1] [,2] [,3] [,4] [,5] [,6] [,7]
##  [1,]      5   19   59   67   79   86   89
##  [2,]    102  108  117  140  150  158  171
##  [3,]    192  193  223  234  251  263  264
```

```
## [4,] 266 268 272 281 283 292 299
## [5,] 307 322 340 345 377 392 400
## [6,] 405 410 433 436 460 483 499
## [7,] 517 532 544 572 588 624 651
## [8,] 652 659 662 665 698 707 712
## [9,] 715 728 734 739 747 749 754
## [10,] 756 768 770 793 797 805 822
## [11,] 849 855 865 919 922 933 977
```

```
# Neutrons created from fission?
nts |> filter(status == "born") |> nrow()
```

```
## [1] 122
```

For instance, here it is clear that many neutrons have been absorbed or have escaped. It is also clear that 122 new neutrons were generated as byproduct of the fission of U_{235} atoms.

As a last exercise in this first demonstration, let us measure the average energy from cycle to cycle. Due to elastic scattering being more likely than absorption, we expect the average energy of all neutrons present in the reactor to be decreasing. The process is, in fact, complicated. Depending on the initial conditions of the mixture and the reactor, sub-criticality, criticality, or super-criticality can occur, where the production of new neutrons dies off quickly, stays constant, or increases without limit. In the current scenario we have obtained super-criticality as $k_{\text{eff}} \approx 1.06 > 1$. This means that the number of neutron will keep increasing, and the reactor will eventually explode.

```
# Start from nts0 again. Copy it to an nts to
# cycle quickly without chaging the variable name
nts <- nts0

# Vector of average energies and neutron mult. factor
# Initialised with values from nts0
aveE <- nts |>
summarise(mean_E=mean(E,na.rm=TRUE)) |>
pull()   # This bit change a tibble into a numeric
totN <- length(nts$id)
Keff <- 1

# 100 cycles
for (i in 1:100) {
nts <- neutron_travel(nts,seed=119)
```

```
nts <- update(nts,seed=120)

# Extract # neutrons and calculate key ratio
totN <- c(totN,length(nts$id))
kk <- totN[length(totN)]/totN[length(totN)-1]
Keff <- c(Keff,kk)

# Extract average energy
aveE <- c(aveE,nts |>
summarise(mean_E=mean(E,na.rm=TRUE)) |>
pull())
}

# Merge results in a tibble
NE <- tibble(Cycle=1:101,Keff=Keff,E=aveE)

# To scale E unto Keff values
scale_factor <-
max(NE$Keff,na.rm=TRUE)/max(NE$E,na.rm=TRUE)

# Plot
ggplot(NE,aes(x=Cycle)) +
geom_line(aes(y=Keff,color="Keff")) +
geom_line(aes(y=E*scale_factor,color="E")) +
scale_y_continuous(
name="",
sec.axis=sec_axis(
transform = ~ . / scale_factor,
name = ""
)
) +
scale_color_manual(values = c("Keff"="blue","E"="red")) +
labs(color="Legend")
```

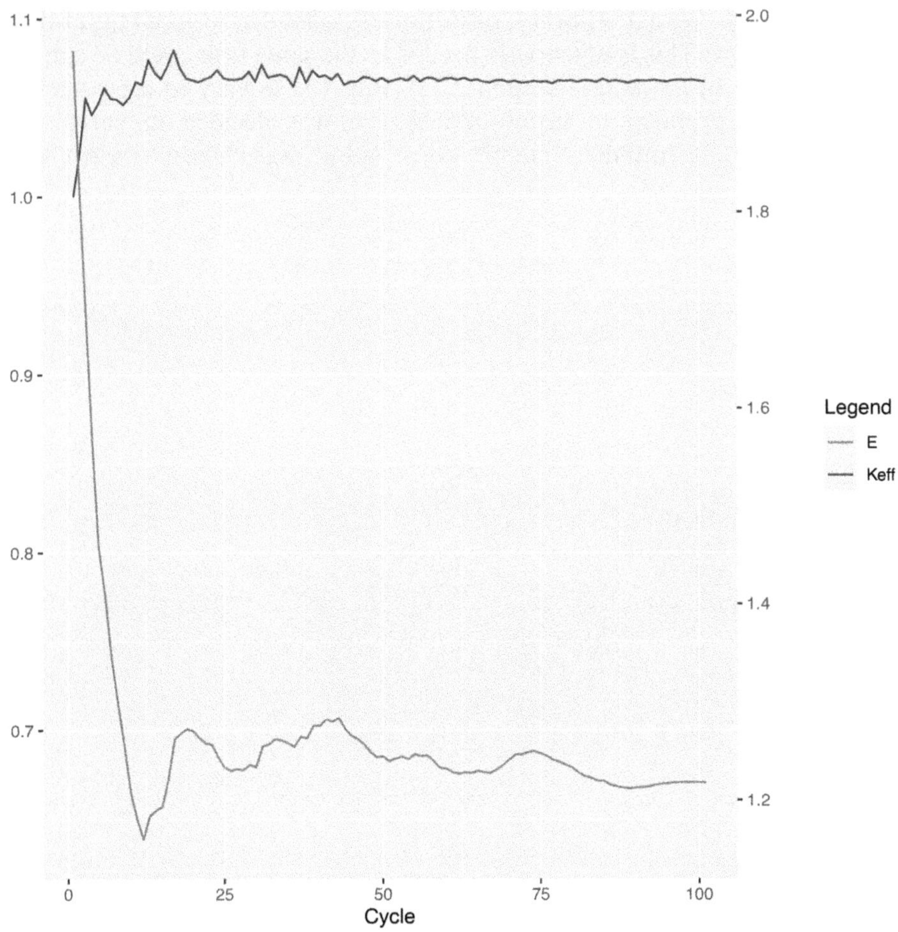

It is instructive to look at how ggplot2 handles the two different scales of energy and K_{eff}. In fact, scaling was not needed in this instance as the two quantities have comparable scales, but we still applied the technique to illustrate how easily different scales can be tamed. The key tool is the mapping enacted by the scale_y_continuous function. A previous scale factor, scale_factor is created and the second quantity, here E, is multiplied by it. The new num,bers are then mapped to the old by scale_y_continuous, simply dividing by the same scale factor. Note also that there has been no need here to add labels to the *y*-axis because they are described in the legend.

9.10.2 2D maps of neutrons in the reactor

This demonstration is relatively straightforward, once the *x*, *y*, *z* columns of the neutron tibble are considered. Let us look at the first production of neutrons, nts0 and create the plots. An extra library, `patchwork`, is loaded to speed the creation of plots in a single row. This is not strictly needed as the same task could be carried out with a proper use of *facets* (see section 2.23). But it is so easy to use `patchwork` (and it is not a problem to install it!), that it was decided to carry out the demonstration with it. Initially, a simple 'black' colour was chosen for each neutron.

```
## 2D plots of neutrons' position inside the reactor
# Uses patchwork, a library not in the tidyverse but
# easy to install. It makes things very easy

# Libraries needed
require(ggplot2)
require(patchwork)

# XY projection
p_xy <- ggplot(nts0,aes(x=x,y=y)) +
geom_point(alpha=0.6) +
labs(title="XY") +
coord_fixed()

# XZ projection
p_xz <- ggplot(nts0,aes(x=x,y=z)) +
geom_point(alpha=0.6) +
labs(title="XZ") +
coord_fixed()

# YZ projection
p_yz <- ggplot(nts0,aes(x=y,y=z)) +
geom_point(alpha=0.6) +
labs(title="YZ") +
coord_fixed()

# Arrange three plots in a row (so easy!)
p_xy + p_xz + p_yz
```

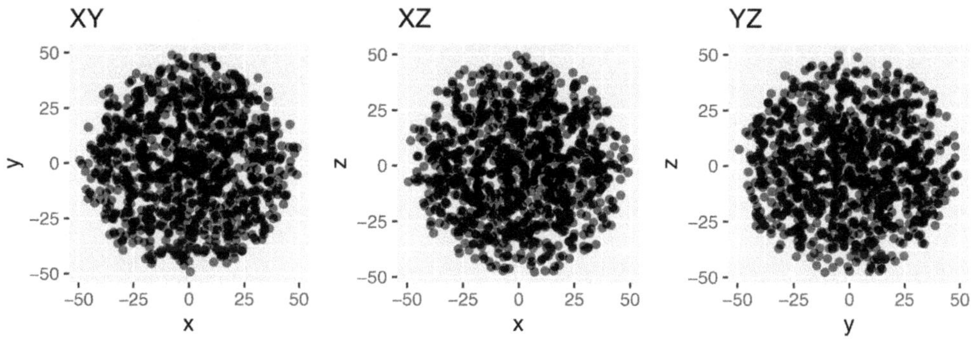

From the three projections, we can see that the initial batch of neutrons is indeed generated uniformly inside a reactor of radius 50 cm. There exist in R tools to visualise 3D plots and some of them are interactive but for presentations and communication purposes it is always better to stick to 2D representations.

Variations from the plots just presented are possible. For example, we could colour the neutrons based on their energy. With `ggplot2` this is not difficult. It can be done by using `colour` inside `aes()`.

```
## 2D plots of neutrons' position inside the reactor
# Uses patchwork, a library not in the tidyverse but
# easy to install. It makes things very easy.
# Neutrons are coloured based on the value of energy

# Libraries needed
require(ggplot2)
```

```
require(patchwork)

# XY projection
p_xy <- ggplot(nts0,aes(x=x,y=y,colour=E)) +
geom_point(alpha=0.6) +
labs(title="XY projection") +
coord_fixed()

# XZ projection
p_xz <- ggplot(nts0,aes(x=x,y=z,colour=E)) +
geom_point(alpha=0.6) +
labs(title="XZ projection") +
coord_fixed()

# YZ projection
p_yz <- ggplot(nts0,aes(x=y,y=z,colour=E)) +
geom_point(alpha=0.6) +
labs(title="YZ projection") +
coord_fixed()

# Arrange three plots in a row (so easy!)
p_xy + p_xz + p_yz
```

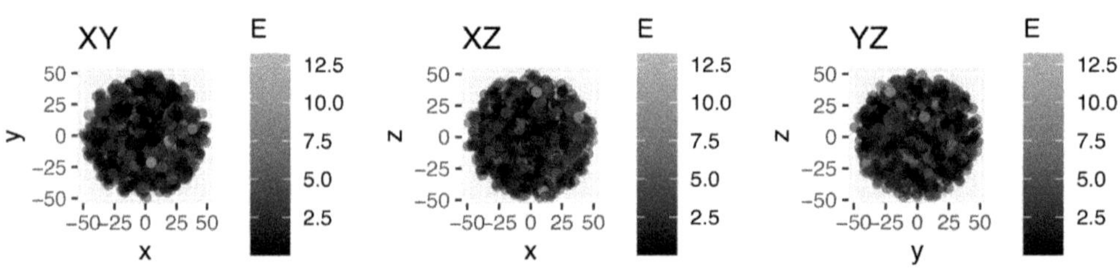

The reason why it is so easy to use colouring and other graphics features in `ggplot2` is that this package (as well as the grammar of graphics the package is based on) has built in optimised and standardised colouring and graphical patterns, making it easier for scientists to visualise and communicate research findings.

9.10.3 Diffusion

In our last demonstration, we monitor how far a group of neutrons generated near the centre of the reactor, travels. It is in general expected that neutrons travel small distances before being absorbed, and have roughly an overall diffusion journey around the spot in which they have been created.

With the Monte Carlo code and the R structures available, the task can be carried out through an initial selection of `id`'s based on the proximity to the centre of the reactor. During cycling, the same neutrons can be tracked via their unique `id`. In the code below, we have executed 20 steps and measured the average distance from the centre of the neutrons monitored.

```
# The initial tibble was created earlier
nts <- nts0

# Identify neutrons near the core (<= 10 cm): new tibble
ids <- nts0 |> filter(sqrt(x^2+y^2+z^2) <= 10) |>
select(id) |> pull()
core0 <- nts0 |> filter(id %in% ids)

# Store other tibbles in a list
Lneutrons <- list()
for (i in 1:20) {
# Free flight + collision
nts <- neutron_travel(nts,seed=119)
nts <- update(nts,seed=120)

# Extract monitored neutrons
core <- nts |> filter(id %in% ids)

# Storage
Lneutrons <- c(Lneutrons,list(core))
}

# Average distances from the centre
Dini <- core0 |>
mutate(D=sqrt(x^2+y^2+z^2)) |>
summarise(mean_D=mean(D,na.rm=TRUE)) |>
pull()
Dfin <- Lneutrons[[20]] |>
mutate(D=sqrt(x^2+y^2+z^2)) |>
summarise(mean_D=mean(D,na.rm=TRUE)) |>
pull()
print(c(Dini,Dfin))
```

```
## [1] 7.468053 6.972500
```

The average distance from the centre, of the neutrons which have not been absorbed after 20 cycles is slightly smaller than what it was at creation time. This essentially means that on average the neutrons have moved very little from their original position. It must also be rememberd that some of the neutrons will have been absorbed and thus it is not possible to measure the average distance travelled by each monitored neutron. A picture taken at the beginning and at the end, further reinforces these findings.

```
# First plot
p00 <- ggplot(core0,aes(x=x,y=y)) +
geom_point(alpha=0.6) +
labs(title="XY projection") +
coord_fixed(xlim=c(-25,25),ylim=c(-25,25)) +
annotate("path",
x = 10*cos(seq(0,2*pi,length.out=100)),
y = 10*sin(seq(0,2*pi,length.out=100)),
color="red")

# Last plot
pLL <- ggplot(Lneutrons[[20]],aes(x=x,y=y)) +
geom_point(alpha=0.6) +
labs(title="XY projection") +
coord_fixed(xlim=c(-25,25),ylim=c(-25,25)) +
annotate("path",
x = 10*cos(seq(0,2*pi,length.out=100)),
y = 10*sin(seq(0,2*pi,length.out=100)),
color="red")

# Put them together
p00 + pLL
```

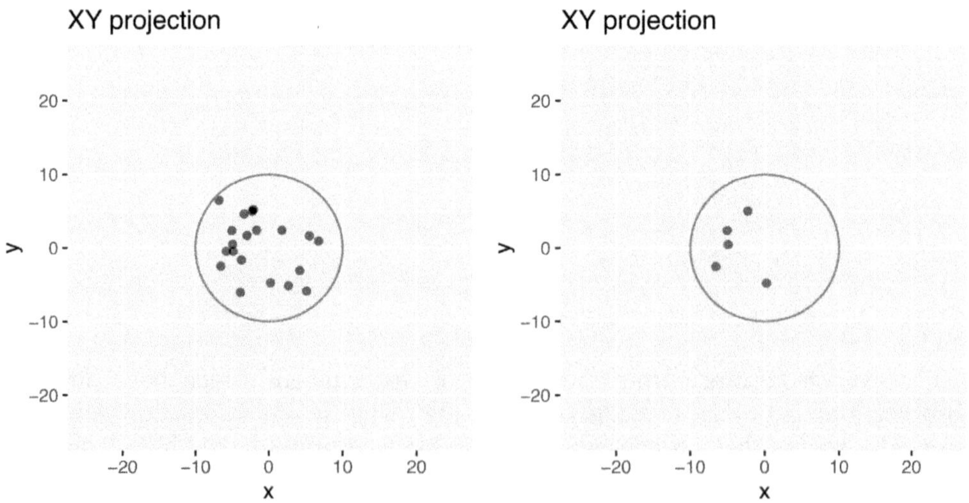

We can also 'zoom in' around a single neutron, one that has survived until cycle 20, and picture its *zig-zag* path. It is easy to find the 'survivor' neutrons (`iAlive` in the code). Then all positions of the selected neutron over all cycles, must be stored into a matrix whose rows contain x, y, z.

```
# The initially-selected group of neutrons, and all
# the tibbles in the subsequent 20 cycles are stored
# in core0 and Lneutrons, generated previously.

# Check which neutron at cycle 20 was present at creation
iAlive <- Lneutrons[[20]] |> select(id) |> pull()

# Neutron 17 has survived until cycle 20. Let us monitor it
nchosen <- 17

# Create matrix of positions at all cycles


Mpos <- matrix(nrow=21,ncol=3)

# Fill matrix
rw <- core0 |>
filter(id == nchosen) |>
select(x,y,z) |>
unlist() |>
as.numeric()
Mpos[1,] <- rw
for (i in 1:20) {
rw <- Lneutrons[[i]] |>
filter(id == nchosen) |>
select(x,y,z) |>
unlist() |>
as.numeric()
Mpos[i+1,] <- rw
}
```

Next, the first two columns of the matrix built are turned into a tibble (`XY`), ready for the plot. The plot clearly shows initial and final position as a green and red circle, respectively.

```
# For the plot, turn x,y columnsof Mpos into  a tibble
# Transpose and keep only x and y
XY <- as_tibble(Mpos[,1:2])
names(XY) <- c("x","y")
XY$step <- seq_len(nrow(XY))  # Add time step

# Plot
ggplot(XY,aes(x=x,y=y)) +
geom_path(color="blue") +
geom_point(size=1,alpha=0.6) +
geom_point(data=XY[1,],aes(x=x,y=y),
color="green",size=1.5) +
geom_point(data=XY[nrow(XY),],aes(x=x,y=y),
color="red",size=1.5) +
coord_fixed() +
labs(title="Neutron 17",x="X",y="Y") +
theme_minimal()
```

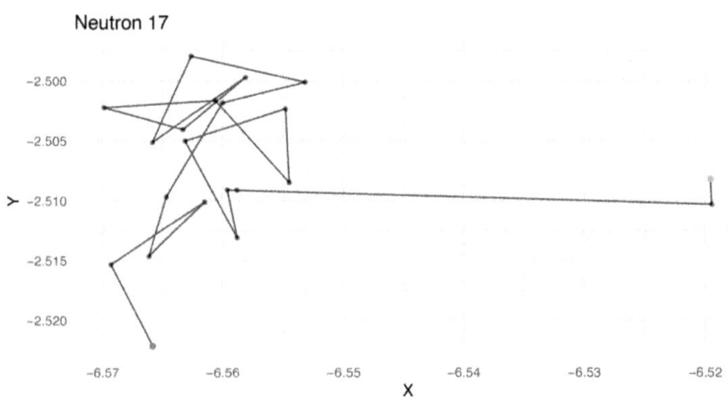

We could repeat the process for the other 4 ‘survivor’ neutrons, but, as it is good practice in R, it is better to prepare a general function to plot a zig-zag path, given the neutron’s id number.

```
# Function to plot a neutron's zig-zag path
ZiggyZaggy <- function(nchosen,core0,Lneutrons) {
# Create container of coordinates
Mpos <- matrix(nrow=21,ncol=3)

# Fill the container
rw <- core0 |>
filter(id == nchosen) |>
select(x,y,z) |>
unlist() |>
as.numeric()
Mpos[1,] <- rw
for (i in 1:20) {
rw <- Lneutrons[[i]] |>
filter(id == nchosen) |>
select(x,y,z) |>
unlist() |>
as.numeric()
Mpos[i+1,] <- rw
}

# Turn to tibble
```

```
XY <- as_tibble(Mpos[,1:2])
names(XY) <- c("x","y")
XY$step <- seq_len(nrow(XY))  # Add time step

# Plot
p <- ggplot(XY,aes(x=x,y=y)) +
geom_path(color="blue") +
geom_point(size=1,alpha=0.6) +
geom_point(data=XY[1,],aes(x=x,y=y),
color="green",size=1.5) +
geom_point(data=XY[nrow(XY),],aes(x=x,y=y),
color="red",size=1.5) +
coord_fixed() +
labs(title = paste("Neutron",n),x="X",y="Y") +
theme_minimal()

return(p)
}
```

We apply, next, the function to the remaining neutrons and tile the four plots together using `patchwork`.

```
# Plot 4 zig-zag paths on a tiled plot
p <- list()
for (n in iAlive[2:5]) {
p <- c(p,list(ZiggyZaggy(n,core0,Lneutrons)))
}

# Use patchwork (easy!)
(p[[1]] | p[[2]]) /
(p[[3]] | p[[4]])
```

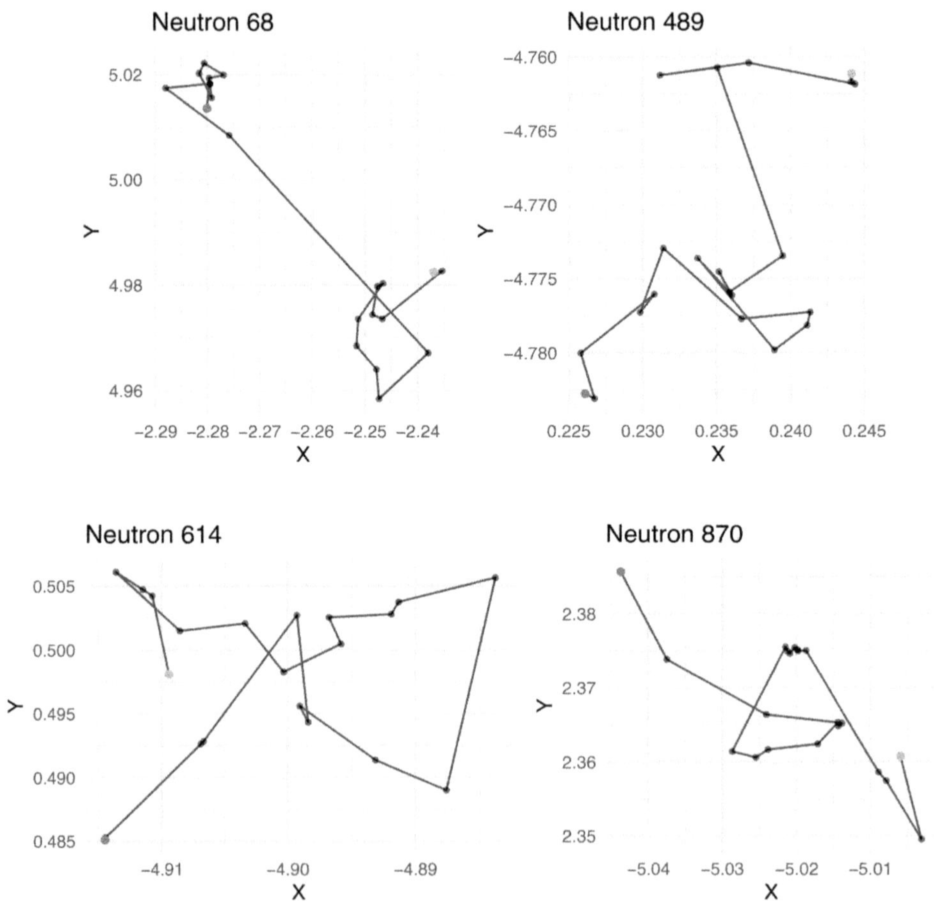

The three demonstrations in this section should give readers enough material to explore and apply the Monte Carlo code with an even larger variety of demonstrations.

References

[1] Rhodes R 1986 *The Making of the Atomic Bomb* (Simon & Schuster) Pulitzer Prize-winning history of the Manhattan Project, including Ulamas Monte Carlo idea
[2] Metropolis N 1980 *The Beginning of the Monte Carlo Method* (Academic) Memoir by Metropolis describing Ulam's insight and the first applications at Los Alamos
[3] Metropolis N and Ulam S 1949 *The Monte Carlo method* **44** 335–41
[4] Neumann J V 1951 Various techniques used in connection with random digits *J. Res. Natl. Bur. Stand. Appl. Math. Ser.* **12** 36–8 Foundational description of pseudo-random number generation for Monte Carlo simulations
[5] Matsumoto M and Nishimura T 1998 Mersenne twister: A 623-dimensionally equidistributed uniform pseudorandom number generator *ACM Trans. Model. Comput. Simul.* **8** 3–30
[6] Box G E P and Muller M E 1958 A note on the generation of random normal deviates *Ann. Math. Stat.* **29** 610–1
[7] Fishman G S 1996 *Monte Carlo: Concepts, Algorithms, and Applications* (Springer)
[8] Woolfson M M and Pert G J 1999 *An Introduction to Computer Simulation* (Oxford Illustrated Edition) (Oxford University Press)

IOP Publishing

Computational Physics with R

James Foadi

Chapter 10

Differential equations with `deSolve`

10.1 The universe of ODE solvers

In chapter 8 we have explored a variety of numerical methods for solving ordinary differential equations (ODEs). We began with explicit initial value methods of fixed step size, then introduced implicit schemes to address stability issues, and finally examined boundary value problems and eigenvalue problems. However, the universe of ODEs is far richer than what we covered in that chapter. In practice, differential equations arise in countless forms and often display behaviours that cannot be efficiently handled by the fixed-step algorithms we have studied. Systems may be stiff, highly oscillatory, constrained, or involve events such as sudden changes in dynamics[1]. For each of these situations, tailored algorithms have been developed, ranging for example from adaptive step-size methods to multistep techniques [1].

It is therefore important to realise that the methods studied in this chapter represent only a small fraction of the available arsenal of ODE solvers. For practical computational physics, one typically relies on well-tested software packages that implement a wide range of such methods, selecting the most appropriate algorithm depending on the problem at hand. In the R environment several packages exist, and in what follows we will focus on one of the most widely used and stable libraries, `deSolve`, which offers a large collection of solvers for ODEs.

10.2 The package `deSolve`

The `deSolve` package is the main library in R for the numerical integration of differential equations. It provides a wide range of solvers for ODEs and other types of differential equations. The package implements both classic algorithms (such as the Runge–Kutta family) and more advanced adaptive solvers (such as the `lsoda`

[1] The reader not familiar with the terminology relevant to solve ODEs should have a quick look at chapter 8.

doi:10.1088/978-0-7503-2632-2ch10

method, which automatically switches between stiff and non-stiff modes). This makes `deSolve` suitable for a broad spectrum of problems in computational physics and beyond. The main reference for `deSolve` is the 2012 book by Soetaert and colleagues [2].

At the core of `deSolve` is the function `ode`, which serves as a general interface to the different solvers. To use it, one must provide:

- the system of differential equations, written as an R function;
- the initial conditions for the state variables;
- the sequence of time points at which the solution is requested;
- and the solver to be employed, specified via the `method` argument.

The output of `ode` is a matrix (or data frame) containing the time points and the corresponding solution values for the system variables. This makes it straightforward to inspect, analyse, and plot results in the usual R workflow.

Example 10.1 *Let us solve numerically the following initial value problem (IVP)*

$$\frac{dy}{dt} = (1 - 2t)y, \qquad y(0) = 1, \; t \in [0, 3]$$

with the Euler method. This is going to be found as `method="euler"` *inside the function* `ode`.

When using the solvers in `comphy`, *the gradient (RHS) of an IVP is typically written as a function of the form* `f(t,y,...)` *that returns a numeric vector. By contrast,* `deSolve`*'s integrator* `ode` *calls the user function with the fixed signature*

```
func <- function(time,state,parms){...}
```

and expects the return value to be a list whose first element is the derivative vector. This is a source of error when a `comphy`*-style function (missing the* `parms` *argument and returning a bare number) is passed to* `ode`. *The minimal input requirements are*:

1. **Arguments**. `time` *(scalar),* `state` *(named numeric vector),* `parms` *(any object; use* `NULL` *if not needed).*
2. **Accessing state variables**. *Either index them, e.g.* `state["y"]`, *or unpack them for readability via* `with(as.list(state), {...})`.
3. **Return type**. `list(deriv)`, *where* `deriv` *is a numeric vector of the same length as* `state` *(even if length 1).*

To keep future options open (e.g. passing extra parameters), define your RHS as

```
function(t,state,parms,...)
```

and pass any extra arguments via `parms` *(a list) or the ellipsis. The key points to remember are the required signature and the* list *return value that* `deSolve` *expects. The following demonstration should clarify this important aspect on the use of* `ode`. *Extra comment lines have been added to further explain the format used for the gradient's definition.*

```
# Don't forget to load deSolve in memory
require(deSolve)

# Gradient (watch out names of gradient variables and the returned list)
#
# The 'state' object arrives here as a numeric vector (e.g. c(y = 1)).
# To make the code cleaner, we convert it into a list so its elements
# can be referred to directly by name:
#
#   as.list(state)  ->  list(y = 1)
#
# Then 'with(...)' creates a local environment where we can use 'y'
# directly instead of writing state["y"] every time.
#
# Inside the block:
#   dy <- (1 - 2*t) * y      # compute the derivative
#
# The function must return the derivatives as a list, even if only
# one equation is present. Hence:
#   list(c(dy))
#
# This structure scales up to multiple variables, in which case we
# would return list(c(dy1,dy2,...)).
#
f <- function(t,state,parms) {
with(as.list(state), {
dy <- (1-2*t)*y
list(c(dy))
})
}

# Integration interval (as grid with fixed step size)
times <- seq(0,3,by=0.2)

# Initial conditions
y0 <- 1

# Generic solver (calls Euler)
Sol <- ode(y=y0,times=times,func=f,parms=NULL,method="euler")

# Sol is a 16 X 2 matrix (t,y)
print(Sol)
```

```
##    time     1
```

```
## 1   0.0  1.00
## 2   0.2  1.20
## 3   0.4  1.32
## 4   0.6  1.36
## 5   0.8  1.32
## 6   1.0  1.20
## 7   1.2  1.00
## 8   1.4  0.72
## 9   1.6  0.36
## 10  1.8 -0.08
## 11  2.0 -0.60
## 12  2.2 -1.20
## 13  2.4 -1.88
## 14  2.6 -2.64
## 15  2.8 -3.48
## 16  3.0 -4.40
```

The output is rather simple: a matrix whose first column contains the grid of the integration variable and whose second column contains the integrated values (t_i and y_i).

Example 10.2 *The second example is a simple linear system,*

$$\begin{cases} \mathrm{d}y_1/\mathrm{d}t &= y_2 \\ \mathrm{d}y_2/\mathrm{d}t &= -y_1 \end{cases}, \qquad y_1(0) = 1,\ y_2(0) = 0,\ t \in [0,\ \pi].$$

The related R code for the use of `ode` *is following the previous one closely, with the difference that now the gradient will be formed for a system and not a single ODE. Also, we will use a different method, the fourth-order Runge–Kutta.*

```
# Gradient
f <- function(t,state,parms) {
with(as.list(state), {
dy1 <- y2
dy2 <- -y1
list(c(dy1,dy2))
})
}

# Integration interval (careful as [0,pi] is incommensurable)
times <- seq(0,pi,length.out=21)

# Initial conditions
y0 <- c(y1=1,y2=0)
```

```
# Generic solver (calls RK4)
Sol <- ode(y=y0,times=times,func=f,parms=NULL,method="rk4")

# Sol is a matrix with columns: time, y1, y2
# Plot
plot(Sol[,1],Sol[,2],type="l",
xlab=expression(t),ylab=expression(y(t)))
points(Sol[,1],Sol[,3],type="l",col="magenta")
```

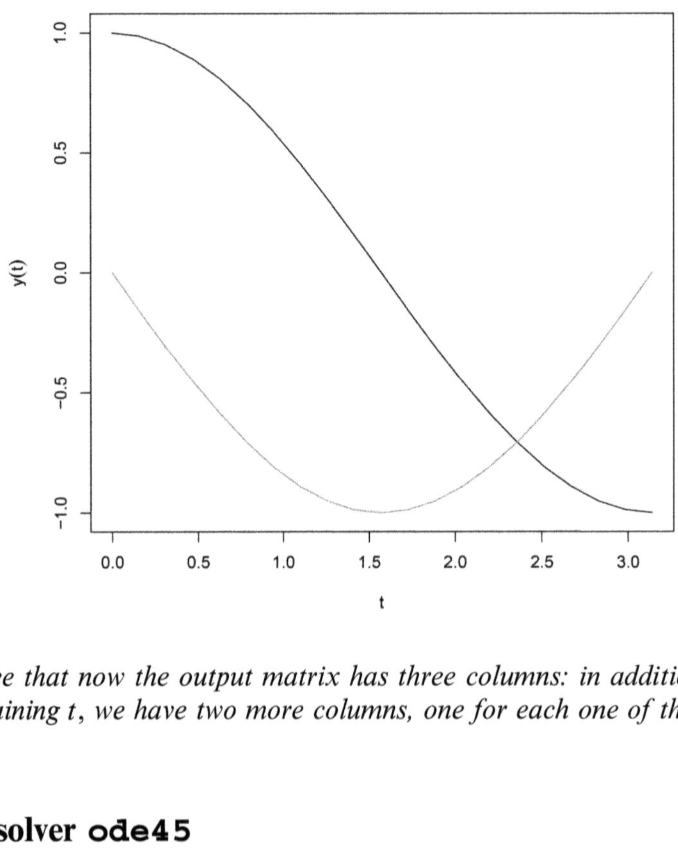

We can see that now the output matrix has three columns: in addition to the first column containing t, we have two more columns, one for each one of the variables y_1 and y_2.

10.3 The solver `ode45`

Among the solvers available in `deSolve`, one of the most convenient is `ode45`. This is often referred to as a 'one-fits-all' solver, and in practice it is usually the first choice when attempting to solve an initial value problem. The name comes from

MATLAB [3], where `ode45` is the default integrator, and the implementation in `deSolve` follows the same idea.

The algorithm behind `ode45` is an explicit Runge–Kutta method of order 5 with an embedded method of order 4 (also known as the *Dormand–Prince pair* [4, 5]). At every step the method produces two approximations of different order, and their difference provides an estimate of the local error. This allows the solver to adjust the step size dynamically: when the error is small the step is increased, leading to efficiency, while when the error grows the step is reduced, ensuring stability and accuracy.

Because of its adaptive mechanism and relatively high order, `ode45` performs well on a wide variety of non-stiff problems without requiring the user to tune many parameters. It is less suitable for stiff equations, where implicit methods or solvers specifically designed for stiffness (such as `lsoda` or `radau`) are recommended. Nonetheless, for many smooth systems of ODEs, `ode45` represents a robust and reliable first attempt.

Example 10.3 *The* Lorenz system *is one of the most famous examples of a low-dimensional nonlinear dynamical system. It was introduced by Edward N Lorenz in 1963 [6] as a simplified model of atmospheric convection. The system consists of three coupled first-order ODEs*:

$$\begin{cases} \dot{x} = \sigma(y - x), \\ \dot{y} = x(\rho - z) - y, \\ \dot{z} = xy - \beta z, \end{cases}$$

where $\sigma > 0$, $\rho > 0$, *and* $\beta > 0$ *are parameters. Here* σ *is the Prandtl number,* ρ *is a scaled Rayleigh number, and* β *is a geometric factor.*

With the classical parameter set $\sigma = 10$, $\rho = 28$, $\beta = 8/3$, *and initial condition* $(x, y, z) = (1, 1, 1)$, *the system exhibits chaotic behaviour. This example is particularly useful to demonstrate adaptive solvers such as* `ode45` *because the trajectories are sensitive to initial conditions and require careful error control.*

Adaptive solvers regulate their step size using two numerical tolerances: the relative tolerance *(*`rtol`*) and the* absolute tolerance *(*`atol`*). Roughly speaking,* `rtol` *controls the number of correct digits requested in each component of the solution, while* `atol` *prevents loss of accuracy when components become very small. For instance, choosing* `rtol = 1e-6` *and* `atol = 1e-9` *requests six correct digits in variables of moderate size, while ensuring accuracy is not lost if one of the variables approaches zero. By adjusting these tolerances, the user can balance precision against computational cost.*

In the numerical experiments presented here, we have chosen to display only the two-dimensional projection of the phase portrait, plotting $y(t)$ *against* $x(t)$. *This avoids the need to call specialised three-dimensional graphics libraries, while still revealing the essential structure of the Lorenz attractor. The 2D projection is sufficient to highlight the system's chaotic oscillations and sensitive dependence on initial conditions. In addition, the following demonstration offers the chance to observe how extra parameters are passed to* `ode` *via the* `parms` *variable.*

```
# Lorenz{63 parameters and RHS
parms  <- c(sigma=10,rho=28,beta=8/3)
lorenz <- function(t,state,parms) {



with(as.list(c(state,parms)), {
dx <- sigma*(y-x)
dy <- x*(rho-z)-y
dz <- x*y-beta*z
list(c(dx,dy,dz))
})
}

# initial condition and time grid (can be named too!)
y0    <- c(x= 1,y=1,z=1)
times <- seq(0,40,by=0.01)

# integrate with Dormand{Prince 4(5) (ode45 analogue)
Sol <- ode(y=y0,times=times,func=lorenz,parms=parms,
method="ode45",rtol=1e-6,atol=1e-9)

# phase portrait: y vs x
# Now can use name, rather than column number
plot(Sol[,"x"],Sol[,"y"],type = "l",
xlab=expression(x(t)),ylab=expression(y(t)),
main="Lorenz{63 phase portrait (y vs x)")
```

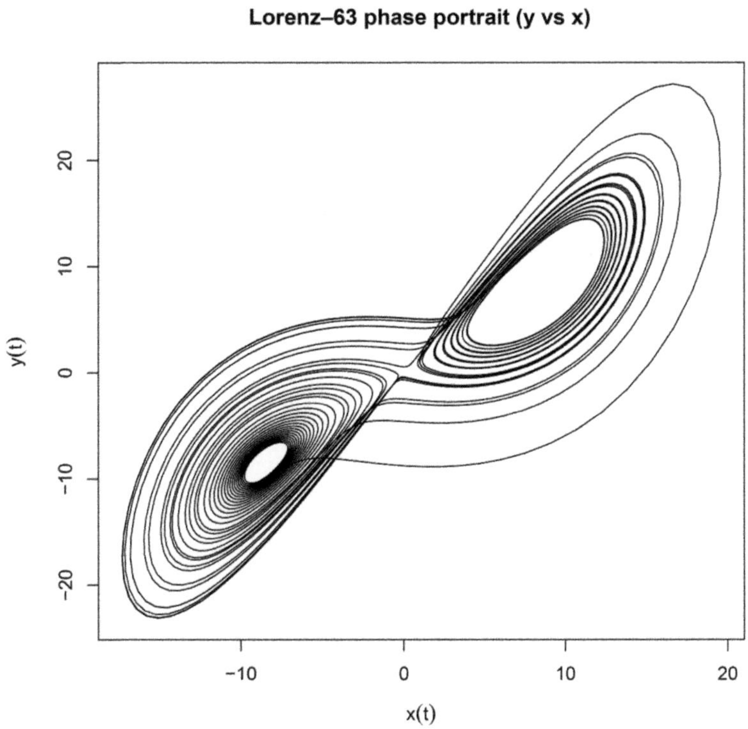

10.4 The solver `lsoda`

Another widely used solver in `deSolve` is `lsoda`. This is in fact the default method called by `ode()` whenever no solver is explicitly specified. Unlike `ode45`, which is designed primarily for non-stiff problems, `lsoda` has the ability to deal effectively with both stiff and non-stiff systems of ODEs. This flexibility makes it one of the most important and robust solvers available.

The strength of `lsoda` lies in its ability to automatically detect stiffness during the integration and to switch methods accordingly. For non-stiff regimes it employs an explicit multistep Adams method[2], which is efficient and accurate. When stiffness is detected, it changes to an implicit method based on backward differentiation formulas (see section 8.10). This dynamic switching means that the user does not need to decide in advance whether the system is stiff or not: the solver adapts during the computation. Because of this dual capability, `lsoda` is often the safest general-purpose choice when there is uncertainty about the nature of the problem. It is particularly well suited to systems where stiffness may emerge intermittently, or when the degree of stiffness is not known *a priori*.

[2] You can read about the method, for instance, in the book by Hairer and Nørsett [1].

Example 10.4 *The nonlinear Van der Pol oscillator has been already introduced as a stiff problem in example 8.9. Transformed as a first-order system of ODEs, it reads*

$$\begin{cases} y' = y_2, \\ y' = \mu\,(1 - y_1^2)\,y_2 - y_1, \end{cases} \qquad y_1(0) = 2, \;\; y_2(0) = 0.$$

As seen, for small μ the dynamics are smooth and non-stiff, but as μ grows the solution develops fast transitions (boundary layers) and becomes stiff. *This makes* `lsoda` *a natural choice: it automatically detects stiffness and switches between non-stiff and stiff modes.*

Below we solve on $t \in [0, 1000]$ for a stiff choice $\mu = 100$, as done in example 8.9. The output grid is uniform, but the internal step size is chosen adaptively by the solver. This point is important as the user only cares to use a comfortable grid, with the bonus of saving on execution time.

```
# Van der Pol in first-order form:
#   y1' = y2
#   y2' = mu*(1 - y1^2)*y2 - y1
f_vdp <- function(t,state,parms) {
with(as.list(c(state,parms)), {
dy1 <- y2
dy2 <- mu*(1-y1^2)*y2-y1
list(c(dy1,dy2))
})
}

# Parameters and problem setup
parms <- list(mu=100)
times <- seq(0,100,by=0.01)
y0  <- c(y1=2,y2=0)

# Solving with timings
t_solve <- system.time({
Sol <- ode(y=y0,times=times,func=f_vdp,parms=parms,method="lsoda")
})
print(t_solve)

##    user  system elapsed
##    0.20    0.00    0.22

# Plot
plot(Sol[,1],Sol[,"y1"],type = "b",pch=16,cex=0.3,
xlab=expression(t),ylab=expression(y(t)))
```

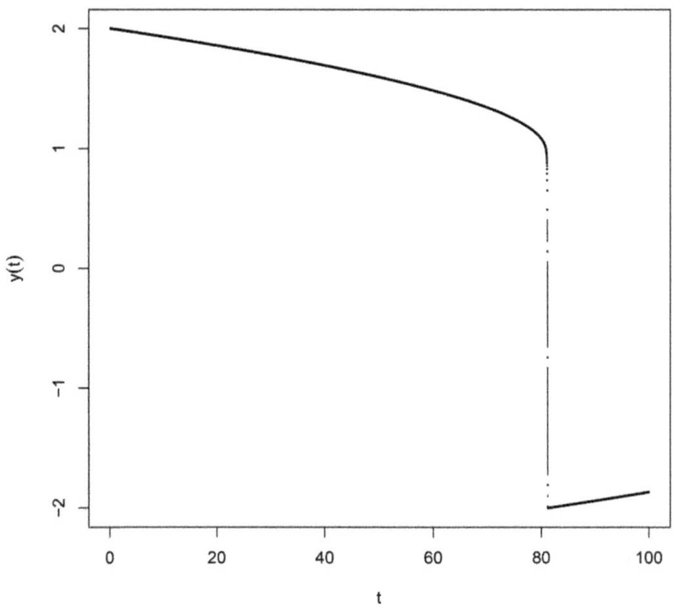

As one can see, the plot of the solution clearly shows details of the curve. But the solution was obtained for a grid with step size $h = 0.01$, *while the standard RK4 algorithm could not converge to a solution with this step size. When it did, with* $h = 0.001$, *execution time was* 0.72 *s, while it is only* 0.20 *seconds with the* `lsoda` *solver, nearly four times shorter. More dramatic time savings can be observed for longer integration intervals and different stiff equations.*

References

[1] Hairer E, Nørsett S P and Wanner G 1993 *Solving Ordinary Differential Equations I: Nonstiff Problem* (Springer Series in Computational Mathematics) vol 8 2nd edn (Springer)
[2] Soetaert K, Cash J R and Mazzia F 2012 *Solving Differential Equations in* R (Springer)
[3] The MathWorks, Inc. 2024 MATLAB (The MathWorks, Inc.)
[4] Prince P J and Dormand J R 1980 A family of embedded Runge-Kutta formulae *J. Comput. Appl. Math.* **6** 19–26
[5] Prince P J and Dormand J R 1981 High order embedded Runge-Kutta formulae *J. Comput. Appl. Math.* **7** 67–75
[6] Lorenz E N 1963 Deterministic nonperiodic flow *J. Atmos. Sci.* **20** 130–41

Chapter 11

A short introduction to machine learning

The year 2022 will be forever associated with artificial intelligence (AI) coming out of obscurity. On 30 November 2022, OpenAI [1] released *chatGPT* [2], an artificial intelligence chatbot that uses GPT (version 3.5), a so-called *large language model*, designed to generate text or language-based output following the input received [3].

chatGPT is a very sophisticated form of *neural network*, one of the topics forming the vast area of *machine learning*. It uses a variant of the so-called *Transformer* architecture, which allows it to process and generate text in a coherent and contextually relevant manner. This is why chatGPT is capable of understanding and generating human-like text based on the input it receives. Indeed, when using chatGPT interactively, the user has a feeling of dialoguing with another intelligent human being.

This chapter touches superficially on the main elements of machine learning, with the inclusion of a description and some easy examples of neural networks. The material is in no way intended as a self-contained introduction to machine learning. Rather, it was felt that some notions to wet the appetite of physicists in this current and popular trend of quantitative research, had to be included in a modern book on computational physics. Serious introductions can be consulted separately, for example in references [4] and [5]. In any case, the knowledge acquired should be enough to understand what a Transformer and tools like chatGPT are. The chapter will also present examples and demonstrations of neural networks and some ways to interface and use chatGPT from within R.

11.1 What is machine learning?

It is not possible to provide a happy single definition of machine learning, as this subject sits at the interface of mathematics, statistics and computer science, so that a definition depends on whether mathematicians rather than statisticians or computer scientists describe it. Furthermore, its use is constantly enriched by the

doi:10.1088/978-0-7503-2632-2ch11

innumerable applications to all fields of human knowledge. In brief, machine learning enables computers to learn from data in an automated or semi-automated fashion. Such learning involves algorithms that analyse and detect patterns, based on direct or indirect validation of specific output. An example in classical mechanics can help to make the above description less abstract. Let us consider a scenario in which a projectile ejected horizontally from a cannon placed at $(x, y) = (0, y_0)$ and with initial speed v_0 (input variable), lands at a position $(d, 0)$ (d is the output variable). Data are gathered by conducting experiments where projectiles with varying initial speeds are shot and the corresponding landing distances are measured. A simple mathematical model can be created to capture the relationship between the distance travelled and the corresponding initial velocity. For example, the linear regression model could be used, presuming that the distance is thought to be linearly dependent from the initial velocity. Although the choice of the mathematical model is important for the accuracy of the prediction, the key point is that the machine learns the appropriate features of the model from the data collected. In the case of linear regression, the two parameters needed come out of the least squares method. Here least squares can be seen as one of the algorithms which are part of machine learning, as the parameters are found automatically by the algorithm, in an attempt to reach the goal of finding the closest straight line to all data points. At this point, if the reader thinks they have already encountered this topic in chapter 5, when discussing data fitting, they are absolutely correct. Machine learning utilises well-established statistical techniques with a particular emphasis on data. In fact, sometimes the term *statistical learning* is used instead of machine learning to emphasise the statistical character of the process. If we indicate with X the input data (in the example of the projectile, these would consist of all, say n, initial speeds, $X = v_{01}, v_{02}, \ldots, v_{0n}$), the output data Y (in the example of the projectile these are the n distances, $Y = d_1, d_2, \ldots, d_n$), the model chosen represents an unknown function, f, and an ineliminable error, ϵ, which quantifies what cannot be modelled:

$$Y = f(X) + \epsilon. \tag{11.1}$$

Statistical learning refers to the vast array of methods used to estimate f. Note that the only accessible learning in trying to model Y is the function f. The error ϵ is not part of the model, it is independent from the model and in principle unknowable. The error yielded to estimate f can be reduced and it is, thus, called *reducible error*. What cannot be reduced is ϵ, which is therefore called *irreducible error*.

The variables X and Y in the example have been introduced, respectively, as *input variables* and *output variables*. But in machine learning jargon (borrowed from modern statistics), input variables are also known as *independent variables*, *explanatory variables* or *predictors*, while output variables are also known as *dependent variables* or *response variables*. These terms can therefore be used interchangeably. In this chapter the number of input variables is denoted by p, while the number of observations is denoted by n. The symbols

$$x_{i1}, x_{i2}, \ldots, x_{ip}$$

represent the value of the p input variables in relation to observation i. If an output variable, y_i, is also available, the set

$$x_{i1}, x_{i2}, \ldots, x_{ip}, y_i$$

is known as *data points*. The n available data points form a group of *training data* because they are used to *train* the algorithm to estimate the function f in formula (11.1).

11.2 Machine learning in physics

The focus on data rather than mathematical laws introduced in the previous section might trigger general questions on its use in physics, a discipline whose central tenet is that the universe is governed by mathematical models, discoverable through reasoning and data analysis. In fact, machine learning and traditional physics approaches often have different focuses when it comes to modelling.

In physics, the emphasis is typically on finding precise mathematical equations or models that describe natural phenomena. These models are often derived from fundamental principles and laws, and they aim to provide a deep understanding of the underlying physics. Physicists seek to uncover the fundamental laws that govern the Universe, and these laws are often expressed in precise mathematical forms, such as differential equations.

In contrast, machine learning does not necessarily aim to find exact mathematical models to describe phenomena. Instead, machine learning approaches are data-driven and focus on learning patterns, relationships, and trends from large datasets. Machine learning models are often used to make predictions or classifications based on data, and they can be highly effective in tasks where the underlying processes are complex and may not be fully understood or described by precise equations.

Machine learning models, such as neural networks, decision trees, or support vector machines, are designed to capture and exploit patterns in data. They excel at tasks like image recognition, natural language processing, and making predictions based on historical data. While they may not provide the same level of fundamental insight into the underlying physics as traditional models, they are powerful tools for tasks where the complexity of the system makes it challenging to derive precise equations.

In some cases, machine learning and physics can complement each other. Physicists can use machine learning to analyse complex experimental data, extract patterns, and make predictions, even when the underlying physics is not fully understood. This can lead to new discoveries or insights. Additionally, machine learning techniques can be used to optimise experiments, accelerate simulations, and solve problems that are computationally intractable with traditional methods.

So, while machine learning and traditional physics modelling have different emphases, they can coexist and be used synergistically in research, especially in areas where the complexity of the system or the amount of data makes traditional modelling approaches challenging.

11.3 Parametric and non-parametric methods

The estimate of function f in equation (11.1) can be done using methods in one of two broad groups known as *parametric* and *non-parametric* methods.

11.3.1 Parametric methods

We have already seen parametric methods in chapter 5, for example in multilinear least squares (section 5.2). The key component of a parametric method is the estimation, *via* the training data, of the parameters characterising the function f. That is, rather than guessing an innumerably large set of functions, a parametric form is imposed to the function and estimates are subsequently performed on the numeric value of these parameters. For example, one could impose (or select in a reasonable way) the linear form

$$f(X) = \beta_0 + \beta_1 X_1 + \beta_2 X_2 + \cdots + \beta_p X_p,$$

where $X_1, \ldots, X_p$ are the p random variables of the model, and use algorithms like least squares to estimate the p parameters. One could then say that parametric methods consist of two steps: the first is the *imposition* of a parametric form to the function, the second is the *estimate* of the parameters used in the function. Least squares, which we have used in chapter 5, is only one of the techniques used to estimate the parameters. There exist other methods which often yield more accurate estimates than least squares like, for instance, *subset selection*, *shrinkage* and *dimension reduction*. There is no room in this introductory chapter to explain these methods, and the reader is directed to the references at the beginning of this chapter for details.

11.3.2 Non-parametric methods

These methods try to discover the shape of f not by imposing a parametric functional form but, rather, by attempting to be as close as possible to the data points, without risking creating a function which is too uneven or erratic. In a certain sense, non-parametric methods have already been described in chapter 3, when describing polynomial interpolation or cubic splines. Those are, in fact, examples of *overfitting*, where the function is estimated so to pass through all the available data points. This amounts to considering an infinite accuracy of the model because all errors are reduced to zero. This is nearly always not the case. Most data are affected by errors, and the criterion of estimating a function that goes through all data points is not sensible and yields, in fact, very erratic results. But methods similar to those of chapter 3, albeit with different guiding principles from what we just described, can produce estimates of f which agree well with the data while still showing features like smoothness which are desirable in a function.

11.3.3 Which method to use?

The adoption a one method rather than another might appear subjective. Quite often researchers stick with whatever method they know best, without asking themselves if a different method would have actually worked better. There are, in fact, guidelines attached to all methods, and an important consideration must be kept in mind when a parametric method is used, rather than a non-parametric one, and vice versa.

A parametric method forces the function f to acquire a specific form. This brings into the method a sort of *rigidity* where one is forced to find agreement with the data even if the model (function) is not really suitable for those data. A typical example could be when data coming from a quadratic are forced to be modelled using a linear model. In such a case the reducible error cannot really be reduced beyond a certain value, and this is mostly to be imputed to the rigidity of the model. On the other hand, a non-parametric method is very flexible, to the extent that a model function can pass through all data points, thus reducing both reducible and, what it is most surprising, the irreducible error to zero. Clearly, the irreducible error should never be zero and thus a non-parametric method producing functions passing through all data points is not a good method, because too flexible.

We get therefore the idea that rigidity and flexibility are two features of a method that need to be monitored. We should also be aware that a propension for rigidity rather than flexibility and vice versa might be related to what it is that one tries to capture with the model. For example, if the correlation between specific variables and the output is what is investigated, then it makes sense to pre-define a model with parameters, some of which are associated with those specific variables. On the other hand, if regions of vicinity to all data points are of interest, then an accurate and flexible model is the preferred choice.

11.4 Supervised, unsupervised and reinforcement learning

Up to this point we have discussed examples of *supervised* learning, where besides the explanatory variables, $x_{i1}, \ldots, x_{ip}$, response variables, y_i, are collected. The reason why algorithms using both explanatory and response variables are said to be part of supervised machine learning is connected with the possibility of changing the model depending on how far the response variable is from a target value. Consider, for instance, the case of linear regression, a supervised machine learning algorithm. Specific values of the parameters $\beta_0, \beta_1, \ldots, \beta_p$ provide a *response* $f(X_i) = \beta_0 + \beta_1 x_{i1} + \cdots + \beta_p x_{ip}$ to the set i of explanatory variables. If the value y_i has been collected, this can be compared with the response $f(X_i)$. Such a comparison, from the point of view of machine learning, *teaches* the model whether the parameter values are acceptable or must be changed. In other words, the presence of response variables provides a way for the algorithm to learn how well or not so well the model justifies the data. Such learning happens in the presence of response data, hence the *supervised* adjective to this type of learning.

When response data are not available, then it is harder to know how the algorithm can learn because there are no means of comparisons. This is the case of *unsupervised* learning. The only available data are input or explanatory variables. Several approaches can be attempted in this case, one of which is data exploration, where input variables are queried to see whether they show presence of unknown patterns.

A classic example of this type of unsupervised learning is cluster analysis where data proximity and the subsequent formation of clusters can reveal relations among variables which were previously unknown. In this case the algorithm discovers the patterns on its own or, equivalently, learns the relations without the assistance of response data.

Reinforcement learning is a branch of machine learning attracting lots of attention in recent years. It revolves around making optimal decisions to maximise rewards in specific situations. This approach is employed to determine the most effective behaviour or path to follow in a given context. Unlike supervised learning, where the training data provides explicit answers, in reinforcement learning there are no pre-defined answers. Instead, a so-called *reinforcement agent* learns by trial and error, figuring out how to accomplish a given task through its experiences and interactions with the environment. To demonstrate the concept, consider a simple game where a computer program controls a virtual robot. The robot doesn't know how to navigate the game world at first. Through repeated attempts and learning from its mistakes, the robot gradually learns to navigate obstacles and reach a goal. It receives positive *rewards* when it makes progress and negative *penalties* for mistakes. Over time, the reinforcement learning algorithm helps the robot develop a strategy to maximise its rewards, effectively learning how to excel at the game without explicit instructions.

While reinforcement learning is not the newest part of machine learning (the concept was introduced in the 1950s), it has gained significant attention and popularity in recent years due to its successes in various applications, including game playing (e.g. AlphaGo and AlphaZero), robotics, autonomous vehicles, and natural language processing. Advances in reinforcement learning algorithms and the availability of powerful computing resources have contributed to its resurgence and widespread use in cutting-edge AI applications.

11.5 The landscape of machine learning

As machine learning is a broad area of research and applications, and with the goal of providing a map to navigate current terminology, we offer here a compact overview. Machine learning is one branch of Artificial Intelligence, and itself subdivides into several main approaches. Among these, *supervised* and *unsupervised* learning are the classical pillars, while *reinforcement learning* covers a different paradigm based on interaction with an environment. *Deep learning* represents a family of methods based on multi-layer neural networks, and underpins many state-of-the-art advances in all three of these areas. Finally, *Large Language Models (LLMs)* are specialised deep learning architectures that have become a dominant application in natural language processing.

The following scheme provides a compact map of these relationships. It should be read from top to bottom, starting with Artificial Intelligence as the broadest concept, and moving down into the specific subfields.

- **Artificial Intelligence (AI)**
  - **Machine Learning (ML)**: systems that learn from data
    - **Supervised Learning**: learning in the presence of response data (e.g. classification, regression)
    - **Unsupervised Learning**: learning without response data, relying only on input variables (e.g. clustering, dimensionality reduction)
    - **Reinforcement Learning**: learning by interaction with an environment through rewards and penalties
      - Deep Reinforcement Learning: reinforcement learning using deep neural networks
    - **Deep Learning (DL)**: multi-layer neural networks, applied in supervised, unsupervised, and reinforcement settings
      - **Large Language Models (LLMs)**: deep learning architectures specialised for natural language

11.6 An example of supervised learning

Let us demonstrate supervised learning as applied to a basic physics example. We are going to explore the relation between the height, y_0, where the cannon is located, the initial horizontal speed, v_0, of the projectile and the distance from the cannon, d, where the projectile lands, i.e. touches the ground (see section 11.1). In a traditional physics approach, one would check agreement with the equations of motion under constant gravity. The landing distance of a projectile launched horizontally with speed v_0 from height y_0 is given exactly by

$$d = v_0\sqrt{\frac{2y_0}{g}},$$

where g is the Earth gravity acceleration (see formula (A.28) of section A.16 in appendix A)). A physicist would therefore test experimental measurements of y_0, v_0, d against this formula, using it as the model to explain the observations. If the data were consistent, the formula would be confirmed; if not, the physicist would refine assumptions or consider additional forces such as air resistance.

A machine learning scientist would act differently. Instead of starting from Newton's laws, they would treat the relation between inputs y_0, v_0 and output d as unknown, and attempt to learn it directly from data. The idea is not to derive or confirm a theoretical formula, but to train an algorithm that can predict the landing distance for new values of the input variables. The tool of choice for this is regression: beginning with a linear model, extending to a quadratic model, and possibly considering testing a cubic model. Each of these models is fitted to data by minimising the discrepancy between predicted and observed distances, and their performance is compared. This is the quantitative route taken in supervised learning, a route structured as a series of steps.

1. **Define the learning problem**. The inputs (or *features*) are the initial conditions of the projectile, $x = (y_0, v_0)$, while the output (or *label*) is the landing distance, d. The aim of supervised learning is to construct an algorithm that

produces a function $\hat{d} = f(x)$ which approximates the true dependence of d on the input variables.

2. **Choose a model class.** We begin with a *linear regression* model

$$\hat{d} = f(y_{0i}, v_{0i}) = \beta_0 + \beta_1 y_0 + \beta_2 v_0,$$

where β_0, β_1, β_2 are coefficients to be determined from the data. If the linear relation proves insufficient, we can extend the class of models to *quadratic regression*, introducing terms such as y_0^2, v_0^2, and $y_0 v_0$. This reflects the trade-off between *rigid* and *flexible* models discussed earlier in the chapter. A cubic model may also be tried, though as we shall see this risks overfitting.

3. **Fit the model (training).** Suppose we have n observed data points

$$\left\{(y_{0i}, v_{0i}), d_i: i = 1, \dots, n\right\}.$$

For a given model, the predicted distances are $\hat{d}_i = f(y_{0i}, v_{0i})$. The coefficients are chosen to minimise the *mean squared error* (MSE)

$$\text{MSE} = \frac{1}{n}\sum_{i=1}^{n}\left(d_i - \hat{d}_i\right)^2,$$

a *loss function* analogous to the sum of squared differences (5.1) introduced in section 5.1. This ensures that the model predictions are, on average, as close as possible to the observed data.

4. **Assess results (test).** To test whether the model captures genuine patterns, data are split into a *training set* and a *test set*. The model is fitted using the training set, while its error is measured on the test set. A good model should achieve a low MSE on both sets. This procedure prevents the machine learning scientist from being misled by models that fit noise, a phenomenon known as *overfitting*. The linear, quadratic, and cubic models can then also be compared by their test-set errors. The linear model is likely to underfit, because it cannot capture the curvature in the true relation $d = v_0\sqrt{2y_0/g}$. The quadratic model should strike a reasonable balance, providing flexibility without excessive complexity. The cubic model is likely to achieve a very low training error, but at the price of a larger test error, illustrating the *bias–variance trade-off*. This is an instance of a general principle in supervised learning: simple models have high bias (that is a systematic error due to their rigidity) but low variance, while very flexible models have low bias but high variance (their predictions fluctuate strongly with the particular training data). Good predictive performance is achieved by balancing these two effects, avoiding both underfitting and overfitting.

11.7 A practical demonstration of supervised learning with R

In this section we put into practice the ideas introduced above by carrying out a complete demonstration of supervised learning using R. The task is to predict the landing distance d of a projectile fired horizontally from a cannon, given its initial height y_0 and initial horizontal speed v_0. Since experimental data are not available, we begin by generating artificial observations through a simulation program. Once the data are available, we proceed through the standard machine learning workflow outlined above.

11.7.1 Define the learning problem

The first step is to define the problem under study. The definition, in this case, is carried out practically through the simulation of a dataset. The function created for this task, called `projectile_data`, randomly selects values of the initial height y_0 (measured in metres) and initial horizontal speed v_0 (measured in metres per second) within physically reasonable ranges, and computes the corresponding landing distance using the exact physics formula

$$d = v_0\sqrt{\frac{2y_0}{g}}.$$

An optional normal error can be added to the output observed d, to simulate unaccounted for effects like air resistance. The result of each simulated launch is therefore a triple (y_0, v_0, d). Code for this function can be inspected in appendix J.

Let us generate $n = 200$ data points, (y_{0i}, v_{0i}, d_i), with the heights y_{0i} coming from a uniform distribution between 10 and 50 metres, the horizontal velocities v_{0i} coming from a uniform distribution between 10 and 50 metres per seconds, and where a random error has been added to d_i from a normal distribution centred at 0 and with a standard deviation equal to 0.1 metres (10 centimetres), so that $d_{\text{obs}\,i} = d_i + \epsilon_i$. The data set generated is contained in a tibble.

```
# Load simulation function in memory
source("projectile_model.R")

# Simulate 200 data points
# y0 ~ Unif(10,50); v0 ~ Unif(10,50); noise_sd=0.1
n <- 200
Data <- projectile_data(n,noise_sd=0.5,seed=867)

# Display tibble
print(Data)
```

```
## # A tibble: 200 x 4
##        y0    v0     d d_obs
##     <dbl> <dbl> <dbl> <dbl>
##  1  38.1  21.5  59.9  59.5
##  2  38.3  16.5  46.1  46.6
##  3  40.0  13.9  39.6  39.0
##  4  28.7  36.1  87.3  88.1
##  5  18.3  44.7  86.4  85.7
##  6  38.0  24.2  67.3  67.1
##  7  45.2  27.6  83.8  84.1
##  8  18.8  35.0  68.6  68.6
##  9  32.6  49.8 128.   128.
## 10  41.4  13.1  38.1  38.6
## # i 190 more rows
```

This dataset provides the inputs (y_{0i}, v_{0i}) and the corresponding outputs d_i that define our supervised learning problem.

11.7.2 Choose a model class

Having generated our dataset, the next step is to decide which models to fit. In our case, three regression models will be considered: a linear model, a quadratic model, and a cubic model. They are described by the following equations:

$$\hat{d} = \beta_0 + \beta_1 y_0 + \beta_2 v_0 \quad \text{Linear,} \tag{11.2}$$

$$\hat{d} = \beta_0 + \beta_1 y_0 + \beta_2 v_0 + \beta_3 y_0^2 + \beta_4 y_0 v_0 + \beta_5 v_0^2 \quad \text{Quadratic,} \tag{11.3}$$

$$\begin{aligned}\hat{d} = {} & \beta_0 + \beta_1 y_0 + \beta_2 v_0 + \beta_3 y_0^2 + \beta_4 y_0 v_0 + \beta_5 v_0^2 + \beta_6 y_0^3 + \beta_7 y_0^2 v_0 + \beta_8 y_0 v_0^2 \\ & + v_0^3 \quad \text{Cubic.}\end{aligned} \tag{11.4}$$

The parameters $\beta_0, \ldots, \beta_9$ define the function $f(y_0, v_0)$ of the model and will be calculated using the available data.

The linear model is deliberately restrictive: it cannot capture curvature or interactions beyond a straight-line dependence. Nevertheless, beginning with a linear model is valuable, since it provides a baseline for comparison and helps us understand whether a more flexible model is justified.

In R, models are prepared using the so-called *grammar of models* explained in section 5.10. For the linear model we have

```
y ~ x,
```

for the quadratic model we have

```
y ~ x + I(x^2),
```

and for the cubic model we have

```
y ~ x + I(x^2) + I(x^3).
```

In the present case, we must deal with variables y_0, v_0, and d. The models, therefore, must be written as

```
d ~ y0 + v0,
```

etc.

11.7.3 Fit the model

Once a candidate model has been chosen, the next step is to estimate its parameters. The goal is always the same: to determine the coefficients so that the mean squared error (MSE) between observed and predicted distances is as small as possible. Different estimation methods are available, such as least squares or maximum likelihood[1]. In R, the function `lm`, which we use to perform the regression, applies maximum likelihood estimation by default. This produces the parameter values that make the observed data most probable under the chosen model.

```
# Fit a linear model
fit_linear <- lm(d_obs ~ y0 + v0,data=Data)

# Fit a quadratic model
fit_quadratic <- lm(d_obs ~ y0 + v0 + I(y0^2) + I(v0^2) + I(y0*v0),
data=Data)

# Fit a cubic model
fit_cubic <- lm(d_obs ~
y0 + v0 +
I(y0^2) + I(v0^2) + I(y0*v0) +
I(y0^3) + I(v0^3) + I(y0^2*v0) + I(y0*v0^2),
data = Data)
```

As in the present case we have two independent variables, y_0 and v_0, the function corresponding to the three models can only be represented as a surface in 3D. This can be done with some R packages that, though, are more useful when used interactively. To communicate results effectivelyit is better to stick to 2D representations. There exist many possibile solutions in 2D; one of them is to plot the 'observed' vs 'predicted' values of d. If the model is accurate, there should be a correlation shown as data points aligning along the 45° diagonal. If the model is inadequate, notable deviations from the line will be visible. One proceeds from the simplest (less parameters) to the most complex (more parameters) model and adopts the first satisfactory one (*Occam's razor*).

[1] Maximum likelihood estimation (MLE) chooses the parameter values that make the observed data most probable under the assumed model. Concretely, one writes down the probability (likelihood) of observing the data as a function of the model parameters and then selects the parameter values that maximise this function.

```
library(dplyr)
library(ggplot2)

# Residues for linear
Data_lin <- Data |>
mutate(pred=predict(fit_linear,newdata=Data))

pLin <- ggplot(Data_lin,aes(x=pred,y=d_obs)) +
geom_point() +
geom_abline(slope=1,intercept=0,linetype=2) +
labs(x="Predicted d", y = "Observed d",
title = "Linear Model") +
coord_equal()

# Residues for quadratic
Data_quad <- Data |>
mutate(pred=predict(fit_quadratic,newdata=Data))

pQuad <- ggplot(Data_quad,aes(x=pred,y=d_obs)) +
```

```
geom_point() +
geom_abline(slope=1,intercept=0,linetype=2) +
labs(x="Predicted d", y = "Observed d",
title = "Quadratic Model") +
coord_equal()

# Residues for cubic
Data_cube <- Data |>
mutate(pred=predict(fit_cubic,newdata=Data))

pCube <- ggplot(Data_cube,aes(x=pred,y=d_obs)) +
geom_point() +
geom_abline(slope=1,intercept=0,linetype=2) +
labs(x="Predicted d", y = "Observed d",
title = "Cubic model") +
coord_equal()

# Tile up plots
library(patchwork)
pLin + pQuad + pCube
```

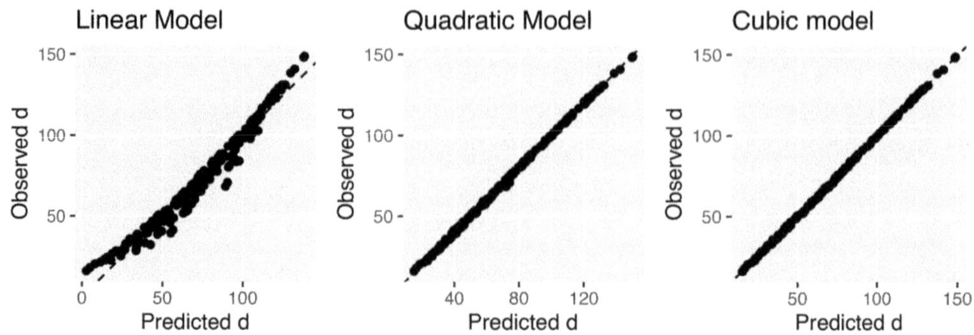

From the plots, it is evident that the linear model is inadequate, while the cubic model does not show any marked improvement, compared to the quadratic one. The quadratic model would therefore be the preferred choice in this case. In relation to the tools used for plotting, the reader can appreciate again the use of the `patchwork` graphics library, that well complements `ggplot2` (see Chapter 9).

The numerical result of all estimations can quickly be compiled into a table using the `tidy` function in package `broom` of the tidyverse.

```
# Load broom for function tidy
library(broom)

# Estimation table for the linear model
tidy(fit_linear)
```

```
## # A tibble: 3 x 5
##   term        estimate std.error statistic   p.value
##   <chr>          <dbl>     <dbl>     <dbl>     <dbl>
## 1 (Intercept)   -37.0     1.58       -23.4 7.31e- 59
## 2 y0              1.27    0.0370      34.2 7.90e- 85
## 3 v0              2.40    0.0356      67.3 6.24e-138
```

```
# Estimation table for the quadratic model
tidy(fit_quadratic)
```

```
## # A tibble: 6 x 5
##   term         estimate std.error statistic   p.value
##   <chr>           <dbl>     <dbl>     <dbl>     <dbl>
## 1 (Intercept) -8.11     0.676       -12.0   3.66e- 25
## 2 y0           0.682    0.0319       21.4   6.87e- 53
## 3 v0           1.11     0.0328       33.8   3.07e- 83
## 4 I(y0^2)     -0.0116   0.000486    -23.8   1.87e- 59
## 5 I(v0^2)      0.000360 0.000486      0.742 4.59e-  1
## 6 I(y0 * v0)   0.0430   0.000438     98.3   1.93e-167
```

```
# Estimation table for the cubic model
tidy(fit_cubic)
```

```
## # A tibble: 10 x 5
##    term             estimate std.error statistic  p.value
##    <chr>               <dbl>     <dbl>     <dbl>    <dbl>
##  1 (Intercept)   -2.44       1.37          -1.79  7.56e- 2
##  2 y0             0.515      0.103          5.00  1.29e- 6
##  3 v0             0.652      0.101          6.46  8.63e-10
##  4 I(y0^2)       -0.0188     0.00322       -5.82  2.42e- 8
##  5 I(v0^2)        0.00686    0.00313        2.19  2.94e- 2
##  6 I(y0 * v0)     0.0657     0.00256       25.7   1.22e-63
##  7 I(y0^3)        0.000206   0.0000342      6.03  8.53e- 9
##  8 I(v0^3)       -0.0000776  0.0000335     -2.32  2.17e- 2
##  9 I(y0^2 * v0) -0.000384    0.0000287    -13.4   3.88e-29
## 10 I(y0 * v0^2)  0.00000369 0.0000309      0.119 9.05e- 1
```

Many interesting comments arise from scrutiny of the above tables. The linear model provides an initial approximation. From the p-values we can see that both y_0 and v_0 are highly significant predictors (v_0 more than y_0), and the fit captures a clear trend in the data. From a machine learning perspective, however, the linear model remains very simple and therefore rigid. Indeed, we have seen in the previous graphical analysis that the quadratic model is more flexible and does a better job than the linear one. The quadratic model represents a substantial improvement. The inclusion of squared and interaction terms allows the regression to adapt more closely to the observed data. Most coefficients are highly significant (with the exception of v_0^2), suggesting that these additional terms contribute meaningfully to the prediction of d. This model offers a good compromise: it provides greater flexibility than the linear model without an excessive number of parameters. With the benefit of knowing the physics related to these data, we can also note that quadratic terms approximate the square-root dependence on y_0 more effectively than a linear expression. The cubic model adds still more flexibility by including third-order terms While several coefficients are statistically significant, others are not, and some estimated effects are extremely small. This indicates that the cubic model is beginning to capture idiosyncrasies of the training data rather than genuine signal. This is essentially overfitting: the model achieves extremely close agreement with the data at hand but introduces terms without clear justification. From the physical viewpoint, the true law involves a square-root, not a cubic, so these extra terms cannot be regarded as meaningful.

11.7.4 Assess results

The immediate question to ask after having chosen and fitted a model is: will it still be a good model outside the data used to construct it? To obtain the parameter estimates, the algorithm relied on the available dataset, which is why these observations are quite rightly called the *training data*. It is therefore no surprise that, when the input variables (y_{0i}, v_{0i}) belong to this training set, the model will produce predictions $\hat{d}_i$ close to the observed outcomes $d_{\text{obs}\,i}$: the parameters were explicitly chosen to make the squared differences $(d_{\text{obs}\,i} - \hat{d}_i)^2$ as small as possible. But what happens when one wishes to predict values for input variables not contained in the original training set? We are basically assessing the ability of a model trained on a given dataset to perform well on new, unseen data. A model that fits the training data perfectly but fails to predict new observations accurately is of little practical use. This problem is connected to *overfitting*, and arises when a model is too flexible and begins to adapt not only to the underlying structure but also to the random variations in the training data (noise). On the other hand, a model that is too simple will not even fit the training data well, which is referred to as *underfitting*.

The assessment starts with the standard practice of splitting the data into two parts: a *training set* and a *test set*. The training set is used to estimate the model parameters, while the test set is kept aside and used only at the end to evaluate predictive performance. In this way the test set plays the role of new, unseen data, providing an unbiased estimate of how general the model is. A common choice is to allocate about 70% of the data to training and the remaining 30% to testing, although the exact proportion can vary. In our projectile example, this means dividing the simulated triples (y_{0i}, v_{0i}, d_i) into two groups. The regression coefficients are fitted on the training group and predictions are made for the test group. The quality of these predictions is measured by comparing them to the true observed values in the test set, using the mean squared error (MSE) or the root mean squared error (RMSE) as metrics. Models with smaller test errors are considered to be more accurate.

As earlier on we did not split data into training and test sets, we need to redo the process. Basically, we will split data as 70% vs 30%, carry out training on only 70% of the data and measure RMSE and R squared on the remaining 30% (test set). To avoid repeating the same calculations for three times, we will build ad hoc functions for doing the analysis. Most specialised R packages have efficient versions of those functions, but it is instructive to actually spell out what they are here.

```
# Create helper functions for testing models
# Some R packages might already have them


# Root Mean Squared Error (RMSE)
rmse <- function(obs,pred) sqrt(mean((obs-pred)^2))

# R squared
r2 <- function(obs,pred) {
ss_res <- sum((obs-pred)^2)
ss_tot <- sum((obs-mean(obs))^2)
rr <- 1-ss_res/ss_tot

return(rr)
}

# Function to compare models
# (using RMSE and R squared)
eval_model <- function(fit,train,test) {
# Training metrics
pred_tr <- predict(fit,newdata=train)
rmse_tr <- rmse(train$d_obs,pred_tr)
r2_tr <- r2(train$d_obs,pred_tr)

# Test metrics
pred_te <- predict(fit,newdata=test)
rmse_te <- rmse(test$d_obs,pred_te)
r2_te   <- r2(test$d_obs,pred_te)

out <- tibble(
RMSE_train=rmse_tr,R2_train=r2_tr,
RMSE_test=rmse_te,R2_test=r2_te
```

```
)
}
```

Let us apply the functions created to the projectile data set. Note how in the code the training-test split is done simply with a sampling action.

```
# Use the same dataset Data. We will need to re-do the
# whole estimate as now we are splitting data into
# training and test sets.

set.seed(2025)  # for reproducible split

# Number of data points
n <- nrow(Data) # We know here n=200

# Typical fraction (70%) for training set
train_frac <- 0.7

# Manual random splitting. idx_train is random integers
idx_train <- sample(1:n,size=floor(train_frac*n),
replace=FALSE)
Data_train <- Data[idx_train, ]
Data_test  <- Data[-idx_train, ]

## Fitting Models

# Linear
fit_linear <- lm(d_obs ~
y0 + v0,
data=Data_train)

# Quadratic
fit_quadratic <- lm(d_obs ~
y0 + v0 + I(y0^2) + I(v0^2) + I(y0*v0),
data=Data_train)

# Cubic
fit_cubic <- lm(d_obs ~
y0 + v0 + I(y0^2) + I(v0^2) + I(y0*v0) +
I(y0^3) + I(v0^3) + I(y0^2*v0) + I(y0*v0^2),
data=Data_train)

## Models comparison (bind three separate tibbles)
```

```
# Recall output of eval_model is a tibble
results <- bind_rows(
eval_model(fit_linear,Data_train,Data_test) |>
mutate(Model="Linear"),
eval_model(fit_quadratic,Data_train,Data_test) |>
mutate(Model="Quadratic"),
eval_model(fit_cubic,Data_train,Data_test) |>
mutate(Model="Cubic")
) |>
select(Model,RMSE_train,RMSE_test,R2_train,R2_test)

print(results)
```

```
## # A tibble: 3 x 5
##   Model     RMSE_train RMSE_test R2_train R2_test
##   <chr>          <dbl>     <dbl>    <dbl>   <dbl>
## 1 Linear         5.91      5.74     0.965   0.969
## 2 Quadratic      0.836     0.731    0.999   1.00
## 3 Cubic          0.528     0.615    1.00    1.00
```

Let us comment on the values of RMSE and R squared. We observe that the RMSE drops dramatically going from linear to quadratic, but the drop is nearly unnoticeable going further to the cubic model. This means that the extra parameters offered by the flexibility of the quadratic model are needed, but more parameters are unnecessary. Similarly, the R squared is high for the linear model, indicating a very good fit. But for the quadratic and cubic models, it is 1, that is perfect. So, we confirm the choice of a quadratic model for these data.

As in this specific example the 'correct' model (the physical model) is known, we can compare the three models with it. Many ways to carry out the comparison are possible. Here we will simply determine the correct and predicted d's on a regular 2D grid of values (y_0, v_0) and represent their absolute difference as a contour map. The key passage, though, is to use a range of values of the input variables, different from the one used to determine the models. The operation and result are containe in the following code. It is relatively straightforward to create a regular gris with `expand.grid`. It is also easy to calculate the values of the true d, of any prediction, and of the absolute error on the same grid. The result is encapsulated, as always when using modern R, in a tibble.

```
# Define function defining true distance
dtrue <- function(y0,v0) {
res <- v0*sqrt(2*y0/9.81)
```

```
return(res)
}

# Tests are on a range different from that
# of the training data
my0 <- 50
My0 <- 100
mv0 <- 50
Mv0 <- 100

# Grid to plot contours of dtrue and prediction
grid <- expand.grid(
y0=seq(my0,My0,length.out=100),
v0=seq(mv0,Mv0,length.out=100)
)

# Correct and predicted values on the grid
Errors <- grid |>
mutate(
d_true=dtrue(y0,v0),
d_lin=predict(fit_linear,newdata=grid),
d_quad=predict(fit_quadratic,newdata=grid),
d_cube=predict(fit_cubic,newdata=grid),
err_lin=abs(d_true-d_lin),
err_quad=abs(d_true-d_quad),
err_cube=abs(d_true-d_cube)
)

# Absolute error for linear
pLin <- ggplot(Errors,aes(x=y0,y=v0,z=err_lin)) +
geom_contour_filled() +
labs(title="Linear",x="y0",y="v0",
fill="Absolute eror") +
coord_equal()

# Absolute error for quadratic
pQuad <- ggplot(Errors,aes(x=y0,y=v0,z=err_quad)) +
geom_contour_filled() +
labs(title="Quadratic",x="y0",y="v0",
fill="Absolute eror") +
coord_equal()

# Absolute error for cubic
pCube <- ggplot(Errors,aes(x=y0,y=v0,z=err_cube)) +
```

```
geom_contour_filled() +
labs(title="Cubic",x="y0",y="v0",
fill="Absolute eror") +
coord_equal()

# The quadratic is best
pQuad
```

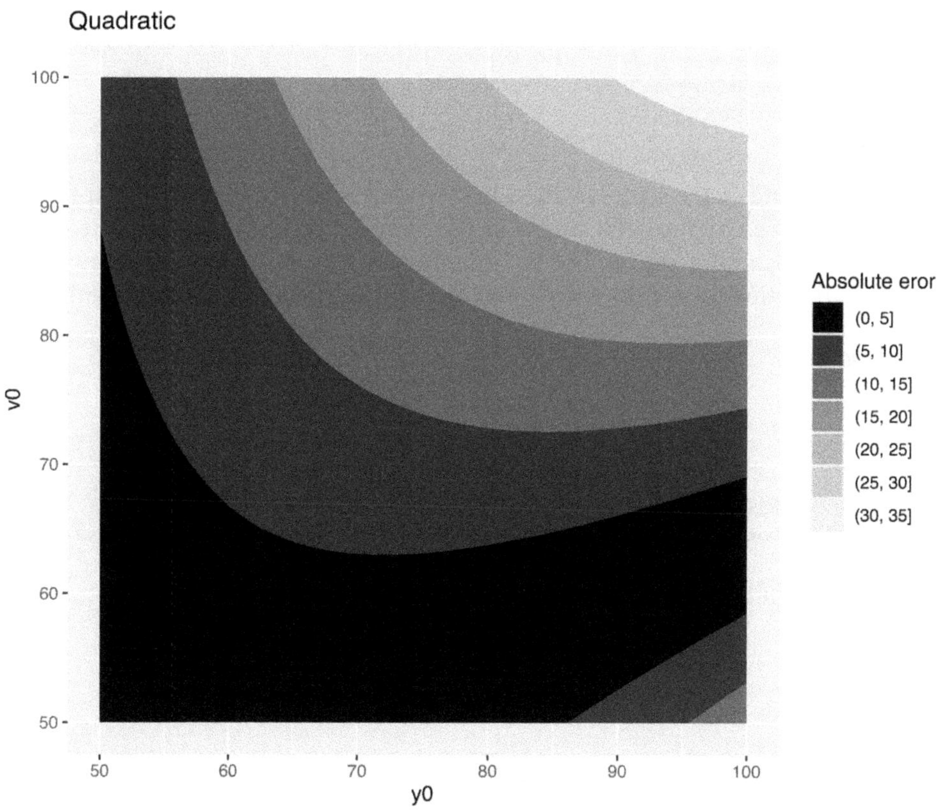


```
# Tile the other two with patchwork
pLin + pCube
```

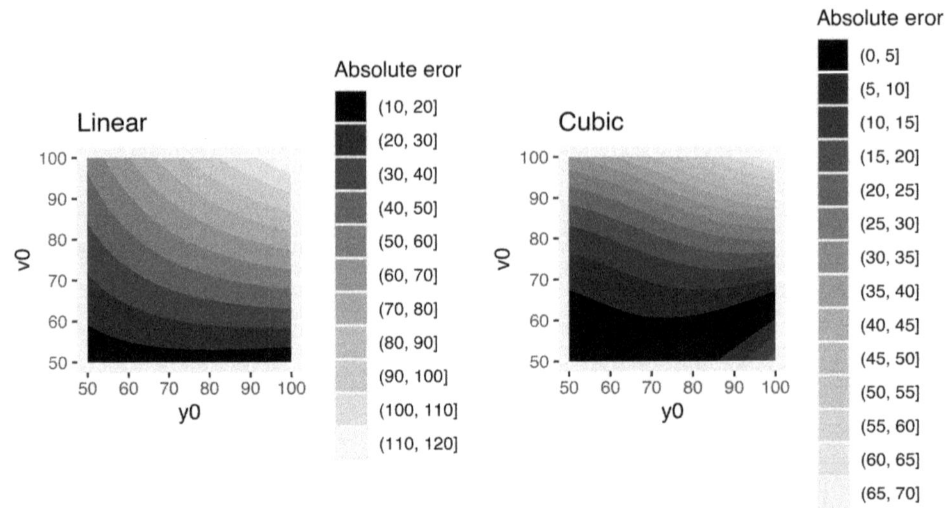

Just by looking at the scales of the contouring we can see that the linear model is the worst, followed by the cubic one. The flexibility of the cubic has made it adapt to the data in the training region, but the adaptation, that is the lack of some rigidity, has meant that this model does not cater well for regions outside the training one.

References

[1] Open AI website https://openai.com/ (Accessed: 2023-06-27)

[2] chatGPT on Wikipedia https://en.wikipedia.org/wiki/ChatGPT#References (Accessed: 2023-06-27)

[3] GPT on Wikipedia https://en.wikipedia.org/wiki/GPT-3 (Accessed: 2023-06-27)

[4] Witten D, Hastie T, James G and Tibshirani R 2021 *An Introduction to Statistical Learning* (Springer)

[5] Kaptein M and Heuvel E V D 2022 *Statistics for Data Scientists* (Springer)

Part IV

Appendices

Appendix A

Mathematical proofs

A.1 Error for the linear interpolation

The analytic expression (3.4) for the linear interpolation's error,

$$\Delta f_{\text{int}}(x) = \frac{f''(\xi)}{2}(x - x_1)(x - x_2)$$

can be proved using a consequence of Calculus' *mean value theorem*, according to which the derivative of a smooth function in the interval between two successive zeros of the function has at least one zero. In other words (and symbols!), if a continuous function $g(z)$ in $[x_1, x_2]$ has at least two zeros in $[x_1, x_2]$ and finite derivatives in (x_1, x_2), then there exists at least one value ξ in (x_1, x_2) such that,

$$g'(\xi) = 0 \tag{A.1}$$

The proof makes use of the following auxiliary function,

$$g(z) \equiv f(z) - f_{\text{int}}(z) - \frac{f(x) - f_{\text{int}}(x)}{(x - x_1)(x - x_2)}(z - x_1)(z - x_2), \tag{A.2}$$

where $x \in (x_1, x_2)$, i.e. $x \neq x_1$ and $x \neq x_2$, and $z \in [x_1, x_2]$. It is straightforward to verify that $g(z)$ has at least three zeros because, given that $f_{\text{int}}(x_1) = f(x_1)$ and $f_{\text{int}}(x_2) = f(x_2)$, then

$$g(x_1) = f(x_1) - f_{\text{int}}(x_1) - \frac{f(x) - f_{\text{int}}(x)}{(x - x_1)(x - x_2)}(x_1 - x_1)(x_1 - x_2) = f(x_1) - f(x_1) = 0,$$

$$g(x_2) = f(x_2) - f_{\text{int}}(x_2) - \frac{f(x) - f_{\text{int}}(x)}{(x - x_1)(x - x_2)}(x_2 - x_1)(x_2 - x_2) = f(x_2) - f(x_2) = 0$$

doi:10.1088/978-0-7503-2632-2ch12

and

$$
\begin{aligned}
g(x) &= f(x) - f_{\text{int}}(x) - \frac{f(x) - f_{\text{int}}(x)}{(x - x_1)(x - x_2)}(x - x_1)(x - x_2) \\
&= f(x) - f_{\text{int}}(x) - f(x) + f_{\text{int}}(x) = 0
\end{aligned}
$$

Using now the consequence of the mean value theorem previously mentioned, we can state that there exist at least two values, one between x_1 and x and the other between x and x_2, for which $g'(z) = 0$. This means that $g'(z)$ has at least two zeros between x_1 and x_2. Using the mean value theorem once more we realise that function $g''(z)$ has at least one zero between x_1 and x_2. Thus, there exists a $\xi \in (x_1, x_2)$ for which condition (A.1) holds. More specifically, in this instance,

$$
g''(\xi) = 0 \quad \Rightarrow \quad f''(\xi) - 0 - \frac{f(x) - f_{\text{int}}(x)}{(x - x_1)(x - x_2)}(2) = 0
$$

from which result (3.4) follows immediately. It is important to remember that the value of ξ changes with the value of x. In other words, ξ is a function of x,

$$
\xi = \xi(x)
$$

A.2 Error for Lagrangian interpolation

The proof follows the same type of reasoning applied when proving the expression for the linear interpolation error (section A.1). The auxiliary function used is now,

$$
g(z) \equiv f(z) - P_n(z) - \frac{f(x) - P_n(x)}{(x - x_1) \cdots (x - x_{n+1})}(z - x_1) \cdots (z - x_{n+1}) \tag{A.3}
$$

This function has at least $n + 2$ zeros because it is easy to verify that $g(x_1) = 0$, $g(x_2) = 0, \ldots, g(x_{n+1}) = 0$ and $g(x) = 0$. Therefore, due to the mean value theorem, its first derivative will be zero in at least $n + 1$ points, the first between x_1 and x_2, the second between x_2 and x_3, etc. This means that $g'(z)$ has at least $n + 1$ zeros between x_1 and x_{n+1}. Once again, due to the mean value theorem, the derivative of $g'(z)$, i.e. $g''(z)$, will have at least n zeros between x_1 and x_{n+1}. Considering derivatives of increasing order we arrive at function $g^{(n+1)}(z)$ which has at least one zero between x_1 and x_{n+1}. Or, more specifically and considering that $P_n^{(n+1)}(z) = 0$, there exists a $\xi \in (x_1, x_{n+1})$ such that,

$$
g^{(n+1)}(\xi) = 0 \quad \Rightarrow \quad f^{(n+1)}(\xi) - 0 - \frac{f(x) - P_n(x)}{(x - x_1) \cdots (x - x_{n+1})}(n + 1)! = 0
$$

from which the result (3.10) readily follows.

This result can also be proved more straightforwardly using the *generalised Rolle theorem* according to which if $g(z)$ is continuous in $[x_1, x_{n+1}]$, has $n + 1$ zeros at $x_1, \ldots, x_{n+1}$ and it is n times differentiable in the interval (x_1, x_{n+1}), then there exist a number ξ in (x_1, x_{n+1}) according to which $g^{(n)}(\xi) = 0$. The generalised Rolle theorem will have here to be applied with $n + 2$ zeros (see above) to a function which is $n + 1$ times differentiable, so that we can have in the end $g^{(n+1)}(\xi) = 0$. For the proof in this appendix it is also important to remember that ξ is a function of x,

$$\xi = \xi(x)$$

Formula (3.10) can be used to show, intuitively, that the magnitude of $\Delta P_n(x)$ is larger the further away x is from the centre of the interval (x_1, x_{n+1}). Let us indicate such a centre with c. The discussed magnitude is essentially determined by the product

$$|(x - x_1)(x - x_2) \cdots (x - x_{n+1})|.$$

Let us re-write x in terms of its distance, $\delta > 0$, from c:

$$x = m - \delta \qquad \text{or} \qquad x = m + \delta.$$

The above product, therefore, becomes

$$|(m - \delta - x_1) \cdots (m - \delta - x_{n+1})|,$$

or

$$|(m + \delta - x_1) \cdots (m + \delta - x_{n+1})|.$$

The larger δ is, the larger the product tends to be as this is the only variable quantity in the product.

A.3 Identical polynomials pass through the same set of points

If two n-degree polynomials $P_n(x)$ and $Q_n(x)$ pass through the same set of $n + 1$ points in an interval $[a, b] \in \mathbb{R}$, they are identical. This means that a polynomial of degree n is completely and uniquely determined by knowledge of $n + 1$ of its points.

To prove this we can proceed by contradiction. Let us assume that both $P_n(x)$ and $Q_n(x)$ pass through the same set of $n + 1$ points, (x_1, f_1), (x_2, f_2),…,(x_{n+1}, f_{n+1}). This means that,

$$P_n(x_1) = Q_n(x_1) = f_1,\ P_n(x_2) = Q_n(x_2) = f_2,\ldots,P_n(x_{n+1}) = Q_n(x_{n+1}) = f_{n+1}$$

Let us now construct a new function $G(x)$ defined as the difference between the two polynomials,

$$G(x) \equiv P_n(x) - Q_n(x)$$

$G(x)$ is a polynomial of at least degree n by construction. Furthermore $G(x_i) = P_n(x_i) - Q_n(x_i) = f_i - f_i = 0$, $i = 1,\ldots,n + 1$. $G(x)$ is, therefore, a polynomial of at least degree n with $n + 1$ zeros. But, due to the fundamental theorem of algebra, a non-constant, n-degree polynomial can have at most n zeros. Therefore, $G(x)$ is a constant function equal to zero, given that $G(x_i) = 0$, $i = 1,\ldots,n + 1$. Recalling the definition of $G(x)$, this means that $P_n(x) = Q_n(x)$.

A.4 Error for divided differences

In order to prove equation (3.20) we are going to make use of the generalised Rolle theorem introduced in appendix A.2, according to which if a function $g(x)$ is continuous in $[x_1, x_{n+1}]$, has $n+1$ zeros at $x_1,\ldots,x_{n+1}$ and it is n times differentiable in the interval (x_1, x_{n+1}), then there exists a number ξ in (x_1, x_{n+1}) such that $g^{(n)}(\xi) = 0$. Let us then define a function $g(x)$ as follows:

$$g(x) \equiv f(x) - P_n(x)$$

As,

$$P_n(x_i) = f(x_i), \quad i = 1,\ldots,n+1,$$

then,

$$g(x_i) = f(x_i) - P_n(x_i) = f(x_i) - f(x_i) = 0, \quad i = 1, \ldots n+1$$

which means that the function $g(x)$ has $n+1$ zeros in the interval $[x_1, x_{n+1}]$. The generalised Rolle theorem can thus be applied. There exists in (x_1, x_{n+1}) a value ξ (remember that ξ is always a function of x) such that $g^{(n)}(\xi) = 0$. Using the definition of $g(x)$ we have,

$$g^{(n)}(\xi) = f^{(n)}(\xi) - P_n^{(n)}(\xi) = 0 \tag{A.4}$$

The leading term of $P_n(x)$ is,

$$f[x_1,\ldots,x_{n+1}](x - x_1) \cdots (x - x_n)$$

Therefore the n-th derivative of $P_n(x)$ is,

$$P_n^{(n)}(x) = n! f[x_1,\ldots,x_{n+1}]$$

Using the above result in expression (A.4) yields the following result,

$$f^{(n)}(\xi) - n! f[x_1,\ldots,x_{n+1}] = 0,$$

which is easily re-arranged into expression (3.20),

$$\frac{f^{(n)}(\xi)}{n!} = f[x_1,\ldots,x_{n+1}],$$

A.5 Singular value decomposition using eigenvalues and eigenvectors

In section 4.16 it was suggested that the singular value decomposition of a matrix M is connected to the calculation of the eigenvalues and eigenvectors of the matrices $MM^\dagger$ and $M^\dagger M$. In order to see how this can be true, let us consider the expression of the singular value decomposition of a $m \times n$ matrix M, where to fix ideas we assume that $m > n$:

$$M = U\Sigma V^\dagger \tag{A.5}$$

U and V are, respectively, $m \times m$ and $n \times n$ unitary matrices, i.e. matrices for which,

$$UU^\dagger = U^\dagger U = I, \qquad VV^\dagger = V^\dagger V = I \tag{A.6}$$

and I is the appropriate identity matrix.

Let us next consider the new matrix $M_u \equiv MM^\dagger$. Using the decomposition (A.5) and the unitarity properties (A.6) yields:

$$M_u = MM^\dagger = (U\Sigma V^\dagger)(U\Sigma V^\dagger)^\dagger = U\Sigma V^\dagger (V^\dagger)^\dagger \Sigma^\dagger U^\dagger = U\Sigma V^\dagger V \Sigma^\dagger U^\dagger = U\Lambda_u U^\dagger$$

The new matrix Λ_u is an $m \times m$ diagonal matrix which, due to Σ singular values σ_i, has n non-negative values,

$$\sigma_i^* \sigma_i = |\sigma_i|^2, \qquad i = 1,\ldots,n \tag{A.7}$$

and $m - n$ zeros along its diagonal. But what written is essentially the spectral decomposition of the Hermitian matrix M_u. Its eigenvalues are the n quantities in equation (A.7) and $m - n$ zeros, while its eigenvectors are the m columns of the matrix U which, due to spectral decomposition, is a unitary matrix.

To find out about V, consider the Hermitian matrix $M_v = M^\dagger M$. Using the expression for the singular value decomposition we have, this time,

$$M_v = M^\dagger M = (U\Sigma V^\dagger)^\dagger (U\Sigma V^\dagger) = (V^\dagger)^\dagger \Sigma^\dagger U^\dagger U \Sigma V^\dagger = V\Sigma^\dagger U^\dagger U \Sigma V^\dagger = V\Lambda_v V^\dagger$$

Λ_v is an $n \times n$ diagonal matrix with $n|\sigma_i|^2$ along the diagonal. Again, using the spectral decomposition for the Hermitian matrix M_v, we see that the n quantities in equation (A.7) are its eigenvalues, with the columns of V being the corresponding eigenvectors. Thus V too is a unitary matrix.

To summarise, the singular value decomposition of a matrix M can be carried out as spectral decomposition of the two matrices $MM^\dagger$ and $M^\dagger M$.

A.6 Condition number for matrices and ill-posedness

The condition number for a matrix A is defined as the product of its norm and the norm of its inverse (see section 4.20). The norm can be defined in any of the possible ways allowed for matrices (some of them are listed in appendix B).

Consider a linear system with n equations,

$$A\mathbf{x} = \mathbf{b} \tag{A.8}$$

and a related system,

$$A\mathbf{x} = \mathbf{b}' \tag{A.9}$$

in which $\mathbf{b}$ is replaced by another column vector, $\mathbf{b}'$, differing from $\mathbf{b}$ of a small quantity $\Delta\mathbf{b}$, that is

$$\Delta\mathbf{b} = \mathbf{b} - \mathbf{b}'$$

We are interested in quantifying the relative change in $\mathbf{x}$, due to the relative change in $\mathbf{b}$. Let us then introduce also $\Delta\mathbf{x}$ as

$$\Delta\mathbf{x} = \mathbf{x} - \mathbf{x}'$$

From the inversion of equations (A.8) and (A.9) the following equations are obtained:

$$\mathbf{x} = A^{-1}\mathbf{b} \tag{A.10}$$

$$\mathbf{x}' = A^{-1}\mathbf{b}' \tag{A.11}$$

Equations mimicking the four equations just introduced can also be derived for the quantities $\Delta\mathbf{x}$ and $\Delta\mathbf{b}$. Indeed, if equation (A.11) is subtracted from equation (A.10), the result is

$$\Delta\mathbf{x} = A^{-1}\Delta\mathbf{b} \tag{A.12}$$

Similarly, if equation (A.9) is subtracted from equation (A.8), the result is

$$\Delta\mathbf{b} = A\Delta\mathbf{x} \tag{A.13}$$

Let us now take the norm of $\Delta\mathbf{x}$, which is also the norm of the product $A^{-1}\Delta\mathbf{b}$. Using the rules for the norm of matrices yields the following inequality:

$$\|\Delta\mathbf{x}\| = \|A^{-1}\Delta\mathbf{b}\| \leqslant \|A^{-1}\|\|\Delta\mathbf{b}\|$$

Doing a similar operation on $\Delta\mathbf{b}$, which is equal to the product $A\Delta\mathbf{x}$, yields

$$\|\Delta\mathbf{b}\| \leqslant \|A\|\|\Delta\mathbf{x}\| \quad \Rightarrow \quad \|\Delta\mathbf{x}\| \geqslant \frac{\|\Delta\mathbf{b}\|}{\|A\|}$$

For $\|\Delta\mathbf{x}\|$ the following double inequality must therefore be satisfied:

$$\frac{1}{\|A\|}\|\Delta\mathbf{b}\| \leqslant \|\Delta\mathbf{x}\| \leqslant \|A^{-1}\|\|\Delta\mathbf{b}\| \tag{A.14}$$

Inequalities for the norm of $\mathbf{x}$ can be obtained in an analogous way, thus yielding,

$$\frac{1}{\|A\|}\|\mathbf{b}\| \leqslant \|\mathbf{x}\| \leqslant \|A\|\|\mathbf{b}\|$$

The reciprocal of the above gives,

$$\frac{1}{\|A^{-1}\|}\frac{1}{\|\mathbf{b}\|} \leqslant \frac{1}{\|\mathbf{x}\|} \leqslant \|A\|\frac{1}{\|\mathbf{b}\|} \tag{A.15}$$

The result we were looking for is obtained simply multiplying equation (A.14) by equation (A.15):

$$\frac{1}{\|A^{-1}\|\|A\|}\frac{\|\Delta\mathbf{b}\|}{\|\mathbf{b}\|} \leqslant \frac{\|\Delta\mathbf{x}\|}{\|\mathbf{x}\|} \leqslant \|A^{-1}\|\|A\|\frac{\|\Delta\mathbf{b}\|}{\|\mathbf{b}\|} \tag{A.16}$$

The product $\|A^{-1}\|\|A\|$ is the *condition number* of matrix A. The result stated above says that when the condition number is high, the variation of the solution, as triggered by a change in the right hand side of the linear system, is also high. This unreasonable behaviour is called *ill-posedness* of the specific linear system.

A.7 Normal equations for multilinear regression

The sum of squared residuals, when the linear model consists of $m + 1$ parameters, is:

$$S(a_1,\ldots,a_m, a_{m+1}) = \sum_{i=1}^{n}(y_i - a_1x_{i1} - \cdots - a_mx_{im} - a_{m+1})^2 \tag{A.17}$$

To find the minimum of S we simply take the $m + 1$ partial derivatives with respect to $a_1,\ldots,a_{m+1}$ and set them to 0:

$$\begin{cases} -2\sum_{i=1}^{n}x_{i1}(y_i - a_1x_{i1} - \cdots - a_mx_{im} - a_{m+1}) & = 0 \\ -2\sum_{i=1}^{n}x_{i2}(y_i - a_1x_{i1} - \cdots - a_mx_{im} - a_{m+1}) & = 0 \\ \cdots & = \cdots \\ -2\sum_{i=1}^{n}x_{im}(y_i - a_1x_{i1} - \cdots - a_mx_{im} - a_{m+1}) & = 0 \\ -2\sum_{i=1}^{n}(y_i - a_1x_{i1} - \cdots - a_mx_{im} - a_{m+1}) & = 0 \end{cases}$$

Simplifying the -2 and re-arranging the terms yields:

$$\begin{cases} \left(\sum_{i=1}^{n}x_{i1}^2\right)a_1 + \left(\sum_{i=1}^{n}x_{i1}x_{i2}\right)a_2 + \cdots + \left(\sum_{i=1}^{n}x_{i1}x_{im}\right)a_m + \left(\sum_{i=1}^{n}x_{i1}\right)a_{m+1} = \left(\sum_{i=1}^{n}x_{i1}y_i\right) \\ \left(\sum_{i=1}^{n}x_{i2}x_{i1}\right)a_1 + \left(\sum_{i=1}^{n}x_{i2}^2\right)a_2 + \cdots + \left(\sum_{i=1}^{n}x_{i2}x_{im}\right)a_m + \left(\sum_{i=1}^{n}x_{i2}\right)a_{m+1} = \left(\sum_{i=1}^{n}x_{i2}y_i\right) \\ \qquad\qquad \cdots \quad \cdots \\ \left(\sum_{i=1}^{n}x_{im}x_{i1}\right)a_1 + \left(\sum_{i=1}^{n}x_{im}x_{i2}\right)a_2 + \cdots + \left(\sum_{i=1}^{n}x_{im}^2\right)a_m + \left(\sum_{i=1}^{n}x_{im}\right)a_{m+1} = \left(\sum_{i=1}^{n}x_{im}y_i\right) \\ \left(\sum_{i=1}^{n}x_{i1}\right)a_1 + \left(\sum_{i=1}^{n}x_{i2}\right)a_2 + \cdots + \left(\sum_{i=1}^{n}x_{im}\right)a_m + na_{m+1} = \left(\sum_{i=1}^{n}y_i\right) \end{cases} \tag{A.18}$$

These are the normal equations for the multilinear regression case. It is possible to show that this system has a unique solution in most cases and that it always has at least one solution.

A.8 Proof of the reciprocal relative error

We would like to prove the result (1.5), i.e. that

$$E_{rr} = \frac{|\tilde{x}-x|}{|\tilde{x}|} \leqslant \frac{E_r}{1 - E_r},$$

where $E_r < 1$ and where

$$E_r = \frac{|x - \tilde{x}|}{|x|} = \left|1 - \frac{\tilde{x}}{x}\right|.$$

We start from $E_r < 1$ which, when the definition is used, yields

$$\frac{|x - \tilde{x}|}{|x|} < 1 \quad \Rightarrow \left| 1 - \frac{\tilde{x}}{x} \right| < 1.$$

Now, using the triangular inequality, we have

$$1 = \left| 1 - \frac{\tilde{x}}{x} + \frac{\tilde{x}}{x} \right| \leqslant \left| 1 - \frac{\tilde{x}}{x} \right| + \left| \frac{\tilde{x}}{x} \right|$$

$$\Downarrow$$

$$\left| 1 - \frac{\tilde{x}}{x} \right| \geqslant 1 - \left| \frac{\tilde{x}}{x} \right|$$

The last two results can be used to work out a lower limit for the expression $E_r/(1 - E_r)$. We start with $1/(1 - E_r)$.

$$\frac{1}{1 - E_r} = \frac{1}{1 - |1 - \tilde{x}/x|} \geqslant \frac{1}{1 - (1 - |\tilde{x}/x|)} = \left| \frac{x}{\tilde{x}} \right|.$$

Then we work out $E_r/(1 - E_r)$.

$$\frac{E_r}{1 - E_r} = E_r \frac{1}{1 - E_r} \geqslant E_r \left| \frac{x}{\tilde{x}} \right| = \left| 1 - \frac{\tilde{x}}{x} \right| \left| \frac{x}{\tilde{x}} \right| = \left| \frac{x - \tilde{x}}{\tilde{x}} \right| = E_{rr},$$

which, when read from right to left, coincides with equation (1.5).

A.9 Derivative of a product with a finite number of terms

Consider the following product of shifted terms:

$$L = \prod_{\ell=0}^{n} (s - \ell) = s(s - 1)(s - 2) \ldots (s - n).$$

Due to the chain rule, the derivative of L yields:

$$\begin{aligned} \mathrm{d}L/\mathrm{d}s = &(s - 1)(s - 2) \ldots (s - n) + s(s - 2)\ldots \\ &(s - n) + \cdots + s(s - 1)(s - 2) \ldots (s - (n - 1)). \end{aligned}$$

This expression can be summarised using the product symbol:

$$\frac{\mathrm{d}L}{\mathrm{d}s} = \frac{\mathrm{d}}{\mathrm{d}s}\left(\prod_{\ell=0}^{n} (s - \ell) \right) = \sum_{k=0}^{n} \prod_{\substack{\ell=0 \\ \ell \neq k}}^{n} (s - \ell). \tag{A.19}$$

A.10 Zeros of the Gaussian quadrature

This appendix provides a proof demonstrating why the zeros (or nodes) used in an n-point Gaussian quadrature rule are precisely the zeros of the nth Legendre polynomial, $P_n(x)$. This connection is fundamental to the high accuracy achieved by Gaussian quadrature.

Recall that an n-point Gaussian quadrature rule approximates a definite integral over the interval $[-1, 1]$ as a weighted sum of function evaluations:

$$\int_{-1}^{1} f(x)\,\mathrm{d}x \approx \sum_{i=1}^{n} w_i f(x_i),$$

where x_i are the nodes and w_i are the corresponding weights. The remarkable property of Gaussian quadrature is that, by a suitable choice of these n nodes and n weights, the formula yields an exact result for any polynomial of degree $2n - 1$ or less. This is the maximum possible degree of exactness for any n-point quadrature rule.

Let $p(x)$ be an arbitrary polynomial of degree at most $2n - 1$. Since the n-point Gaussian quadrature rule is exact for such polynomials, we have:

$$\int_{-1}^{1} p(x)\,\mathrm{d}x = \sum_{i=1}^{n} w_i p(x_i). \tag{A.20}$$

Now, let us define a polynomial $\omega(x)$ whose roots are exactly the n nodes x_i of the quadrature rule:

$$\omega(x) = \prod_{i=1}^{n} (x - x_i)$$

This polynomial $\omega(x)$ is of degree n. Consider any polynomial $q(x)$ of degree at most $n - 1$. We can express the polynomial $p(x)$ (of degree at most $2n - 1$) using polynomial division by $\omega(x)$:

$$p(x) = \omega(x)q(x) + r(x)$$

Here, $r(x)$ is the remainder polynomial, and its degree is at most $n - 1$. This decomposition is valid for any polynomial $p(x)$ of degree less than $2n$. Substitute, next, this expression for $p(x)$ into equation (A.20):

$$\int_{-1}^{1} (\omega(x)q(x) + r(x))\,\mathrm{d}x = \sum_{i=1}^{n} w_i(\omega(x_i)q(x_i) + r(x_i)).$$

Let us analyse both sides of this equation.

1. **Right-hand side**. Since x_i are the roots of $\omega(x)$, we have $\omega(x_i) = 0$ for all $i = 1,\ldots,n$. Therefore, the right-hand side simplifies to

$$\sum_{i=1}^{n} w_i(0 \cdot q(x_i) + r(x_i)) = \sum_{i=1}^{n} w_i r(x_i)$$

2. **Left-hand side**. We can split the integral:

$$\int_{-1}^{1} \omega(x)q(x)\,\mathrm{d}x + \int_{-1}^{1} r(x)\,\mathrm{d}x.$$

Equating the simplified left and right sides yields:

$$\int_{-1}^{1} \omega(x)q(x)\ \mathrm{d}x + \int_{-1}^{1} r(x)\ \mathrm{d}x = \sum_{i=1}^{n} w_i r(x_i)$$

Since $r(x)$ is a polynomial of degree at most $n-1$, and the Gaussian quadrature rule is exact for all polynomials up to degree $2n-1$ (which includes degree $n-1$), it must be exact for $r(x)$. This means that

$$\int_{-1}^{1} r(x)\ \mathrm{d}x = \sum_{i=1}^{n} w_i r(x_i).$$

Substituting this back into the main equation, we find:

$$\int_{-1}^{1} \omega(x)q(x)\ \mathrm{d}x = 0.$$

This crucial result implies that the polynomial $\omega(x)$ is orthogonal to any polynomial $q(x)$ of degree less than n over the interval $[-1, 1]$. We know from the definition of Legendre polynomials (see appendix F) that $P_n(x)$ is a polynomial of degree n and is orthogonal to all polynomials of degree less than n over the interval $[-1, 1]$ with respect to a weight function of 1. Since $\omega(x)$ also has degree n and satisfies this same orthogonality condition, it must be proportional to $P_n(x)$:

$$\omega(x) = C\ P_n(x)$$

for some non-zero constant C.

Since the roots of $\omega(x)$ are precisely the nodes x_i of the Gaussian quadrature rule, and $\omega(x)$ is proportional to $P_n(x)$, it follows that the nodes x_i must be the roots of the nth Legendre polynomial, $P_n(x)$.

A.11 Weights of the Gaussian quadrature

This appendix provides a proof demonstrating how the weights w_i in an n-point Gaussian quadrature rule are determined. It builds upon two key facts:

1. The nodes x_i $(i = 1,\ldots,n)$ are the roots of the n-th Legendre polynomial, $P_n(x)$ (as established in appendix F).
2. An n-point Gaussian quadrature rule is designed to be exact for any polynomial $p(x)$ of degree up to $2n-1$.

Let us begin with the exactness property of a Gaussian quadrature: for any polynomial $p(x)$ of degree less or equal than $2n-1$:

$$\int_{-1}^{1} p(x)\ \mathrm{d}x = \sum_{i=1}^{n} w_i p(x_i). \tag{A.21}$$

Next, we introduce a specific set of polynomials from chapter 3, the Lagrange basis polynomials. For the n distinct nodes $x_1, x_2,\ldots,x_n$ (which are the roots of $P_n(x)$), we define the k-th Lagrange basis polynomial, denoted here $L_k(x)$, as:

$$L_k(x) = \prod_{j=1, j \neq k}^{n} \frac{x - x_j}{x_k - x_j}$$

for $k = 1, 2, \ldots, n$. As discussed in chapter 3, each $L_k(x)$ is a polynomial of degree $n - 1$. These n polynomials form a basis for the space of all polynomials of degree at most $n - 1$, meaning any polynomial of degree $n - 1$ can be uniquely expressed as a linear combination of these $L_k(x)$. The crucial 'cardinality' property of $L_k(x)$ means that

$$L_k(x_i) = \begin{cases} 1 & \text{if } i = k \\ 0 & \text{if } i \neq k, \end{cases}$$

or concisely, using the Kronecker delta, $L_k(x_i) = \delta_{ki}$. Since $L_k(x)$ is a polynomial of degree $n - 1$, and the Gaussian quadrature rule is exact for all polynomials of degree up to $2n - 1$ (which include all polynomials of degree $n - 1$ for $n \geqslant 1$), we can substitute $p(x) = L_k(x)$ into equation (A.21):

$$\int_{-1}^{1} L_k(x)\, \mathrm{d}x = \sum_{i=1}^{n} w_i L_k(x_i).$$

Now, applying the cardinality property $L_k(x_i) = \delta_{ki}$ to the sum on the right-hand side, yields

$$\sum_{i=1}^{n} w_i \delta_{ki} = w_k.$$

This fundamental step reveals that the weight w_k associated with the node x_k is simply the integral of the corresponding Lagrange basis polynomial. Specifically:

$$w_k = \int_{-1}^{1} L_k(x)\, \mathrm{d}x. \tag{A.22}$$

To obtain an explicit formula for w_k, we use a sensible expression for $L_k(x)$ when the nodes x_i are roots of an orthogonal polynomial $P_n(x)$. Since x_k is a simple root of $P_n(x)$, we can write $P_n(x) = (x - x_k)q(x)$, where $q(x)$ is a polynomial such that $q(x_k) = P'(x_k)$. From the definition of Lagrange basis polynomials, it is relatively straightforward to show that $L_k(x)$ can be expressed as

$$L_k(x) = \frac{P_n(x)}{(x - x_k)P'(x_k)}.$$

Substituting this expression for $L_k(x)$ into equation (A.22) yields

$$w_k = \int_{-1}^{1} \frac{P_n(x)}{(x - x_k)P'(x_k)}\, \mathrm{d}x.$$

Since $P'(x_k)$ is a constant with respect to the integration variable x, we can factor it out of the integral:

$$w_k = \frac{1}{P'(x_k)} \int_{-1}^{1} \frac{P_n(x)}{x - x_k}\, \mathrm{d}x.$$

When the integral identity (F.3) is employed, the above expression becomes

$$w_k = \frac{2}{nP_n'(x_k)P_{n-1}(x_k)}.$$

Finally, we can use formula (F.2) which relates $P_n(x)$ and $P_{n-1}(x)$ at the roots of $P_n(x)$, and obtain the expression we were looking for:

$$w_k = \frac{2}{(1 - x_k^2)(P'(x_k))^2}.$$

This formula demonstrates that the weights w_k for Gaussian quadrature are uniquely determined by the nodes x_k and the derivative of the Legendre polynomial at those nodes.

A.12 Calculation of the local error for the Heun method

This calculation relates to what was described in section 8.6. The numerical approximation, y_i, obtained with the Heun method for ordinary differential equations is different, at each step, from the correct value, $y(t_i)$. The local error, $\epsilon = y(t_i) - y_i$, can be calculated using the results and assumptions reported in that section. The goal of this appendix is to carry out and illustrate that calculation in full.

While the general method to calculate local errors is similar to the one used to derive Euler's local error, different and often challenging levels of difficulty can arise, depending on the method. The main problem for this new calculation (Heun) is represented by the intermediate term, $f(t_i, \tilde{y}_i)$, as this involves the predicted value $\tilde{y}_i$ and the next time point t_i. As we know, we assume $y_{i-1} = y(t_{i-1})$ and $f(t_{i-1}, y_{i-1}) = \mathrm{d}y(t_{i-1})/\mathrm{d}t$, and this is all we can use to make the comparison and the calculation with Taylor expansion. Therefore, we need to express $f(t_i, \tilde{y}_i)$ in terms of t_{i-1} and y_{i-1} using a two-variable Taylor expansion around the point (t_{i-1}, y_{i-1}).

Recall that $t_i = t_{i-1} + h$ and $\tilde{y}_i = y_{i-1} + hf(t_{i-1}, y_{i-1})$. To make the notation less heavy, let us use the following symbols:

$$f \equiv f(t_{i-1}, y_{i-1}), \quad f_t \equiv \frac{\partial f(t_{i-1}, y_{i-1})}{\partial t}, \quad f_y \equiv \frac{\partial f(t_{i-1}, y_{i-1})}{\partial y},$$

$$f_{tt} \equiv \frac{\partial f^2(t_{i-1}, y_{i-1})}{\partial t^2}, \quad f_{ty} \equiv \frac{\partial f^2(t_{i-1}, y_{i-1})}{\partial t \partial y}, \quad f_{yy} \equiv \frac{\partial f^2(t_{i-1}, y_{i-1})}{\partial y^2}.$$

Then the Taylor expansion of $f(t, y)$ around (t_{i-1}, y_{i-1}) is:

$$\begin{aligned} f(t, y) &= f + (t - t_{i-1})f_t + (y - y_{i-1})f_y \\ &\quad + (1/2!)[(t - t_{i-1})^2 f_{tt} + 2(t - t_{i-1})(y - y_{i-1})f_{ty} + (y - y_{i-1})^2 f_{yy}] \\ &\quad + O(h^3). \end{aligned}$$

The substitutions

$$t = t_i = t_{i-1} + h \quad \Rightarrow t - t_{i-1} = h, \quad y = \tilde{y}_i = y_{i-1} + hf \quad \Rightarrow y - y_{i-1} = hf,$$

yield

$$f(t_i, \tilde{y}_i) = f + hf_t + (hf)f_y + \frac{1}{2!}[h^2 f_{tt} + 2h(hf)f_{ty} + (hf)^2 f_{yy}] + O(h^3)$$

$$\Downarrow$$

$$f(t_i, \tilde{y}_i) = f + h(f_t + ff_y) + \frac{h^2}{2}(f_{tt} + 2ff_{ty} + f^2 f_{yy}) + O(h^3)$$

Now, substitute this expansion of $f(t_i, \tilde{y}_i)$ back into the corrector equation for y_i:

$$y_i = y_{i-1} + \frac{h}{2}\left(f + \left[f + h(f_t + ff_y) + \frac{h^2}{2}(f_{tt} + 2ff_{ty} + f^2 f_{yy}) + O(h^3)\right]\right)$$

$$y_i = y_{i-1} + \frac{h}{2}\left(2f + h(f_t + ff_y) + \frac{h^2}{2}(f_{tt} + 2ff_{ty} + f^2 f_{yy}) + O(h^3)\right),$$

or, collecting equal powers of h:

$$y_i = y_{i-1} + hf + \frac{h^2}{2}(f_t + ff_y) + \frac{h^3}{4}(f_{tt} + 2ff_{ty} + f^2 f_{yy}) + O(h^4)$$

This expression represents the numerical approximation y_i obtained with the Heun's method, expanded in terms of h and values at (t_{i-1}, y_{i-1}), and ready for the comparison with the Taylor expansion of $y(t_i)$ (i.e. around t_{i-1}):

$$y(t_i) = y(t_{i-1}) + hy'(t_{i-1}) + \frac{h^2}{2!}y''(t_{i-1}) + \frac{h^3}{3!}y'''(t_{i-1}) + O(h^4)$$

Since we assumed $y_{i-1} = y(t_{i-1})$, we can substitute y_{i-1} for $y(t_{i-1})$. We also know that this implies $y'(t_{i-1}) = f(t_{i-1}, y(t_{i-1})) = f$. But we are not finished yet as we need the second derivative, $y''(t)$. Using the chain rule:

$$y''(t) = \frac{\mathrm{d}}{\mathrm{d}t}f(t, y(t)) = \frac{\partial f}{\partial t}(t, y(t)) + \frac{\partial f}{\partial y}(t, y(t))\frac{\mathrm{d}y}{\mathrm{d}t}$$

$$y''(t) = f_t(t, y(t)) + f_y(t, y(t))f(t, y(t))$$

At t_{i-1}:

$$y''(t_{i-1}) = f_t(t_{i-1}, y(t_{i-1})) + f_y(t_{i-1}, y(t_{i-1}))f(t_{i-1}, y(t_{i-1})) = f_t + ff_y$$

For the third derivative, $y'''(t)$:

$$y'''(t) = \frac{\mathrm{d}}{\mathrm{d}t}y''(t) = \frac{\mathrm{d}}{\mathrm{d}t}(f_t + ff_y)$$

$$y'''(t) = \frac{\partial}{\partial t}(f_t + ff_y) + \frac{\partial}{\partial y}(f_t + ff_y)\frac{\mathrm{d}y}{\mathrm{d}t}$$

$$y'''(t) = (f_{tt} + f_t f_y + f_y f_t + ff_{yy}f) + (f_{ty} + ff_{yy} + f_y f_y)f$$

$$y'''(t) = f_{tt} + 2f_t f_y + f^2 f_{yy} + f(f_{ty} + ff_{yy} + f_y^2)$$

This expression for $y'''(t_{i-1})$ is complex, but its exact form is not strictly necessary for determining the order of the method, only that it is $O(1)$. Substituting these into the exact Taylor series expansion of $y(t_i)$:

$$y(t_i) = y_{i-1} + hf + \frac{h^2}{2}(f_t + ff_y) + \frac{h^3}{6}y'''(t_{i-1}) + O(h^4)$$

We are now in the position of determining the local error,

$$\epsilon = y(t_i) - y_i.$$

We start by substituting the expanded forms of $y(t_i)$ and y_i:

$$\begin{aligned} \epsilon &= \left(y_{i-1} + hf + \frac{h^2}{2}(f_t + ff_y) + \frac{h^3}{6}y'''(t_{i-1}) + O(h^4)\right) \\ &\quad -\left(y_{i-1} + hf + \frac{h^2}{2}(f_t + ff_y) + \frac{h^3}{4}(f_{tt} + 2ff_{ty} + f^2 f_{yy}) + O(h^4)\right) \end{aligned}$$

We observe that the terms up to $O(h^2)$ cancel out, so that:

$$\epsilon = \left(\frac{h^3}{6}y'''(t_{i-1}) - \frac{h^3}{4}(f_{tt} + 2ff_{ty} + f^2 f_{yy})\right) + O(h^4)$$

$$\epsilon = h^3\left(\frac{1}{6}y'''(t_{i-1}) - \frac{1}{4}(f_{tt} + 2ff_{ty} + f^2 f_{yy})\right) + O(h^4)$$

The coefficient of h^3 is generally non-zero. Therefore, the local truncation error for Heun's method is $O(h^3)$.

A.13 Centred difference for $p(x)\,\mathrm{d}y/\mathrm{d}x$

This part of the appendix explains how one arrives at formula (8.21). In section 7.3, we introduced the centred difference formula for approximating the second derivative of a function when the coefficient is constant:

$$\frac{\mathrm{d}^2 y}{\mathrm{d}x^2}(x_i) \approx \frac{y_{i-1} - 2y_i + y_{i+1}}{h^2}.$$

We now consider the more general differential operator

$$-\frac{\mathrm{d}}{\mathrm{d}x}\left(p(x)\frac{\mathrm{d}y}{\mathrm{d}x}\right),$$

where $p(x)$ is a variable coefficient. This expression involves the derivative of a product and cannot be discretised using the same formula. Instead, we proceed in two steps.

First, we approximate the first derivative $\frac{\mathrm{d}y}{\mathrm{d}x}$ at the midpoints $x_{i+1/2}$ and $x_{i-1/2}$ using forward and backward differences, respectively:

$$\left.\frac{\mathrm{d}y}{\mathrm{d}x}\right|_{x_{i+1/2}} \approx \frac{y_{i+1} - y_i}{h}, \qquad \left.\frac{\mathrm{d}y}{\mathrm{d}x}\right|_{x_{i-1/2}} \approx \frac{y_i - y_{i-1}}{h}.$$

Next, we approximate the outer derivative at the interior point x_i by applying the standard difference formula to the two midpoint expressions:

$$-\frac{\mathrm{d}}{\mathrm{d}x}\left.\left(p(x)\frac{\mathrm{d}y}{\mathrm{d}x}\right)\right|_{x_i} \approx -\frac{1}{h}\left[p_{i+1/2}\cdot\left(\frac{y_{i+1} - y_i}{h}\right) - p_{i-1/2}\cdot\left(\frac{y_i - y_{i-1}}{h}\right)\right],$$

where $p_{i\pm1/2} = p(x_i \pm h/2)$. This simplifies to:

$$-\frac{1}{h^2}[p_{i+1/2}(y_{i+1} - y_i) - p_{i-1/2}(y_i - y_{i-1})].$$

Although the individual approximations of the first derivative are only first-order accurate, their combination yields a second-order accurate expression for the full differential operator. Furthermore, as we do not have grid points at $x_{i\pm1/2}$, the quantities $p_{i+1/2}$ and $p_{i-1/2}$ are approximated by

$$p_{i+1/2} \approx \frac{p(x_i) + p(x_{i+1})}{2}, \qquad p_{i-1/2} \approx \frac{p(x_{i-1}) + p(x_i)}{2}.$$

This is still a second-order accurate approximation, provided that $p(x)$ is sufficiently smooth.

A.14 Stability bound for RK4

This note derives the explicit stability interval on the real axis for the classical fourth-order Runge–Kutta method (RK4), starting from the stability condition described at formula (8.12). If we introduce the non negative variable $\xi = \lambda h$, the inequality can be re-written as follows:

$$-1 < R(\xi) < 1,$$

where

$$R(\xi) = 1 - \xi + \frac{1}{2}\xi^2 - \frac{1}{6}\xi^3 + \frac{1}{24}\xi^4.$$

Considering the left inequality first, a direct check shows that $R(\xi)$ attains its minimum at $\xi \approx 1.596$ with $R_{\min} \approx 0.270 > -1$. Hence the left inequality $-1 < R(\xi)$ holds for all $\xi \geqslant 0$ and imposes no restriction.

Let us next consider the right inequality. This is equivalent to

$$R(\xi) - 1 = -\xi + \frac{1}{2}\xi^2 - \frac{1}{6}\xi^3 + \frac{1}{24}\xi^4 < 0.$$

Factoring $\xi > 0$ out yields:

$$\xi\left(-1 + \frac{1}{2}\xi - \frac{1}{6}\xi^2 + \frac{1}{24}\xi^3\right) < 0.$$

Thus the boundary occurs when the cubic in parentheses vanishes:

$$-1 + \frac{1}{2}\xi - \frac{1}{6}\xi^2 + \frac{1}{24}\xi^3 = 0 \iff \xi^3 - 4\xi^2 + 12\xi - 24 = 0.$$

This cubic has a unique positive real root

$$\gamma = \frac{4}{3} + \sqrt[3]{\frac{172 + 36\sqrt{29}}{27}} + \sqrt[3]{\frac{172 - 36\sqrt{29}}{27}} \approx 2.785\,293\,563.$$

Since $\xi = \lambda h$ with $\lambda > 0$, the RK4 stability interval on the real axis is

$$0 < \xi < \gamma \iff 0 < h < \frac{\gamma}{\lambda} \approx \frac{2.785\,293\,563}{\lambda}. \tag{A.23}$$

A.15 Variance of Monte Carlo estimators

In the main text (section 9.3) we considered three different Monte Carlo approaches for estimating the integral

$$I = \int_0^1 \mathrm{e}^{-x^2}\,\mathrm{d}x.$$

We derived that in each case the estimator is unbiased, i.e. its expectation equals I. Here we compute the variances of these estimators. Although in practice one estimates the variance numerically from the samples, it is useful to know their exact analytic forms.

1. **Hit-or-miss Monte Carlo**. In this method we generate random points (X, Y) uniformly in the unit square $\mathcal{S}$. Each point is classified as a 'hit' if it falls below the curve $y = \mathrm{e}^{-x^2}$, and as a 'miss' otherwise. The estimator is the fraction of hits out of the total number of sampled points.

   The probability of a hit is exactly the area under the curve:

   $$p = \int_0^1 \mathrm{e}^{-x^2}\,\mathrm{d}x = \frac{\sqrt{\pi}}{2}\,\mathrm{erf}(1).$$

   Since each point is a hit with probability p and a miss with probability $1 - p$, the variance of the fraction of hits is

$$\mathrm{Var}(\hat{I}_N) = \frac{1}{N}\, p(1 - p).$$

That is,

$$\mathrm{Var}(\hat{I}_N) = \frac{1}{N}\left(\frac{\sqrt{\pi}}{2}\,\mathrm{erf}(1)\right)\left(1 - \frac{\sqrt{\pi}}{2}\,\mathrm{erf}(1)\right) \approx \frac{0.189\,077\,8}{N}. \tag{A.24}$$

2. **Plain Monte Carlo (uniform sampling in [0,1])**. Here we sample $X \sim U[0, 1]$ and set

$$\hat{I}_N = \frac{1}{N}\sum_{i=1}^{N} \mathrm{e}^{-X_i^2}.$$

The variance is

$$\mathrm{Var}(\hat{I}_N) = \frac{1}{N}\,(\mathbb{E}(\mathrm{e}^{-2X^2}) - (\mathbb{E}(\mathrm{e}^{-X^2}))^2).$$

We compute

$$\mathbb{E}(\mathrm{e}^{-X^2}) = \int_0^1 \mathrm{e}^{-x^2}\,\mathrm{d}x = \frac{\sqrt{\pi}}{2}\,\mathrm{erf}(1),$$

$$\mathbb{E}(\mathrm{e}^{-2X^2}) = \int_0^1 \mathrm{e}^{-2x^2}\,\mathrm{d}x = \frac{\sqrt{\pi}}{2\sqrt{2}}\,\mathrm{erf}(\sqrt{2}).$$

Thus

$$\mathrm{Var}(\hat{I}_N) = \frac{1}{N}\left(\frac{\sqrt{\pi}}{2\sqrt{2}}\,\mathrm{erf}(\sqrt{2}) - \left(\frac{\sqrt{\pi}}{2}\,\mathrm{erf}(1)\right)^2\right) \approx \frac{0.040\,397\,7}{N}. \tag{A.25}$$

3. **Importance sampling** Suppose we choose importance density $g(x)$ on $[0, 1]$ and form

$$\hat{I}_N = \frac{1}{N}\sum_{i=1}^{N}\frac{\mathrm{e}^{-X_i^2}}{g(X_i)}, \qquad X_i \sim g.$$

Since the summands are i.i.d.,

$$\mathrm{Var}(\hat{I}_N) = \frac{1}{N}\,\mathrm{Var}\left(\frac{\mathrm{e}^{-Y^2}}{g(Y)}\right) = \frac{1}{N}\left\{\int_0^1 \frac{\mathrm{e}^{-2y^2}}{g(y)}\,\mathrm{d}y - I^2\right\}.$$

For the special case $g(x) = 1$ (uniform distribution), we recover the variance computed in case 2. With $g(y) = \lambda \mathrm{e}^{-\lambda y}/(1 - \mathrm{e}^{-\lambda})$ we have

$$\frac{1}{g(y)} = \frac{1 - \mathrm{e}^{-\lambda}}{\lambda}\,\mathrm{e}^{\lambda y},$$

hence

$$\int_0^1 \frac{e^{-2y^2}}{g(y)}\,dy = \frac{1-e^{-\lambda}}{\lambda}\int_0^1 e^{-2y^2+\lambda y}\,dy.$$

Complete the square:

$$-2y^2 + \lambda y = -2\left(y - \frac{\lambda}{4}\right)^2 + \frac{\lambda^2}{8},$$

so that

$$\int_0^1 e^{-2y^2+\lambda y}\,dy = e^{\lambda^2/8}\int_0^1 e^{-2(y-\lambda/4)^2}\,dy = \frac{\sqrt{\pi}}{2\sqrt{2}}\,e^{\lambda^2/8}\left(\mathrm{erf}\left(\sqrt{2}\left(1-\frac{\lambda}{4}\right)\right) - \mathrm{erf}\left(-\sqrt{2}\,\frac{\lambda}{4}\right)\right).$$

Therefore,

$$\mathrm{Var}(\hat{I}_N) = \frac{1}{N}\left\{\frac{1-e^{-\lambda}}{\lambda}\cdot\frac{\sqrt{\pi}}{2\sqrt{2}}\,e^{\lambda^2/8}\left(\mathrm{erf}\left(\sqrt{2}\left(1-\frac{\lambda}{4}\right)\right) + \mathrm{erf}\left(\sqrt{2}\,\frac{\lambda}{4}\right)\right) - I^2\right\},$$

where

$$I = \int_0^1 e^{-x^2}\,dx = \frac{\sqrt{\pi}}{2}\,\mathrm{erf}(1).$$

For practical purposes we can pick $\lambda = 1$ so that

$$\begin{aligned}\mathrm{Var}(\hat{I}'_N) &= \frac{1}{N}\left\{\left(1-\frac{1}{e}\right)\frac{\sqrt{\pi}}{2\sqrt{2}}\,e^{1/8}\left(\mathrm{erf}\left(\frac{3\sqrt{2}}{4}\right) + \mathrm{erf}\left(\frac{\sqrt{2}}{4}\right)\right) - \frac{\pi}{4}\mathrm{erf}^2(1)\right\} \\ &\approx 0.003\,026\,68/N.\end{aligned} \tag{A.26}$$

A.16 Formulas for the motion of a projectile

Consider a projectile launched with initial position (x_0, y_0) and initial velocity components (v_{x0}, v_{y0}). Neglecting air resistance, the equations of motion under constant gravity g are

$$x(t) = x_0 + v_{x0}t, \qquad y(t) = y_0 + v_{y0}t - \frac{1}{2}gt^2. \tag{A.27}$$

In the scenario treated in Chapter 11 we take

$$v_{x0} = v_0, \quad v_{y0} = 0, \quad x_0 = 0,$$

as the projectile is launched horizontally from height y_0 with speed v_0. The projectile touches the ground when $y(t) = 0$, i.e.

$$0 = y_0 - \frac{1}{2}gt^2,$$

which gives the flight time (only the positive root is physical)

$$t = \sqrt{\frac{2y_0}{g}}.$$

Substituting this into the horizontal displacement yields the landing distance

$$d = v_0\sqrt{\frac{2y_0}{g}}, \tag{A.28}$$

used to explain *supervised learning* in Section 11.6.

IOP Publishing

Computational Physics with R

James Foadi

Appendix B

A short introduction to matrices

A matrix M is a rectangular array of real or complex numbers. Such an array has two *dimensions*; the first dimension concerns the rows of the matrix and the second dimension its columns. If the matrix consists of m rows and n columns, it is said to be an $m \times n$ matrix. An example of a real 3×4 matrix is,

$$M = \begin{pmatrix} 7 & 2 & -1 & 0 \\ 1 & -1 & 0 & 1 \\ 0 & 1 & 0 & 1 \end{pmatrix}$$

This matrix can be defined once all its elements are defined. Each element can be indicated using the same symbol representing the matrix, with two integer indices attached that correspond to the row and column in which the element is contained. So, using the example just introduced, $M_{23} = 0$ and $M_{13} = -1$, for example. Row and column vectors like,

$$\mathbf{v} = \begin{pmatrix} v_1 & v_2 & \cdots & v_n \end{pmatrix} \qquad \text{and} \qquad \mathbf{w} = \begin{pmatrix} w_1 \\ w_2 \\ \cdots \\ w_m \end{pmatrix}$$

are special types of matrix of size $1 \times n$ and $m \times 1$, respectively.

Matrices with an equal number of rows and columns are called *square matrices* and play an important role in all matrix calculations. Two important types of square matrices are the *null* matrix, 0 and the *identity* matrix, I. They are simply defined as,

$$0 \equiv \begin{pmatrix} 0 & 0 & \cdots & 0 \\ 0 & 0 & \cdots & 0 \\ \cdots & \cdots & \cdots & \cdots \\ 0 & 0 & \cdots & 0 \end{pmatrix}, \qquad I \equiv \begin{pmatrix} 1 & 0 & \cdots & 0 \\ 0 & 1 & \cdots & 0 \\ \cdots & \cdots & \cdots & \cdots \\ 0 & 0 & \cdots & 1 \end{pmatrix}$$

While the null matrix has all its elements equal to 0, the identity matrix has all its elements equal to 0, with the exception of the elements on its diagonal, which are

doi:10.1088/978-0-7503-2632-2ch13

equal to 1. As the symbols chosen here to represent matrices can be mixed with other objects, it will be made clear explicitly or through the context, when a given symbol does actually indicate a matrix.

B.1 Basic matrix operations

Certain binary operations are valid for real and complex numbers, like addition, subtraction and multiplication, can be defined for matrices too, provided the correct matrix size is used. Two matrices of equal size, $m \times n$, can be added or subtracted; the result of this operation is another matrix of size $m \times n$. In detail, if a_{ij} is an element of matrix A and b_{ij} is the corresponding element of matrix B, the element c_{ij} of $C = A \pm B$ is

$$c_{ij} = a_{ij} \pm b_{ij} \tag{B.1}$$

and the size of C is $m \times n$. For example,

$$\begin{pmatrix} 2 & 1 & -1 & 0 \\ 1 & 3 & 3 & 1 \\ 0 & -1 & 1 & 2 \end{pmatrix} + \begin{pmatrix} 3 & -1 & 1 & 2 \\ -1 & 2 & 3 & 0 \\ 1 & 1 & 1 & 3 \end{pmatrix} = \begin{pmatrix} 5 & 0 & 0 & 2 \\ 0 & 5 & 6 & 1 \\ 1 & 0 & 2 & 5 \end{pmatrix},$$

and

$$\begin{pmatrix} 2 & 1 & -1 & 0 \\ 1 & 3 & 3 & 1 \\ 0 & -1 & 1 & 2 \end{pmatrix} - \begin{pmatrix} 3 & -1 & 1 & 2 \\ -1 & 2 & 3 & 0 \\ 1 & 1 & 1 & 3 \end{pmatrix} = \begin{pmatrix} -1 & 2 & -2 & -2 \\ 2 & 1 & 0 & 1 \\ -1 & -2 & 0 & -1 \end{pmatrix},$$

But, for instance, $A \pm B$ cannot be calculated for the matrices,

$$A = \begin{pmatrix} 2 & 1 & -1 & 0 \\ 1 & 3 & 3 & 1 \\ 0 & -1 & 1 & 2 \end{pmatrix} \quad \text{and} \quad \begin{pmatrix} 1 & 2 & 3 \\ 4 & 5 & 6 \\ 5 & 4 & 3 \\ 2 & 1 & 0 \end{pmatrix}$$

because their sizes are different (the first is a 3×4 matrix and the second a 4×3 matrix).

Matrix multiplication is a somewhat more complicated operation than addition and subtraction. It is typically said that matrix multiplication involves one row of the first matrix and one column of the second matrix. More specifically, given matrix A of size $m \times n'$ and matrix B of size $n' \times n$, the product matrix C has size $m \times n$. The product is possible only if the second size of the first matrix is equal to the first size of the second matrix (in the example, this common size is n'). The element c_{ij} of the product matrix $C = AB$ is given by the sum

$$c_{ij} = \sum_{k=1}^{n'} a_{ik} b_{kj} \tag{B.2}$$

If row i of matrix A is indicated as the vector of length n', $\mathbf{a}$, and the column j of matrix B is indicated as the vector of identical length n', $\mathbf{b}$, the element c_{ij} of the matrix product is equivalent to the inner product (dot product) between $\mathbf{a}$ and $\mathbf{b}$:

$$c_{ij} = \mathbf{a} \cdot \mathbf{b}$$

It can also be pedagogically useful to single out visually the row and column needed for matrix multiplication as follows:

$$\begin{pmatrix} c_{11} & c_{12} & \cdots & c_{1j} & \cdots & c_{1n} \\ c_{21} & c_{22} & \cdots & c_{2j} & \cdots & c_{2n} \\ \cdots & \cdots & \cdots & \cdots & \cdots & \cdots \\ c_{i1} & c_{i2} & \cdots & \mathbf{c_{ij}} & \cdots & c_{in} \\ \cdots & \cdots & \cdots & \cdots & \cdots & \cdots \\ c_{m1} & c_{m2} & \cdots & c_{mj} & \cdots & c_{mn} \end{pmatrix} =$$

$$\begin{pmatrix} a_{11} & a_{12} & \cdots & a_{1n'} \\ a_{21} & a_{22} & \cdots & a_{2n'} \\ \cdots & \cdots & \cdots & \cdots \\ \mathbf{a_{i1}} & \mathbf{a_{i2}} & \cdots & \mathbf{a_{in'}} \\ \cdots & \cdots & \cdots & \cdots \\ a_{m1} & a_{m2} & \cdots & a_{mn'} \end{pmatrix} \begin{pmatrix} b_{11} & b_{12} & \cdots & \mathbf{b_{1j}} & \cdots & b_{1n} \\ b_{21} & b_{22} & \cdots & \mathbf{b_{2j}} & \cdots & b_{2n} \\ \cdots & \cdots & \cdots & \cdots & \cdots & \cdots \\ b_{n'1} & b_{n'2} & \cdots & \mathbf{b_{n'j}} & \cdots & b_{n'n} \end{pmatrix}$$

Given the complicated nature of matrix multiplication, it should be of no surprise to learn that, differently to the multiplication between real numbers, matrix multiplication does not obey the multiplication rule. This means that, given two (square) matrices A and B, then in general

$$AB \neq BA$$

This prescription applies only to square matrices because it has to be possible to multiply both AB and BA.

There are other operations that are applicable to matrices and that are straightforward, like the transpose and the conjugate transpose of a matrix. The *transpose* of a matrix M is the matrix M^T obtained exchange M's rows with columns. This means that,

$$M_{ij}^T = M_{ji} \tag{B.3}$$

The *adjoint* (or complex conjugate transpose) of a matrix M is the matrix $M^\dagger$ for which all its elements have been changed into their complex conjugates and where rows have been exchanged with columns. This means that,

$$M_{ij}^\dagger = M_{ji}^* \tag{B.4}$$

where c^* is the complex conjugate of c. For real matrices, the adjoint is equivalent to the transpose.

B.2 Determinant and trace

The *determinant*, $|M|$, of a square $n \times n$ matrix M, is the complex number defined recursively through the relation,

$$|M| = \sum_{j=1}^{n}(-1)^{i+j}M_{ij}|R_{ij}|, \quad i = 1, 2,\ldots,n \tag{B.5}$$

where the $(n-1) \times (n-1)$ matrix R is obtained from M by suppressing the ith row and the jth column. The determinant $|R_{ij}|$ is known as *first minor* or simply *minor* of M. Therefore, the calculation of the determinant of, say, a 3×3 matrix is formed by the sum of determinants of three 2×2 matrices. Each 2×2 minor matrix is, in turn, composed by the sum of determinants of two 1×1 matrices. To uniquely define a determinant, thus, we must establish how to calculate the determinant of a 1×1 matrix. This is, very simply, the numerical value of the only element of the 1×1 matrix. It is possible to perform the expansion given using any value of i; in standard jargon the determinant is said to be calculated using a specific row, i. It is, in fact, possible to calculate the determinant with an expansion that makes use of columns, rather than rows. In such a case, the sum will be obtained changing the index i in the summation symbol, and selecting the appropriate column j. Consider, for example, the 3×3 matrix,

$$A = \begin{pmatrix} 1 & 1 & 0 \\ 0 & 1 & 1 \\ 1 & 1 & 1 \end{pmatrix}$$

Its determinant can be expressed as the sum of three of its minors, using the first row, i.e. index $i = 1$:

$$|A| = 1\begin{vmatrix} 1 & 1 \\ 1 & 1 \end{vmatrix} - 1\begin{vmatrix} 0 & 1 \\ 1 & 1 \end{vmatrix} + 0\begin{vmatrix} 0 & 1 \\ 1 & 1 \end{vmatrix}$$

Each one of the three 2×2 determinants can also be calculated using their first row. This means that,

$$\begin{vmatrix} 1 & 1 \\ 1 & 1 \end{vmatrix} = 1(1) - 1(1) = 0$$

$$\begin{vmatrix} 0 & 1 \\ 1 & 1 \end{vmatrix} = 0(1) - 1(1) = -1$$

$$\begin{vmatrix} 0 & 1 \\ 1 & 1 \end{vmatrix} = 0(1) - 1(1) = -1$$

Therefore,

$$|A| = 1(0) - 1(-1) + 0(-1) = 1$$

From this example it emerges clearly that it is convenient to carry out an expansion using the row or column containing the maximum number of zeros, because in that case it will not be necessary to calculate some of the minors.

As the determinant can be calculated with an expansion using any row or column, it is not difficult to realise that the determinant of the transpose of a matrix is equal to the determinant of the matrix itself:

$$|M^T| = |M| \tag{B.6}$$

In addition to determinants, the *trace*, Tr(M), of a square matrix M of size n is another important quantity related to matrices. It is defined as,

$$\mathrm{Tr}(M) = \sum_{i=1}^{n} m_{ii} \tag{B.7}$$

An interesting consequence of this definition is, for instance, that the trace of all identity matrices is equal to the size of the matrix, while the trace of all null matrices is equal to zero. It is also clear that, given any square matrix A,

$$\mathrm{Tr}(A^T) = \mathrm{Tr}(A) \tag{B.8}$$

because the diagonal of a transpose is equal to the unchanged matrix.

B.3 Inverse of a square matrix

Another complicated operation when dealing with matrices is the *inverse of a square matrix* of size n. If the matrix is indicated by M, its inverse is indicated by M^{-1}. The components of M^{-1} can be calculated solving the following system of n^2 equations, here expressed in matrix form:

$$MM^{-1} = M^{-1}M = I \tag{B.9}$$

where I is the identity matrix of size n. For example, the inverse A^{-1} of

$$A = \begin{pmatrix} 1 & 1 & 0 \\ 0 & 1 & 1 \\ 1 & 1 & 1 \end{pmatrix}$$

is the matrix whose components x_{ij} can be found using the nine equations,

$$\begin{pmatrix} 1 & 1 & 0 \\ 0 & 1 & 1 \\ 1 & 1 & 1 \end{pmatrix} \begin{pmatrix} x_{11} & x_{12} & x_{13} \\ x_{21} & x_{22} & x_{23} \\ x_{31} & x_{32} & x_{33} \end{pmatrix} = \begin{pmatrix} 1 & 0 & 0 \\ 0 & 1 & 0 \\ 0 & 0 & 1 \end{pmatrix}$$

Symbolically, it is possible to show that the elements x_{ij} of an inverse M^{-1} can be expressed as,

$$x_{ij} = (-1)^{i+j} \frac{|R_{ij}|}{|M|} \tag{B.10}$$

In the above expression, $|R_{ij}|$ is the minor which we have defined earlier, and $|M|$ is the determinant of M. The inverse of a matrix involves lots of calculations and thus it is in general a time-consuming operation. For this reason, several algorithmic approaches are devoted to fast calculations of a matrix inverse.

When the determinant of a square matrix is zero, the matrix is known as a *singular matrix*, and its inverse does not exist.

B.4 Eigenvalues and eigenvectors

Consider an $m \times n$ matrix M, an $n \times 1$ column vector $\mathbf{v}$ and a real or complex variable λ. If the so-called *eigenvector equation* (or eigenvalue equation),

$$M\mathbf{v} = \lambda\mathbf{v} \tag{B.11}$$

is satisfied by any $\mathbf{v}$, then this solution is called an *eigenvector* of matrix M and λ is known as the corresponding *eigenvalue*. An example can demonstrate the idea in practical terms. Matrix

$$M = \begin{pmatrix} -2 & 1 \\ 1 & -2 \end{pmatrix}$$

yields the following eigenvector equation:

$$M\mathbf{v} = \lambda\mathbf{v} \Leftrightarrow \begin{pmatrix} -2 & 1 \\ 1 & -2 \end{pmatrix}\begin{pmatrix} v_x \\ v_y \end{pmatrix} = \lambda\begin{pmatrix} v_x \\ v_y \end{pmatrix}$$

The above matrix equation can be turned into the system,

$$\begin{cases} (-2-\lambda)v_x + v_y &= 0 \\ v_x + (-2-\lambda)v_y &= 0 \end{cases}$$

The system has solutions different from the trivial solution $\mathbf{v}^T = (0 \quad 0 \quad 0)$ only if the determinant,

$$\begin{vmatrix} -2-\lambda & 1 \\ 1 & -2-\lambda \end{vmatrix}$$

is equal to zero. This leads to the following equation in the unknown λ:

$$(2+\lambda)^2 - 1 = 0$$

which has solutions $\lambda = -1$ and $\lambda = -3$. These are the two eigenvalues of M (a square matrix of size n can have at most n eigenvalues). Let's see what eigenvectors are associated, for instance, with the eigenvalue $\lambda = -1$. It suffices to replace this value in the original system, to obtain,

$$\begin{cases} -v_x + v_y &= 0 \\ v_x - v_y &= 0 \end{cases}$$

which has parametric solution,

$$\mathbf{v} = \begin{pmatrix} \alpha \\ \alpha \end{pmatrix}, \qquad \alpha \in \mathbb{R}$$

More than one eigenvector corresponds to a same eigenvalue. In fact, all vectors parallel to one of the eigenvectors are also eigenvectors of the same eigenvalue. To see this it suffices to consider the eigenvector equation for $\mathbf{v}$ and a vector $\mathbf{w}$ parallel to $\mathbf{v}$, so that $\mathbf{w} = k\mathbf{v}$. Then,

$$M\mathbf{w} = M(k\mathbf{v}) = kM\mathbf{v} = k\lambda\mathbf{v} = \lambda(k\mathbf{v}) = \lambda\mathbf{w}$$

which clearly shows that $\mathbf{w}$ is an eigenvector of M corresponding to the eigenvalue λ.

An important class of matrix transformation, in connection to eigenvectors and eigenvalues, is the class of *similarity transformations*. They are represented by invertible square matrices, like S, for which two matrices A and B are related (transformed) as follows:

$$B = S^{-1}AS \tag{B.12}$$

It is relatively easy to show that A and B have the same eigenvalues. Suppose λ is an eigenvalue of A, corresponding to the eigenvector $\mathbf{v}$. Consider, next, the vector,

$$\mathbf{w} \equiv S^{-1}\mathbf{v} \quad \Rightarrow \quad S\mathbf{w} = \mathbf{v} \tag{B.13}$$

This vector is an eigenvector of B, corresponding to the same eigenvalue λ. Indeed, using equations (B.12) and (B.13), in addition to the eigenvector equation,

$$B\mathbf{w} = S^{-1}AS\mathbf{w} = S^{-1}A\mathbf{v} = S^{-1}\lambda\mathbf{v} = \lambda S^{-1}\mathbf{v} = \lambda\mathbf{w}$$

which means that $\mathbf{w}$ is an eigenvector of B, corresponding to the eigenvalue λ.

More properties and applications inherent to matrices and matrix calculations can be found in chapters 1 and 2 of [1].

B.5 Special types of matrices

Some matrices enjoy properties that make them stand out from generic matrices. We have already met some of these matrices, like the identity matrix, the null matrix and the diagonal matrix. There are many more special matrices and in this section we will review a few of them.

B.5.1 Symmetric, Hermitian and unitary matrices

An $n \times n$ matrix, A, is *symmetric* if

$$A^T = A \tag{B.14}$$

i.e. if the matrix is equal to its transpose. In other words, a matrix is symmetric if it does not change when its rows are swapped with its columns.

When the elements of the matrix are complex numbers, the concept of symmetric matrix turns out to be not useful. It is, in fact, replaced with the related idea of Hermitian matrix. A matrix, A, is said to be *Hermitian* if,

$$A^\dagger = A, \tag{B.15}$$

where $A^\dagger$ indicates the adjoint of A. Clearly, an Hermitian matrix is equivalent to a symmetric matrix, when the matrix elements are real numbers.

A matrix U for which the following relation is valid

$$U^{\dagger}U = I \qquad \Rightarrow \qquad U^{\dagger} = U^{-1}, \tag{B.16}$$

is called *unitary*.

B.5.2 Positive definite matrices

A real and symmetric $n \times n$ matrix A is said to be *positive definite* if,

$$\mathbf{x}^T\, A\, \mathbf{x} > 0 \tag{B.17}$$

for all non-zero vectors $\mathbf{x}$ of length n. This definition is not very helpful to build up what positive definiteness is, intuitively. Luckily, the definition leads to results which are more intelligible. One of them (perhaps the better known of them) is that a positive definite matrix has all its eigenvalues real and positive. Similarly to what defined for symmetric matrices, an Hermitian matrix A is positive definite if

$$\mathbf{z}^*\, A\, \mathbf{z} > 0, \tag{B.18}$$

for all $\mathbf{z} \in \mathbb{C}$, that is, $\mathbf{z}$ is a vector of complex numbers and $\mathbf{z}^*$ the corresponding vector of complex conjugates. The eigenvalues of a positive definite Hermitian matrix are, too, real and positive.

B.6 The norm of a matrix

The norm of a vector is a non-negative real number that characterises the vector magnitude. As the vector is in general completely defined by several numeric components, the norm does not fully characterise it, but it obeys a set of properties that make it very useful in many ways. If the norm of an abstract vector x (or, in general, of a generic geometric object x) is denoted by $||x||$, then the following are the wanted properties for it:

- $||x|| \geqslant 0$ with $||x|| = 0$ if and only if $x = 0$, where 0 is the null element.
- $||kx|| = |k|\,||x||$, where $k \in \mathbb{C}$.
- $||x + y|| \leqslant ||x|| + ||y||$ triangle inequality.
- $||xy|| \leqslant ||x||\,||y||$.

The *Euclidean norm* for vectors, for example, defined by the operation,

$$||x||_2 = \left(\sum_{i=1}^{n} x_i^2\right)^{1/2}$$

in which the x_i's are the components of the vector $x \in \mathbb{R}^n$, satisfies all of the above properties and it is, therefore, a valid definition of norm. Following this definition, and using other and equally valid definitions, four types of norm can be used for matrices. If A is an $m \times n$ complex matrix, we have:

$$1\text{−norm:} \qquad ||A||_1 \equiv \max_{1 \leqslant j \leqslant n} \left\{\sum_{i=1}^{m} |a_{ij}|\right\} \tag{B.19}$$

$$\text{infinity–norm:} \qquad \|A\|_\infty \equiv \max_{1\leqslant i\leqslant m}\left\{\sum_{j=1}^{n}|a_{ij}|\right\} \tag{B.20}$$

$$\text{Frobenius–norm:} \qquad \|A\|_f \equiv \left(\sum_{i=1}^{m}\sum_{j=1}^{n}a_{ij}^2\right)^{1/2} \tag{B.21}$$

$$\text{spectral–norm:} \qquad \|A\|_2 \equiv \lambda_{\max}^{1/2} \tag{B.22}$$

In the last definition, $\lambda_{\max}$ is the largest eigenvalue of the matrix $A^\dagger A$.

Let us demonstrate the four definitions just introduced using the following matrix:

$$A = \begin{pmatrix} 5 & -5 & -7 \\ -4 & 2 & -4 \\ -7 & -4 & 5 \end{pmatrix}$$

We have:

- $\|A\|_1 = 16$ because the sums of the absolute values in each column are 16, 11, 16 and therefore $\max\{16, 11, 16\} = 16$.
- $\|A\|_\infty = 17$ because the sums of the absolute values in each row are 17, 10, 16 and therefore $\max\{17, 10, 16\} = 17$.
- $\|A\|_f = 15$ because the sum of the squares of all the nine elements of A is 225, and $(225)^{1/2} = 15$.
- $\|A\|_2 = 12.030$ because the three eigenvalues of $A^\dagger A$ are 144.722, 44.511, 35.767 and the square roor of the largest one is 12.030.

Reference

[1] Heading J 1958 *Matrix Theory for Physicists* (Longmans Green)

IOP Publishing

Computational Physics with R

James Foadi

Appendix C

Some statistical concepts and theory

Some important ideas and results from statistics are collected here without a rigorous and complete proof. The aim is just one of aiding recalling to mind what is needed to understand the more statistically-oriented parts of the book.

C.1 The functions of probability

When a random variable is continuous, it is trickier to work out probabilities, especially if we think that the probability of an event associated with a single value of the variable, is zero. The following functions are found to be useful, when working with continuous functions.

1. The *probability density function*, or **PDF** in short. This function returns the probability of a given outcome consisting of an interval of the continuous variable, as the area underneath. If the probability density function of a random variable X is indicated as $f(x)$, with x any outcome of X, then:

$$P(x_1 \leqslant X \leqslant x_2) = \int_{x_1}^{x_2} f(x)\, \mathrm{d}x \tag{C.1}$$

   in R the PDF is implemented as `dtype`, where `type` must be replaced by the appropriate type of probability function. For example, the PDF for the normal distribution is `dnorm`, while the one for the t-Student distribution is `dt`.

2. The *cumulative distribution function*, or **CDF** in short, returns the probability of a value less than or equal to a given outcome of the random variable. At value x, this function, indicated as $F(x)$, is calculated as the integral of the PDF from $-\infty$ to x:

$$F(x) = \int_{-\infty}^{x} f(t)\, \mathrm{d}t \tag{C.2}$$

   The cumulative distribution function is therefore a monotonic function rising from 0 to 1. In R, the CDF is implemented as `ptype`, where `type` must be

doi:10.1088/978-0-7503-2632-2ch14

replaced by the appropriate type of probability function. For example, the CDF for the normal distribution is `pnorm`, while the one for the t-Student distribution is `pt`.

3. The *quantile function*, or *percent-point function*, **PPF** in short, is the inverse of the cumulative distribution function. If the probability that the outcome of a random variable X is less or equal than x is,

$$P(X \leqslant x) \equiv F(x) = p,$$

then the quantile function, $Q(p)$, is x:

$$Q(p) = x \qquad \text{if} \qquad p = F(x) \tag{C.3}$$

In R, the PPF is implemented as `qtype`, where `type` must be replaced by the appropriate type of probability function. For example, the PPF for the normal distribution is `qnorm`, while the one for the t-Student distribution is `qt`.

C.2 Linear regression

Consider a functional relation between an independent variable x and a dependent variable $y = g(x)$. If randomness is not involved, at each value of x, say x_i, corresponds a unique value of y, $y_i = g(x_i)$. When randomness plays a role, though, y_i will not necessarily be $g(x_i)$, but will rather have a probability distribution around $g(x_i)$. In the following, we will assume that this distribution is a normal distribution with mean $g(x_i)$ and variance σ^2. Using statistical notation, the same assumption can be formulated as

$$E(Y_i) = g(x_i), \qquad \text{Var}(Y_i) = \sigma^2, \tag{C.4}$$

where Y_i indicates the random variable whose outcomes yield the possible values that y_i can assume. We have not introduced the same random variable to represent the values of x because these are not supposed to be random.

An important assumption about σ^2 in the theory of linear regression here studied is that σ^2 is a constant, i.e. the distributions of Y_i have different means, but the same variance. When this requirement is relaxed, the theory will need to be slightly modified. Another important assumption relates to the so called *residual*, ϵ_i, defined as,

$$\epsilon_i = Y_i - g(x_i), \qquad i = 1,\ldots,n \tag{C.5}$$

where we are considering n known values of x. Each residual is defined in terms of the random variable Y_i and it is therefore itself a random variable, following a normal distribution. Its expectation is,

$$E(\epsilon_i) = E(Y_i - g(x_i)) = E(Y_i) - g(x_i) = g(x_i) - g(x_i) = 0,$$

while its variance is,

$$\text{Var}(\epsilon_i) = \text{Var}(Y_i - g(x_i)) = \text{Var}(Y_i) = \sigma^2$$

Therefore, each residual varies according to a normal distribution with mean 0 and variance σ^2. In the following, and this is key to the standard theory of linear regression, all residuals will not be correlated with each other, i.e.

$$\mathrm{Cov}(\epsilon_i, \epsilon_j) = 0, \qquad i, j = 1,\ldots,n,\ i \neq j \tag{C.6}$$

In the theory of linear regression, the functional form of $g(x)$ is linear in the parameters of the function. More specifically, the general expression of $g(x)$ is,

$$g(x;\ a_1,\ a_2,\ldots,a_m)$$

where $a_1,\ a_2,\ldots,a_m$ are the possible parameters needed to describe g's functional form. Some examples of functions of this type are:

- $g(x) = a_1 x + a_2$
- $g(x) = a_1 \sin(2\pi x) + a_2 \cos(2\pi x)$
- $g(x) = a_1 x^3 + a_2 x^2 + a_3 x + a_4$

The first two lines describe functional forms with two parameters ($m = 2$), while the last line describes a functional form with four parameters ($m = 4$). Notice that although there are nonlinear expressions of x in the above functions, still their full expression is linear in the parameters.

The main goal of linear regression is the estimate of the m parameters of the function involved, based on the knowledge of the values y_i available. Several estimators can be designed for the estimate, but particularly successful ones are the *least squares estimators*, based on the minimisation of the following expression,

$$Q = \sum_{i=1}^{n}(Y_i - g(x_i))^2 \equiv \sum_{i=1}^{n}\epsilon^2 \tag{C.7}$$

The solution of the minimisation results in expressions for estimators of each one of the parameters a_i. For example, if the function $g(x)$ is a straight line with two parameters,

$$y = a_1 x + a_2,$$

the least squares estimators for the two parameters are,

$$\hat{a}_1 = \bar{Y} - \hat{a}_2, \qquad \hat{a}_2 = \left(\sum_{i=1}^{n}(x_i - \bar{x})(Y_i - \bar{Y})\right) \Big/ \sum_{i=1}^{n}(x_i - \bar{x})^2 \tag{C.8}$$

Besides the parameters, also the variance σ^2 is unknown and needs to be estimated. The estimator used is,

$$s^2 = \sum_{i=1}^{n}(Y_i - \hat{g}(x_i))^2/(n - (m + 1)), \tag{C.9}$$

where $\hat{g}(x)$ is the function $g(x)$ in which the parameters have been estimated with the least squares estimators. When $g(x)$ is the straight line, it is possible to show, starting from the above expression, that the unbiased estimator for σ^2 is,

$$s^2 = \frac{1}{n-2}\left(\sum_{i=1}^{n}(Y_i - \bar{Y})^2 - \hat{a}_2^2\sum_{i=1}^{n}(x_i - \bar{x})^2\right) \tag{C.10}$$

In the case of a straight line, $\hat{a}_1$ and $\hat{a}_2$ as given by equation (C.8) are unbiased estimators. This means that,

$$E(\hat{a}_1) = a_1 \quad \text{and} \quad E(\hat{a}_2) = a_2$$

The same goes for s^2, as given by equation (C.9) or (C.10),

$$E(s^2) = \sigma^2$$

Formulas for the variances of $\hat{a}_1$ and $\hat{a}_2$ and for their covariance, can also be calculated:

$$\text{Var}(\hat{a}_1) \equiv \sigma_{\hat{a}_1}^2 = \left(\sigma^2\sum_{i=1}^{n}x_i^2\right)\Big/\left(n\sum_{i=1}^{n}(x_j - \bar{x})^2\right) \tag{C.11}$$

$$\text{Var}(\hat{a}_2) \equiv \sigma_{\hat{a}_2}^2 = \sigma^2\Big/\left(\sum_{i=1}^{n}(x_j - \bar{x})^2\right) \tag{C.12}$$

$$\text{Cov}(\hat{a}_1, \hat{a}_2) = -\sigma^2\bar{x}\Big/\left(\sum_{i=1}^{n}(x_j - \bar{x})^2\right) \tag{C.13}$$

Estimators are also possible for these statistics, once σ^2 is replaced with its estimator, s^2. They will be indicated by $s_{\hat{a}_1}^2$ and $s_{\hat{a}_2}^2$.

An important result in the case of $g(x) = a_1x + a_2$, relates to the two following statistics:

$$\frac{a_1 - \hat{a}_1}{s_{\hat{a}_1}}, \qquad \frac{a_2 - \hat{a}_2}{s_{\hat{a}_2}} \tag{C.14}$$

The are distributed according to a t-Student's distribution with $n - 1$ degrees of freedom. The importance of this result stems from the fact that the t-Student's distribution makes it possible to calculate confidence intervals for both $\hat{a}_1$ and $\hat{a}_2$.

To conclude this sketchy revision of linear regression, it is worth noticing that if the assumption of equal variances of the Y_i is dropped, i.e. if Y_i has variance $\sigma^2c_i^2$, with c_i a given constant, then the results previously found will still be valid, as long as the various formulae are changed to take the difference into account. For instance, the expression (C.7) to find least squares estimators has to be replaced with the following:

$$Q = \left(\sum_{i=1}^{2}(Y_i - g(x_i))^2\right)\Big/c_i^2 \tag{C.15}$$

IOP Publishing

Computational Physics with R

James Foadi

Appendix D

The IEEE 754 standard for floating-point arithmetic

Numbers processed by a computer have binary format. They are stored in *bits*, where each *bit* can accommodate two values, 0 and 1. Different computer architectures carry out the storing in different ways, but more and more they follow the *IEEE 754 standard*, created and managed by the Institute of Electrical and Electronics Engineers.

There are mainly systems with 32-bits or 64-bits processors. This means that a number in a 32-bits system can be stored in a string containing 32 bits (or, equivalently, 4 bytes), while a number in a 64-bits system can be stored in a string containing 64 bits (equivalent to 8 bytes). In addition, each number is composed of *sign*, *significand* (or *mantissa*) and *exponent*. For example, the base 2 number 1000.1 (equivalent to 8.5 in base 10) can be written as

$$1000.1 = 01.0001 \times 2^3$$

(verify this is true). In this expression, the first 0 is the sign ('0' corresponds to '+' and '1' corresponds to '−'), 1.0001 is the significand and 3 (the power of 2) is the exponent. In a 32-bits system, one bit is reserved to the sign, 8 bits to the exponent and 23 bits to the significand. Furthermore, the following prescriptions are needed to take into account negative exponents and to increase the accuracy of the representation:

- The point dividing the digits corresponding to positive powers of 2 from those corresponding to negative powers (*floating point*) is always shifted immediately to the right of the leftmost available 1 (*normalisation*).

- 127 is always added to the exponent and the obtained integer is transformed into a base 2 number. The reason has to do with the possibility of having negative exponents. The 127 is the result of $2^{(8-1)} - 1$, where 8 are the number of bits for the exponent. The number to add for 64-bits systems is 1023,

doi:10.1088/978-0-7503-2632-2ch15

corresponding to $2^{(11-1)} - 1$, because 11 bits are reserved for the exponent in a 64-bits system. The smalles and largest usable powers of 2 are fixed as -126 and 127, respectively. The integers -127 and 128 are reserved for special cases (see table below).

- The leading 1 of the significand (now immediately at the left of the floating point) is dropped as there will always be a 1 at the left of the floating point and this redundant piece of information does not need to be stored.

- The order of storage for a 32-bits system is: 1 bit for the sign, followed by 8 bits for the exponent, followed by 23 bits for the significand. For a 64-bits system we have, instead: 1 bit for the sign, followed by 11 bits for the exponent and 52 bits for the significand.

- Both exponent and significand strings have to be padded with zeros, if necessary.

An example can help to see how the standard works. Consider, first, the number 8.5, which in binary notation becomes 1000.1 (it is easy to see this is true because the notation means $2^3 + 2^{-1} = 8 + 0.5 = 8.5$). The sign is positive, so that the first bit of the stored number will be 0. Next, we shift the floating point of three positions towards the left and, at the same time, multiply the number by 2^3:

$$8.5 \rightarrow 1000.1 = 1.0001 \times 2^3$$

The exponent is 3 and, in a 32-bits system, we need to add 127 to it, thus obtaining 130. In binary notation, 130 is 10000010. This is also the expression of the exponent. The significand, neglecting the initial 1, is 0001. We need to add 19 zeros to fill the 23-bits part reserved to the mantissa. In summary, the standard 32-bits string for this number is

$$0 \quad 10000010 \quad 00010000000000000000000$$

When the number to be represented is periodical or has an infinite number of digits after the floating point, the computer will carry out an appropriate type of rounding. Quite often, to avoid a bias caused by the building up of round-off errors deriving from rounding always in a given direction, a stochastic rounding is applied so that the accumulated error is random, rather than systematic (see appendix E)

The IEEE 754 standard includes four special cases, associated with specific mathematical numbers and objects. For the 32-bits system these are:

Exponent value	Significand	Mathematical object
11111111	All zeros	Infinity (∞)
11111111	Not all zeros	Not a number (NaN)
00000000	All zeros	Zero (0)
00000000	Not all zeros	Subnormal numbers

The *subnormal numbers* are numbers smaller than the smallest number representable with 2^{-126}, but greater than 0. This is one way of encoding information optimally. More can be learnt by reading the IEEE 754 standard directly on Wikipedia [1].

Single and double precision

The 32-bits and 64-bits standard are also known as *single precision* and *double precision* systems. The reason is connected to the number of decimals with which a number can be represented. This can be worked out with the following simple reasoning. What matters, when discussing precision, is the number of significant digits with which a number is represented. The significand of this number can always be transformed into an integer, provided the exponent is modified accordingly. So, for instance, the number 2.345 has a precision of 4 digits, and can be re-written as 2345×10^{-3}. The precision can be now observed in the number of digits of the integer. In a system with m bits dedicated to the significand, the largest possible integer (and the one with the highest number of digits) is 2^m. We can write, for such a number

$$x = 2^m,$$

where x is the integer whose number of digits signifies the highest possible precision. The base-10 logarithm of x returns an approximation to the number of possible digits in x[1]. This means that the number, n, of digits in x is equal to the floor function of $\log_{10}(x)$[2] plus 1. Therefore, taking the base-10 logarithm of both sides of the equation above, we obtain

$$\log_{10}(x) = \log_{10}(2^m) \;\Rightarrow\; \log_{10}(x) = m\log_{10}(2).$$

So,

$$n = \lfloor \log_{10}(x) \rfloor + 1 = \lfloor m\log_{10}(2) \rfloor + 1. \tag{D.1}$$

For a single-precision number we have $m = 23$ so that

$$n = \lfloor 23\log_{10}(2) \rfloor + 1 = 7.$$

For a double-precision number we have $m = 52$ and hence

$$n = \lfloor 52\log_{10}(2) \rfloor + 1 = 16.$$

Therefore, the maximum precision obtainable with the single-precision system is 7 digits, while with the double-precision system is 16 digits.

[1] Consider for example $x = 1000$. In this case the logarithm yields the exact number of digits after 1, as $\log_{10}(1000) = 3$. The value is not exact when the number is not an exact power of 10, but it is close to the exact number. For example, when $x = 9760$, we have $\log_{10}(9760) = 3.989$.

[2] The *floor function* (indicated by $\lfloor\rfloor$) of a real number x, is the closest integer immediately below x. For example, $\lfloor 1.3 \rfloor = 1$.

Reference

[1] Wikipedia 2022 IEEE 754 https://en.wikipedia.org/wiki/IEEE_754 (Accessed: 2022-09-13)

Appendix E

The IEEE standard to binary rounding

Rounding of binary numbers follows rules similar to the rounding of decimal numbers. There exist, though, multiple methods [1] and only the *nearest, tie to even* method will be illustrated here.

Before explaining this type of rounding, it is important to remember that a binary number has only digits 0 and 1. When rounding down is executed on a binary digit, the digit does not change. But when rounding up is applied, a 0 becomes a 1, while a 1 becomes a 10. Considering for example the first digit after the floating point of 0.11, this means that rounding down yields 0.1, while rounding up yields 1.0.

We can now turn to the set of rules forming the nearest-tie-to-even method. The rules concern the digit at the nth place in the number. The rounding is based on the value of the digit immediately following the nth digit:

- If this is 0, then the nth digit is rounded down.
- If this is 1 and any of following digits is also a 1, then the nth digit should be rounded up (it becomes 0 and the $(n-1)$th digit is added a 1).
- If this is 1, but all the following digits are 0, then the *round to even* rule is applied. This means that rounding up or rounding down will be chosen according to which one causes the nth digit to become a 0.

Let us demonstrate the rules with a few examples in which we wish to round the third digit after the floating point. Consider, first, the number 10.110 010 1. This is rounded down to 10.110 because the digit following the third digit after the point is 0. Next, consider 10.110 100 1; the fourth digit is a 1 and the digits following it are not all zeros. Therefore, we round up to 10.111. Now we explain the round-to-even rule with two more examples. The fourth digit of 10.111 100 0 is a 1, but all following digits are zeros. This is where the round-to-even rule comes in. Rounding up yields 11.000; rounding down yields 10.111. The third digit for the first approximation is 0, while that for the second is 1; therefore we round up to

doi:10.1088/978-0-7503-2632-2ch16

11.000. One more example should be enough to understand the rule. The number to round is 10.110 100 0. Here, again, the fourth digit is 1, but the following ones are all zeros, and therefore we need to apply the round-to-even rule. The number rounded up is 10.111, the one rounded down is 10.110. In this case the rule selects the number rounded down because it has a 0 at the third digit.

The round-to-even rule yields both rounded up and rounded down values in a random fashion, thus making the building up of systematic errors less likely.

Reference

[1] Wikipedia 2022 IEEE 754 https://en.wikipedia.org/wiki/IEEE_754 (Accessed: 2022-09-13)

IOP Publishing

Computational Physics with R

James Foadi

Appendix F

Legendre Polynomials

Legendre polynomials, denoted $P_n(x)$, constitute a family of special functions that are polynomials defined on the interval $[-1, 1]$. They arise in various areas of physics and mathematics, particularly in problems involving spherical symmetry (e.g. in electromagnetism, quantum mechanics, and celestial mechanics) and in the field of numerical analysis.

They can be generated using *Rodrigues' formula*:

$$P_n(x) = \frac{1}{2^n n!}\frac{\mathrm{d}^n}{\mathrm{d}x^n}(x^2 - 1)^n \tag{F.1}$$

F.1 Key properties of Legendre polynomials

A number of properties of interest for the material covered in this book are listed in what follows.

1. **Orthogonality**. A defining characteristic of Legendre polynomials is their orthogonality over the interval $[-1, 1]$ with respect to 1 as weight function. This means that the integral of the product of two different Legendre polynomials over this interval is zero:

$$\int_{-1}^{1} P_m(x)P_n(x)\mathrm{d}x = \frac{2}{2n + 1}\delta_{mn}$$

   where δ_{mn} is the Kronecker delta. In simpler terms, just as two vectors are orthogonal if their dot product is zero, two functions (in this case, polynomials) are orthogonal if the integral of their product over a specified interval is zero. This property is fundamental for their use in approximating functions and constructing efficient numerical schemes.
2. **Real and distinct roots**. The nth Legendre polynomial, $P_n(x)$, has exactly n distinct real roots, all of which lie within the open interval $(-1, 1)$. These roots possess unique mathematical properties that make them particularly useful for certain approximation and integration techniques.

doi:10.1088/978-0-7503-2632-2ch17

3. **Relationship between $P_n(x)$ and $P_{n-1}(x)$ at roots.** For any root x_k of $P_n(x)$, there exists a specific relationship between the value of the $(n-1)$th Legendre polynomial at that root and the derivative of the nth Legendre polynomial at that root. More specifically:

$$P_{n-1}(x_k) = \frac{(1 - x_k^2)P_n'(x_k)}{n} \tag{F.2}$$

This identity can be verified by direct calculation for the first few polynomials and roots. The expression turns out to be particularly important when deriving the weights for Gaussian quadrature.

4. **Integral identity at roots.** For any root x_k of $P_n(x)$, the following integral identity holds:

$$\int_{-1}^{1} \frac{P_n(x)}{x - x_k}\,\mathrm{d}x = \frac{2}{nP_{n-1}(x_k)}. \tag{F.3}$$

As the previous formula, this identity is a crucial step in deriving the explicit formula for the weights in Gaussian quadrature.

F.2 The first few Legendre polynomials

These could be generated using Rodrigues' formula (please verify).

- $P_0(x) = 1$
- $P_1(x) = x$
- $P_2(x) = \frac{1}{2}(3x^2 - 1)$
- $P_3(x) = \frac{1}{2}(5x^3 - 3x)$

IOP Publishing

Computational Physics with R

James Foadi

Appendix G

The eigenvalue problem in ordinary differential equations

G.1 Introduction

This short appendix is not intended as a full theoretical treatment, but simply to set the stage for the numerical approach to eigenvalue problems, which will be presented in chapter 8. We aim to explain what kind of boundary value problems give rise to eigenvalues and eigenfunctions, and why this matters in physical and computational settings.

In the context of differential equations, an *eigenvalue problem* arises when we look for non-trivial solutions of an equation involving a parameter λ, under the constraint that the solution must satisfy certain boundary conditions. These solutions exist only for specific values of λ, called *eigenvalues*, and each such value has one or more associated *eigenfunctions*. To illustrate the idea, consider the simple differential equation

$$\frac{d^2y}{dx^2} + \lambda y = 0,$$

defined on the interval $[0, \pi]$, with boundary conditions

$$y(0) = 0, \qquad y(\pi) = 0.$$

We are looking for values of λ such that there exists a function $y(x) \neq 0$ satisfying both the differential equation and the boundary conditions. Let us solve the equation explicitly. The general solution of

$$y'' + \lambda y = 0$$

depends on the sign of λ. If $\lambda > 0$, say $\lambda = \mu^2$, then the general solution is

$$y(x) = A\cos(\mu x) + B\sin(\mu x).$$

doi:10.1088/978-0-7503-2632-2ch18

Now impose the boundary conditions:

- At $x = 0$: $y(0) = A = 0$, so $A = 0$.
- At $x = \pi$: $y(\pi) = B\sin(\mu\pi) = 0$.

To obtain a non-trivial solution ($B \neq 0$), we must have $\sin(\mu\pi) = 0$, which occurs when $\mu = n$ for integers $n = 1, 2, 3,\ldots$ Therefore, the allowed values of λ are

$$\lambda_n = n^2, \qquad n = 1, 2, 3,\ldots$$

and the corresponding eigenfunctions are

$$y_n(x) = \sin(nx).$$

These functions are linearly independent, orthogonal over $[0, \pi]$, and form a complete basis for square-integrable functions on that interval (for a refresher on square-integrable functions and all that, see for instance reference [1]). The key point is that the boundary conditions determine the spectrum of eigenvalues.

G.2 The Sturm–Liouville form

The example above is a special case of a broader class of problems called *Sturm–Liouville problems*, which have the general form

$$\frac{\mathrm{d}}{\mathrm{d}x}\left(p(x)\frac{\mathrm{d}y}{\mathrm{d}x}\right) + [\lambda w(x) - q(x)]y = 0,$$

with appropriate boundary conditions at the endpoints of an interval $[a, b]$. This form appears in many areas of applied mathematics and physics. The function $w(x)$ is called the *weight function*, and the eigenfunctions $y_n(x)$ corresponding to different eigenvalues λ_n are orthogonal with respect to it. Here *orthogonality* between two functions $y_n(x)$, $y_m(x)$ means:

$$\int_a^b y_n(x)y_m(x)w(x)\,\mathrm{d}x = 0 \qquad \text{if } n \neq m.$$

In general, analytical solutions are available only for a limited number of Sturm–Liouville problems. In chapter 8 we explore how to solve them numerically. The key idea is that, by discretising the differential operator (for example using finite differences), the continuous eigenvalue problem is transformed into a matrix eigenvalue problem (see appendix B), which can then be solved using standard numerical methods. The resulting eigenvalues and eigenvectors provide approximations to the true eigenvalues and eigenfunctions of the original problem.

G.3 Classification of boundary conditions

In boundary-value problems, the behaviour of the solution is constrained at two or more points in the domain. These constraints are called, as we have seen, *boundary conditions* and play a fundamental role in determining the nature of the solution. Several standard types of boundary conditions are encountered in physical and mathematical problems:

- **Dirichlet boundary conditions**. These specify the value of the unknown function at the boundary:

$$y(a) = \alpha, \quad y(b) = \beta.$$

If $\alpha = \beta = 0$, the conditions are said to be *homogeneous*. Otherwise, they are *nonhomogeneous*.

- **Neumann boundary conditions**. These specify the value of the derivative of the function at the boundary:

$$y'(a) = \gamma, \quad y'(b) = \delta.$$

Again, the conditions are homogeneous if $\gamma = \delta = 0$.

- **Robin (or mixed) boundary conditions**. These involve a linear combination of the function and its derivative:

$$a_1 y(a) + b_1 y'(a) = \eta, \quad a_2 y(b) + b_2 y'(b) = \zeta.$$

This form includes both Dirichlet and Neumann conditions as special cases.

- **Periodic boundary conditions**. These arise in problems defined on a periodic domain. They require the function and its derivative to match at the endpoints:

$$y(a) = y(b), \quad y'(a) = y'(b).$$

Homogeneous boundary conditions often simplify the mathematical structure of the problem and are easier to handle numerically. However, nonhomogeneous conditions arise naturally in many physical applications and require more elaborate techniques, such as function shifting or superposition, as discussed in chapter 8. Boundary conditions also play a crucial role in determining whether the differential equation leads to a well-defined eigenvalue problem. In particular, homogeneous Dirichlet conditions are frequently used in Sturm–Liouville problems and lead to symmetric, positive-definite matrix formulations in numerical methods.

Reference

[1] Arfken G B, Weber H J and Harris F E 2013 *Mathematical Methods for Physicists* 7th edn (Academic)

IOP Publishing

Computational Physics with R

James Foadi

Appendix H

List of functions in package comphy

Function	Description
BVPlinshoot2	Linear shooting method for second-order linear BVPs
BVPshoot2	Solves a second-order BVP using the shooting method
EPSturmLiouville2	Sturm–Liouville eigenproblem with homogeneous Dirichlet boundary conditions
EulerODE	Euler method for systems of ODEs
GSeidel	The Gauss–Seidel algorithm
Gquad	Numerical integration using n-point Gaussian quadrature.
HeunODE	Heun method for systems of ODEs
LUdeco	LU decomposition
PJacobi	The Jacobi method
RK4ODE	Runge–Kutta fourth order method for systems of ODEs
backdif	Backward differences
condet	Determinant of a square matrix
decidepoly_n	Degree of best-interpolating polynomial
deriv_irr	First derivative for an irregular grid
deriv_reg	First derivative on a regular grid
divdif	Divided differences
forwdif	Forward differences
gauss_elim	Gaussian Elimination
illcond_sample	Ill-conditioned sampling
linpol	1D linear interpolation
nevaitpol	Neville–Aitken algorithm for polynomial interpolation
numint_reg	Numerical integration using the trapezoid or simpson's rule
oddity	Parity of a permutation
polydivdif	Approximating polynomial for divided differences
polysolveLS	Polynomial Least Squares
roots_bisec	Bisection method for roots

(*Continued*)

doi:10.1088/978-0-7503-2632-2ch19

(*Continued*)

Function	Description
roots_newton	Newton method for roots
roots_secant	Secant method for roots
solveLS	Multilinear Least Squares
solve_tridiag	Tridiagonal linear system
transform_upper	Transform to upper triangular
which_poly	Find optimal polynomial model

IOP Publishing

Computational Physics with R

James Foadi

Appendix I

R code for the reactor simulation

In Chapter 9 we have described a simplified version of a nuclear reactor in which the dynamics of neutron production and absorption were illustrated with the help of R code. The code is divided into nine separate files with extension `.R`, each file containing one function, with the exception of the file `all.R` which must be sourced to load all functions in the working memory. The list of the nine files and the functions they contain is presented in table I.1.

Table I.1. Files containing the R code for the Monte carlo simulation of section 9.9.

Name of file	Main function	Description
`all.R`	R script	Load all functions in memory.
`FixedData.R`	Data container	Contains numeric data and parameters to be loaded in memory.
`neutronProduction.R`	`neutron_production()`	Produces the initial neutrons.
`neutronTravel.R`	`neutron_travel()`	Neutron's free flight.
`update.R`	`update()`	Update neutron's status after free flight.
`selectEvent.R`	`select_event()`	Decide on neutron's interaction after free flight.
`removeNeutrons.R`	`remove_neutron()`	Remove one row (a neutron) from the main tibble.
`elasticScattering.R`	`elastic_scattering()`	Deals with a neutron's elastic collision.
`fission.R`	`fission()`	Deals with a neutron's absorption by U_{235} and following fission.

doi:10.1088/978-0-7503-2632-2ch20

Each file includes lines of explanation (using roxygen2 [1]) and a commented-out runnable example. The full content of each file is included in this Appendix.

File `all.R`

```
###
## Monte Carlo reactor
###

# Load all functions in working memory
source("FixedData.R")
source("neutronProduction.R")
source("neutronTravel.R")
source("update.R")
source("removeNeutrons.R")
source("elasticScattering.R")
source("selectEvent.R")
source("fission.R")
```

File `FixedData.R`

```
###
## Monte carlo simulation of a nuclear reactor
###


###
## Parameters likely to change
###

# Energy of ignition neutrons
meanE <- 2

# Radius of spherical reactor (cm)
R <- 50
```

```
# Importance of presence of moderator and uranium
# in mixture (will be normalised)
phi_C <- 1.8
phi_U <- 0.9


###
## Parameters unlikely to change
###

# Natural uranium atomic fractions (U235/U238 = 1/138)
f235 <- 1/139
f238 <- 138/139

# Fast neutrons (created by fission)
meanE_fission <- 2

# Densities (kg/m^3)
rho_C  <- 1.65e3
rho_U  <- 1.905e4

# Atomic weights (g/mol)
A_C    <- 12.01
A_235  <- 235.00
A_238  <- 238.07

# Microscopic cross sections at thermal (barns)


 sig_s_C    <- 4.8
 sig_s_235  <- 10.0
 sig_s_238  <- 8.3

 sig_a_C    <- 3.2e-3
 sig_a_235  <- 694.0
 sig_a_238  <- 2.37

 sig_f_235  <- 582.0  # fission part of 235 absorption
 # U238 fission at thermal ~ 0

 # Energies (MeV)
 E_therm <- 0.025
 E0_res  <- 2.0
```

```
# Conversion factor:
# Sigma [1/m] approx 0.06025 *
#   (rho [kg/m^3] / A [g/mol]) * sigma [barns]
K <- 0.06025

# Fission multiplicity
nu_bar <- 2.47

# Parameters for U238 resonance bump
c_max <- 4
sigma_res <- 0.4
H <- 4

# Cross section (function depending on neutron's energy)
#Sigma_t <- 0.1
Sigma_t <- function(E) {
scale_a <- pmin(sqrt(E_therm/pmax(E,1e-12)),c_max)
bump_238 <- 1+H*exp(-(E-E0_res)^2/(2*sigma_res^2))
scale_a_238 <- scale_a*bump_238

Sig_s_C <- K*(rho_C/A_C)*sig_s_C*phi_C
Sig_a_C <- K*(rho_C/A_C)* sig_a_C*phi_C*scale_a

Sig_s_235 <- K*rho_U*(f235/A_235)*sig_s_235*phi_U
Sig_s_238 <- K*rho_U*(f238/A_238)*sig_s_238*phi_U

Sig_a_235 <- K*rho_U*(f235/A_235)*sig_a_235*phi_U*scale_a
Sig_a_238 <- K*rho_U*(f238/A_238)*sig_a_238*phi_U*scale_a_238
   Sig_s_total <- Sig_s_C+Sig_s_235+Sig_s_238
   Sig_a_total <- Sig_a_C+Sig_a_235+Sig_a_238

   sig_t <- Sig_s_total+Sig_a_total

   return(sig_t)
   }
```

File neutronProduction.R

```
#' Produce initial neutrons as a tibble (tidyverse-friendly)
#'
#' @param N Integer, number of neutrons to generate.
#' @param seed Optional integer for reproducibility.
#'
#' @return A tibble with columns:
#'   id, x, y, z (cm), dx, dy, dz (unit direction),
#'   E (MeV), status ("born")
#'
#' Notes:
#'   These are:
#' - Positions are uniform in the ball of
#'     radius R (r = R * U^(1/3)).
#' - Directions are isotropic
#'     (cos theta ~ Unif[-1,1], phi ~ Unif[0, 2Pi)).
#' - Energies are Exp(rate = 1/mean_energy);
#'     swap the sampler if needed later.
#'
#' - !!! Must source "FixedData.R" to have hard-coded
#'        parameters available.
neutron_production <- function(N,seed=NULL) {

# Source fixed parameters
source("FixedData.R")

# Some checks
stopifnot(N >= 0, R > 0, meanE > 0)
if (!is.null(seed)) set.seed(seed)

# sample isotropic directions
u  <- runif(N,-1,1)
phi <- runif(N,0,2*pi)
s  <- sqrt(pmax(0,1-u^2)) # sin(theta), safe against
# tiny negatives
dx <- s * cos(phi)
dy <- s * sin(phi)
dz <- u

# sample uniform points in sphere of radius R
r   <- R*runif(N)^(1/3)
# reuse independent isotropic directions for placement
u2  <- runif(N,-1,1)
```

```
phi2 <- runif(N,0,2*pi)
s2  <- sqrt(pmax(0,1-u2^2))
x <- r*s2*cos(phi2)
y <- r*s2*sin(phi2)
z <- r*u2

# energies (exponential with mean = meanE)
E <- rexp(N,rate=1/meanE)

tibble::tibble(
id     = seq_len(N),
x = x, y = y, z = z,
dx = dx, dy = dy, dz = dz,
E = E,
# flight placeholders (filled later by free-flight step)
s_int = as.numeric(NA),  # distance to next interaction
s_bnd = as.numeric(NA),  # distance to boundary
s     = as.numeric(NA),  # actual step: min(s_int, s_bnd)
status = "born"
)
}
```

File neutronTravel.R

```
#' Advance neutrons to the next event (interaction or boundary)
#'
#' @param neutrons Tibble from neutron_production()
#'                 (must have x,y,z, dx,dy,dz, E).
#' @param seed     Optional integer for reproducibility of the
#'                 free-flight sampling.
#' @return Updated tibble with new positions (x,y,z)
#'         and step columns: s_int (to interaction),
#'         s_bnd (to boundary), s = pmin(s_int, s_bnd).
#'
neutron_travel <- function(neutrons,seed=NULL) {

# Source fixed parameters
source("FixedData.R")

# Checks
stopifnot(R > 0)
if (!is.null(seed)) set.seed(seed)

n <- nrow(neutrons)
if (n == 0L) return(neutrons)
```

```
needed <- c("x","y","z","dx","dy","dz","E")
missing <- setdiff(needed, names(neutrons))
if (length(missing)) {
stop("neutron_travel(): missing columns: ",
paste(missing, collapse=", "))
}

dnorm <- sqrt(neutrons$dx^2 + neutrons$dy^2 + neutrons$dz^2)
bad_dir <- which(!is.finite(dnorm) | dnorm <= 0)
if (length(bad_dir))
stop("neutron_travel():
 found non-finite/zero direction vectors.")

dxu <- neutrons$dx / dnorm
dyu <- neutrons$dy / dnorm
dzu <- neutrons$dz / dnorm

rv   <- neutrons$x*dxu + neutrons$y*dyu + neutrons$z*dzu
r2   <- neutrons$x^2 + neutrons$y^2 + neutrons$z^2
disc <- pmax(rv^2 - (r2 - R^2), 0)
s_bnd_new <- -rv + sqrt(disc)

# Repeat Sigma_t into a vector
#sig_t <- rep_len(as.numeric(Sigma_t), n)
sig_t <- Sigma_t(neutrons$E)

if (any(!is.finite(sig_t) | sig_t <= 0)) {
warning("neutron_travel():
 non-positive/invalid Sigma_t; using Inf for s_int there.")
}
u <- runif(n)
s_int_new <- ifelse(is.finite(sig_t) & sig_t > 0,
-log(u)/sig_t, Inf)

s_new <- pmin(s_int_new, s_bnd_new)

x_new <- neutrons$x + dxu * s_new
y_new <- neutrons$y + dyu * s_new
z_new <- neutrons$z + dzu * s_new

neutrons |>
dplyr::mutate(
x = x_new, y = y_new, z = z_new,
s_int = s_int_new,
s_bnd = s_bnd_new,
s     = s_new
)
}
```

File update.R

```r
#' Update neutron status after a flight step
#'
#' @param neutrons Tibble after neutron_travel()
#' @param seed     Optional integer for reproducibility
#'
#' @return neutrons changed by collision or absorption
#'
update <- function(neutrons,seed=NULL) {

# Source fixed parameters
source("FixedData.R")

# Checks
if (!is.null(seed)) set.seed(seed)

# Remove escaped neutrons
idx <- which(abs(neutrons$s-neutrons$s_bnd) < 1e-12)
neutrons <- remove_neutrons(neutrons,idx)

### Probability of events
# Select events (scattering, absorption) based
# on cross sections and mixtures
neutrons <- select_event(neutrons,seed)

# Remove absorbed (not fissioned) neutrons
idx <- which(neutrons$status == "absorbed238" |
neutrons$status == "absorbedM")
neutrons <- remove_neutrons(neutrons,idx)

# Current number of neutrons
N <- length(neutrons$id)

# Act on scattering events
# Elastic scattering
idx <- which(neutrons$status == "elasticM" |
neutrons$status == "elasticU")
if (length(idx) > 0) neutrons <-
elastic_scattering(neutrons,seed)

# Fission, at last!
idx <- which(neutrons$status == "absorbed235")
if (length(idx) > 0) neutrons <- fission(neutrons,seed)

return(neutrons)
}
```

File removeNeutrons.R

```
#' Remove neutrons by id
#'
#' @param neutrons Tibble of neutrons (must have column `id`).
#' @param ids Integer vector of ids to remove.
#'
#' @return Tibble without the specified neutrons.
#'
remove_neutrons <- function(neutrons,ids) {
if (!"id" %in% names(neutrons)) {
stop("remove_neutrons(): tibble must contain column 'id'")
}
neutrons <- neutrons |> dplyr::filter(!id %in% ids)

return(neutrons)
}
```

File elasticScattering.R

```
#' Elastic scattering at selected indices (A inferred from status)
#'
#' Uses status to choose the target nucleus:
#'   - "elasticM" -> Graphite moderator (A = 12.01)
#'   - "elasticU" -> Uranium (A = 238.07 by default)
#' Also accepts (future-proof):
#'   - "elasticU235" -> A = 235
#'   - "elasticU238" -> A = 238.07
#'
#' Energy update: g = E'/E ~ Unif[alpha, 1],
#'  with alpha = ((A-1)/(A+1))^2
#' Direction update: new isotropic LAB direction
#'  (teaching simplification)
#' Clears s_int, s_bnd, s so the next neutron_travel()
#'  recomputes flight.
#'
#' @param neutrons Tibble of neutrons (must have columns:
#'        E, dx, dy, dz,
#'                  s_int, s_bnd, s, status)
```

```
#' @param seed     Optional RNG seed for reproducibility.
#'
#' @return Updated tibble.
#'
elastic_scattering <- function(neutrons,seed=NULL) {

# Source fixed parameters
source("FixedData.R")

# Find indices of scattered neutrons
idx <- which(neutrons$status == "elasticM" |
neutrons$status == "elasticU")

# basic checks
if (length(idx) == 0L) return(neutrons)
if (!all(idx %in% seq_len(nrow(neutrons))))
stop("elastic_scattering(): some idx out of range.")

# Map status -> mass number A
stat <- neutrons$status[idx]
A_vec <- numeric(length(idx))

A_vec[stat == "elasticM"]    <- A_C
A_vec[stat == "elasticU"]    <- A_238

if (any(!is.finite(A_vec) | A_vec <= 0)) {
bad <- unique(stat[!is.finite(A_vec) | A_vec <= 0])
stop("elastic_scattering():
 unrecognised status for A mapping: ",
paste(bad, collapse = ", "))
}

# Energy fraction bounds from isotropic CM scattering
alpha <- ((A_vec - 1)^2) / (A_vec + 1)^2

# Sample g = E'/E ~ Uniform[alpha, 1]
g <- alpha + (1 - alpha) * runif(length(idx))

# Update energies
E_new <- neutrons$E
E_new[idx] <- g * E_new[idx]

# New isotropic LAB directions
udir <- runif(length(idx), -1, 1) # cos(theta)
phi  <- runif(length(idx), 0, 2*pi)
sdir <- sqrt(pmax(0, 1 - udir^2))
```

```
dx_n <- sdir * cos(phi)
dy_n <- sdir * sin(phi)
dz_n <- udir

dx_new <- neutrons$d;
dy_new <- neutrons$dy
dz_new <- neutrons$dz
dx_new[idx] <- dx_n
dy_new[idx] <- dy_n
dz_new[idx] <- dz_n

# Clear step fields for scattered rows
s_int_new <- neutrons$s_int
s_int_new[idx] <- NA_real_
s_bnd_new <- neutrons$s_bnd
s_bnd_new[idx] <- NA_real_
s_new      <- neutrons$s
s_new[idx]      <- NA_real_

# Keep status as is ("elasticM"/"elasticU"/...)
status_new <- neutrons$status
# (Optional) store A used in this scatter
scatter_A <- if (!"scatter_A" %in% names(neutrons))
rep(NA_real_, nrow(neutrons)) else neutrons$scatter_A
scatter_A[idx] <- A_vec

neutrons <- neutrons |>
dplyr::mutate(
E          = E_new,
dx         = dx_new,
dy         = dy_new,
dz         = dz_new,
s_int      = s_int_new,
s_bnd      = s_bnd_new,
s          = s_new,
status     = status_new,
scatter_A = scatter_A
)

return(neutrons)
}
```

File selectEvent.R

```
#' Select event type for each neutron after a flight
#'  step (hard-coded data)
#'
#' Uses thermal cross sections you provided and simple
#'  energy scaling:
#'   - absorption ~ 1/sqrt(E) below Ethermal (clamped)
#'   - crude resonance bump for U-238 absorption around E0
#' You can refine these later; the structure stays the same.
#'
#' @param neutrons Tibble after neutron_travel()
#'                 (needs s, s_bnd, E).
#' @param seed     Optional RNG seed for reproducibility.
#'
#' @return Tibble with updated `status` and `scatter_target`
#'         (NA if not elastic).
select_event <- function(neutrons,seed=NULL) {

# Source fixed parameters
source("FixedData.R")

# Checks
if (!is.null(seed)) set.seed(seed)
n <- nrow(neutrons)
if (n == 0L) return(neutrons)

# normalise mixing weights (allow arbitrary positives)
s <- phi_C + phi_U
if (!is.finite(s) || s <= 0)
stop("select_event(): phi_C and phi_U must be nonnegative
 with phi_C+phi_U > 0")
phi_C <- phi_C / s
phi_U <- phi_U / s

req <- c("s","s_bnd","E")
miss <- setdiff(req, names(neutrons))
if (length(miss)) stop("select_event(): missing columns: ",
paste(miss, collapse=", "))

# ---------- Energy scaling (simple, pedagogical) ----------
E <- pmax(neutrons$E, 1e-12)  # guard

# 1/v for absorption relative to thermal
```

```
# (cap near 1 below thermal)
scale_a <- sqrt(E_therm / pmax(E, 1e-12))
scale_a <- pmin(scale_a,c_max)

# U-238 resonance bump near 2 MeV
# width ~0.4 MeV, height +4
bump_238 <- 1+H*exp(-((E-E0_res)^2)/(2*sigma_res^2))
scale_a_238 <- scale_a*bump_238

# Scattering weakly energy dependent here → leave as constant
# (you can introduce small tweaks later if desired)

# - Macroscopic cross sections (per neutron energy) -
# Graphite
Sigma_s_C <- K*(rho_C/A_C)*sig_s_C* phi_C
Sigma_a_C <- K*(rho_C/A_C)*sig_a_C*phi_C*scale_a

# Uranium mixture, isotope-weighted
# Scatter
Sigma_s_U235 <- K*rho_U*(f235/A_235)*sig_s_235*phi_U
Sigma_s_U238 <- K*rho_U*(f238/A_238)*sig_s_238*phi_U
# Absorption (total); split U-235 into
# fission vs capture for later use
Sigma_a_U235 <- K*rho_U*(f235/A_235)*sig_a_235*phi_U*scale_a
Sigma_f_U235 <- K*rho_U*(f235/A_235)*sig_f_235*phi_U*scale_a
Sigma_c_U235 <- pmax(Sigma_a_U235-Sigma_f_U235,0)

Sigma_a_U238 <- K*rho_U*(f238/A_238)*sig_a_238*phi_U*scale_a_238

# Totals needed for branching
Sigma_s_total <- Sigma_s_C+Sigma_s_U235+Sigma_s_U238
Sigma_a_total <- Sigma_a_C+Sigma_a_U235+Sigma_a_U238
Sigma_t_total <- Sigma_s_total+Sigma_a_total

# ---------- Event probabilities at a collision
#            (conditional on not escaping) ----------
# Per-neutron vectors
P_scatter   <- ifelse(Sigma_t_total >
0,Sigma_s_total/Sigma_t_total,0)
P_abs235    <- ifelse(Sigma_t_total >
0,Sigma_a_U235/Sigma_t_total,0)
P_abs238    <- ifelse(Sigma_t_total > 0,
Sigma_a_U238/Sigma_t_total,0)
P_absM      <- ifelse(Sigma_t_total > 0,
```

```
Sigma_a_C/Sigma_t_total,0)

# Among scatters, choose target
P_sC   <- ifelse(Sigma_s_total > 0,
Sigma_s_C   / Sigma_s_total, 0)
P_s235 <- ifelse(Sigma_s_total > 0,
Sigma_s_U235/ Sigma_s_total, 0)
P_s238 <- ifelse(Sigma_s_total > 0,
Sigma_s_U238/ Sigma_s_total, 0)

# ---------- Assign events ----------
status <- neutrons$status
scatter_target <- rep(NA_character_, n)

# Collisions: sample by probabilities
coll_idx <- which(is.finite(Sigma_t_total)
& Sigma_t_total > 0)
if (length(coll_idx)) {
u <- runif(length(coll_idx))

# cumulative bins
c1 <- P_scatter[coll_idx]
c2 <- c1 + P_abs235[coll_idx]
c3 <- c2 + P_abs238[coll_idx]
# remainder goes to moderator absorption

# decide primary event
e <- character(length(coll_idx))

# For elastic scattering choose phiC,phiU probs
idx <- which(u < c1)
jdx <- partition_vector(idx,phi_C,phi_U)
e[jdx]                         <- "elasticM"
e[!jdx]                        <- "elasticU"
e[u >= c1 & u < c2]            <- "absorbed235"
e[u >= c2 & u < c3]            <- "absorbed238"
e[u >= c3]                     <- "absorbedM"
status[coll_idx] <- e
```

```
# for nucleus
elast_idx <- coll_idx[e == "elastic"]
if (length(elast_idx)) {
ue <- runif(length(elast_idx))
# cumulative for targets

ce1 <- P_sC[elast_idx]
ce2 <- ce1 + P_s235[elast_idx]
t <- character(length(elast_idx))
t[ue < ce1]                  <- "C"
t[ue >= ce1 & ue < ce2]      <- "U235"
t[ue >= ce2]                 <- "U238"
scatter_target[elast_idx] <- t
}
}

# Write back; keep a note of nu_bar for later fission
# spawning if needed
neutrons$status <- status
neutrons$scatter_target <- scatter_target
neutrons$nu_bar <- nu_bar

return(neutrons)
}

###------------------------------------------------------------------
## Auxiliary functions
##-------------------------------------------------------------------

# Bernoulli split of indices: TRUE with prob p = a/(a+b)
partition_vector <- function(v, a, b) {
if (a < 0 || b < 0)
stop("partition_vector(): a, b must be nonnegative.")
N <- length(v)
if (N == 0L) return(logical(0))
p <- if ((a + b) > 0) a/(a + b) else 0
as.logical(stats::rbinom(N, 1, p))
}
```

File fission.R

```
#' Fission: handle U-235 absorption by removing parents
#'  and spawning newborns
#'
#' For every neutron with status == "absorbed235", remove
#'  the parent and spawn k newborn neutrons at the same
#'  position, with isotropic directions and fast energies.
#'  Here k is 2 or 3 with 50%-50% probability.
#'
#' Energies are sampled from an exponential with mean
#'  `meanE_fission` (if defined in FixedData.R),
#'  otherwise default 2 MeV.
#'
#' Newborns are appended with status = "born" and fresh,
#'  consecutive ids.
#'
#' @param neutrons Tibble of neutrons.
#' @param seed     Optional RNG seed for reproducibility.
#'
#' @return neutrons: likely to include new ones
#'
fission <- function(neutrons, seed = NULL) {
# Source fixed parameters (may define meanE_fission,
# nu_bar, etc.)
source("FixedData.R")
if (!is.null(seed)) set.seed(seed)

# Parents: those that underwent U-235 absorption
parents_rows <- which(neutrons$status == "absorbed235")
if (length(parents_rows) == 0L) return(neutrons)

# Multiplicity: 2 or 3 with equal probability
k_vec <- sample(c(2L, 3L), size = length(parents_rows),
replace = TRUE)
Ktot  <- sum(k_vec) # Total number of neutrons produced

# Defaults for fast fission spectrum (simple placeholder)
meanE_fast <- if (exists("meanE_fission"))
meanE_fission else 2.0  # MeV

# Snapshot parent geometry before removal
parents <- neutrons[parents_rows, , drop = FALSE]
```

```
# Remove parents (by id)
if (!"id" %in% names(neutrons))
stop("fission(): tibble must contain column 'id'")
neutrons <- remove_neutrons(neutrons, parents$id)

# If no newborns to add
# (shouldn't happen with k in {2,3}), return
if (Ktot == 0L) return(neutrons)

# Build newborns at parents' positions
rep_idx <- rep(seq_along(parents_rows), times = k_vec)
x <- parents$x[rep_idx]
y <- parents$y[rep_idx]
z <- parents$z[rep_idx]

# Isotropic directions
u  <- runif(Ktot, -1, 1)          # cos(theta)
ph <- runif(Ktot, 0, 2*pi)
s  <- sqrt(pmax(0, 1 - u^2))
dx <- s * cos(ph)
dy <- s * sin(ph)
dz <- u

# Fast energies (placeholder exponential spectrum)
E  <- rexp(Ktot, rate = 1/meanE_fast)

# Next ids
next_id <- if (nrow(neutrons)) max(neutrons$id) + 1L else 1L
new_ids <- seq.int(next_id, length.out = Ktot)

# Create newborns with core columns
newborns <- tibble::tibble(
id = new_ids,
x = x, y = y, z = z,
dx = dx, dy = dy, dz = dz,
E  = E,
s_int = NA_real_, s_bnd = NA_real_, s = NA_real_,
status = "born"
)

# Add optional columns only if they exist in 'neutrons'
if ("scatter_A" %in% names(neutrons))
newborns$scatter_A <- NA_real_
if ("scatter_target" %in% names(neutrons))
```

```
newborns$scatter_target <- NA_character_
if ("nu_bar" %in% names(neutrons))
newborns$nu_bar <- if (exists("nu_bar")) nu_bar else NA_real_

# Append newborns
neutrons <- dplyr::bind_rows(neutrons, newborns)

return(neutrons)
}
```

Reference

[1] Wickham H, Danenberg P, Csardi G, Eugster M and RStudio 2024 roxygen2: In-line documentation for r. R package version 7.3.2

IOP Publishing

Computational Physics with R

James Foadi

Appendix J

R code for the projectile simulation

The data used to demonstrate supervised learning in section 11.7 have been generated with the following simulator, the function `projectile_data`.

```
#' Generate supervised-learning data for projectile distance
#'
#' Simulates n launches with inputs (y0,v0) and output d, where
#'    d = v0 * sqrt(2*y0/g).
#' Optionally adds Gaussian measurement noise to d to produce d_obs.
#'
#' Gravity acceleration is hard-coded as 9.81 m/s^2
#'
#' @param n         Integer, number of observations.
#' @param y0_range Numeric length-2, range for y0 in metres.
#' @param v0_range Numeric length-2, range for v0 in m/s.
#' @param noise_sd Standard deviation of measurement noise on d
#'                  (default 0).
#' @param seed      Optional RNG seed for reproducibility.
#'
#' @return A tibble with columns: y0, v0, d (true), d_obs (measured)
#'         If noise_sd = 0, then d_obs == d.
#'
projectile_data <- function(n,
y0_range = c(1,5), # m
v0_range = c(1,5), # m/s
noise_sd = 0,      # m
seed = NULL) {
# Checks
stopifnot(length(y0_range) == 2, length(v0_range) == 2)
if (!is.null(seed)) set.seed(seed)

# Gravity acceleration (hard coded)
g <- 9.81
```

doi:10.1088/978-0-7503-2632-2ch21

```
# Draw inputs uniformly in the specified physical ranges
y0 <- runif(n,min=y0_range[1],max=y0_range[2])
v0 <- runif(n,min=v0_range[1],max=v0_range[2])

# True distance from the physics formula
d_true <- v0*sqrt(2*y0/g)

# Measured distance with optional Gaussian noise
if (noise_sd > 0) {
d_obs <- d_true+rnorm(n,mean=0,sd=noise_sd)

# guard against negative due to noise
# (distance cannot be negative)
d_obs[d_obs < 0] <- 0
} else {
d_obs <- d_true
}

# Return as tibble
Data <- tibble::tibble(
y0 = y0,
v0 = v0,
d  = d_true,
d_obs = d_obs
)

return(Data)
}
```

Appendix K

Solutions to exercises and downloadable R code

Solutions to the exercises and most R code presented in the book can be accessed at the following link: https://www.jfoadi.uk/CPR.html.

The link redirect the browser to a web page containing further links to both solutions and R code:

- The *solutions* are included as individual files, one per chapter[1]. They can be downloaded both in PDFand HTML formats.

- The *R code* is presented as compressed folders (ZIP), each folder containing `.R` files relevant to each chapter[2].

Readers do not in principle need to download the included R code, as in the book this is presented as ready–to–copy–and–paste text. But it is certainly useful to have it also available as a set of ready `.R` files.

[1] Only chapters 1 to 8 have exercises.
[2] Only chapters 2 to 11 have code included in `.R` files.

doi:10.1088/978-0-7503-2632-2ch22

www.ingramcontent.com/pod-product-compliance
Ingram Content Group UK Ltd.
Pitfield, Milton Keynes, MK11 3LW, UK
UKHW051923100926
8556IPUK00021B/1343